Biomolecules from Natural Sources

Biomolecules from Natural Sources

Advances and Applications

Edited by

Vijai Kumar Gupta
Scotland's Rural College (SRUC),
Edinburgh, Scotland, UK

Satyajit D. Sarker
Liverpool John Moores University
Liverpool, UK

Minaxi Sharma
University of Science and Technology,
Meghalaya, India &
Haute Ecole Provinciale du Hainaut- Condorcet,
ATH, Belgium

María Elida Pirovani
Universidad Nacional del Litoral
Santa Fe, Argentina

Zeba Usmani
University of Science and Technology,
Meghalaya, India

Chelliah Jayabaskaran
Indian Institute of Science
Bangalore, India

WILEY

This edition first published 2022

Registered Office(s)
John Wiley & Sons, Inc., 111 River Street, Hoboken, NJ 07030, USA
John Wiley & Sons Ltd, The Atrium, Southern Gate, Chichester, West Sussex, PO19 8SQ, UK

Editorial Office
The Atrium, Southern Gate, Chichester, West Sussex, PO19 8SQ, UK

For details of our global editorial offices, customer services, and more information about Wiley products visit us at www.wiley.com.

Wiley also publishes its books in a variety of electronic formats and by print-on-demand. Some content that appears in standard print versions of this book may not be available in other formats.

Library of Congress Cataloging-in-Publication Data
Names: Gupta, Vijai Kumar, editor. | Sarker, Satyajit D., editor. | Sharma, Minaxi, editor. | Pirovani, María Elida, editor. | Usmani, Zeba, editor. | Jayabaskaran, Chelliah, editor.
Title: Biomolecules from natural sources: : advances and applications / edited by Vijai Kumar Gupta, Satyajit D. Sarker, Minaxi Sharma, María Elida Pirovani, Zeba Usmani, Chelliah Jayabaskaran.
Description: Hoboken, NJ : John Wiley & Sons, [2022] | Includes bibliographical references and index.
Identifiers: LCCN 2021061387 (print) | LCCN 2021061388 (ebook) | ISBN 9781119769576 (hardback) | ISBN 9781119769590 (pdf) | ISBN 9781119769613 (epub) | ISBN 9781119769620 (ebook)
Subjects: LCSH: Biomolecules. | Natural resources.
Classification: LCC QP514.2 .B577 2022 (print) | LCC QP514.2 (ebook) | DDC 612/.015--dc23/eng/20220125
LC record available at https://lccn.loc.gov/2021061387
LC ebook record available at https://lccn.loc.gov/2021061388

Cover image: © CHRISTOPH BURGSTEDT/Getty Images
Cover design by Wiley

Set in 9.5/12.5pt STIXTwoText by Integra Software Services Pvt. Ltd, Pondicherry, India

10 9 8 7 6 5 4 3 2 1

Contents

Preface

Developing strategies for the use of natural resources and turning them into useful biomolecules and products at an economical cost is the need of the hour. Biomolecules have huge commercial and industrial applications due to their wide bioactive potential and biological activities. In regard to this, biomolecules obtained from natural resources, having biological activity can be used in the development of novel products and applications. Biomolecules have unique properties such as biodegradability, non-toxicity, biocompatibility, bioavailability, they are eco-friendly, and have diverse applications. These compounds can be defined as secondary metabolites and are produced as the primary biosynthetic pathway and metabolic routes of chief biological compounds associated with growth and development. Biomolecules can be extracted from several organisms ranging from cyanobacteria, bacteria, fungi, algae, viruses, plants, corals, and fish. In nature, biomolecular interactions are mostly associated with biochemical and biological diversity, especially in light of the discovery of associated biomolecules from various bioresources. These biologically active molecules derived from nature have novel applications which contribute to the wide range of natural product-based preparations and formulations currently being developed on an industrial/pilot scale. The understanding of the bioprocessing and applications of these biomolecules will help its value addition in the sectors of agriculture, food, biopharma, nutraceuticals, cosmeceuticals, therapeutic, and the environment. Thus, this book deals with the recent breakthrough in the technological progress and applications of naturally derived biomolecules. It provides students and researchers with important information about the natural sources of biomolecules, strategies of production, their pharmacological and health promoting properties, their advance use in drug therapy, and other possible future applications. This book also throws light on emerging techniques for studying naturally bioactive molecules.

Chapters from some renowned scientists working in this area provide recent and valuable information in the fields of screening, valorization, characterization, and new applications of natural biomolecules. Biological activities of natural glycosides, natural biopolymer types and applications, mushrooms pigments and their applications, pharmacological potential of pigments, protein nanoparticles to encapsulate bioactive compounds, protein structure, function, and specificity: PhaC synthases type I, II, III and IV as a model, extremozyme-based technology: Bioactivity and stability performances, the role of divalent cations in antibiotic sensitivity: The molecular position, biomolecules from vegetal wastes, natural bioactive compounds of berry fruits during surface decontamination, biomolecules from basil: Pharmacological significance, Himalayan peony (Paeonia emodi Royle): Enlightening bioactive compounds and biological applications towards sustainable use, health properties of dietary monoterpenes, Biomolecules derived from whey, strategies for production and biological properties, exopolysaccharides (EPS) produced by lactic acid bacteria, characterization (technological and molecular) of bacteriocins produced by lactic acid bacteria of industrial interest are all described in the chapters. So, this publication is targeted at bringing forward an exhaustive collection of research and information used to investigate various sources of biomolecules, their products and their inter- and transdisciplinary applications that represent recent advancements made in this regard and their relevance to scientists and researchers investigating natural biomolecules.

Editors

Vijai Kumar Gupta
Satyajit D. Sarker,
Minaxi Sharma,
María Elida Pirovani,
Zeba Usmani,
Chelliah Jayabaskaran

List of Contributors

Amro Abd Al Fattah Amara
Protein Research Department,
City of Scientific Research and
Technological Applications,
Alexandria, Egypt

Elisa C. Ale
Instituto de Lactología Industrial,
Santa Fe, Argentina

Begoña de Ancos
Department of Characterization,
Quality and Safety,
Spanish National Research Council,
Madrid, Spain

Ilian Badjakov
Department of Agrobiotechnology,
Agricultural Academy, Sofia,
Bulgaria

C. V. Bergamini
Instituto de Lactología Industrial,
Universidad Nacional del Litoral, Santa
Fe, Argentina

Ana G. Binetti
Instituto de Lactología Industrial,
Facultad de Ingeniería Química, Santa Fe,
Argentina

Rafael Chelala Moreira
Department of Food Science and
Nutrition, University of Campinas,
Campinas – SP, Brazil

Ameur Cherif
University of Manouba, ISBST, Sidi
Thabet, Tunisia

E. J. A. Corrêa
Empresa de Pesquisa Agropecuária de
Minas Gerais EPAMIG-URECO,
Pitangui, MG, Brazil

Natália Cruz-Martins
Faculty of Medicine, University of Porto,
Porto, Portugal and
Institute for Research and Innovation in
Health, University of Porto, Porto,
Portugal and
Toxicology Research Unit, University
Institute of Health Sciences, Gandra,
Portugal

Andrea del Luján Quiberoni
Instituto de Lactología Industrial,
Facultad de Ingeniería Química,
Santa Fe, Argentina

María Laura Deseta
Área de Biocoloides y Nanotecnología,
Universidad Nacional del Litoral,
Santa Fe, Argentina and
Consejo Nacional de Investigaciones
Científicas y Técnicas, Argentina

Gerardo Díaz-Godínez
Research Center for Biological Sciences, Autonomous University of Tlaxcala, Tlaxcala, México

Ivayla Dincheva
Department of Agrobiotechnology, Agricultural Academy, Sofia, Bulgaria

N. F. Duarte
Federal University of Minas Gerais, Belo Horizonte, Minas Gerais, Brazil

Evaa Fahr
Department of Pharmaceutical Sciences and Medicines, Universidade Lisboa, Lisbon, Portugal

Joana B. Ferrado
Área de Biocoloides y Nanotecnología, Universidad Nacional del Litoral, Santa Fe, Argentina and
Consejo Nacional de Investigaciones Científicas y Técnicas, Argentina

Afwa Ghorrab
University of Manouba, ISBST, Sidi Thabet, Tunisia

Daniela Marta Guglielmotti
Instituto de Lactología Industrial (INLAIN, UNL-CONICET), Santa Fe, Argentina

Khouloud Hammami
University of Manouba, ISBST, Sidi Thabet, Tunisia

Juliano Lemos Bicas
Department of Food Science and Nutrition, University of Campinas, Campinas – SP, Brazil

Sara Lopes
Department of Pharmaceutical Sciences and Medicines, Universidade Lisboa, Lisbon, Portugal

Mario Roberto Marostica Junior
Department of Food Science and Nutrition, University of Campinas, Campinas – SP, Brazil

María P. Méndez-Galarraga
Instituto de Tecnología de Alimentos, Universidad Nacional del Litoral, Santa Fe, Argentina and
Consejo de Investigaciones Científicas y Técnicas, Godoy Cruz, Argentina

Gustavo Molina
Institute of Science and Technology, Federal University of Jequitinhonha and Mucury Valleys,
Diamantina – MG, Brazil

Mohamed Neifar
University of Manouba, ISBST, Sidi Thabet, Tunisia

M. C. Pagano
Federal University of Minas Gerais, Minas Gerais, Brazil

Sakshi Painuli
Department of Biotechnology, Graphic Era University, Uttarakhand, India
and Himalayan Environmental Studies and Conservation Organization, Uttarakhand, India

G. H. Peralta
Instituto de Lactología Industrial, Universidad Nacional del Litoral, Santa Fe, Argentina

Adrian A. Perez
Área de Biocoloides y Nanotecnología, Universidad Nacional del Litoral, Santa Fe, Argentina and Consejo Nacional de Investigaciones Científicas y Técnicas, Argentina

M. C. Perotti
Instituto de Lactología Industrial, Universidad Nacional del Litoral, Santa Fe, Argentina

Andrea M. Piagentini
Instituto de Tecnología de Alimentos, Universidad Nacional del Litoral, Santa Fe, Argentina

María Elida Pirovani
Instituto de Tecnología de Alimentos, Universidad Nacional del Litoral, Santa Fe, Argentina

Silvina Alicia Pujato
Instituto de Lactología Industrial, Facultad de Ingeniería Química, Santa Fe, Argentina

Melisa A. Puntillo
Instituto de Lactología Industrial, Facultad de Ingeniería Química, Santa Fe, Argentina

A. Quiberoni
Instituto de Lactología Industrial, Universidad Nacional del Litoral, Santa Fe, Argentina

Maria H. Ribeiro
Department of Pharmaceutical Sciences and Medicines, Universidade Lisboa, Lisbon, Portugal

María F. Rojas
Instituto de Lactología Industrial, Facultad de Ingeniería Química, Santa Fe, Argentina

Concepción Sánchez-Moreno
Department of Characterization, Spanish National Research Council, Madrid, Spain

Lilana G. Santiago
Área de Biocoloides y Nanotecnología, Universidad Nacional del Litoral, Santa Fe Argentina

Prabhakar Semwal
Department of Biotechnology, Graphic Era University, India and
Uttarakhand State Council for Science and Technology, Uttarakhand, India

Ahmed Slaheddine Masmoudi
University of Manouba, ISBST, Sidi Thabet, Tunisia

Amal Souii
University of Manouba, ISBST, Sidi Thabet, Tunisia

Maura Téllez-Téllez
Biological Research Center, Autonomous University of the State of Morelos, Morelos, México

Ashish Thapliyal
Department of Biotechnology, Graphic Era University, Uttarakhand, India

Franco Van de Velde
Instituto de Tecnología de Alimentos, Universidad Nacional del Litoral, Santa Fe, Argentina and
Consejo de Investigaciones Científicas y Técnicas, Godoy Cruz, Argentina

M. A. Vélez
Instituto de Lactología Industrial, Universidad Nacional del Litoral, Santa Fe, Argentina

C. I. Vénica
Instituto de Lactología Industrial, Universidad Nacional del Litoral, Santa Fe, Argentina

Kele A. C. Vespermann
Institute of Science and Technology, Federal University of Jequitinhonha and Mucury Valleys, Diamantina – MG, Brazil

Flavia F. Visentini
Área de Biocoloides y NanotecnologíaUniversidad Nacional del Litoral, Santa Fe, Argentina and Consejo Nacional de Investigaciones Científicas y Técnicas, Argentina

I. V. Wolf
Instituto de Lactología Industrial, Universidad Nacional del Litoral, Santa Fe, Argentina

B. K. Yelikbayev
Kazakh National Agrarian University, Almaty, Kazakhstan

1

Glycolipids

From Biosynthesis to Biological Activity toward Therapeutic Application

Maria H. Ribeiro, Eva Fahr, and Sara Lopes

Research Institute for Medicines (iMed.ULisboa), Department of Pharmaceutical Sciences and Medicines, Faculty of Pharmacy, Universidade Lisboa, Lisbon, Portugal
**Corresponding author: mhribeiro@ff.ulisboa.pt*

1.1 Introduction

Biosurfactants are surface active biomolecules, mostly produced by microorganisms, with a wide range of industrial applications. Biosurfactants are usually designed with a hydrophilic moiety composed of amino acids or peptides, anions, or cations; mono-, di-, or polysaccharides; and a hydrophobic moiety consisting of unsaturated, saturated, or fatty acids (Banat et al. 2010).

Since biosurfactants are a wide group of biocompounds, there are different methods of classification. The most usual is classification according to the nature, chemical composition and microbial origin of the biosurfactants. They can be divided into five major categories: glycolipids, fatty acid/phospholipid, lipopeptide/lipoprotein, polymeric and surfactant particles (Cortés-Sánchez et al. 2013).

Among the biosurfactants, glycolipids have been intensively studied and are one of the most promising categories for commercial production and utilization (Warnecke and Heinz 2010). Glycolipids with one or two sugar residues attached to different lipid backbones can be found in cell membranes of bacteria, fungi, plants and animals in the form of sterylglycosides, glycosylceramides, and diacylglycerolglycosides (Warnecke and Heinz 2010). The most well-known glycolipids are sophorolipids (Bogaert et al. 2007; Oliveira et al. 2015), mannosylerythritol lipids (Im et al. 2001), rhamnolipids and trehalose lipids (Figure 1.1). The glycolipods that this chapter will focus on is a case study of trehalose lipids, also known as trehalolipids.

The amphiphilic character triggers them to aggregate at liquid interfaces with different degrees of polarity and hydrogen bridges, giving them the ability to reduce surface- and interfacial-tension between solids, liquids and gases.

Furthermore most biosurfactants exhibit characteristics such as tolerance to pH, temperature and ionic strength, biodegradability, low toxicity, detergency, emulsification, de-emulsification and foaming. There is considerable interest in potential applications,

Biomolecules from Natural Sources: Advances and Applications, First Edition. Edited by Vijai Kumar Gupta, Satyajit D. Sarker, Minaxi Sharma, María Elida Pirovani, Zeba Usmani, and Chelliah Jayabaskaran.

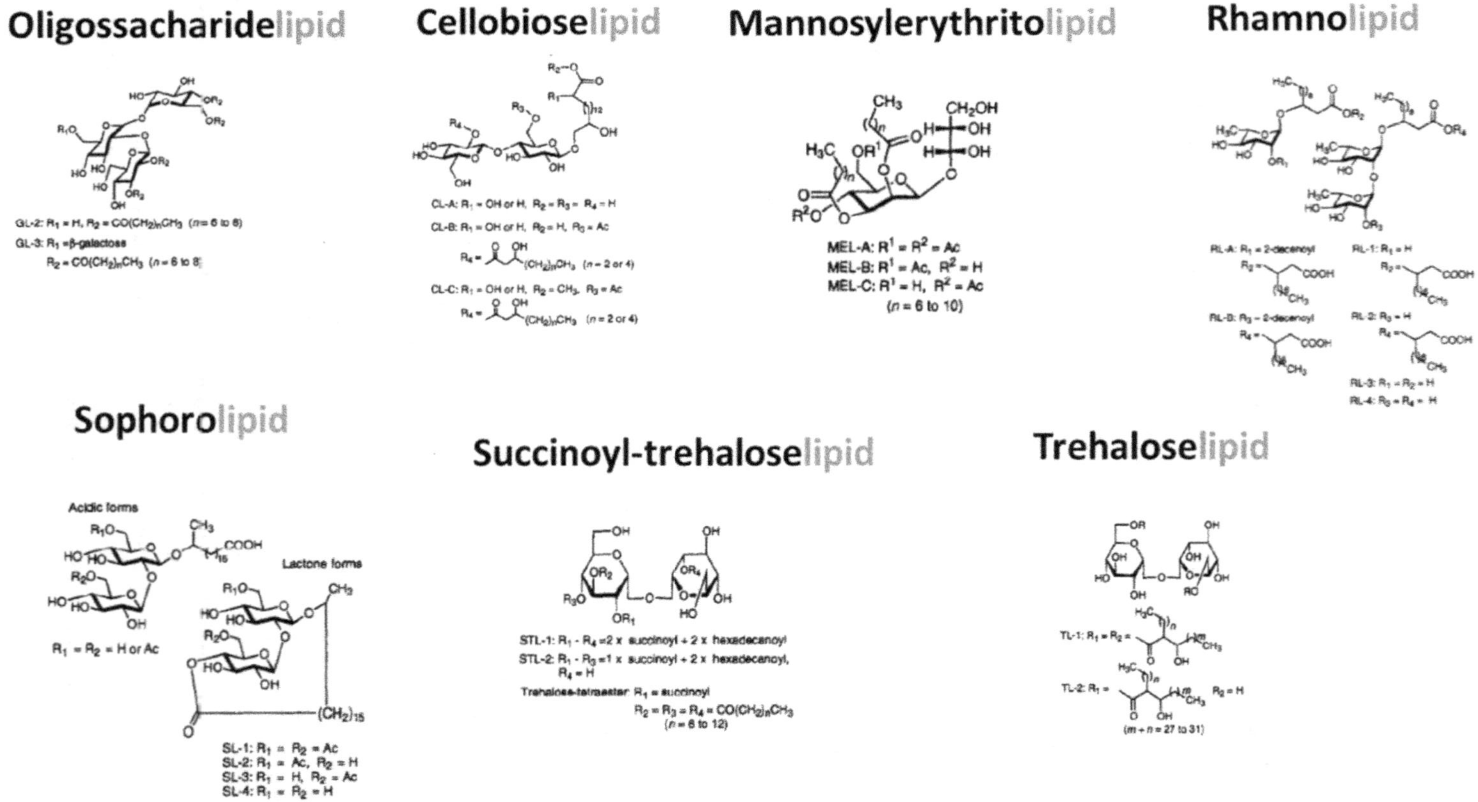

Figure 1.1 The chemical structures of the most common glycolipids.

due to their environmental friendly character and sustainability (Geys et al. 2014; Makkar et al. 2011; Santos et al. 2016; Smyth et al. 2010). Nowadays the preservation of the Earth as a sustainable planet is one of humanities greatest concerns. In line with this concern about the environment, many industries are changing to a global viewpoint on the future of manufacturing. In fact, they have recognized the potential of living cells in the pre-treatment of raw materials, processing operations, product modifications, selective waste management, energy recycling and conservation.

Biosurfactants are quite adaptable, their performance is versatile in a wide range of applications, in different areas, such as pharmaceutics, cosmetics, agronomy, food, beverages, metallurgy, agrochemicals, organic chemicals, petrochemicals, fertilizers, and others (Abdel-Mawgoud and Stephanopoulos 2018). The main applications in the pharmaceutical field are as anti-microbial, anti-cancer, anti-viral and anti-adhesive agents, immunological adjuvants, and in drug and gene delivery (Abdel-Mawgoud and Stephanopoulos 2018).

1.1.1 Application of Biosurfactants

The most commonly used surfactants are chemically produced from petroleum, these synthetic derived agents are generally toxic and not biodegradable. This problem motivates the search for more environmentally friendly surfactants, such as biosurfactants produced by microorganisms, which provides a wide range of applications (Santos et al. 2016; Vijayakumar and Saravanan 2015).

Biosurfactants have several advantages over petroleum based surfactants, such as structural diversity, low toxicity, greater biodegradability, the ability to function in wide ranges of pH, temperature and salinity, as well as greater selectivity, lower protein denaturing potency and lower critical micellar concentration (CMC). The wide range of industrial applications include the field of petroleum industry as well as bioremediation, agricultural, food processing, health, chemical, and cosmetic industries (Abdel-Mawgoud and Stephanopoulos 2018; Santos et al. 2016). The following sections present an overview of the most investigated application fields for biosurfactants.

1.1.1.1 Petroleum Industry

The major accruing market for biosurfactants is currently the petroleum industry, which offers different applications for them (Santos et al. 2016).

Petroleum, also known as crude oil is a natural energy source found beneath the earth's surface. It is a resource that is in great demand, and has become the leading raw material for development and the economy in the past century. It basically consists of two to three phases (liquid/solid and gas), the industry uses several mechanisms to separate these (Almeida et al. 2016). Biosurfactants have shown promising applications in this industry, such as extraction, transportation or petrochemical manufacturing (Makkar et al. 2011).

Approximately 50–65% of oil residues are found in porous rocks, caused by high forces of capillarity, interfacial tension between the aqueous and the hydrocarbon phases and the high viscosity of the crude oil. These residues can often not be recovered by conventional oil recovery methods (Almeida et al. 2016; Santos et al. 2016). Biosurfactants can lower the interfacial tension between oil/rock and oil/water interfaces, in that way the oil-recovery can be enhanced. A mixture containing the producing microorganism can be injected into the porous rock, which will be sealed for a certain time to allow microbial growth. Therefore, the production of biosurfactants allows the breakdown of the oil film in the rocks and oil flow can be re-established. Experiments have shown that the use of indigenous microorganisms *Arthobacter sp, Pseudomonas sp* and *Bacillus sp* resulted in a reduction of paraffin from 29.8 to 25.5 % in 9 months (Bachmann et al. 2014).

Because of their amphipilic structure, biosurfactants have highly emulsifying properties which are important for the extraction of oil, to form stable water–oil emulsions (Bachmann et al. 2014).

1.1.1.2 Bioremediation

Oil spills during the transport, exploration or refining of petroleum products cause huge environmental hazards. The primary transportation method of oil products is via ship, which increases marine oil contamination due to the routine operations of ship washing, and accidents during exploration and transportation (Souza et al. 2014). Conventionally the spilled oil is removed via physicochemical methods, which doesn't solve the problem in the long term, since the contaminants are simply removed from one environment to another, which often results in the development of even more toxic byproducts. For this reason, the search for biological alternatives, such as biosurfactants, is important to contribute to environmental health (Silva et al. 2014).

Generally, bioremediation is a process where toxic compounds are fully or partially degraded through living organisms such as bacteria or plants.

The production of biosurfactants generally requires a hydrophobic and hydrophilic carbon source in the culture medium. This process is economically and environmentally friendly when using waste products as substrates (Silva et al. 2014). The biodegradation of the oil-derived compounds is based on different mechanisms. Through the production of biosurfactants the bioavailability of the hydrophobic substrate for the producing bacterium increases, whereby the surface tension of the medium around the bacteria reduces, which results in a lower interfacial tension. Another mechanism is a membrane modification through an interaction of the cell surface and the biosurfactant, which increases the hydrophobicity of the cell wall by reducing the lipopolysaccharide index through an adhesion of the hydrocarbons, without damaging the membrane. Through these mechanisms the formation of hydrogen bonds is blocked and the surface/interfacial tension is reduced, enhancing the dispersion of the hydrocarbon into micelles (which breaks down the biomass into drops) and amplifies the bioavailability and biodegradability (Aparna et al. 2011; Santos et al. 2016).

Regarding the environmental industry trehalose lipids are used as microbial-enhanced oil recovery, biodegradation of polycyclic aromatic hydrocarbons or oil-spill treatments, in the cosmetics industry and most importantly in the biomedical field with biologic properties, like anti-microbial, anti-viral (Azuma et al. 1987) and anti-tumor activities (Franzetti et al. 2010; Gudiña et al. 2013; Kadinov et al. 2020). Moreover, they can act as therapeutic agents due to their functions in cell membrane interactions (Franzetti et al. 2010).

1.1.1.3 Agriculture

Great agricultural damage can be caused by plant pathogens, which results in the use of chemical pesticides that have a number of negative effects on the health of humans and the environment. This is why the requirement and search for natural and environmental friendly pesticides is highly relevant. Biosurfactants, show anti-microbial properties against a large group of plant pathogens, or they are able to stimulate the plant immune system. In particular rhamnolipids have been broadly studied in terms of crop protection (Crouzet et al. 2020).

1.1.1.4 Food Industry

Due to their emulsifying and anti-bacterial activities biosurfactants show great potential in the food industry. Because of their low toxicity and biodegradability their use can be explored as additives in food.

An emulsion describes a dispersion of two or more phases that do not blend with each other, resulting in a liquid–liquid separation. Emulsions are usually unstable and sensitive to agglomeration of the inner phase, through the addition of biosurfactants the system can be stabilized by the reduction of the surface and interfacial tension, thereby preventing the coalescence through steric and electrosteric barriers (Santos et al. 2016). Examples of foods that are defined as emulsions would be milk, heavy cream or mayonnaise.

Another use in the food industry is caused by the anti-biofilm properties of biosurfactants. Biofilms are a significant issue since they can lead to faster food spoilage, distribution of diseases or contamination. Biosurfactants have been shown to have strong anti-adhesive activities against bacterial and yeast strains, through modifying the physicochemical properties of surfaces or changing bacterial interactions. Surfaces can therefore be pre-conditioned using biosurfactants (Mnif and Ghribia 2016).

There are several further applications in food, such as controlling fat globule agglomeration, improving texture and consistency of fat-based products or stabilizing aerated systems (Santos et al. 2016).

1.1.1.5 Biomedicine

Biosurfactants show great potential in therapeutical applications, due to their biological activities. They have the great advantage of being mostly non-toxic and stable in extreme conditions. In particular the anti-microbial, and anti-biofilm properties of biosurfactants make them attractive agents for pharmaceutical and medical applications. Anti-bacterial properties of biosurfactants (Sana et al. 2018; Solaiman et al. 2016), anti-fungal (Sarwar et al. 2018), and anti-viral activities (Borsanyiova et al. 2016) have been reported.

Biosurfactants are well known by their membrane permeabilization properties as they can induce pore and ion channel formation in lipid bilayer membranes. Moreover, they are able to destabilize membranes disturbing their integrity and permeability. Also, pore formation in membranes may cause transmembrane ion influxes, including Na^+ and K^+, which result in membrane disruption and cell death (Fracchia et al. 2015; Matsumoto et al. 2016).

The anti-microbial property (Rodrigues et al. 2006) of the biosurfactant might be due to their capacity to disturb the integrity of the membranes, destabilizing them and leading to cell lysis by increasing the permeability of the membrane, leading to leakage of metabolites. This effect is based on the interruption of the protein conformation that signals important membrane functions leading to changes in the physical structure of the membrane (Banat et al. 2010; Cortés-Sánchez et al. 2013; Fracchia et al. 2015).

Biofilms describe groups of bacteria that colonize or accumulate on a given surface. They often contain bacterial strains that can become highly resistant. Biosurfactants can interfere with microbial adhesion and biofilm formation, by modifying the physico-chemical properties of surfaces or changing bacterial interactions (Fracchia et al. 2015).

One example for a widely studied biosurfactant in the medical and pharmaceutical field is trehalose lipid. Studies have shown that trehalose lipid contains growth inhibiting properties against several resistant pathogen. Trehalose lipid (TL1) shows anti-fungal activity by which it inhibits the *Chlamydospore* germination of the fungus *Glomerella cingulata* at a concentration of 300 mg L^{-1}. It also shows 30% inhibition of *Candida albicans* growth (Janek et al. 2018). Furthermore succinoyl-trehalose (STL-1 and -2) presents virucidal activity by inhibiting the herpes simplex virus and influenza virus at concentrations from 11–33 mg L^{-1}. Trehalose lipids were investigated to present high anti-adhesive properties against several microorgansims, such as *Candida albicans* and *Escherichia coli*, on polystyrene surface and silicone urethral catheters, therefore it could be a promising coating agent (Janek et al. 2018; Kitamoto et al. 2002).

1.2 Biosynthesis of Glycolipids

The exact synthetic pathways of the majority of glycolipids are not yet fully known. Various pathways are involved in the biosynthesis of precursors for biosurfactant production, depending on the main carbon source used in the fermentation medium.

Generally the last step of the glycolipid-synthesis implicates the linking of glycosyl and lipid precursors. The linking is mostly via o-glycosidic or ester bonds formed by glycosyltransferases or acyltransferases. Glycosyltransferases are highly valuable glycosylation catalysts, inducing the transfer of the glycosyl residue from an activated glycosyl donor (mostly sugar nucleotides or phosphates) to a lipid acceptor, by making glycosidic bonds between the nucleophilic hydroxyl groups of the acceptor and the anomeric carbon of the sugar donor (Abdel-Mawgoud and Stephanopoulos 2018; Bungaruang et al. 2013).

Acyltransferases catalyze the transfer of the lipid residue from an activated acyl donor, such as acyl-coA, to a glycosyl acceptor, by forming an ester bond between the nucleophilic hydroxyl group of the glycosyl residue (acceptor) and the donor's carbonyl group (Abdel-Mawgoud and Stephanopoulos 2018).

Glycolipids could also be theoretically hydrolyzed, as is shown in Figures 1.2 and 1.3. Glycoside hydrolase can hydrolyze the sugar–lipid or sugar–sugar glycosidic bonds (Figure 1.2), while carbohydrate esterase can hydrolyze the sugar–lipid ester bonds (Figure 1.3) (Abdel-Mawgoud and Stephanopoulos 2018).

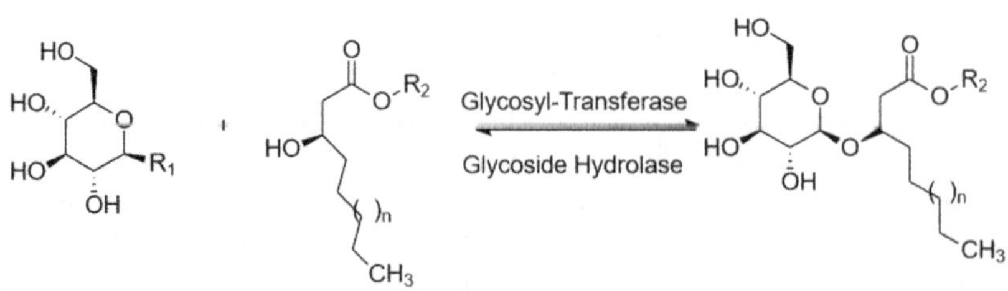

Figure 1.2 Hydrolysis of the sugar–lipid or sugar–sugar glycosidic bonds from glycolipids. R_1: Nucleotides or phosphates groups of the glycosyl residue; R_2: any substitution that could be glycosyl, lipid, or glycolipid units.

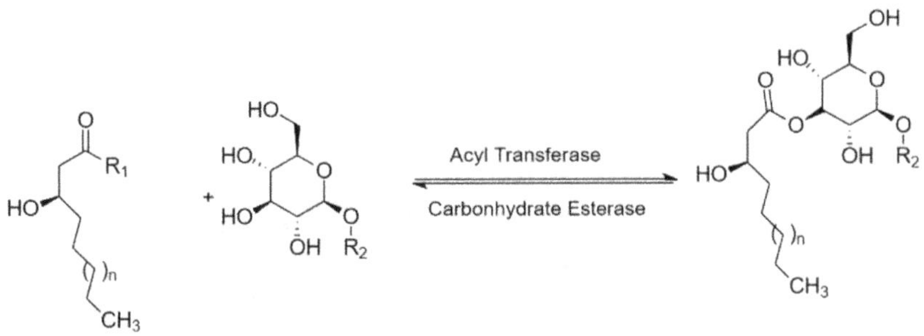

Figure 1.3 Hydrolysis of sugar–lipid ester bonds from glycolipids. R_1: Coenzyme A or Acyl Carrier Protein groups of the acyl donor; R_2: any substitution that could be glycosyl, lipid, or glycolipid units.

1.3 Biosynthesis of Trehalose Lipids

Trehalose lipids are glycolipids containing trehalose hydrophilic moiety. In fact, structurally, trehalose lipids consist of a hydrophilic moiety (trehalose) formed by two glucose units linked through the α,α-1, 1-glycosidic linkage and a hydrophobic moiety represented by chains of fatty acids, such as succinic, octanoic, decanoic, and mycolic acids (Paulino et al. 2016). Trehalose is a non-reducing disaccharide, it is the carbohydrate group of the cell wall glycolipids in *Mycobacteria* and *Corynebacteria* (Franzetti et al. 2010). Trehalose displays thermostability, resistance to acid hydrolysis and non-reactivity to the Maillard reaction.

Trehalose lipids were discovered in 1933 (Anderson and Newman 1933) and purified in 1956 (Soberón-Chávez 2011). They are among the best known biosurfactants, as rhamnolipids and sophorolipids, in both composition and activity (Soberón-Chávez 2011). In comparison to other microbial glycolipids, trehalose lipids have shown contrasting results and achievements in both cases of inhibition and enhancement of biodegradation rates (Kügler et al. 2014; Silva et al. 2014). They can reduce the surface tension of aqueous solutions and the interfacial tension between aqueous and oil phases to levels observed with synthetic surfactants, and have low critical micelle concentrations.

A wide variety of trehalose lipids can be found in several forms (Figure 1.4), namely: trehalose monomycolates, dimycolates, trimycolates, non-ionic acylated trehalose derivatives, anionic trehalose tetraesters, and succinoyl trehalolipids (Franzetti et al. 2010; Kügler et al. 2014; Paulino et al. 2016).

Therefore the different types of trehalose lipids can be classified into two general subclasses:The first group is the 6,6-trehalose diesters such as fatty acid trehalose diesters (TDEs, 1), trehalose dicorynomycolates (TDCMs, 2) and trehalose dimycolates (TDMs, 3). The most quantified trehaloselipid is trehalose 6,6′-dimycolate which is an a-branched chain mycolic acid esterified to the C6 position of each glucose (Franzetti et al. 2010). Although TDEs are the simplest glycolipids found in the trehalose 6,6′-diester series, they can be differentiated by lipid length and branching of the fatty acids, which can be divided into anionic or non-anionic. It is known that anionic surface-active TDEs have a higher surface activity than non-ionic (Silva et al. 2014).The second group is the

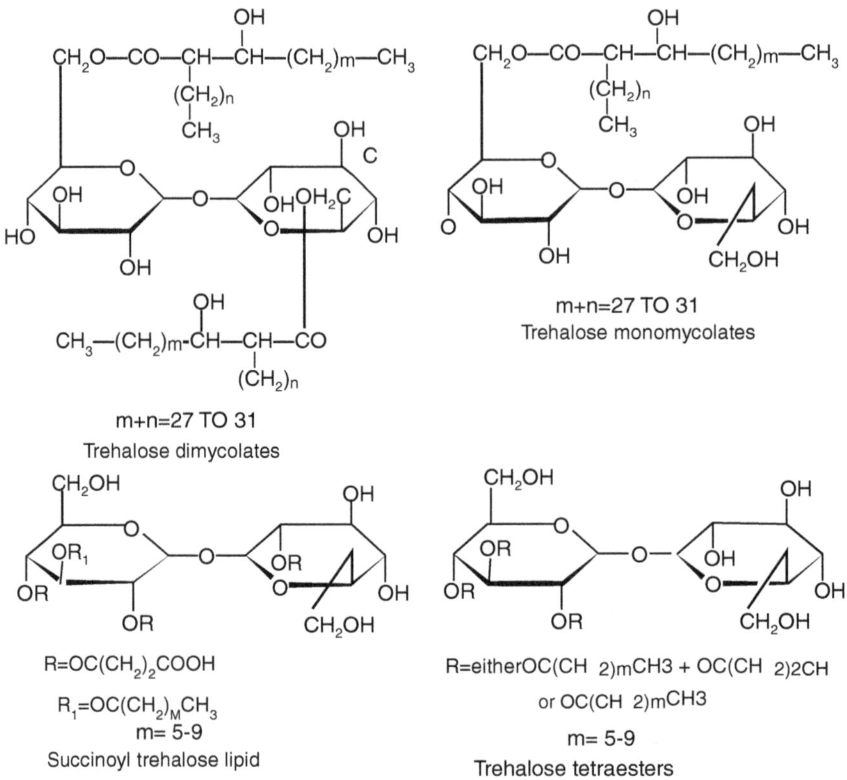

Figure 1.4 The main chemical structure of trehalose lipids (adaped from Franzetti et al. 2010).

2,3-trehalose diesters such as diacyl trehalose sulfates (STL) and sulfolipid-1 (SLL) (Kügler et al. 2014). One of the studies showed that succinoyl trehalose lipids (SCTLs) are promising biosurfactants since they can be efficiently produced from *n*-alkanes by *Rhodococcus* bacteria and recovered by precipitation. Among SCTL producers, *Rhodococcus* sp. SD-74 provided the best yield (40 g L^{-1}) from *n*-hexadecane under high osmotic conditions. These SCTLs also have also been demonstrated to show growth inhibition against the influenza virus and cell-differentiation induction towards human leukemia cells (Isoda et al. 1997; Sudo et al. 2000).

Mycolic acids are 2-alkyl-3-hydroxy fatty acids of high molecular mass, present exclusively in the cell envelope of bacteria of the mycolata taxa, including *Rhodococcus* species (Paulino et al. 2016).

The synthesis of mycolic residues, such as trehalose lipids, is believed to be a Claisen condensation. The key reaction for the formation of trehalose-6-phosphate is catalyzed by trehalose-6-phosphate synthetase (TPS). TPS links the two D-glycopyranosyl units at C1 and C1′ (Franzetti et al. 2010). Uridine diphosphate-glucose and glucose-6 phosphate participate in that reaction, forming trehalose phosphate (Franzetti et al. 2010):

$$\text{UDP-Glucose} + \text{Glucose-6-phosphate} \longrightarrow \text{UDP} + \text{trehalose phosphate}$$

In alkanotrophic *Rhodococci*, TPS is induced by *n*-alkanes. Further reactions for the formation of trehalose lipids have been clearly elucidated for trehalose dimycolates (TDMs) in *M. tuberculosis*. The mycobacterial glycolipid has been proposed to play a key role in the immunopathogenesis of tuberculosis (Hoq et al. 1997; Jain and Roy 2009; Paulino et al. 2016). The production of threalose lipids established during the final stages of the cell wall synthesis of *M.tuberculosis*. In this step newly synthesized mycolic acids are transported and attached to the peptidoglycan-arabinogalactan complex of the cell wall, followed by the formation of trehalose dimycolates. This formation occurs in four different reactions (Franzetti et al. 2010, p. 25; Takayama et al. 2005).

The first reaction of the synthesis, is the transfer of the mycolyl group to D-mannopyranosyl-1-phosphoheptaprenol by an as-yet unidentified cytoplasmic mycolyltransferase, forming 6-O-mycolyl-β-D-mannopyranosyl-1-phosphoheptaprenol (Franzetti et al. 2010; Takayama et al. 2005).

The mycolyl group is then transferred to trehalose 6-phosphate by a membrane-associated mycolyltransferase II to form trehalose mono mycolate (TMM)-phosphate. After a dephosphorylation, the resulting product is TMM (Franzetti et al. 2010) (reaction 2) (Figure 1.5).

The third reaction consists of an extracellular transportation of TMM via an ABC-transporter.

In the fourth reaction trehalose dimycolates(TDM) (Figure 1.5) and arabinogalactan-mycolate are formed from TMM through an extracellular mycolyltransferase (Ag85/Fbp/PS1) (Franzetti et al. 2010; Takayama et al. 2005).

The key reaction for synthesis of the final resulting sugar residue, trehalose-6-phosphate, is catalyzed by a trehalose-6-phosphate synthetase (TPS) which links the two D-glycopyranosyl units at C1 and C10. UDP-glucose and glucose-6-phosphate act as the immediate precursors. In fact, the trehalose moiety and the fatty (mycolic) acid moiety of trehaloselipid molecules are synthesized independently and are subsequently etherified. Trehalose monomycolate was claimed to be an intermediate of TDM biosynthesis. The formation of the mycolate is considered to be a Claisen-type condensation of two fatty acids, a carboxylated acyl-coenzyme A and an activated acyl chain to yield a 3-oxo intermediate, which would then be reduced to form mycolic acid (Kuyukina and Ivshina 2010)

The additional reactions involved in the synthesis of THL have been elucidated for TDMs in which the production happens in the final stages of the synthesis of the cell wall. In this phase, the newly synthesized mycolic acids are transported and attached to the peptidoglycan-arabinogalactan complex of the cell wall, followed by the formation of TDM which occurs by four different reactions.

The biosynthesis proceeds when the first reaction occurs through the transmission of the mycolyl group to D-mannopyranosyl-1-phosphoheptaprenol by a proposed cytoplasmic

1) Mycolyl-S-Pks13 + Man-P-heptaprenol ⟶ Myc-PL — Inside the cell
2) Myc-PL + Treh 6-P ⟶ TMM — Inside the cell
3) TMM(inside) + ATP ⟶ TMM(inside) + ADP + P_i — Transfer out of the cell
4) TMM + TMM ⟶ TDM + Treh — Outside the cell

Figure 1.5 Synthesis of trehalose dimycolates, in four different reactions, the first two reactions are performed inside the cell, the third reaction forms the transport out of the cell and the fourth is outside the cell. Pks13: polyketide synthase 13; Man-P-heptaprenol: D-mannopyranosyl-1-phosphoheptaprenol; Myc-PL: 6-O-mycolyl-β-D-mannopyranosyl-1-phosphoheptaprenol; Treh 6-P: trehalose 6-phosphate; TMM: trehalose mono mycolate; TDM: trehalose dimycolate.

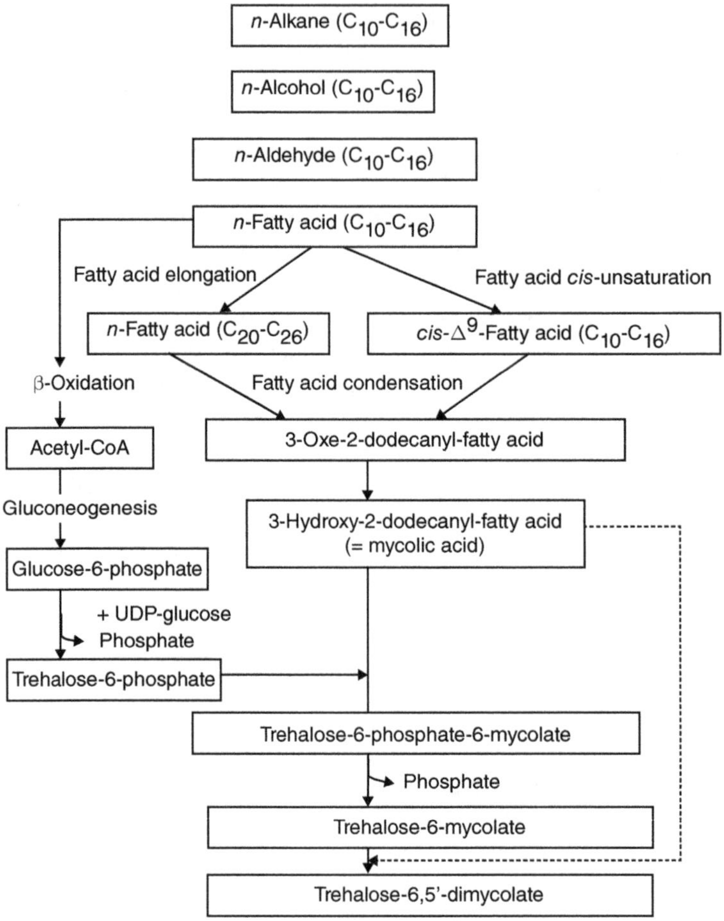

Figure 1.6 Scheme of trehalose-mono- and -dimycolate synthesis from n-alkanes (adapted from Kuyukina and Ivshina 2010).

mycolyltransferase I to form 6-O-mycolyl-b-D-mannopyranosyl-1-phosphoheptaprenol (Myc-PL) (Figure 1.6). For the second reaction the mycolyl group is transferred to trehalose 6-phosphate by a membrane-associated mycolyltransferase II to form trehalose mono mycolate (TMM)-phosphate and, after dephosphorylation, results in the formation of TMM. The third reaction happens when TMM is transported outside the cell by an ABC transporter. A rapid and efficient transfer of TMM from the inside to the outside of the cell is necessary for the synthesis of cell wall arabinogalactan-mycolate and TDM. The fourth and last reaction occurs by the action of the extracellular mycolyltransferase called Ag85/Fbp/PS1, the final products of the cell wall arabinogalactan-mycolate and TDM are formed from TMM (Franzetti et al. 2010).

Production of Glycolipids

The kinetics of glycolipids production has considerable variations among various systems. Only a few generalizations can be made, by grouping kinetic parameters into the following types: (1) growth-associated production, (2) production under growth-limiting conditions, (3) production by resting or immobilized cells, and (4) production with precursor supplementation (Desai and Banat 1997; Santos et al. 2016).

Examples for the growth-associated production (1) can be the synthesis of rhamnolipids by some *Pseudomonas spp* strains. The biosurfactant production is in direct relation to growth and substrate utilization. The excreted emulsan-like substance accumulates on the cell surface during the exponential phase of growth (Desai and Banat 1997).

The production under growth-limiting conditions (2) is characterized by an accentuated increase in the biosurfactant level as a result of the limitation of one or more medium components. An example is the production of rhamnolipids by *Pseudomonas aeruginosa*, particularly when the cells become limited for nitrogen or iron. The limitation results in an overproduction of biosurfactant when the culture reaches the stationary phase of growth (Desai and Banat 1997; Kumari et al. 2010).

Production by resting or immobilized cells (3), describes a biosurfactant production with no cell multiplication; whereby the cells nevertheless continue to utilize the carbon source for biosurfactant synthesis. For example the gram-positive bacterium *Arthobacter crystallopoietes* was shown to directly produce trehalose from maltose by resting cell reaction (Desai and Banat 1997; Seo and Shin 2011).

Santos et al. (2016) report that the addition of biosurfactant precursors (4) to the growth medium can cause qualitative and quantitative changes in the final product (Ashby et al. 2008; Santos et al. 2016).

Furthermore different factors, such as the carbon source, nitrogen source or environmental conditions can influence the glycolipids production and will be focus on next points in the case study of trehalose lipids.

1.4 Production of Trehalose Lipids

Trehalose lipids are the basic component of the cell wall glycolipids in *Mycobacteria* and *Corynebacteria* and are known to be produced by Gram-positive bacteria, as *Actinomycetales* such as *Mycobacterium, Nocardia* or *Corynebacterium* and they differ in the structure, size and degree of saturation (Cappelletti et al. 2020; Franzetti et al. 2010).

Rhodococcus erythropolis DSM43215 was reported for the first time as a producer of trehalose lipids with chain length (C20–C90) of the esterified fatty acids in 1982 (Kretschmer et al. 1982) and in 1983. These trehalose lipids were characterized as trehalose-6-mono-corynomycolates, trehalose 6,6′-diacylates and trehalose-6-acylates (Kretschmer et al. 1982). A non-ionic trehalose lipid, consisting of one major and ten minor components was produced using *Rhodococcus* strain H13-A (Bryant 1990). Other types of trehalose lipids, including mono-, di- and tri-corynomycolates, mono-, di-, tetra-, hexa- and octa-acylated derivatives of trehalose, trehalose tetraesters and succinoyl trehalose lipids were produced, in the following years, using *R. erythropolis* and *R. ruber* (Esders and Light 1972; Uchida et al. 1989; White et al. 2013). The large-scale production of trehalose lipids is very challenging. The effective use of biosurfactants is limited by the high cost of production and complex downstream processing (Franzetti et al. 2010). In addition, when *Rhodococcus* strains are used for this purpose, the major problem is the fact that trehalose lipids are associated with the cell walls leading to an increase in the costs of downstream processing and recovery (Espuny et al. 1996). Furthermore, several studies have shown that the production of trehalose lipids can either be extracellular, or cell-bound, depending on the growth conditions. Experiments presented that *R. erythropolis ATCC 4277* was able to produce extracellular trehalose lipids, which were all released into the medium, using glycerol as the sole carbon source, while the production was partially cell-bound when cells were grown on *n*-hexadecane (Ciapina et al. 2006; Franzetti et al. 2010).

The yields of trehalose lipids appear to be very low compared to sophorolipids, rhamnolipids and mannosylerythritol lipids. They are often bound to cell surfaces, which reduces the production yield and increases downstream costs. Three basic strategies have been adopted to make the fermentation process cost-competitive: (i) using cheap and waste substrates, (ii) development of efficient bioprocesses, such as optimization of fermentative conditions, (iii) development of overproducing mutant or recombinant strains (Franzetti et al. 2010; Uchida et al. 1989). One study has shown that a high phosphate buffer concentration and neutral pH conditions optimize the production of succinoyl trehalose lipids in *R. erythropols SD-74* up to 40 g L^{-1} (Franzetti et al. 2010; Uchida et al. 1989).

In the bioproduction of trehalose lipids many factors need to be considered. Firstly, the microorganism must be carefully chosen to produce the trehalose lipid required, with special attention to the purpose of its application. The microbial production can be influenced by different factors such as media composition (e.g., carbon and nitrogen sources, salt composition or use of extract in culture broth), bioreactor and environmental conditions (e.g. temperature, pH, oxygen, speed). As well as sophorolipids, for trehalose lipids, there is the possibility of further chemical modifications to obtain novel analogues with diverse properties.

1.4.1 Microorganisms

Trehalose lipids are made up of a disaccharide, trehalose, linked by an ester bond to a-branched b-hydroxy fatty acids (Lang and Philp 1998). The a-branched b-hydroxy fatty acids are connected at the C6 and C60 of the carbohydrate structure in the case of the trehalose dimycolates and at C6 for the monomycolates; other structure types have also been reported (Lang and Philp 1998). The production of trehalose lipids is associated with most species of *Mycobacterium*, *Rhodococcus* and *Corynebacterium*.

Trehalose lipids are usually produced by Gram-positive bacteria. Many studies display use of a different microorganisms, such as *Arthrobacter*, *Rhodococcus*, *Gordonia* (Lang and Philp 1998), *Mycobacterium*, *Nocardia* and C*orynebacterium*, which allowed trehalose lipids which differ in structure, size, degree of saturation and mycolate group (Franzetti et al. 2010).

Nevertheless, some of the producing bacteria are pathogenic mycobacteria (including pathogens in the *Mycobacterium avium*, M. *intracellulare* group, *nocardia (Nocardia asteroids)*, and *corynebacteria (Corynebacterium diphtheria, C. matruchotii*, and *C. xerosis*) and the observed or potential pathogenicity of these producer strains, as well as the high toxicity of the biosynthesized glycolipids, significantly restricted their use, specifically in the biomedical field, in which is mandatory they are GRAS (Generally Regarded As Safe). For this purpose, non-pathogenic producing bacteria have been studied, in particular, representatives of *Rhodococcus*, *Gordonia*, *Dietzia*, *Tsukamurella*, *Skermania*, *Williamsia*, among others (Kuyukina and Ivshina 2019; Kuyukina et al. 2020). Although all of these microorganisms can be used for the production of trehalose lipids, *Nocardia* and *Rhodococcus* are the most widely explored (Patil and Pratap 2018).

In the case of *Rhodococcus* and related genera, the trehalose lipids biosynthesized are usually bound to the cell envelope and are produced mainly when the microorganisms are grown on hydrocarbon sources. This ability to access hydrocarbons is related to the microorganism's cell surface hydrophobicity. Cells with high hydrophobicity directly contact oil drops and solid hydrocarbons, while low hydrophobicity allows the adhesion of microbial cells to the micelles or emulsified oils, formed due to the presence of extracellular biosurfactants (Franzetti et al. 2010). White et al. (2013) evaluated the production of trehalose lipids by a marine bacterium *Rhodococcus* sp. PML026 using sunflower oil as a hydrophobic substrate.

The capacity to produce a metabolite is bestowed by the genes of the microrganism, affecting the yield of biotechnological products. A production process is commercially viable and profitable after the yield of the final product by the producer organisms is high, in fact the bioindustrial production process is often dependent on the use of hyperproducing microbial strains, even with cheap raw materials, optimized medium and culture conditions, and efficient recovery processes (Bouassida et al. 2018). The genetic manipulation to improve production of biosurfactants was highlighted in 2016 by Paulino et al. (2016).

Moreover, when good yields from natural producer strains are lacking, the industrial production process will depend on the availability of recombinant and mutant organisms. To economize the production process further and to obtain products with better commercially important properties, recombinant hyperproducers have been developed. However, only a few mutant and recombinant varieties with enhanced biosurfactant production characteristics are reported in the literature (Mukherjee et al. 2006). Transposons (Koch et al. 1991), chemical mutagens (e.g., N-methyl-N0 nitro-Nnitrosoguanidine (Lin et al. 1998), radiation (Iqbal et al. 1995) or by selection are strategies used to produce mutant varieties.

The first report about the use of genetic engineering was the insertion and expression of the *Vitreoscilla* hemoglobin gene (vgb) in *Gordonia amarae*. This resulted in enhancement of trehalose lipid production in a medium supplemented with 1% hexadecane (Dogan et al. 2006). The elucidation of biosynthetic pathways through the identification of putative acyl coenzyme A transferase, fructose-biphosphate aldolase, and alkane monooxygenase genes and genetic manipulation allowed an increase in succinoyl trehalolipid production by *Rhodococcus* sp. strain SD-74 (Inaba et al. 2013).

The screening of biosurfactant producers can be carried out using several methods (qualitative and/or quantitative) with variable precisions and aims. Some examples of these methods are the following: (i) hemolytic assay (HA) (Carrillo et al. 1996), (ii) the N-Cetyl-N,N,N-trimethylammoniumn bromide (CTAB) agar plate method (Pinzon and Ju 2009), (iii) cell surface hydrophobicity methods, that is to say bacterial adherence to hydrocarbons (BATH) assay (Desai and Banat 1997) and microbial adherence to hydrocarbons (MATH) assay (Siegmund and Wagner 1991), (iv) the oil spreading method (Morikawa et al. 2000), (v) emulsification activity (Varjani et al. 2014), (vi) tilted glass slide (Satpute et al. 2010), (vii) hydrocarbon overlay agar (HOA) (Satpute et al. 2008), (viii) aximetric drop shape analysis (ADSA) (Satpute et al. 2010), (ix) direct colony thin layer chromatographic, turbidity assay (Satpute et al. 2008), (x) tensiometric measurements, like surface and interfacial tension (Varjani et al. 2014; Varjani and Upasani 2016). The primary screening method for biosurfactant producers were hemolytic assay and CTAB agar test methods (Mulligan et al. 1984; Satpute et al. 2010).

1.5 Factors Affecting Trehalose Lipid Production

The factors affecting trehalose lipids production can be divided in two major groups: (i) nutritional and (ii) environmental.

i) Nutritional factors

The carbon source can influence the biosurfactant synthesis either by induction or repression, it has considerable bearing on the produced type of biosurfactants. Furthermore nitrogen can play an important role in the regulation of biosurfactant synthesis (Cameotra and Makkar 1998; Ribeiro et al. 2013; Santos et al. 2016).

1.5.1 Carbon Source

The growth of the microorganisms for trehalose lipid production requires a carbon source. This carbon source influences not only the quantity but also the quality of the formed trehalose lipids. When carbon sources and insoluble substrates are combined the intracellular diffusion and production of different substances are simplified (Santos et al. 2016). For instance, in 1996 a study (Espuny et al. 1996) proceeded to test the effect of different carbon sources where two different mixtures were used, P-120 and P-147. P-120 was composed of *n*-decane, *n*-undecane, dodecane, *n*-tridecane, *n*-tetradecane. The second mixture, P-147, had in its composition *n*-tetradecane, *n*-pentadecane, *n*-hexadecane, *n*-heptadecane. In this study, the higher yield was accomplished using *n*-tridecane alone (Espuny et al. 1996).

In several studies, *n*-hexane was used as the carbon source (Christova et al. 2015; Janek et al. 2018; Kuyukina et al. 2016; Tokumoto et al. 2009). Another study showed that *Rhodococcus erithropolis*, using *n*-hexane as the carbon source was able to produce mono- and di-succinyl trehalose lipids (Kitamoto et al. 2002; Uchida et al. 1989).

Kuyukina et al. (2015) used as the carbon source, *n*-dodecane and *n*-hexadecane and obtained 6.5 and 9.4 g L^{-1} of trehalose lipids, showing that the best method uses hydrocarbons in the bioprocess. *Rhodococcus wratislaviensis* BN38 grown in *n*-hexadecane produced 3.1 g L^{-1} of 2,3,4,2′-trehalose tetraesters containing succinic acid. These trehalose lipids showed high surface activity reducing the surface tension to 28.6 mN m^{-1} in mixture and 24.4 mN m^{-1} when purified at a CMC value of 5 mg L^{-1}.

In a different study, Marqués et al. (2009) carried out a study with *Rhodococcus erythropolis* 51T7 in a medium containing 2% of tetradecane, enabled the production of trehalose tetraester in a concentration of 0.48–1.12 g L^{-1} and a reduction in the surface tension to 27.9 mN m^{-1} at a CMC value of 37 mg L^{-1}.

Nevertheless, other sources have been used such as *n*-alkanes (Inaba et al. 2013), which have been used for the production of TL-1 and TL-2, and *n*-decane to produce trehalose-tetraesters (Kitamoto et al. 2002), glucose or vegetable oil (Santos et al. 2016), glyceryltrioleate (Kügler et al. 2014).

Furthermore, using sugars as carbon sources directly may affect the carbohydrate moiety of trehalose lipids, which might have a good effect on the production and be less expensive.

Different carbon sources, like glycerol, diesel oil, and vegetable oil, showed feasibility as trehalose lipids producers (Bajaj et al. 2014). Additionally the authors (Bajaj et al.

2014) correlated the possibility of using trehalose lipid bioproducers as potential agents in bioremediation process.

A further target for production is the use of low-cost substrates. Experiments on *R. erythropolis 16 LM.USTHB* showed the conversion of used sunflower frying oil, a cheap renewable source, into extracellular glycolipids with the ability to lower the surface tension of crude broth to 31 mN m^{-1} (23 Franzetti et al. 2010).

Kügler et al. (2014) performed a study using non-pathogenic actinomycetes *Tsukamurella spumae* and *Tsukamurella pseudospumae* using sunflower oil as the carbon source in a fermentation medium. They found trehalose lipids with C4–C6 and C16–C18 short acyl chains in hydrophobic moiety, in a low concentration of product (approximately 1.3 g L^{-1}). Kundu et al. (2013) performed a study using the *Rhodococcus pyridinivorans* NT2 strain isolated from effluent sediment contaminated with pesticides and evaluated it for biodegradation of 4-nitrotoluene. The biosurfactant produced, a trehalose lipid, exhibited surface activity, which allowed the reduction of surface tension of the media from 71 to 29 mN m^{-1}, with a CMC value of the 30 mg L^{-1}. The most abundant trehalose lipid derivatives produced by the mentioned strain were trehalose-succinic acids and 2,3,4,2′-trehalose tetraester analogues. Moreover, an emulsification index of 90–95% was obtained with long-chain hydrocarbons (diesel, liquid paraffin, motor oil, groundnut oil, and soybean oil) while shorter-chain alkanes resulted in a lower emulsification index (50–80%). The microbial characteristics of *Rhodococcus pyridinivorans* NT2 contributed to its potential use in bioremediation field, despite the low concentration in the production of trehalose lipids (45 mg L^{-1}) (Kundu et al. 2016a, 2016b).

1.5.2 Nitrogen Source

In fermentative processes, the carbon/nitrogen (C/N) ratio usually affects the accumulation of metabolites. High carbon/nitrogen (C/N) ratios limit bacterial growth. On the other hand, low carbon/nitrogen ratios lead to the synthesis of cellular material and limit the accumulation of products (Santos et al. 2016).

In the specific case of trehalose lipids, different nitrogen sources can be used.

In 1988, Ramsay et al. (1988) used sodium nitrate instead of ammonium sulfate, as a nitrogen source, with good results.

In another study (Uchida et al. 1989) the effect of the different nitrogen sources was evaluated. The cell growth was not affected considerately by ammonium sulfate, ammonium dihydrogen phosphate, ammonium nitrate, or urea. Nevertheless, potassium nitrate allowed a higher yield (Uchida et al. 1989). In another work, making use of limited nitrogen conditions allowed trehalose mycolates to be formed by *Rhodococcus erythropolis*, to produce anionic trehalose tetraesters (Lang and Philp 1998).

In fact, nitrogen sources played a key role in biosurfactant production. Wang et al. (2019) noticed that when organic nitrogen urea was used as the nitrogen source, in the *Rhodococcus qingshengii* strain FF growth, the yield of trehalose lipids was always higher compared to the use of inorganic nitrogen, ammonium nitrate. When *n*-hexadecane was used as the sole carbon source, the yield of trehalose lipids was 1.62 g L^{-1} with urea as the nitrogen source, which was 2.36 times the yield obtained when inorganic nitrogen ammonium nitrate was used (Wang et al. 2019). Through screening different types of carbon and nitrogen sources, Wang et al. (2019) identified, hexadecane:oleic acid ($m : m = 1 : 1$) and urea, as the best carbon and nitrogen sources respectively.

ii) Environmental factors

The environmental parameters for the fermentative process are of great importance. Environmental factors like pH, temperature, agitation and oxygen availability, play vital roles in microbial growth and glycolipid production as they show their effects on cellular growth and activity (Varjani and Upasani 2017). For trehalose lipid production in a bioreactor with control systems (e.g. digital), these parameters were stable during bacterial growth (Pacheco et al. 2010).

Although trehalose lipids produced by *Rhodococcus* sp. formed emulsions that were stable at pH 2–10 and temperatures of 20–100 °C (Mnif and Ghribi 2015; White et al. 2013), the temperatures were stabilized between 28 °C and 30 °C during growth and the pH was neutral and stable as well, leading to higher yields (Janek et al. 2018; Kügler et al. 2014; Kuyukina et al. 2016).

In order to avoid the problem of low concentration of trehalose lipids in biotechnological processes, the use of statistical methods is an alternative approach to optimize the factors that affect growth and production, increasing yields and reducing process costs. Through response surface methodology (RSM), Mutalik et al. (2008) achieved an increase from 3.2 to 10.9 g L^{-1} in the concentration of trehalose lipids, using *Rhodococcus* spp. MTCC2574 as a biocatalyst and *n*-hexadecane as a substrate (Mutalik et al. 2008).

1.6 Downstream Process

The downstream process is probably the major problem to overcome in bioprocesses, as normally the purification of the target biological compound can account for over half of the manufacturing cost in many biotechnology processes (Desai and Banat 1997). In fact, even if production is optimized related to media and culture conditions, the manufacturing process is incomplete without an efficient and economical strategy for the recovery of the bioproducts. For many biotechnological products, the downstream processing costs account for 60% of the total production costs.

In the recovery of biosurfactants several conventional methods have been used over time, like acid precipitation, solvent extraction, crystallization, ammonium sulfate precipitation and centrifugation (Desai and Banat 1997). One of the main problems in these recovery processes is the toxicity of the solvents (e.g. acetone, methanol, chloroform) in nature and which are harmful to the environment. In recent years an alternative to those solvents was the use of methyl tertiary-butyl ether (MTBE), a cheap and less toxic solvent, mainly used in the recovery of biosurfactants produced by *Rhodococcus* (Kuyukina et al. 2001; Philp et al. 2002).

In the case-study of the trehalose lipid downstream process, although a variety of methods are available, the most commonly used is solvent extraction. In fact, extraction is one of the most used to obtain a crude extract free from the aqueous culture medium. One of the main problems is the number of impurities, often co-extracted during extraction along with several structural types of the target biosurfactant. Afterwards, they need to be evaluated by separating and removing the impurities. There are different mixtures of solvent systems that can be used, the three most frequently used are chloroform-methanol, methyl tert-butyl ether (MTBE), or a mixture of ethyl acetate-methanol (Franzetti et al. 2010). When the goal is the production of trehalose lipids to be applied in the biomedical field, some of the solvent/extraction systems may represent a problem, due to their toxicity.

Other interesting methods have been reported, such as foam fractionation (Davis et al. 2001), ultrafiltration (Sen and Swaminathan 2005), adsorption–desorption on polystyrene resins and ion exchange chromatography and adsorption–desorption on wood-based activated carbon (Dubey et al. 2005). These methods are based on the surface activity of trehalose lipids or their ability to form micelles and/or vesicles and are particularly useful for large-scale continuous recovery of extracellular biosurfactants, with high purity from culture broth. In particular in the pharmaceutical industry, but also in the food and cosmetic industries, individually purified trehalose lipids are required. In these applications the trehalose lipids can be obtained performing multistep recovery by further purification steps. Silica gel column chromatography is a relatively inexpensive method that can be combined with other methods to purify trehalose lipids. Using this technique milligram to kilogram quantities can be obtained free from impurities and can also be used to separate structural types of trehalose lipids for further analysis.

1.7 Identification and Characterization

Glycolipids can be identified and characterized using a broad range of techniques from simple colorimetric assays to sophisticated techniques, like mass spectrometry (MS) and nuclear magnetic resonance (NMR).

The choice of the chemical and structural analysis of glycolipids depend on the purpose. The experimental procedure mainly rely on the following key steps: (i) extraction of glycolipids from the culture medium, (ii) detection; (iii) separation and purification of the crude product (iv) structural analysis.

In the first step colorimetric methods can be used to evaluate the presence of the biosurfactants in either the culture medium or the extract. Detection can be carried out using assays based on the detection of sugar moieties, such as anthrone (Hodge and Hofreiter 1962) or orcinol assay without the need for extraction. The anthrone method is a colorimetric technique based on the reaction of anthrone with the sugar, forming a coloured complex that can be quantified in the visible region, by spectrophotometry. However, interferences from chemicals and carbon sources can result in inaccurate results and, therefore, these methods should only be used as a rough indicator of glycolipid production.

Thin layer chromatography (TLC) is a simple method that allows the detection of glycolipids and can also provide information on possible structural types of glycolipids present in the sample. TLC detection should be carried out before purification procedures to evaluate the presence of glycolipids and should be used afterwards to determine purity.

Further analyses are required for both quantification and/or identification of structural features, mainly using high-performance liquid chromatography-mass spectrometry (HPLC-MS) (Marqués et al. 2009) and nuclear magnetic resonance (NMR) (Kügler et al. 2014). Among these techniques, mass spectrometry offers the greatest amount of information with regard to purity and structural conformation of carbohydrate and fatty acid components of the glycolipids.

In fact, HPLC is a method that allows the separation of glycolipids and when coupled with an evaporative light scattering detector (ELSD) or mass spectrometry provides valuable information related to the identification and quantification of glycolipids. HPLC-UV can also be used for analysis when the test compounds have been derivatized to p-bromophenacyl esters (Mata-Sandoval et al. 1999).

In the specific case of trehalose lipids many techniques can be used for the identification and characterization. Similar to other glycolipids, trehalose lipid content can be estimated using the colorimetric anthrone method. Also TLC has been extensively used in the detection of trehalose lipids. A few solvent systems have been used to separate trehalose lipids, although the most frequently used is chloroform:methanol:water (65:15:2) for the mobile phase (Christova et al. 2015) and to reveal them, using a chemical developer, frequently made of acetic acid: anisaldehyde: sulfuric acid, under 150°C airstream for 2–4 min. Trehalose lipids appear in the form of spots (Kügler et al. 2014), with the monomycolates near the point of application of the sample and the mark of dimycolates a little further away.

High-performance liquid chromatography (HPLC) and reverse phase (HPLC-RC) have been used for the identification of trehalose lipids (Janek et al. 2018), especially with mass spectrometry HPLC (HPLC-MS) it is possible to disclose the glycoside linkage between two of the sugar moieties (Patil and Pratap 2018). Moreover reversed-phase HPLC (HPLC-RC) was also used for the identification of these compounds.

Mass spectrometry (MS) provides detailed structural information on the molecular mass of the compounds under investigation. Tandem MS (MS/MS) results in the fragmentation of structures thus allowing the identification of individual isomers without the need for separation. Moreover, when combined with HPLC, it provides the most sensitive method for identification and quantification of trehalose lipids. A drawback is that it requires a high level of purification as salts and free non-polar lipids can induce suppression of ion signals under MS conditions. Glycolipids can be analyzed on all types of mass spectrometers, with electrospray ionization (ESI-MS) and matrix assisted laser desorption ionization (MALDI) (Yagi-Utsumi 2019).

Electrospray ionization provides excellent trehalose lipid ionization when used for direct infusion or HPLC-MS. Using this technique virtually no fragmentation occurs in the primary molecules under investigation. Ionized molecules are detected by a mass analyzer according to their mass to charge ratio (m/z) and can be fragmented using collision-induced dissociation (CID) to provide valuable information about each structure and their isomers (Yagi-Utsumi 2019).

MALDI is a soft ionization mass spectrometry technique that allows the identification of intact compounds. Basically, samples to be analyzed are mixed with a matrix and dried on a platform, onto which a laser is fired with various degrees of energy, thus forming gaseous ions, which can then be observed in a time of flight analyzer (Yagi-Utsumi 2019).

The characterization can be done by breaking down the structure separating the fatty acids from the carbohydrate or just by analysing the entire molecule. Mass spectrometry or mass spectrometry in conjugation gas chromatography (GS) (Christova et al. 2015) can be used.

To achieve a full structural determination, nuclear magnetic resonance spectroscopy (NMR) is utilized and is the most powerful method able to identify functional groups as well as the position of linkages within the carbohydrate and lipid molecules. Using a series of NMR experiments the exact location of each functional group can be obtained and information about the structural isomers is also possible (Sato et al. 2019).

Fourier-transform infrared spectroscopy (FTIR) is applied for the characterization of trehalose lipids, once these molecules form polymeric aggregates in aqueous media above the critical micellar concentration (CMC). The type of aggregate structure can play a biological role and is determined by the shape of the contributing molecules which is determined by their primary chemical structure and is influenced by pH, concentration of mono- and divalent cations, among others (Brandenburg and Seydel 1988).

1.8 Surface-Active Properties

Biosurfactants are surface-active compounds, capable of reducing surface and interfacial tension at the interfaces between liquids, solids and gases, allowing them to mix or disperse readily as emulsions in water or other liquids.

Surfactant-related parameters are surface tension, maximum surface excess value, minimum surface area occupied by the surfactant molecule at the air–water interface at saturated adsorption, the minimum concentration of the surfactant required to reduce surface tension and the Gibbs free energy of adsorption (Marqués et al. 2009). The interfacial tension, the emulsification index, and the hydrophilic lipophilic balance (HLB) are other important surface active parameters. The interfacial tension is usually measured by tensiometry. The emulsification index is the direct indication of the amount of surfactant. Afterwards the emulsification index can be monitored during the course of fermentation to evaluate the production of trehalose lipids, by mixing an equal volume of cell free broth and liquid paraffin (Cooper and Goldenberg 1987).

Moreover, surface tension has a tendency to decrease with the increase of glycolipid concentration. The structural diversity of glycolipids structures originate different surfactant properties. Generally they are able to reduce the surface tension of water to 43–24.1 mN m^{-1} and present CMC values of 0.7–37 mg L^{-1} (Marqués et al. 2009; Tuleva et al. 2009; Yakimov et al. 1999).

The trehalose lipids are able to reduce the surface tension of the water, for instance from 72 to 34 mN m^{-1} (Janek et al. 2018). This is in agreement with most trehalose lipids produced from *Rhodococcus*, with strong surface activity by lowering water surface below 30 mN m^{-1}, lowering the interfacial tension against hexadecane up to 1 mN m^{-1}. An example, is a purified trehalose lipid, produced by a novel marine bacterium *Rhodococcus* sp. PML026 using sunflower oil as a hydrophobic substrate, being able to reduce the surface tension of water to 29 mN m^{-1} with a CMC value of 250 mg L^{-1} (White et al. 2013).

Marqués et al. (2009) analyzed the chemical and physical properties of a glycolipid synthesized by *Rhodococcus* sp. 51T7. They demonstrated that this biosurfactant was a trehalose tetraester (THL) consisting of six components: one major and five minor. The hydrophobic moieties ranged in size from 9 to 11 carbons. The critical micelle concentration (CMC) was 37 mg L^{-1} and the interfacial tension against hexadecane was 5 mN m^{-1}. At pH 7.4 the trehalose lipid CMC/critical aggregation concentration (CAC) was 50 mg L^{-1} and at pH 4 it was 34 mg L^{-1} (Marqués et al. 2009). A break at around 40 mg L^{-1} was consistent with the CMC/CAC obtained from surface tension measurements. An increase in pH (4–7.4), CAC values increased as the more negatively charged carboxylate group of trehalose lipid tends to complicate aggregation. When NaCl was added at pH 7 trehalose lipid behaved in a non-ionic way and the CAC was unaffected. A phase diagram showed effective emulsification with water and paraffin or isopropyl myristate. A composition of 11.3–7.5–81.8% (isopropyl myristate–trehalose lipid–water) was stable for at least three months. The HLB was 11 and the phase behaviour of the glycolipid revealed the formation of lamellar and hexagonal liquid-crystalline textures (Marqués et al. 2009).

Depending on the *Rhodococcus sp.* the different trehalose lipids produced changes in interfacial tensions, critical micelle concentrations and emulsifier index and biologic activity.

1.9 Biologic Activity

As mentioned before glycolipids have a wide range of applications. Since trehalose lipids are able to reduce the surface tension of aqueous solutions and the interfacial tension between aqueous and oil phases and have low CMCs, they are especially used as emulsifiers, foaming, wetting (Mujumdar et al. 2019), solubilizers anti-adhesive and anti-microbial agents, among others.

Therefore, trehalose lipids show great potential in different fields (Kuyukina et al. 2001; Mutalik et al. 2008). They are able to act in diverse industries, including (i) the environmental industry with applications such as microbial-enhanced oil recovery, biodegradation of polycyclic aromatic hydrocarbons or oil-spill treatment, (ii) the food industry, where trehalose lipids are especially used as emulsifiers, foaming, wetting, solubilizers anti-adhesive, and anti-microbial agents, (iii) the cosmetics industry and most importantly in (iv) the biomedical field with properties like anti-microbial (Sen et al. 2017), anti-viral (Azuma et al. 1987), anti-tumor activities (Franzetti et al. 2010; Gudiña et al. 2013) (e.g. inhibitory activity on calcium-dependent protein kinase C of human promyelocytic leukemia HL60 cells, inhibitory effects in growth and differentiation-induced against human leukemia cells (Baeva et al. 2014; Kuyukina et al. 2007; Paulino et al. 2016) and immunomodulation activity (Paulino et al. 2016).

Moreover, they can act as therapeutic agents due to their functions in cell membrane interactions. Many studies showed the influence of trehalose lipid interaction with membranes, proteic models, and enzymes, demonstrating the role and hypothetical action site of these biosurfactants (DeBosch et al. 2016).

In the work of Mclaughlin et al. 1980, a tumor regression was seen in guinea pigs bearing transplantable, line-10 hepatocellular carcinoma when synthetic muramyl dipeptides combined with trehalose dimycolate in oil-in-water emulsions were injected directly into the tumors.

In 1996 succinoyl trehalose lipid extracted from *n*-hexadecane culture of *Rhodococcus erythropolis* SD-74 remarkably inhibited the growth of a human monocytoid leukemic cell line (Isoda et al. 1997). In a recent study, trehalose lipids proved to be potent in reducing breast cancer cell viability and ineffective on the contractility of rat mesenteric arteries in vitro (Kadinov et al. 2020).

The work of Natsuhara et al. (1990) revealed an anti-tumor effect of trehalose dimycolates (TDM) from *Rhodococcus ruber* M1 on a subcutaneously implanted sarcoma-180 and allogeneic sarcoma of mice, which was accompanied by significant granuloma formation in lungs, spleen and liver and elevated levels of TNF-*a* (Natsuhara et al. 1990). The abilities of trehalolipids from *Rhodococcus* to induce TNF-*a* determine their anti-tumor activities (Natsuhara et al. 1990).

The biological activity of several succinoyl trehalose lipids (STL) from *Rhodococcus erythropolis* SD-74 was investigated by Isoda et al. (1996). They found that STL induced cell differentiation into monocytes instead of cell proliferation in human myeloid (HL60), monocytoid (U937), and erythroid (K562 and KU812) leukemia cell lines. Additionally they found that biological effects of STL were dependent on the structure of hydrophobic moiety (Isoda et al. 1997; Sudo et al. 2000). Pretreatment of U937 cells with STL evolved their phagocytic activity (Isoda et al. 1997), by increasing the presence of cells with Fc receptors for bacterial recognition and, after the phagocytic uptake of PK-2 dye or opsonized yeast particles (Groves et al. 2008; Matsumoto et al. 2016; Ribeiro et al. 2012). Gein et al. (2011) showed that lipids, in the forms mono-acyl trehalose and di-acyl trehalose, produced by *Rhodococcus ruber* IEGM 23 in a culture medium containing *n*-dodecane, prevented the adhesion of human monocytes to polystyrene surfaces and inhibited their cytokine production without any cytotoxic effects in vitro. This was assessed measuring the inhibition of proliferative activity of cultured human peripheral blood lymphocytes. Trehalose lipids were also studied in emulsions as immunomodulatory and anti-tumor agents (Matsumoto et al., 2013; Kuyukina et al., 2015). Cytokine stimulating activity and an increased tumor necrosis factor-α (TNF-α), interleukin (IL)-1β, and IL-6 production were observed with trehalose lipids, when applied to the adherent human peripheral blood monocyte culture (Gein et al. 2011).

Kuyukina et al. (2016) described the production of trehalose lipid from *Rhodococcus ruber* IEGM 231 using 3% (v/v) *n*-hexadecane at 160 rpm, 28°C for 48 h, and the anti-adhesive and biofilm-preventing effects against Gram-positive and Gram-negative bacteria strains. Additionally, interesting anti-adhesive effects were obtained using the trehalose lipid at 10 mg L^{-1} against actively growing *B. subtilis* ATCC 6613, *Corynebacterium glutamicum* IEGM 1861, *E. coli* K-12, *Micrococcus luteus* IEGM 401, and *Pseudomonas fluorescence* NCIMB 9046 cells with different percentages of inhibition (30–76%) (Kuyukina et al. 2016). Moreover, these authors suggested that anti-adhesive properties were dependent on hydrophobicity/surface characteristics of the strains tested and their physiological stage and not strongly dependent upon the concentration of trehalose lipid (Kuyukina et al. 2016).

A trehalose lipid biosurfactant secreted by *Rhodococcus fascians* BD8 was investigated as an anti-microbial and anti-adhesive against pathogenic bacteria and *Candida albicans* to polystyrene, silicone, and glass surfaces (Janek et al. 2018). Up to 95% prevention of *Candida albicans* adhesion to a polystyrene surface was achieved with 0.5 mg mL^{-1} trehalose lipid. The authors (Janek et al. 2018) concluded that the exploration of trehalose lipid interaction with medical surfaces using quantum chemical calculations and due to its surface tension properties, trehalose lipids are interesting as surface coating agent against microbial colonization of various surfaces (e.g., implants and urethral catheters).

1.10 Conclusions

Glycolipid biosurfactants are surface-active natural compounds produced by several microorganisms with biological activities and potential applications in environmental, medical, cosmetic, pharmaceutical, and food industries.

Microbial glycolipid biosurfactants have many advantages over chemically synthesized surfactants, such as lower toxicity, they are environmentally friendly, have similar surface activity. Downstream processing is probably the most expensive process in the production of microbial glycolipids. To obtain pure glycolipids from production medium requires several operations and purification steps, with extraction being still the most commonly used.

The use of economically feasible renewable substrates, the optimization of growth and production conditions and efficient multi-step downstream processing will enhance the manufacturing and application of glycolipids, and be more profitable.

Novel recombinant varieties, especially beyond the development of novel recombinant microorganism hyperproducers may potentially bring the required development in these biosurfactant production process.

With ever increasing reports regarding the therapeutic and biomedical properties of glycolipids (e.g. trehalose lipids) as biosurfactants, these molecules will surpass the realm of surfactants and might emerge as highly valued molecules with relevance to health in the near future. The future application of glycolipids (e.g. trehalose lipids) in drugs or medicines will make it really interesting for industry. Therefore, future glycolipid research should be focused on making the production process economical with the potential use of hyperproducers in addition to novel cost-effective bioprocesses.

In the study of trehalose lipids, future work should be focussed on the use of inexpensive (when adequate) carbon substrates, optimization of C/N and enviromental conditions, leading to the highest yields, combined with cost effective downstream processing methods. A large group of biosurfactant producers belonging to the generas *Rhodococus*, *Gordonia* or *Torulopsis* have not been exploited extensively for the economical production of trealose lipids.

Additionally there is the possibility of further chemical modifications of trehalose lipids, to obtain novel analogues with diverse and improved properties.

References

Abdel-Mawgoud, A.M. and Stephanopoulos, G. (2018). Simple glycolipids of microbes: chemistry, biological activity and metabolic engineering. *Synthetic and Systems Biotechnology* 3 (1): 3–19.

Almeida, D.G., Soares Da Silva, R.C.F., Luna, J.M., Rufino, R.D., Santos, V.A., Banat, I.M., and Sarubbo, L.A. (2016). Biosurfactants: promising molecules for petroleum biotechnology advances. *Frontiers in Microbiology* 7: 1–14.

Anderson, R.J. and Newman, M.S. (1933). The chemistry of the lipids of tubercle bacilli: XXXIII. Isolation of trehalose from the acetone-soluble fat of the human tubercle *bacillus*. *The Journal of Biological Chemistry* 101: 499–504.

Aparna, A., Srinikethan, G., and Hedge, S. (2011). Effect of addition of biosurfactant produced by Pseudomonas ssp. on biodegradation of crude oil. *International Proceedings of Chemical, Biological & Environmental Engineering* 6: 71.

Ashby, R.D., Solaiman, D.K.Y., and Foglia, T.A. (2008). Property control of sophorolipids: influence of fatty acid substrate and blending. *Biotechnology Letters* 30 (6): 1093–1100.

Azuma, M., Suzutani, T., Sazaki, K., Yoshida, I., Sakuma, T., and Yoshida, T. (1987). Role of interferon in the augmented resistance of trehalose 6,6'-dimycolate-treated mice to influenza virus infection. *The Journal of General Virology* 68: 835–843.

Bachmann, R.T., Johnson, A.C., and Edyvean, R.G.J. (2014). Biotechnology in the petroleum industry: an overview. *International Biodeterioration and Biodegradation* 86: 225–237.

Baeva, T.A., Gein, S.V., Kuyukina, M.S., Ivshina, I.B., Kochina, O.A., and Chereshnev, V.A. (2014). Effect of glycolipid *Rhodococcus* biosurfactant on secretory activity of neutrophils in vitro. *Bulletin of Experimental Biology and Medicine* 157 (2): 238–242.

Bajaj, A., Mayliraj, S., Mudiam, M.K.R., Patel, D.K., and Manickam, N. (2014). Isolation and functional analysis of a glycolipid producing *Rhodococcus* sp. strain IITR03 with potential for degradation of 1, 1,1-trichloro-2,2-bis(4-chlorophenyl)ethane (DDT). *Bioresource Technology* 167: 398–406.

Banat, I.M., Franzetti, A., Gandolfi, I., Bestetti, G., Martinotti, M.G., Fracchia, L., Smyth, T.J., and Marchant, R. (2010). Microbial biosurfactants production, applications and future potential. *Applied Microbiology and Biotechnology* 87 (2): 427–444.

Bogaert, I.N.A.V., Saerens, K., Muynck, C., Develter, D.M., Soetaert, W., and Vandamme, E.J. (2007). Microbial production and application of sophorolipids. *Applied Microbiology and Biotechnology* 76 (1): 23–34.

Borsanyiova, M., Patil, A., Mukherji, R., Prabhune, A., and Bopegamage, S. (2016). Biological activity of sophorolipids and their possible use as antiviral agents. *Folia Microbiologica* (Praha). 61 (1): 85–89.

Bouassida, M., Ghazala, I., Ellouze-Chaabouni, S., and Ghribi, D. (2018). Improved biosurfactant production by *Bacillus subtilis* SPB1 mutant obtained by random mutagenesis and its application in enhanced oil recovery in a sand system. *Journal of Microbiology and Biotechnology* 28 (1): 95–104.

Brandenburg, K. and Seydel, U. (1988). Infrared spectroscopy of glycolipids. *Chemistry Physical Lipids* 96 (1–2): 23–40.

Bryant, F.O. (1990). Improved method for the isolation of biosurfactant glycolipids from Rhodococcus sp. strain H13A. *Applied Environmental Microbiology* 56: 1494–1496.

Bungaruang, L., Gutmann, A., and Nidetzky, B. (2013). Leloir glycosyltransferases and natural product glycosylation: biocatalytic synthesis of the C-glucoside nothofagin, a major antioxidant of redbush herbal tea. *Advanced Synthesis & Catalysis* 355 (14–15): 2757–2763.

Cameotra, S.S. and Makkar, R.S. (1998). Synthesis of biosurfactants in extreme conditions. *Applied Microbiology and Biotechnology* 50 (5): 520–529.

Cappelletti, M., Presentato, A., Piacenza, E., Firrincieli, A., Turner, R.J., and Zannoni, D. (2020). Biotechnology of *Rhodococcus* for the production of valuable compounds. *Applied Microbiology and Biotechnology* 104: 8567–8594.

Carrillo, P.G., Mardaraz, C., Pitta-Alvarez, S.I., and Giuliett, A.M. (1996). Isolation and selection of biosurfactant producing bacteria. *World Journal of Microbiology & Biotechnology* 12 (1): 82–84.

Christova, N., Lang, S., Wray, V., Kaloyanov, K., Konstantinov, S., and Stoineva, I. (2015). Production, structural elucidation and in vitro antitumor activity of trehalose lipid biosurfactant from *Nocardia farcinica* strain. *Journal of Microbiology and Biotechnology* 25: 439–447.

Ciapina, E.M.P., Melo, W.C., Santa Anna, L.M.M., Santos, A.S., Freire, D.M.G., and Pereira, N. (2006). Biosurfactant production by *Rhodococcus erythropolis* grown on glycerol as sole carbon source. *Applied Biochemistry and Biotechnology* 131 (1–3): 880–886.

Cooper, D.G. and Goldenberg, B.G. (1987). Surface-active agents from two bacilllus species. *Applied and Environmental Microbiology* 53: 224–229.

Cortés-Sánchez, A.J., Hernández-Sánchez, H., and Jaramillo-Flores, M.E. (2013). Biological activity of glycolipids produced by microorganisms: new trends and possible therapeutic alternatives. *Microbiological Research* 168 (1): 22–32.

Crouzet, J., Arguelles-Arias, A., Dhondt-Cordelier, S., Cordelier, S., Pršić, J., Hoff, G., Mazeyrat-Gourbeyre, F., Baillieul, F., Clément, C., Ongena, M., and Dorey, S. (2020). Biosurfactants in plant protection against diseases: rhamnolipids and lipopeptides case study. *Frontiers in Bioengineering and Biotechnology* 8: 1–11.

Davis, D.A. et al. (2001). The application of foaming for recovery of surfactin from *B. subtilis* ATCC 21332. *Enzyme Microbiology Technology* 28: 346–354.

DeBosch, B.J., Heitmeier, M.R., Mayer, A.L., Higgins, C.B., Crowley, J.R., Kraft, T.E., Chi, M., Newberry, E.P., Chen, Z., Finck, B.N., Davidson, N.O., Yarasheski, K.E., Hruz, P.W., and Moley, K.H. (2016). Trehalose inhibits solute carrier 2A (Slc2A) proteins to induce autophagy and prevent hepatic steatosis. *Science Signaling* 9 (416): 1.

Desai, J.D. and Banat, I.M. (1997). Microbial production of surfactants and their commercial potential. *Microbiology and Molecular Biology Reviews : MMBR* 61 (1): 47–64.

Dogan, I., Pagilla, K.R., Webster, D.A., and Stark, B.C. (2006). Expression of Vitreoscilla haemoglobin in *Gordonia amarae* enhances biosurfactant production. *Journal of Industrial Microbiology & Biotechnology* 33: 693–700.

Dubey, K.V. et al. (2005). Adsorption–desorption process using wood–based activated carbon for recovery of biosurfactant from fermented distillery wastewater. *Biotechnology Progress* 21: 860–867.

Esders, T.W. and Light, R.J. (1972). Glucosyl and acetyltransferases involved in the biosynthesis of glycolipids from *Candida bogoriensis*. *The Journal of Biological Chemistry* 10, 247 (5): 1375–1386.

Espuny, M.J., Egido, S., Rodón, I., Manresa, A., and Mercadé, M.E. (1996). Nutritional requirements of a biosurfactant producing strain *Rhodococcus* sp 51T7. *Biotechnology Letters* 18 (5): 521–526.

Fracchia, L., Banat, J., Cavallo, J., Ceresa, M., and Banat, C.I.M. (2015). Potential therapeutic applications of microbial surface-active compounds. *AIMS Bioengineering* 2 (3): 144–162.

Franzetti, A., Gandolfi, I., Bestetti, G., Smyth, J.P.T., and Banat, I.M. (2010). Production and applications of trehalose lipid biosurfactants. *European Journal of Lipid Science and Technology* 112 (6): 617–627.

Gein, S.V., Kuyukina, M.S., Ivshina, I.B., Baeva, T.A., and Chereshnev, V.A. (2011). In vitro cytokine stimulation assay for glycolipid biosurfactant from *Rhodococcus ruber*: role of monocyte adhesion. *Cytotechnology* 63: 559–566.

Geys, R., Soetaert, W., and Bogaert, I.V. (2014). Biotechnological opportunities in biosurfactant production. *Current Opinion in Biotechnology* 30: 66–72.

Groves, E., Dart, A.E., Covarelli, V., and Caron, E. (2008). Molecular mechanisms of phagocytic uptake in mammalian cells. *Cellular and Molecular Life Sciences* 65: 1957–1976.

Gudiña, E.J., Rangarajan, V., Sen, R., and Rodrigues, L.R. (2013). Potential therapeutic applications of biosurfactants. *Trends in Pharmacological Sciences* 34 (12): 667–675.

Hoq, M.M., Suzutani, T., Toyoda, T., Horiike, T., Yoshida, G., and Azuma, I.M. (1997). Role of gamma delta TCR + lymphocytes in the augmented resistance of trehalose 6,6'-dimycolate-treated mice to influenza virus infection. *The Journal of General Virology* 78: 1597–1603.

Im, J.H., Nakane, T., Yanagishita, H., Ikegami, T., and Kitamoto, D. (2001). Mannosylerythritol lipid, a yeast extracellular glycolipid, shows high binding affinity towards human immunoglobulin. *BMC Biotechnology* 1: 5.

Inaba, T., Tokumoto, Y., Miyazaki, Y., Inoue, N., Maseda, H., Nakajima-Kambe, T., Uchiyama, H., and Nomura, N. (2013). Analysis of genes for succinoyl trehalose lipid production and increasing production in *Rhodococcus* sp. strain SD-74. *Applied and Environmental Microbiology* 79: 7082–7890.

Iqbal, S., Khalid, Z.M., and Malik, K.A. (1995). Enhanced biodegradation and emulsification of crude oil and hyperproduction of biosurfactants by a gamma ray induced mutant of *Pseudomonas aeruginosa*. *Letters in Applied Microbiology* 21: 176–179.

Isoda, H., Kitamoto, D., Shinmoto, H., Matsumura, M., and Nakahara, T. (1997). Microbial extracellular glycolipid induction of differentiation and inhibition of the protein kinase C activity of human promyelocytic leukemia cell line HL60. *Bioscience, Biotechnology, and Biochemistry* 61 (4): 609–614.

Isoda, H., Shinmoto, H., Matsumura, M., and Nakahara, T. (1996). Succinoyl trehalose lipid induced differentiation of human monocytoid leukemic cell line U937 into monocyte–macrophages. *Cytotechnology* 19: 79–88.

Jain, N.K. and Roy, I. (2009). Effect of trehalose on protein structure. *Protein Science* 18 (1): 24–36.

Janek, T., Krasowska, A., Czyznikowska, Z., and Łukaszewicz, M. (2018). Trehalose lipid biosurfactant reduces adhesion of microbial pathogens to polystyrene and silicone surfaces: an experimental and computational approach. *Frontiers in Microbiology* 9 2441

Kadinov, B., Nikolova, B., Tsoneva, I., Semkova, S., Kabaivanova, L., and Dimitrova, D. (2020). Trehalose lipid biosurfactant reduced cancer cell viability but did not affect the isometric contraction of rat mesenteric arteries in vitro. *International Journal Bioautomation* 24 (1): 79–86.

Kitamoto, D., Isoda, H., and Nakahara, T. (2002). Functions and potential applications of glycolipid biosurfactants. *Journal of Bioscience and Bioengineering* 94 (3): 187–201.

Koch, A.K., Käppeli, O., Fiechter, A., and Reiser, J. (1991). Hydrocarbon assimilation and biosurfactant production in *Pseudomonas aeruginosa* mutants. *Journal of Bacteriology* 173: 4212–4219.

Kretschmer, A., Bock, H., and Wagner, F. (1982). Chemical and physical characterization of interfacial-active lipids from *Rhodococcus erythropolis* grown on n-alkanes. *Applied and Environmental Microbiology* 44 (4): 864–870.

Kügler, J. H. et al. (2014). Trehalose lipid biosurfactants produced by the actinomycetes *Tsukamurella spumae* and *T. pseudospumae*. *Applied Microbiology and Biotechnology* 98 (21): 8905–8915.

Kumari, S., Sekar, K.V., Nagasathya, A., Palanivel, S., and Nambaru, S. (2010). Effective biosurfactants production by Pseudomonas aeruginosa and its efficacy on different oils. *Journal of Advanced Laboratory Research in Biology* 1 (1): 31–34.

Kundu, D., Hazra, C., and Chaudhari, A. (2016a). Biodegradation of 2,6- dinitrotoluene and plant growth promoting traits by *Rhodococcus pyridinivorans* NT2: identification and toxicological analysis of metabolites and proteomic insights. *Biocatalysis and Agricultural Biotechnology* 8: 55–65.

Kundu, D., Hazra, C., and Chaudhari, A. (2016b). Bioremediation potential of *Rhodococcus pyridinivorans* NT2 in nitrotoluene-contaminated soils: the effectiveness of natural attenuation, biostimulation and bioaugmentation approaches. *Soil and Sediment Contamination: An International Journal* 25: 637–651.

Kundu, D., Hazra, C., Dandi, N., and Chaudhari, A. (2013). Biodegradation of 4-nitrotoluene with biosurfactant production by *Rhodococcus pyridinivorans* NT2: metabolic pathway, cell surface properties and toxicological characterization. *Biodegradation* 24: 775–793.

Kuyukina, M.S. and Ivshina, I.B. (2010). *Rhodococcus* biosurfactants: biosynthesis, properties, and potential applications. In: Biology of Rhodococcus (ed. H. Alvarez), Microbiology Monographs, vol 16, 291–313. Springer.

Kuyukina, M.S. and Ivshina, I.B. (2019). Production of trehalolipid biosurfactants by *Rhodococcus*. In: Biology of Rhodococcus (ed. H. Alvarez), Microbiology Monographs, vol 16, 271–298. Cham: Springer.

Kuyukina, M.S., Ivshina, I.B., Baeva, T.A., Kochina, O.A., Gein, S.V., and Chereshnev, V.A. (2015). Trehalolipid biosurfactants from nonpathogenic *Rhodococcus actinobacteria* with diverse immunomodulatory activities. *New Biotechnology* 25; 32 (6): 559–568.

Kuyukina, M.S., Ivshina, I.B., Gein, S.V., Baeva, T.A., and Chereshnev, V.A. (2007). In vitro immunomodalating acitivity of biosurfactant glycolipid complex from *Rhodococcus rubber*. *Bulletin of Experimental Biology and Medicine* 144 (3): 326–330.

Kuyukina, M.S., Ivshina, I.B., Korshunova, I.O., Stukova, G.I., and Krivoruchko, A.V. (2016). Diverse effects of a biosurfactant from *Rhodococcus ruber* IEGM 231 on the adhesion of resting and growing bacteria to polystyrene. *AMB Express* 6 (14): 1–12.

Kuyukina, M.S., Ivshina, I.B., Philp, J.C., Christofi, N., Dunbar, S.A., and Ritchkova, M.I. (2001). Recovery of *Rhodococcus* biosurfactants using methyl tertiary-butyl ether extraction. *Journal of Microbiological Methods* 46: 149–156.

Kuyukina, M.S., Varushkina, A.M., and Ivshina, I.B. (2020). Effects of electroporation on antibiotic susceptibility and adhesive activity to *n*-hexadecane in *Rhodococcus ruber* IEGM 231. *Applied Biochemistry and Microbiology* 56: 729–735.

Lang, S. and Philp, J.C. (1998). Surface-active lipids in *rhodococci*. *Antonie van Leeuwenhoek* 74 (1/3): 59–70.

Lin, S.-C., Lin, K.-G., and Lin, Y.M. (1998). Enhanced biosurfactant production by a *Bacillus licheniformis* mutant. *Enzyme and Microbial Technology* 23: 267–273.

Mclaughlin, C.A., Schwartzman, S.M., Horner, B.L. et al. (1980). Regression of tumors in guinea pigs after treatment with synthetic muramyl dipeptides and trehalose dimycolate. *Science (80)* 208 (4442): 415–416.

Makkar, R.S., Cameotra, S.S., and Banat, I.M. (2011). Advances in utilization of renewable substrates for biosurfactant production. *AMB Express* 1 (1): 1–19.

Marqués, A.M., Pinazo, M., Farfana, F.J., Aranda, J.A., Teruel, A., Ortiz, A., Manresa, A., and Espunya, M.J. (2009). The physicochemical properties and chemical composition of trehalose lipids produced by *Rhodococcus erythropolis* 51T7. *Chemistry and Physics of Lipids* 158: 110–117.

Mata-Sandoval, J.C., Karns, J., and Torrents, A. (1999). High-performance liquid chromatography method for the characterization of rhamnolipid mixtures produced by Pseudomonas aeruginosa UG2 on corn oil. *Journal of Chromatography A* 864: 211–220.

Matsumoto, Y., Cao, E., and Ueoka, R. (2013). Novel liposomes composed of dimyristoylphosphatidylcholine and trehalose surfactants inhibit the growth of tumor cells along with apoptosis. *Biological & Pharmaceutical Bulletin* 36 (8): 1258–1262.

Matsumoto, Y., Kuwabara, K., Ichihara, H., and Kuwano, M. (2016). Therapeutic effects of trehalose liposomes against lymphoblastic leukemia leading to apoptosis *in vitro* and *in vivo*. *Bioorganic & Medicinal Chemistry Letters* 15, 26 (2): 301–305.

Mnif, I., Ghribi, D. (2015). Microbial derived surface active compounds: properties and screening concept. *World Journal of Microbiology and Biotechnology* 31 (7): 1001–1020.

Mnif, I. and Ghribia, D. (2016). Glycolipid biosurfactants: main properties and potential applications in agriculture and food industry. *Journal of the Science of Food and Agriculture* 96 (13): 4310–4320.

Morikawa, M., Hirata, Y., and Imanaka, T. (2000). A study on the structure-function relationship of lipopeptide biosurfactants. *Biochimica et Biophysica Acta* 1488: 211–218.

Mujumdar, S., Joshi, P., and Karve, N. (2019). Production, characterization, and applications of bioemulsifiers (BE) and biosurfactants (BS) produced by *Acinetobacter* spp.: a review. *Journal of Basic Microbiology* 59 (3): 277–287.

Mukherjee, S., Das, P., and Sen, R. (2006). Towards commercial production of microbial surfactants. *Trends in Biotechnology* 24 (11): 509–514.

Mulligan, C.N., Cooper, D.G., and Neufeld, R.J. (1984). Selection of microbes producing biosurfactants in media without hydrocarbons. *Journal of Fermentation Technology* 62: 311–314.

Mutalik, S.R., Vaidya, B.K., Joshi, R.M., Desai, K.M., and Nene, S.N. (2008). Use of response surface optimization for the production of biosurfactant from Rhodococcus spp. MTCC 2574. *Bioresource Technology* 99: 1875–7880.

Natsuhara, Y., Oka, S., Kaneda, K., and Yano, I. (1990). Parallel antitumor, granuloma forming and tumor-necrosis-factor-priming activities of mycoloyl glycolipids from *Nocardia rubra* that differ in carbohydrate moiety: structure–activity relationships. *Cancer Immunology, Immunotherapy* 31: 99–106.

Oliveira, M.R., Magri, A., Baldo, C., Camilios-Neto, D., Minucelli, T., and Celligo, M.A.P.C. (2015). Review: sophorolipids a promising biosurfactant and its applications. *International Journal of Advanced Biotechnology and Research* 6 (2): 161–174.

Pacheco, G.J., Ciapina, E.M.P., Gomes, E.B., and Pereira-Junior, N. (2010). Biosurfactant production by *Rhodococcus erythropolis* and its application to oil removal. *Brazilian Journal of Microbiology* 41 (3): 685–693.

Patil, H.I. and Pratap, A.P. (2018). Production and quantitative analysis of trehalose lipid biosurfactants using high-performance liquid chromatography. *Journal of Surfactants and Detergents* 21 (4): 553–564.

Paulino, B.N., Pessôa, M.G., Mano, M.C.R., Molina, G., Neri-Numa, I.A., and Pastore, G.M. (2016). Current status in biotechnological production and applications of glycolipid biosurfactants. *Applied Microbiology and Biotechnology* 100: 10265–10293.

Philp, J.C., Kuyukina, M.S., Ivshina, I.B., Dunbar, S.A., Christofi, N., Lang, S., and Wray, V. (2002). Alkanotropic *Rhodococcus* ruber as a biosurfactant producer. *Applied Microbiology and Biotechnology* 59: 318–324.

Pinzon, N. and Ju, L.K. (2009). Improved detection of rhamnolipid production using agar plates containing Methylene blue and cetyltrimethyl ammonium bromide. *Biotechnology Letters* 31: 1583–1588.

Pinzon, N.M., Zhang, Q., Koganti, S., and Ju, L.K. (2009). Advances in bioprocess development of rhamnolipid and sophorolipid production. In: Biobase Surfactants and Detergents—Synthesis, Properties, and Applications (ed. D.G. Hayes, D. Kitamoto, D.K.Y. Solaiman, and R.D. Ashby), 77–105. Urbana, IL: AOCS Press.

Ramsay, B., McCarthy, J., Guerra-Santos, L., Kappeli, O., Fiechter, A., and Margaritis, A. (1988). Biosurfactant production and diauxic growth of Rhodococcus aurantiacus when using n -alkanes as the carbon source. *Canadian Journal of Microbiology* 34 (11): 1209–1212.

Ribeiro, I.A., Bronze, M.R., Castro, M.F., and Ribeiro, M.H.L. (2012). Design of selective production of sophorolipids by *Rhodotorula bogoriensis* through nutritional requirements. *Journal of Molecular Recognition* 25 (11): 630–640.

Ribeiro, I.A., Bronze, M.R., Castro, M.F., and Ribeiro, M.H.L. (2013). Sophorolipids: improvement of the selective production by *Starmerella bombicola* through the design of nutritional requirements. *Applied Microbiology and Biotechnology* 97 (5): 1875–1887.

Rodrigues, L., Banat, I.M., Teixeira, J., and Oliveira, R. (2006). Biosurfactants: potential applications in medicine. *Journal of Antimicrobial Chemotherapy* 57 (4): 609–618.

Sana, S., Datta, S., Biswas, D., and Sengupta, D. (2018). Assessment of synergistic antibacterial activity of combined biosurfactants revealed by bacterial cell envelop damage. *Biochimica et Biophysica Acta - Biomembranes* 1860 (2): 579–585.

Santos, D.K.F., Rufino, R.D., Luna, J.M., Santos, V.A., and Sarubbo, L.A. (2016). Biosurfactants: multifunctional biomolecules of the 21st century. *International Journal of Molecular Sciences* 18, 17 (3): 1–31.

Sarwar, A., Brader, G., Corretto, E., Aleti, G., Corretto, E., Abaidullah, M., Sessitsch, A., and Hafeez, F.Y. (2018). Qualitative analysis of biosurfactants from *Bacillus* species exhibiting antifungal activity. *PLoS One* 13 (6): 1–15.

Sato, S., Fukuoka, T., Saika, A., Koshiyama, T., and Morita, T. (2019). A New Screening Approach for Glycolipid-type Biosurfactant Producers Using MALDI-TOF/MS. *Journal of Oleo Science* 68 (12): 1287–1294.

Satpute, S.K., Banat, I.M., Dhakephalkar, P.K., Banpurkar, A.G., and Chopade, B.A. (2010). Biosurfactants, bioemulsifiers and exopolysaccharides from marine microorganisms. *Biotechnology Advances* 28 (4): 436–450.

Satpute, S.K., Bhawsar, B.D., Dhakephalkar, P.K., and Chopad, B.A. (2008). Assessment of different screening methods for selecting biosurfactant producing marine bacteria. *Indian Journal of Marine Sciences* 37 (3): 243–250.

Sen, R. and Swaminathan, T. (2005). Characterization of concentration and purification parameters and operating conditions for the small-scale recovery of surfactin. *Process Biochemistry* 40: 2953–2958.

Sen, S., Borah, S.N., Bora, A., and Deka, S. (2017). Production, characterization, and antifungal activity of a biosurfactant produced by *Rhodotorula babjevae* YS3. *Microbial Cell Factories* 16 (95): 1–14.

Seo, Y.S. and Shin, K.S. (2011). Optimal conditions and substrate specificity for trehalose production by resting cells of arthrobacter crystallopoietes n-08. *Journal of Food Science and Nutrition* 16 (4): 357–363.

Siegmund, I. and Wagner, F. (1991). New method for detecting rhamnolipids excreted by *Pseudomonas* species during growth on mineral agar. *Biotechnology Techniques* 5: 265–268.

Silva, R.C.F.S., Almeida, D.G., Rufino, R.D., Luna, J.M., Santos, V.A., and Sarubbo, L.A. (2014). Applications of biosurfactants in the petroleum industry and the remediation of oil spills. *International Journal of Molecular Sciences* 15 (7): 12523–12542.

Smyth, T.J.P., Perfumo, A., Marchant, R., and Banat, I.M. (2010). Isolation and analysis of low molecular weight microbial glycolipids. In: Handbook of Hydrocarbon and Lipid Microbiology (ed. K.N. Timmis), 3705–3723. Berlin, Heidelberg: Springer.

Soberón-Chávez, G. (ed.) (2011). Biosurfactants, Microbiology monographs 20. eBook from Heidelberg: Springer, 127.

Solaiman, D., Ashby, R., Birbir, M., and Caglayan, P. (2016). Antibacterial activity of sophorolipids produced by *Candida bombicola* on gram-positive and gram negative bacteria isolated from salted hides. *Journal of the American Leather Chemists Association*111: 358–364.

Souza, E.C., Vessoni-Penna, T.C., and Oliveira, R.P.S. (2014). Biosurfactant-enhanced hydrocarbon bioremediation: an overview. *International Biodeterioration & Biodegradation* 89: 88–94.

Sudo, T., Zhao, X., Wakamatsu, Y., Shibahara, M., Nomura, N., Nakahara, T., Suzuki, A., Kobayashi, Y., Jin, C., Murata, T., and Yokohama, K.K. (2000). Induction of the differentiation of human HL60 promyelocytic leukemia cell line by succinoyl trehalose lipids. *Cytotechnology* 33 (1–3): 259–264.

Takayama, K., Wang, C., and Besra, G.S. (2005). Pathway to synthesis and processing of mycolic acids in *Mycobacterium tuberculosis*. *Society* 18 (1): 81–101.

Tokumoto, Y., Nomura, N., Uchiyama, H. et al. (2009). Structural Characterization and Surface-Active Properties of a Succinoyl Trehalose Lipid Produced by *Rhodococcus* sp. SD-74. *Journal of Oleo Science* 102 (2): 97–102.

Tuleva, B., Christova, N., Cohen, R., Antonova, D., Todorov, T., and Stoineva, I. (2009). Isolation and characterization of trehalose tetraester biosurfactants from soil strain *Micrococcus luteus* BN56. *Process Biochemistry* 44: 135–141.

Uchida, Y., Misawa, S., Nakahara, T., and Tabuchi, T. (1989). Factors affecting the production of succinoyl trehalose lipids by *Rhodococcus erythropolis* SD-74 grown on n-alkanes. *Agricultural and Biological Chemistry* 53 (3): 765–769.

Varjani, S.J., Rana, D.P., Bateja, S., Sharma, M.C., and Upasani, V.N. (2014). Screening and identification of biosurfactant (bioemulsifier) producing bacteria from crude oil contaminated sites of Gujarat, India. *International Journal of Innovative Research in Science Engineering and Technology* 3 (2): 9205–9213.

Varjani, S.J. and Upasani, V.N. (2016). Biodegradation of petroleum hydrocarbons by oleophilic strain of *Pseudomonas aeruginosa* NCIM 5514. *Bioresource Technology* 222: 195–201.

Varjani, S.J. and Upasani, V.N. (2017). Critical review on biosurfactant analysis, purification and characterization using rhamnolipid as a model biosurfactant. *Bioresource Technology* 232: 389–397.

Vijayakumar, S. and Saravanan, V. (2015). Biosurfactants-types, sources and applications. *Research Journal of Microbiology* 10 (5): 181–192.

Wang, Y., Nie, M., Diwu, Z., Lei, Y., Li, H., and Bai, X. (2019). Characterization of trehalose lipids produced by a unique environmental isolate bacterium *Rhodococcus qingshengii* strain FF. *Journal of Applied Microbiology* 127: 1442–1453.

Warnecke, D. and Heinz, E. (2010). Glycolipid headgroup replacement: A new approach for the analysis of specific functions of glycolipids in vivo. *European Journal of Cell Biology* 89 (1): 53–61.

White, D.A., Hird, L.C., and Ali, S.T. (2013). Production and characterization of a trehalolipid biosurfactant produced by the novel marine bacterium *Rhodococcus* sp., strain PML026. *Journal of Applied Microbiology* 115: 744–755.

Yagi-Utsumi, M. (2019). NMR characterization of conformational dynamics and molecular assemblies of proteins. *Biological and Pharmaceutical Bulletin* 42 (6): 867–872.

Yakimov, M.M., Giuliano, L., Scarfi, B.V., and Golyshin, P.N. (1999). Characterization of antarctic hydrocarbon-degrading bacteria capable of producing bioemulsifiers. *The New Microbiologica* 22: 249–256.

Zhang, J., Saerens, K.M.J., Bogaert, I.N.A.V., and Soetaert, W. (2011). Vegetable oil enhances sophorolipid production by *Rhodotorula bogoriensis*. *Biotechnology Letters* 33 (12): 2417–2423.

2

Natural Polymer Types and Applications

Amro Abd Al Fattah Amara[1,2]

[1]*Protein Research Department, Genetic Engineering and Biotechnology Research Institute, City of Scientific Research and Technological Applications, Alexandria, Egypt*
[2]*Head of Scientific Publishing office, City of Scientific Research and Technological Applications, Alexandria, Egypt*
**Corresponding author: amroamara@web.de*

2.1 Introduction

The first to use biopolymers (non-fibrous biopolymers) in a technical application were the ancient Egyptians (The Pharaohs), who used gum (later named gum Arabic or Acacia gum) (Figure 2.1) before the second millennium BC [1–3]. The Pharaohs were known to use the Arabian Gum and other plant glue in preparing inks used in writing, to install and fix colors and to prepare papyrus (the material upon which the ancient Egyptians wrote) (Figure 2.2). Papyrus was formed by cutting the stem a tall rushlike plant (*Cyperus Papyrus*) of the Sedge family, growing in Egypt, Abyssinia, Syria, Sicily, etc.), into thin longitudinal slices, which were gummed together and pressed. Gum Arabic was also used to prepare different treatments used in many therapeutic and protective combinations. The resin was used in the mummification process, which included soft tissue removal, dehydration of the body, embalming, sealing inside and outside with resin, covering with bandages and preserving in an evacuated sealed chamber and in a dry place [4]. Apparently, they recognized that gums and resins protect plants from spoilage and so they invented a gum/resin sealing strategy. They used it in most of the applications where an adhesive was required. Gum Acacia or gum Arabic was also used by the ancient Egyptians as a food. The ancient Indian, Chinese, Greeks, Romans, and many other ancient civilizations have also used the gum and resin in different applications. Biopolymers have unique properties, which made them the best choice in various applications. They are in most cases, biodegradable, bioavailable, biocompatible, non-toxic, environmentally friendly, applicable, diverse, and have many other useful properties [5–14]. They also degrade to safe structures. Most of the biopolymers are produced by plants but some are produced by animals, algae and microbes.

Biomolecules from Natural Sources: Advances and Applications, First Edition. Edited by Vijai Kumar Gupta, Satyajit D. Sarker, Minaxi Sharma, María Elida Pirovani, Zeba Usmani, and Chelliah Jayabaskaran.

Figure 2.1 Gum Arabic; the first known biopolymer, its name comes from the Arabic merchants who sold it in Europe. It is still sold nowadays in local markets as big colored transparent granules.

Figure 2.2 Acacia tree as drawn and memorized by ancient Egyptian civilization.

2.1.1 The Monomer, Polymer and Biopolymer

Polymer is the generic name for a species of macromolecules which have the unique properties of repeated monomers, a linear or branched backbone or a naturally occurring or synthetic compound consisting of large molecules made up of a linked series of repeated simple monomers. In contrast, a monomer is a simple compound whose molecules are joined to form polymers. The monomeric constituents of the polymer are responsible for their properties. They either have the same chemical structures, in such cases the polymer is named a homopolymer, or they are different in their chemical structures and the polymer is named heteropolymer. The more monomers with different properties the polymer has the the wider its range of physicochemical and biological activity might be. For example, different proteins consist of different amounts and sequences of 20 amino acids. Repetition of different constituents of amino acids gives each protein its unique specificity.

2.1.2 The Monomeric Structure

Why are some polymers nearly inert while other are so dynamic? [15, 16] Why are they different? Why are some of them grouped in a certain ways? It is, of course, because of their structure which depends on their monomeric subunits [17–21]. As an example, protein varies in their monomeric types (amino acids), numbers and location. So, each protein is unique in its structure/function/specificity [22–24].

2.1.3 Enzymes (Protein Polymers) Building Polymers

The sequence of polymer building in the cells starts from the DNA and the protein. Life must be started by the existence of many biological and chemical forms including both DNA and protein which are highly complicated polymers, not only in their components, but also in their design and the large amount of information installed which gives one the code and the other the dynamicability. Proteins alone are inactive structures, but if other elements exist (e.g., ions, water, pH, etc.) the requrement that enable them to react as biologically active and dynamic macromolecules with high specific reaction are satisfied.

Accumulation of mutants causes change in the protein function. Mutants in the cell cycle, repairing or apoptosis genes might turn normal cells to cancer cells. One important example is DNA and the RNA polymerases. DNA is a long linear polymer; found in the nucleus of a eukaryotic cell or in the cytoplasm of the prokaryotic cells and formed from nucleotides and shaped like a double helix; associated with the transmission of genetic information. RNA is a single strand long linear polymer of nucleotides found in the nucleus but mainly in the cytoplasm of the eukaryotic cell. The polymerases which work on them are different. The polymers represent essential and vital parts of the cells such as the DNA, RNA and protein. Additionally, they can be used as storage components such as starch, polyhydroxyalkanoates, etc. They can also be used by the cell for different purposes, for example as a protective agent (e.g., alginate, gums, and resin).

Biopolymers are different from the synthetic polymers in two main ways in that they are produced by living cells (produced naturally) and that they can be used by their main producer or by related or different kinds of the living cells (naturally biodegradable: capable of being decomposed). Generally, they belong to the biological system and their polymeric structure, the polymerization steps, and their degradation is done through various enzymatic activities. In other words they are a globally essential part of the biological system, produced by it and also degraded through it. For that it is normal to find a polymer produced by a microbe and degraded by the same microbe. More simply, it is a polymer for us, but it is a food or a part of the cell's different structures for their producer. Their presence is governed by the biological aspects [25, 26]. As they from the biological system, their elements in most cases (except structures like native foreign protein and the LPS) are compatible with the human immune system [25, 27–31]. In some instances they are named white or green to demonstrate their compatibility with the biological system or their safety to nature [8, 28, 32, 33].

2.1.4 The Synthetic Polymers are Non-homogenized with Nature

The synthetic polymer produced chemically through specific chemical reactions or from petroleum oil and their monomers are linked together to form large molecules. They are macromolecules made of linked series of repeated monomers joined by chemical bonds through chemical reactions, mainly polymerization, polycondensation and polyaddition. The polymerization process, in most cases, contains compounds toxic to the live cells. While the biopolymer is compatible with nature, the synthetic polymer is incompatible. This incompatibility was first considered a revolution. The synthetic polymer, in most cases, is undegradable. For example, the invention of durable plastic was considered a revolution at the time of its discovery; however, this image has changed over time. Plastic (and its products) is either a synthetic or semisynthetic material that can be molded or extruded into objects, films, filaments or be used to make structures such as coatings and adhesives. Their accumulation as waste is causing great concern. Millions of tons each year are discarded and accumulated in the earth, which causes affects from the ground through to the ozone layer (the plastic byproducts and the gases produced either through production processes or after its burning). Today nobody could guarantee that 100% of polymerization step(s) are free from toxic chemical compounds and stable against the natural, physical or chemical degradation.

2.1.5 The Competition between Biopolymers and Chemically Synthetic Polymers

The plastic revolution brings wealth and happiness for many and allows nearly all industrial sectors to flourish, but nowadays it is a subject for continuous evaluation and validation. However, the image is not entirely black, in fact there are many positive things that compete strongly and direct us toward better management of our resources: are we are able to reduce the amount of synthetic polymers? Are we are able to fill the market demand? In fact, biopolymers are not able to do that today. Some points could be summarized:

1) The synthetic polymers are mostly hydrocarbons. Some powerful microbes are able to degrade them successfully such as *Pseudomonas aeruginosa* and the different *Gordonia spp* [34].
2) There are a huge number of biopolymer products and types that could be used as alternatives to synthetic polymers. An example of biopolymers that could be polymerized, which have similar or better properties to the natural products are the polyhydroxyalkanoates. They is considered to be alternatives to plastic. Another example is natural rubber which is an alternative to synthetic rubbers.
3) The recycling of synthetic polymers is under continuous quality control and validation and has reduced the global demand for the amount produced annually.
4) There is an increasing worldwide awareness concerning the plastic accumulation problem and global pollution.
5) Investment in natural materials is profitable.
6) There is a rising political awareness in the problems caused by synthetic materials after the many side effects.
7) Not all synthetic materials have the same issues at the same level. Some forms are beneficial and their side effect could be avoided.

2.1.6 The Plastic Success

The most well know polymers are plastic(s); this is a generic name for synthetic, semi-synthetic or natural materials that can be molded or extruded into objects, films, filaments or be used to make, for example, coatings and adhesives. Synthetic plastic is mainly derived from petroleum oil or through chemical reactions. But there are a considerable number of plastics that have a biological origin. Because of their perfect mechanical properties, different types of plastic were formulated to match different applications. Plastics were first used in packaging and housing materials. Later plastics find their way into medicinal, pharmaceutical and industrial applications. Today, plastics applications have either totally or partly substituted the other materials used previously in industrial applications (on all or some of their parts) such as wood, mud, metals, glass and other materials [35, 36]. Plastic is the best choice in many applications because of its low cost, stability, durability, good mechanical and thermal properties. Those who are interested in the materials produced by biopolymers and are investing funds and arranging resources aimed at commercializing species of biopolymers should identify the areas which lead to the success of the plastic applications. That will enable a successful start for any of the biopolymer application species still out of the market because of the value of the synthetic polymer. Some of the important biopolymer species, such as the bioplastc polyhydroxyalkanoates, gum Arabic, agar, alginate and so on, already exist in the market [28, 35, 37].

2.1.7 Biopolymer Commercialization

As well as plastics and bioplastics, the other types of biopolymer were given more opportunities and many of them were commercialized. Biopolymers have unique properties (there are some exceptions); they are produced and degraded through the biological system, therefore they are non-toxic (mostly); they are bioavailable (mostly); They are diverse, applicable, and renewable. There are many types of classifications which are based on their chemical, biological and physical properties; their source (plant, animal and microbes); their applications (medicinal, pharmaceutical, agriculture, industrial); their economic value; their biodegradability (biodegradable, non-degradable); their bioavailability; their cost and their mechanical properties. In this chapter, the classification which is based on the polymer chemical structure will be used. The main limiting factor in commercializing biopolymers is their production cost. For that this chapter will use the classification which is based on the polymer chemical structure [35].

2.1.8 The Eight Different Biopolymers

The eight types of biopolymers are: (1) nucleic acids (DNA and RNA); (2) polyamides which are polymers containing repeated amide groups (protein poly-(amino acids) such as, gelatine, casein, wheat gluten, silk and wool); (3) polysaccharides, any of a class of carbohydrates whose molecules contain chains of monosaccharide molecules (such as, starch, cellulose, lignin, chitin); (4) organic polyoxoesters (such as poly(hydroxyalkanoic acids), poly(malic acid) and cutin); (5) polyisoprenoides (such as natural rubber or gutta-percha [a whitish rubber derived from the coagulated milky latex of gutta-percha trees; used for insulation of electrical cables]); (6) inorganic polymers such as inorganic polyesters with polyphosphate, (7) polyphenols (such as lignin or humic acids), and (8) polythioesters, for example, poly(3-mercaptopropionate). Polymers from bioderived monomers could be polymerized and might be added as group nine. Additionally, some inorganic elements might show accumulation in the microbial cells in repeated forms but due to their nature they form crystals which are usually different to those made in labs. For example, magnetotactic bacteria show Fe_3O_4 chains of similar crystals which are unique in their structures. The helical twist of the Fe_3O_4 series of crystals are not cubes. It might be interesting to report that similar structures are found in goethite in the strengthening of limpet teeth. Other examples are iron, sulphides, pyrite crystals found in some anaerobic bacteria. In fact, more research should be conducted on the nature of the inorganic structures which might be finally classified as biopolymer because they are not crystalline spontaneously but due to the effect of proteins and enzymes. The amazing structure of different diatoms might be a good example [35, 38, 39].

2.2 Biopolymer Type Number 1: Nucleic Acids

Nucleic acids are the genetic code in living cells. DNA and RNA are the most important biopolymers that are located in the nucleotides of the eukaryotic cells and in the cytoplasm of the prokaryotic cells [40, 41]. DNA and RNA are polymers in their nature. They are usually used in various genetic manipulation tools, as well as in some fine technical applications and nanoapplications. DNA is used to generate new protein through mutagenesis, which gives new protein and thus new function or products (protein engineering). RNA is made from polymers of ribose sugar, phosphate and nitrogenous base.

DNA is made from polymer of deoxyribose sugar, phosphate and nitrogenous base. DNA was used as a platform based on self-assembled DNA biopolymer for high-performance cancer therapy [42]. DNA novel nanomaterial is designed for applications in photonics and in electronic [43, 44].

2.2.1 Tissue Engineering

Different types of polymers are used in tissue engineering [45]. The existence of the genetic elements in the cells is an essential part of their viability and productivity. Cells aggregate to form tissue: growing free cells in a particular space made of a biodegradable polymer enables their injection or cultivation in a tissue. In a successful process the cells then will differentiate to form or to fill in the target tissue. Upon degrading the polymer and replicating the cells, the new cells become a part of the oriignal repaired tissue, a new technology named "tissue engineering" which offers the possibility to help in regenerating tissues damaged by disease or trauma and, in some cases, to create new tissues. Usually this is achieved through using degradable biomaterials to either induce the surrounding tissue and cell to grow or to serve as temporary scaffolds for transplanted cells to attach, grow, and maintain differentiated functions. In some trials the polymer improves the growth of the cultivated cells. Leucocytes show improvement growth on PHA polymer surface [28].

2.2.2 Gene Therapy and Delivery

Biopolymers were designed in many formulations to either react by themselves or to be used in gene therapy [46–48]. They are used widely in tissue engineering and in genetic cell engineering [49]. They enable the cells to provide biochemical signals [50] which direct cell proliferation and differentiation [51]. The unique criteria in the polymers used in gene therapy, is their biocompatibility, mucoadhesive character, and biodegradability. The biocompatibility of natural polymers allows cells to infiltrate the matrix and transfection can occur as these cells come into contact with the embedded DNA. The biodegradability of the matrices obtained from natural polymers may also assist the release of gene transfer agents into the surrounding environment and thus affect nearby cells [48, 52, 53].

2.2.3 As Biosensor

A biosensor is a molecular device that converts a biological response into a detectable measurable signal. A biosensor is formed from a sensor which could use different types of materials including the polymer and the bioreceptor, such as antibodies, enzymes, nucleic acid, etc., [54], is coupled with the sensor through different immobilizing techniques [54]. Different polymers are used to produce biosensors such as in antibacterial electrospun, dual fluorescence "turn-on" sensors of cysteine and silver ions, chitin nanofiber paper toward optical (bio)sensing applications, click off colorimetric detection of peroxide and glucose etc. [55].

2.3 Biopolymer Type Number 2: Polyamides

2.3.1 Protein (πρώτειος)

The word protein comes from the Greek word πρώτειος (*proteios*) "primary". Proteins were first described and named by the Swedish chemist Jöns Jakob Berzelius in 1838. However, the involvement of proteins in living system organisms was correctly understood in 1926, when James B. Sumner showed that urease was a protein defined by the sequence of its related nucleotides and amino acids [56]. The genetic code can include selenocysteine and in certain cases (such as in archaea) pyrrolysine. The residues in a protein are often observed to be chemically modified by post-translational modification, which can happen either before the protein is used in the cell, or as a part of control mechanisms. Proteins can also work together to achieve a particular function, and they often associate to form stable complex functions, such as actin and myosin in muscles and proteins in the cytoskeleton, which form a system of scaffolding that maintains cell shape. Other proteins are important in cell signaling, immune responses, cell adhesion, and the cell cycle [57].

Proteins are natural chains of amino acids joined by amide linkages. They are degraded by enzymes (proteases). For thousands of years people used natural proteins such as wool [58], silk [45] and hair (keratin) for clothes, decoration or to display their wealth. After understanding their composition, they were reformulated and their properties changed to match certain demands. Numerous proteins were studied for developing natural bio-based materials such as keratin, collagen, albumin, gelatin, and fibroin [59]. They are degraded by enzymes. Nowaday some old techniques are still used to produce products from some types protein–polymer such as wool [60], silk [61–64], and hair [65–67]. However, some modern applications have been invented such as the use of Keratin [68–70] in hydrogel. Silk is used in tissue engineering, in drug delivery for musculoskeletal therapeutics. The first industrial applications of protein as polymer were in the early 1930s and 1940s with casein and with soy protein. Protein biopolymers can be classified with animal proteins (e.g. casein [71–75], whey [76–79], keratin [58, 68, 80, 81], collagen [82–85] and gelatine [86], polyglutamic) and in plant proteins (wheat, corn, soy, pea and potato proteins) and microbial protein such as polyglutamic [87–89], cyanophycin [90–93] protein biopolymers remained present in some niche markets such as encapsulates (pharmaceutical), coatings (food industry), adhesives or surfactants. They are used in the packaging industry for breweries, wineries and essential oil composite film for refrigerated products; microcapsules based on biodegradable polymers [72, 90, 93]. Protein biopolymers remain in some niche markets such as encapsulates (pharmaceutical), coatings (food industry), adhesives or surfactants. They are used in the packaging industry for breweries, wineries and refrigerated products essential oil composite film, microcapsules based on biodegradable polymers [57, 72].

Mammalians are not able to produce all 20 amino acids (essential amino acids). They obtain those essential amino acids through their diet. The first protein to be fully sequenced was insulin, by Frederick Sanger, who won the Nobel Prize for this achievement in 1958. It is important to highlight that Sanger succeed in solving the insulin sequence using insulin itself rather using the related genes. It was a complicated process especially in the presence of the disulfide bond between A and B fragments. Sanger discovered early the importance of the 3D structure of the protein. The first protein structures to be solved using data obtained from x-ray diffraction analysis included hemoglobin and myoglobin, by Max Perutz and Sir John Cowdery Kendrew, respectively, in 1958 [94, 95]. The 3D structures of both proteins were determined; Perutz and Kendrew shared the 1962 Nobel Prize in Chemistry for these discoveries. Rapid advances in site-directed mutagenesis and total gene synthesis combined with new expression systems in prokaryotic and eukaryotic cells have provided the molecular biologist with tools for modification of existing proteins to improve catalytic activity, stability and selectivity, for construction of chimeric molecules and for synthesis of completely novel molecules that may be endowed with some useful activity. The results are used in the improvement of the design by using knowledge-based procedures that exploit facts, rules and observations about proteins of known 3D structure [96].

2.3.2 The Biology of the Protein

Most of organic compounds have applications that were successfully synthesized chemically after solving their structures. Even though the basic concept of the protein structure and composition was solved, proteins resist chemical synthesis, this is due mainly to its long variant polymeric chain. Proteins cannot be synthesized by organic chemists in large quantities and they cannot be manipulated *in vitro* to modify single amino acids in a protein and leave all other amino acids of that variety unchanged. However, the majority of proteins can be manipulated using *in vivo* genetic engineering. Once a gene coding for a protein has been cloned from the original wild-type genome into a vector it can be manipulated by using synthetic oligonucleotides to produce site specific mutations in the cloned material. This is specific and can alter any side chain of a particular amino acid to any other of the 20 naturally occurring amino acids. The technique of site-directed mutagenesis can alter any number of amino acids in a protein and can be used to build proteins from scratch. The position of the amino acid is decided by inspection of the tertiary structure of the primary structure (using conserved amino acids and site directed mutagenesis experiments) and the tertiary structure of the protein (using protein modeling), and the interaction of the amino acid with the substrate or with other parts of the main protein is evaluated mathematically. Conserved amino acids can be determined from the protein primary structure using alignment. Conserved amino acids are usually responsible for important functions. It is, of course, necessary to have a reasonable idea of what property one is trying to enhance in the target protein. Wrong manipulation of protein could lead to fatal problems. Proteins produced through biological systems (genetically unmodified protein) are the safest choice [57, 97, 98].

There are many concepts controlling the uses of protein in medicinal applications such as is purity. In technical applications (e.g., technical enzyme) the boundaries of purity are different. Regarding activity and stability, the protein must match perfectly the purpose of its use. The importance of protein engineering in industry continues to grow as the number of applications of proteins expands, and the technology used to discover proteins efficiently with useful properties is better able to address industrially relevant problems. Recent advances in directed evolution are implemented in many established industrial laboratories as well as in start-up companies, augmenting the rational design approach. Additionally, organisms from extreme environments are becoming an important source of new backbones for engineering proteins with significantly different properties. The successfully engineered protein generally requires a proper combination of properties. For example, a detergent protease would require, at minimum, stability in the presence of detergent and activity against certain protein stains. Nevertheless, the control of a few basic properties is a recurring theme in many applications. Properties such as sufficient stability, high activity (in the case of enzymes), and the ability to interact correctly with surfaces are necessary for a variety of industrially important proteins [99].

2.3.3 Engineered Proteins

2.3.3.1 Technical Enzymes: e.g. Proteases and Lipases

The demand for technical enzymes corresponded to a market size equal to 1 billion USD in 1999 [100]. Some of these enzymes are the thermostable enzymes, which are well represented in different industrial processes and constitute more than 65% of the worldwide market [101]. Enzymes were implemented in many important industrial products and applications such as in the paper industry, detergents, drugs, degradation of different wastes, textiles, food, pharmaceuticals, leather, degumming of silk goods, manufacture of liquid glue, cosmetics, meat tenderization, cheese production, growth promoters, etc. Enzymes used with detergent are the most important and profitable applications with a market size equal to 0.6 billion USD in 2000 (Novozymes data) [100]. The first use of enzymes in detergents occurred in 1913 when Röhm and Haas introduced crude trypsin into their detergent Burnus based on a German patent issued to Otto Röhm (1913) [100]. Enzymes used with detergent must be stable and function well in the presence of a variety of potentially unfriendly detergent ingredients (e.g., anionic/ non-ionic/ cationic surfactants, chelators (e.g. EDTA), builders, polymers, bleaches) and in various forms of detergent products (i.e., liquids and powders) [100]. Thermostable enzymes are active and stable at temperatures higher than optimal growth of their producer strains. Bacilli strains isolated from diverse sources with diverse properties have made these organisms the focus of attention in biotechnology. Thermostable enzymes can be produced by both thermophilic and mesophilic microbes. The use of high temperature has many significant applications due to solubility and reducing viscosity [102, 103].

2.3.3.1.1 Proteases

Up to the 1980s, proteases were considered the only commercially relevant enzymes. Today, many laundry-detergent products contain at least a protease, and many contain cocktails of enzymes including proteases, amylases, cellulases, and lipases [100]. The major source of proteases is microorganisms while proteases from plant origin have not been well investigated. Based on their catalytic mechanisms, proteases can be classified into Ser, Cys, Aspartic and metalloproteases. In nature, proteases have valuable biochemical and physiological functions. They can be very specific, not only cleaving proteins into amino acids or short peptides but also can cleaved specifically to produce useful peptides. *Bacillus licheniformis*, *Bacillus subtilis* and *Bacillus pumilus* are the most well-known species used in industry for alkaline protease production [104]. Proteases were used widely in many industrial applications included detergent, wool quality improvement, meat tenderization, leather, etc. Ideally, proteases used in detergent formulations should have high activity and stability within a broad range of pH and temperatures, and should be compatible with various detergent components along with oxidizing and sequestering agents [105].

Protein engineering was used to improve the stability of BPN' from *Bacillus amyfoliquefaciens* in the chelating environment of the detergent by deleting the strong calcium-binding site (residues 75–83) and re-stabilizing the enzyme through interactions not involving metal ion binding. Stability increases of greater than 1000-fold in EDTA were reported for this protease [106]. The surface properties of BPN' have also been engineered. It was found that variants containing mutations that produce negative charges in the active site region of the molecule adsorbed less strongly and gave better laundry performance.

2.3.3.1.2 Lipases

Lipases were characterized by their ability to hydrolyze long chain triglycerides [107]. Lipase catalyzes the hydrolysis (or synthesis) of insoluble esters. The primary use of lipase is in cleaning applications, although its use in the chiral synthesis of high value chemicals is also important. A comparison of the experimental results of several site-directed variants with structural modeling has provided much insight into the catalytic mechanism of a fungal lipase from *Rhizopus oryzae* at the molecular level [108]. In order to understand lipase activity fully one must also take into account its ability to interact with a macroscopic substrate, such as a triglyceride surface. Most lipases are activated at the oil(substrate)–water interface by a conformational change to adapt the enzyme–substrate interaction [109]. Changes at Glu87 and Trp89 were reported to alter activity of the lipase from Humicola lanuginosa (Lipolase) [110]. Surfactant and calcium sequestering agents, such as sodium tripolyphosphate, reduce the activity of current lipases 100–1000-fold in laundry detergents [111, 112]. Some progress in designing variants that reduce this inhibition by creating favorable surfactant–enzyme interactions were reported to give improved laundry performance. The commercial applications of lipases include, detergents such as in dishwashing, clearing of drains clogged by lipids in food processing or domestic/industrial effluent treatment plants [96].

2.3.3.2 Pharmaceutical Applications

The estimate of about 350 biotechnology drugs currently undergoing development, including vaccines, gene therapy, antisense technology and antibodies derived from "humanized" transgenic mice. Protein engineering was used to produce therapeutic proteins with improved properties such as increased solubility and stability. Many of the early protein drugs derived from biotechnology failed because they were primary molecules with suboptimal affinity or poor half-life *in vivo*, leading to poor efficacy. In other cases, many of the original protein drug molecules were non-human and caused immune responses against the drug itself. Affinity, half-life and dosing are all interrelated and play a role in determining the clinical efficacy and financial viability of protein-based drugs. This increased understanding of the issues affecting success in drug development was paralleled by increased capabilities in protein engineering and selection/screening technologies. These were used to improve the effectiveness of a number of protein drug candidates.

2.3.3.3 Reducing the Immunogenicity of Protein Drug Molecules

Many early attempts at introducing protein therapeutic molecules failed because the protein drug molecules were recognized as non-human and led to an immune response against the drug itself. As a result, most proteins used in clinical trials now are primarily human or are humanized, even if the original "proof of concept" work was done with non-human proteins. For example, Pulmozyme (Genentech) is a drug based on human DNAse which was developed for use in managing cystic fibrosis, following successful "proof of principle" studies with bovine pancreatic DNAse I [113]. The immunogenicity of mouse antibodies in humans was one of the major reasons why early monoclonal antibodies did not deliver the anticipated therapeutic benefits. This led to the development of chimaeric antibodies, created by fusing mouse variable domains to human constant domains to retain binding specificity while reducing the proportion of mouse sequence. TNFα-neutralizing chimaeric monoclonal antibody, was approved for use in treating Crohn's disease and rheumatoid arthritis [114]. The reduction in monoclonal antibody immunogenicity was taken a stage further by complementarity-determining region (CDR) grafting, where the 34 CDRs of mouse antibodies were grafted onto human frameworks to reduce the proportion of mouse sequences in the drug still further while retaining binding specificity [115].

2.3.3.3.1 Insulin

Insulin was engineered through mutagenesis to create monomeric forms that are fast acting (insulin lispro and insulin aspart). Conversely, another form (insulin glargine) was created by mutagenesis to precipitate upon injection and give a sustained release of insulin. More research was done on insulin. Whittingham et al. 1997 reported a crystal structure of a prolonged-acting insulin with albumin-binding properties [116].

2.3.3.3.2 Catalytic Antibody

A catalytic antibody is a variant of an antibody. Antibodies are proteins that normally bind to a specific molecule but do not alter the bound molecule in any way. A catalytic antibody is one which was changed by mutations to have a novel sequence that folds into a structure that catalyzes a specific reaction (such as amide bond formation, ester hydrolysis, and decarboxylation). Catalytic antibodies function like enzymes, and are created to catalyze reactions for which there are no naturally occurring enzymes. Fifty or more different reactions have been catalyzed by the action of catalytic antibodies that were obtained individually by methods of protein engineering [117].

2.3.3.3.3 Polyketide Synthases

Antibiotics such as erythromycin are made by large multidomain proteins called polyketide synthases. Site-directed mutagenesis was used to modify the substrate specificity of one polyketide synthase reaction so that the product contains a malonate unit, whereas the product of the original enzyme contained a methylmalonate unit. In addition to site-directed mutagenesis, the order of the polyketide synthase domains was shuffled to create proteins that could catalyze the synthesis of new antibiotics. An extension of site-directed mutagenesis allows non-natural amino acids to be incorporated into proteins. Non-natural amino acids are not naturally encoded by the genome, but instead include a wide variety of amino acids that are present in cells or produced by synthetic methods [117].

2.3.4 Traditional Protein

2.3.4.1 Casein

Casein is a natural polymer extracted from skimmed milk proteins. Casein protein is used in many industrial and technical applications [71, 118, 119], such the manufacture of adhesives and the packaging industry for breweries, wineries and refrigerated products and it can also be used as a plasticizer for concrete. Casein is also used as microcapsules and in synthetic peptides [120]. Caseins evolved from members of a group of secreted calcium (phosphate)-binding phosphoproteins. The first industrial applications of protein as polymer were in the early 1930s and 1940s with casein and with soy protein. Casein is also used as microcapsules and in synthetic peptides [75, 121].

2.3.4.2 Keratin

Keratin derived materials have shown potential to transform the world of bio-based materials because of their intrinsic biocompatibility, biodegradability, mechanical durability, and natural abundance [58]. keratin is the most abundant structural protein in epithelial cells [122] and a most important biopolymer in animals along with collagen. Keratin is a polypeptide consisting of amino acids having intermolecular bonding of cysteine and few intramolecular bonding of polar and non-polar groups. The cysteine residues have thiol groups which produce strong disulfide bonds leading to the cross-linking of the matrix molecule. Keratins can be exited as delicate keratins (such as stratum corneum) generally poorly united and with a lower measure of sulfur and lipids, and hard keratins found in hair [123], nails, paws, noses, feathers [124], plumes, which have a more rational structure and a higher measure of sulfur.

2.3.4.3 Worm and Spider Silk

Silk is a polymer made from the fine threads produced by certain insect larvae. Silk was traditionally known to be produced from the silk worm [125]. Silk is, by nature, a protein biopolymer produced by polymer producer organisms. Examples of silk producers are spiders, silkworms [126], mites [127], scorpions and flies [128]. There is a rise in interest in the silk produced by spiders [129]. Spider silk is an interesting biomaterial, elastic and strong that is comparable to the best fibers synthetized by new technology in terms of mechanical properties. It is also a biodegradable material and environmentally safe. Because of the limited amount of spider silk, silk fibroin as a natural polymer produced by silkworms is a good alternative. Sericin and fibroin are the major components of it. Fibroin, a fibrous protein creating the silk core, is composed of a fibroin light chain, fibroin heavy chain and fibrohexamerin. Excellent mechanical properties, biocompatibility and slow degradability make this material interesting.

2.3.4.4 Collagen, Gelatin, Elastin, Albumine and Fibrin

Collagen and gelatine are animal polymers found in skin and connective tissues. Collagen degraded to high molecular weight polypeptide, called gelatine, can be obtained by thermal denaturing of collagen. Gelatin is a water-soluble proteinaceous substance [130]. Gelatin is an important high molecular weight polypeptide hydrocolloid. It is commonly used in a wide range of food, medicinal, pharmaceutical, and polymeric materials. Most hydrocolloids are polysaccharide, whereas gelatin is a protein containing all the amino acids except tryptophan [131]. It *was* fabricated to different forms to match different applications [132]. It is essential in drug caps, X-rays, photographic film development and food processing. Gelatin grades used in drug delivery and tissue engineering are also available in a wide range of viscosities. It does not show antigenity and is resorbable *in vivo*. Its physico-chemical properties can be suitably modulated. Gelatine can be plasticized thanks to the addition of water or of glycerol. There is, however, a limit to the use of this interesting material because there is a risk of viral animal contamination. Blends of polyvinyl alcohol and gelatine are the object of studies and research. Elastin, albumine and fibrin are other proteins from animal sources. They have been investigated especially for various biomedical applications. Elastin is used as a biopolymer in enhancing cellular uptake in the tumor cells [82, 83, 133–136].

2.3.4.5 Wheat Gluten

Wheat gluten is a protein by-product of the starch fabrication. In addition to wheat, grain sources of gluten are barley, rye, triticale, spelt, einkorn, emmer and kumut. It is available in high quantity and at low cost [137]. Gluten is a part of our food and is contained in pasta, bread, cereals, soups, deserts, soy sauce, hydrolyzed wheat proteins, wheat bran hydro lysate, wheat protein isolate, wheat starch, glucose syrups, wheat maltodextrin, sorbitol, lactitol, maltitol, caramel, glucan, alcohol/ ethanol, vinegar, wheat germ oil, medications, and so on. They are relatively impermeable to oxygen and to CO_2 but are sensitive to humidity. Potential applications are producing soluble receptacles for the controlled release of a chemical product (such as toilet detergent). Wheat gluten contains two main groups of proteins, gliadin and glutenin [138]. Gliadins are protein molecules with disulphide bonds. They have low molecular weight and a low level of amino acids with charged side groups. Gliadin has antimicrobial activity and is used in food packaging and coating applications [130, 139].

The molecular weight of glutenins is at least ten times higher than that of gliadins. Wheat gluten materials have the fastest degradation rates. Gluten is fully biodegradable and the products obtained are non-toxic. Wheat gluten has proved to be an excellent film-forming agent [18, 140]. In practice, the term “gluten” refers to the proteins, because they play a key role in determining the unique baking quality of wheat by conferring water absorption capacity, cohesivity, viscosity and elasticity to dough. It is also used for improving solubility, emulsification, and film-forming properties [22, 141, 142]. The amino acid compositions of glutenins are similar to those of gliadins, with high levels of glutamine and proline and low levels of charged amino acids. Glutenins can be broadly classified into two groups, the high molecular weight (HMW) and the low molecular weight (LMW) subunits.

2.3.4.6 Soy Protein

Soy protein has been used since 1959 as an ingredient in a variety of foods for its functional properties, which include emulsification and texturizing. Soy protein is used in many applications [143]. Recently the popularity of soy protein has risen, mainly because of its health benefits. It has been proved that soy protein can help to prevent heart problems. Soy protein films do not have as good mechanical and barrier properties as most protein films, due to their hydrophilic nature. They are used to produce flexible and edible films.

2.4 Biopolymer Type Number 3: Polysaccharides

Polysaccharides are among the most widespread organic compounds in the plant kingdom and used in many applications [144]. Polysaccharides play essential roles in the life processes of all plants. They can be divided into several broad groups according their functions, i.e., structural polymers (cellulose), protective polysaccharides (pectin and hemicelluloses) and reserve polysaccharides (starch). Further, polysaccharides can form glycoconjugates with proteins and lipids resulting in biological macromolecules in the cell wall and cell membranes, and play important roles in many physiological and biochemical processes.

2.4.1 Starch

Starch is a complex carbohydrate and one of the largest molecules in nature found chiefly in seeds, fruits, tubers, roots and stem pith of plants, notably in corn, potatoes, wheat, and rice. Starch is a polymer of D-glucose organized in two major constituents of huge molecular weights. Amylose contains amorphous and crystalline regions [145, 146]. Amylopectin is used in high-performance flocculents and as non-ionic surfactant [145]. To improve their resistance to shock and moisture, polyolefins were added in small quantities (about 10–15%) or in large proportions up to 85–95% to starch. Those polymer mixtures disappeared during the biodegradation process leaving small fragments whose degradation time was a function of their carbon chain length. Supol (Supol, Germany) Potato flour is submitted to a thermal treatment under pressure. Pellets can be injected to produce single-use dishes which are microwavable and which are compostable or can be added to animal food. Evercom (Comstarch, Japan) plasticized maize starch can be injected to make small parts for catering or for horticultural applications. This product is compatible with other biopolymers such as PHBV, PLA, PCL polyesters. Native starches differ in the amylose/amylopectin ratio depending on their botanic source, such as native starches are composed of 20–30% amylose and an additional amount of amylopectin, amylose-enriched starch may contain up to 84% amylose while waxy starches consist of nearly pure amylopectin. The main applications are for producing mulch films, shopping bags, food packaging (yogurts), nappies, medicinal and personal hygiene products [147, 148].

2.4.2 Cellulose and Cellulose Derivative

Cellulose is a polysaccharide that is the main constituent of all plant tissues and fibers. The name cellulose, from *Latin cellula*, was coined in 1838 by French chemist Anselme Payen. Special forms of cellulose fiber are found in most plants in the leaves and stalks, with cotton fiber (95% cellulose) [149]; wood (about 50%) being the principle industrial sources of cellulose [150]. Because of its ubiquity in the plant kingdom, cellulose is arguably the most abundant polymer on earth, with billions of tons produced annually through photosynthesis. The strong glucosidic bonds ensure the stability of the cellulose in various media. Cellulose is mostly insoluble and highly crystalline. Chemical reactions such as esterification are conducted on the free hydroxyl groups to improve its thermoplastic behavior. The main uses of cellulose are for paper, membranes, dietary fibers, explosives and textiles [151]. Cellophane™is a blend of cellulose and diaphane [152]. A transparent paper-like product that is impervious to moisture and used to wrap candy, cigarettes, etc. The name was coined in 1912 by Swiss chemist Jacques Edwin Brandenberger, the invention of the material dates from 1908. Cellophane films are obtained by dissolution of cellulose in a sodium hydroxide and carbon disulphide solution (Xanthation) and then by recasting in a sulphuric acid bath. Degradation takes place after six weeks of composting. Cellophane films are mainly used in food packaging where they are valuable for their barrier properties against microorganisms, gases and smells. Cellulose acetate is mainly used in the synthesis of membranes for reverse osmosis.

Cellulose is the principle part of most plant cell walls, and is currently of interest because of processes for paper making, and as a major structural component of textile fibers such as (*Gossypium spp.*), ramie (*Boehmeria nivea*), jute (*Corcorus capsularis*), flax (*Linum usitatissiumum*) and sisal (*Agare sisalana*). Surgical cotton is actually made from wood. Cellulose used as biopolymer, biomaterial, drug delivery, bacterial microparticles [153, 154].

Bacterial cellulose is synthesized in a process whereby the polymer material is extruded from the bacterial cells. Most cellulose-producing bacteria (i.e., Acefobacter) extrude cellulose as a ribbon-like product from one fixed site on the cell surface. This results in a network of interlocking fibers forming. Bacterial cellulose is produced under conditions of agitated fermentation. Bacterial cellulose is a water-insoluble material that has a large surface area because of its large network of fibers; bacterial fibers have roughly 200 times the surface area of fibers from wood pulp. This, coupled with their ability to form hydrogen bonds, makes them unique when used as suspensions, they have pseudoplastic thickening properties. The bacterial cellulose is used for dressing chronic wounds [155], nanocomposites [156], cancer treatment [157], natural rubber latex formation [158], antimicrobial food packaging applications [77], biocomposite [159], gravity-driven oil/water separation [160] and many others.

2.4.3 Hemicellulose

Hemicelluloses are a group of substances which occur in association with cellulose, and are usually a mixture of xylans, xyloglucans, arabinogalactans, glucomannans and galactoglucomannans. Hemicellulose is one of the most abundant polysaccharides after cellulose. Hemicelluloses were used as feedstock for producing sugars. Hemicelluloses are mostly heteropolysaccharides classified according to the sugar residues present, namely xylans, mannans, arabinans and galactans, and they are either linear or branched polymers. Xyloglucan is used as a barrier film [161].

2.4.4 Chitin and Chitosan

Chitin was discovered in 1811 by Braconnot and was initially termed fungine because it was discovered in mushrooms [25], it is derived from shrimp, crab, Antarctic krill, and cultivated fungi. It was later in 1823 that Odier gave the name chitin to the same material discovered now in the elytrum of the cock chafer beetle based on the Greek term chitos, meaning coat. Chitin as a natural polysaccharide is a tough, semitransparent horny substance; the principal component of the exoskeletons of arthropods and the cell walls of certain fungi.

Chitosan is known as soluble chitin [162–167]. Chitosan is produced commercially by deacetylation of chitin. It has many uses: diluent, binder, drug carrier, drug release, site specific drug delivery, absorption, enhancer, carrier, anticancer, anticoagulant, antiviral, antioxidant, tissue engineering and food technology. Chitosan is soluble in water and in some organic solvents. The difference between chitin and chitosan is defined by their solubility in a dilute solution of weak acids. Chitosan dissolves in dilute acetic acid. It presents a unique combination of properties, brought about by its polysaccharide structure, large molecular weight, and cationic character. Chitin and chitosan are biocompatible and present antithrombogenic and hemostatic properties. These polymers can be extruded to make films for packaging applications. Chitosan is used in many applications including: hydrogel, tissue engineering, drug delivery, tissue repair.

Chitosan is the partially or fully deacetylated form of chitin. The chitosan deacetylation degree is usually in the range between 70% and 95%, and the molecular weight is also between 10 to 1000 kDa. Its application in the tissue engineering and drug delivery fields is wide ranging from cartilage, bone, vascular grafts and skin to substrates for cell culture. Biologically renewable, biocompatible, biodegradable, non-toxic and non-antigenic properties of chitosan make it a bio-functionally useful biomaterial. In addition, hydroxyl and amino groups of chitosan can be modified chemically to provide a high chemical diversity. It also has bio-adhesive properties. Chitosan exhibits different behaviors at various pH levels. It does not dissolve at high pH while it is soluble at lower pH ranges. This property makes chitosan a suitable tool for delivery applications. Chitosan, and their derivatives have found a number of pharmaceutical or biomedical applications. Although chitosan was mostly used as a diluent in tablet manufacturing, it was also proposed as a binder, lubricant, or potential disintegrating agent. The mucoadhesive properties of chitosan make it an attractive material for the local delivery of drugs in the oral cavity.

2.4.5 Xanthan

Xanthan gum, a complex copolymer produced by a bacterium, one of the first commercially successful bacterial polysaccharides to be produced by fermentation [168, 169]. In terms of production volume, xanthan gum is the most widely used microbial polysaccharide. Worldwide production is currently in the range of 10,000 to 20,000 tons. Xanthans have many applications such as in hydrogel, antibacterial and catalytic applications, in silver nano-particles, in medicinal and pharmaceutical applications [170]. It is used in drug release [171], wound healing [172], blended hydrogels for connective tissue regeneration [173], immobilized biocatalyst [174], edible coatings and many others. Xanthan gum is a HMW extracellular polysaccharide produced by the fermentation of the gram-negative bacterium *Xanthomonas campestris*. Companies such as ADM and Merck have recently announced the expansion of their xanthan production facilities. About 60% of the xanthan produced is used in foods, with the remaining 40% used in industrial applications. Food-grade xanthan costs about $8 to $10 per pound, while non-food grades sell for about $5 per pound. So far, only experimental samples of genetically modified xanthan have been produced.

2.4.6 Dextran

Dextran is the generic name of a large family of microbial polysaccharides that are assembled or polymerized outside the cell by enzymes called dextran sucrases. This class of polysaccharides is composed of building blocks (monomers) of the simple sugar glucose [169, 175–177]. It can be found as storage material in yeasts and bacteria. Dextrans are produced by enzymatic conversion of the feedstock sucrose. Most commercial dextran production uses the microorganism *Leuconstoc mesenteroides*. Cyclodextrins is used as a stabilizer and in edotoxin removal [178]. Dextran polymers have some medical applications. Dextrans were used for wound coverings, in surgical sutures, as blood volume expanders, to improve blood flow in capillaries in the treatment of vascular occlusion, and in the treatment of iron deficiency anemia in humans and animals.

Chemically modified dextrans such as dextran sulfate [179] have antiulcer and anticoagulant properties. Other modified dextrans such as Sephadex are used extensively in the separation of biological compounds. In the industrial area, dextrans are being incorporated into x-ray and other photographic emulsions. Dextran is used as a food-grade biopolymer, nanoparticles, hydrogel, and in transdermal delivery [180].

Dextran-hemoglobin compounds may be used as blood substitutes that have oxygen delivery potential and can also function as plasma expanders.

2.4.7 Pullulan

Pullulan is a non-toxic exopolysaccharide of fungi origin. It is non-toxic, non-immunogenic, non-mutagenic and non-carcinogenic [181]. It produced from *Aureobasiduim pullulans* [182]. Pullulan is biodegradable impermeable to oxygen, non-hygroscopic and non-reducing. Pullulan possesses oxygen barrier properties, good moisture retention and it inhibits fungal growth so it enhances the shelf-life. It is used in tissue engineering, molecular chaperones, plasma expender and surface modification. Pullulan is a linear polymer made of monomers that contain three glucose sugars linked together. For more than a decade, a Japanese firm, Hayashibara Biochemical Laboratories, has used a simple fermentation process to produce pullulan. Some feedstocks are used for this process, including the streams containing simple sugars. Pullulan can be chemically modified to produce a polymer that is either less soluble or insoluble in water. The thermal and ionic (electrical) properties of pullulans can also be altered.

It can be used as a food additive, providing bulk and texture. It is tasteless, odorless, and non-toxic. It does not break down in the presence of naturally occurring digestive enzymes and therefore has no caloric content. So, it can be used as a food additive in low-calorie foods and drinks, in place of starch or other fillers. In addition, pullulan inhibits fungal growth and has good moisture retention, and thus can be used as a preservative. Pullulan can also be used as a water-soluble, edible film for the packaging of food products. It is transparent, impermeable to oxygen, and oil- and grease-resistant. Foods can be either immersed in a solution of pullulan or coated by a polymer spray. After the pullulan coating is dried, an airtight membrane is formed. Pullulan is used as gold nanoparticles for cancer treatment, drug delivery and in controlled release of biopharmaceutical [183, 184].

2.4.8 Glucan

A common source for glucan is baker's yeast, *Saccharomyces cerevisiae*, although it is also found in some other sources (bacteria, fungi, lichen, and higher plants, such as, barley). Large supplies of inexpensive yeast are available from both the baking and the brewing (brewer's yeast) industries. Glucans are the most abundant polymers in yeast, making up approximately 12–14% of the total dry cell weight. Glucan is readily purified from yeast cells by using hot alkali treatment to remove all other cellular materials, thereby allowing recovery of the insoluble glucan material. Yeast glucan particles purified by this method contain both HMW and lower molecular weight polymers. Glucan is used as an immunosuppressive [185]. Glucan has been modified and applied in different applications such as hydrogels [32, 186], and a potential prebiotic [187], in supporting the treatment of viruses [188] and many others. Although glucans are being exploited mainly for their antitumor, anti-infectious, and radioprotective properties, they also have non-medical applications. Glucans resist breakdown when attacked by digestive enzymes, and thus can be used as non-caloric food thickeners. Other possible applications include use in sustained-release tablets, encapsulation of oxygen for mass transfer in fermentation reactions, and as a solid support material for chromatographic separations [189–193].

2.4.9 Gellan

Gellan is a complex polysaccharide having a four-sugar repeat unit (glucose–glucuronic acid–glucose–rhamnose). It is produced by the bacterium *Pseudomonas elodea*, which is derived from plant tissue. The properties of gellan can be easily modified. A hot caustic treatment of gellan yields a polymer that has the desirable characteristic of low viscosity at high temperature. Cooling gellan in various cations (such as, calcium) results in forming strong gels. Gellan is used to inhibit *Plasmodium falciparum* growth in the food industry, personal care products [194], pharmacy and medicine [194], hard capsule [195], drug delivery [196], microspheres [197], cell differentiation [198], microbeads [199], nanocomposites [200] and many others.

2.4.10 Pectin

Pectin is a water-soluble colloidal carbohydrate that occurs in ripe fruit and vegetables. Pectin is a family of complex polysaccharides present in the walls that surround growing and dividing plant cells. It is also present in the junctional zone between cells within secondary cell walls including xylem and fiber cells in woody tissue. Its traditional application is in making fruit jellies and jams. Pectin is an essential component in the initial growth and ripening process of fruit and is often a waste material from the food and fruit processing industry. Pectins are made either from apple pulp, byproducts of cider manufacture, or from the peel of citrus fruits such as limes, the preferred source, lemons or oranges, byproducts of fruit juice manufacture. Pectins have long been recognized as the main gel-forming agents in jams and fruit-based preserves. It was known for a time that polysaccharides can stabilize colloidal sols and suspensions. The simple addition of water-soluble pectate for example: usually the sodium salt will prevent aggregation and precipitation. Gels can be obtained by mixing gelatin and pectin into lipid in water emulsions. The results are low-fat margarines and other spread type products. Pectins are used in drug delivery and as a colloidal [201].

2.4.11 Gums

Gums are plant substances which can be dissolved or dispersed in water to form more or less viscous colloidal solutions or dispersions. The gums are naturally occurring substances, mainly carbohydrate in nature and are being used since the beginning of civilization for various purposes e.g., a food ingredient (for humans and cattle) and manufacturing domestic items. The properties of natural gums have been known for a long time including its gelling, thickening and binding properties. The mannans are commonly referred to as "gums". Guar gum comes from the endosperm of the seed of the legume plant *Cyamopsistetragonolobus*. The gum is commercially extracted from the seeds essentially by a mechanical process of roasting, differential attrition, sieving and polishing [202]. Khaya gum is a polysaccharide obtained from the incised trunk of the tree *Khaya grandifoliola* from the family Meliaceae. Acacia gum (GA) is the exudate from the *Acacia senegal* and *Acacia seyal* trees, belonging to Leguminosae family [203]. Most of the GA is harvested from the arid lands of Sudan, Chad, Nigeria, Senegal, and Ethiopia. Sudan is the largest exporter, accounting for up to 80% of the trade, followed by Nigeria. The use of GA dates back to 5000 years ago. GA has also found use in the textile industry for its ability to enhance tensile strength of the yarns [204]. Glactomannan is one of the principle biotechnologically interesting molecules. There is a particular interest in the synergistic interactions between both galactomnnans and glucomannans with xanthan, carrageenans and agarose. Glucomannan is used as antioxidant [205].

2.4.12 Hyaluronic Acid

Hyaluronic acid capsular component by the bacteria, Staphylococcus and some Streptococci. Hyaluronic is used in drug delivery, as an antioxidant, as a cationic biopolymer, and in nanotube technology [40, 206–210].

2.4.13 Fructans

Inulin and levan fructans (which used to be called fructosans, and are described as "inulin" in commercial practice) are found in large amounts in only a few plants but are widely distributed [211].

2.4.14 Marine Polysaccharides

Seaweed contains different types of polysaccharides such as alginates, carrageenans, agar (and its agarose constitutent) shells, crabs, and lobsters. Shrimps contain chitin and its soluble derivative,

chitosan.

2.4.14.1 Alginate

The naturally occurring alginate polymers have great potential in drug formulation because of their extensive application as food additives and their recognized lack of toxicity. Alginate is a historic term used in many applications such as the cosmetic and pharmaceutical industries. As this group of polymers possess numerous characteristics that makes it useful as a formulation aid, both as a conventional excipient and more specifically as a tool in polymeric-controlled drug delivery. The alginates were discovered by a British Pharmacist, E.C.C. Stanford; commercial production started in 1929. The annual production of alginates in the world is about 30 000 tonnes. Alginic acid and it's salts (Ca, Mg, Na and K) are abundantly present in brown algae (pheophyta) of the genera macrocystis, laminaria, ascophyllum, alario, ecklonia, eisenia, nercocystis, sargassum, cystoseira, and fucus. Acetylated alginates are also isolated from some bacteria genera pseudomonas and acetobacter. Red algae belonging to the family coralenacease also contain these substances. Alginate are used in nanoparticles, grafted donor sites in burns, wound dressing and in recycling of textile dye [212].

2.4.14.2 Carrageenans and Red Seaweed

Carrageenans are marine hydrocolloids obtained by extraction from some members of the class rhodophyceae. The carrageenans are a group of related linear and sulfated polymers. The most important members of this class are *Chondrus crispus* and *Gigartina stellata*. Carrageenans were mainly used as gelling and thickening agents. Only a few studies have dealt with carrageenans for controlled release tablets. These studies dealt only with drug delivery from tablets on a hydraulic press or from tablets that contain the carrageenans in a mixture with other excipients. A polysaccharide obtained from edible red seaweeds (*Chondrus crispus*). It is used in pharmaceutical formulations and cosmetics. As an anticoagulant, antithrombotic, antiviral, antitumor, immunomodulatory, immobiliser, and cleaning of industrial effluents. Unlike alginate, carrageenan use is dominated by food applications, particularly in concentration with milk like products. Carrageenans are used in gelling and thickening toothpaste, microencapsulation and immobilization. Carrageenan is used as a grafted copolymer in food packaging and coating applications [213].

2.4.14.3 Agar and Agarose

A polysaccharide produced commercially from red algae blonging to the family Rhodophyceae particullay from *Gelidium* and *Gracilaria*. Agar polysaccharide is composed of neutral polysaccharide (agarose), charged polysaccharide (agaropectin), and highly sulfated galactans. It is used as a food additive in icings, glazes, processed cheese, jelly, sweets and marshmallows. Agar is used for growing microorganisms while most species are unable to degrade it. Agar is resistant to high temperature, can form brittle gels and can hold a large number of soluble solids. It used as gelling, thickening, stabilizing and viscosity controlling agent for jellies, candies and jams.

It shows also many medical, pharmaceutical and industrial applications such as a laxative, an anti-rheumatic agent and for making dental impressions. Agarose is a neutral polysaccharide in agar and is used mainly in the separation of biological macromolecules. Also, it has antioxidative, antibacterial, antimutagenic and immune modulating activities.

2.5 Biopolymer Type Number 4: Organic Polyoxoesters

Poly(3-hydroxyalkanoates) (PHAs) are structurally simple macromolecules. PHAs accumulate as discrete granules to levels as high as 90% of the cell dry weight and are generally believed to play a role as a sink for carbon and reducing equivalents [35, 37]. Most of the well identified PHAs are linear, head-to-tail polyesters composed of 3-hydroxy fatty acid monomers (Figure 2.3). In these polymers, the carboxyl group of one monomer forms an ester bond with the hydroxyl group of another monomer. In 1976, Imperial Chemical Industries (ICI) in England started to produce P(3HB) by fermentation. In 1993, Zeneca Bioproducts started their production business, later in 1996 Monsanto bought the production business from Zeneca. Using *R. eutropha* about 800 tons per year of P(3HB-co-3HV) produced under the trademark Biopol™. Monsanto terminated its activities in this area by the end of 1998. Today many other companies remain still active in research and development of PHAs like Procter and Gamble as well as Metabolix (USA) and others all over the world. Biodegradable materials are often used within the biomedical field as implants or as drug carrier systems. Early inteest in environmentally friendly bioplastics was by the European environmental legislation in 2005

[37, 214]. PHA was proved to be biocompatible and can be used in tissue engineering, implantations, and so on. Retinal pigment epithelium cells grow well on P(3HB-*co*-3HV) as a monolayer for their subretinal transplantation. PHA can be melted or solution processed into a variety of forms. Salt leaching, dip coating and thermally induced phase separation were used to produce scaffolds for cardiovascular tissue engineering. When seeded with cells and cultured *in vitro*, these scaffolds were used to create living tissue implants [215].

The hydroxyl-substituted carbon atom has *R* configuration in all characterized PHAs. At the C-3 atom (β position), an alkyl group length can vary from methyl (C1) to tridecyl (C13) [216]. This alkyl side chain is not necessarily saturated. Unsaturated, aromatic, epoxidized, halogenated, and branched monomers were reported as well [217]. Cross-linking of unsaturated bond substations in the side chains of PHAs can be added

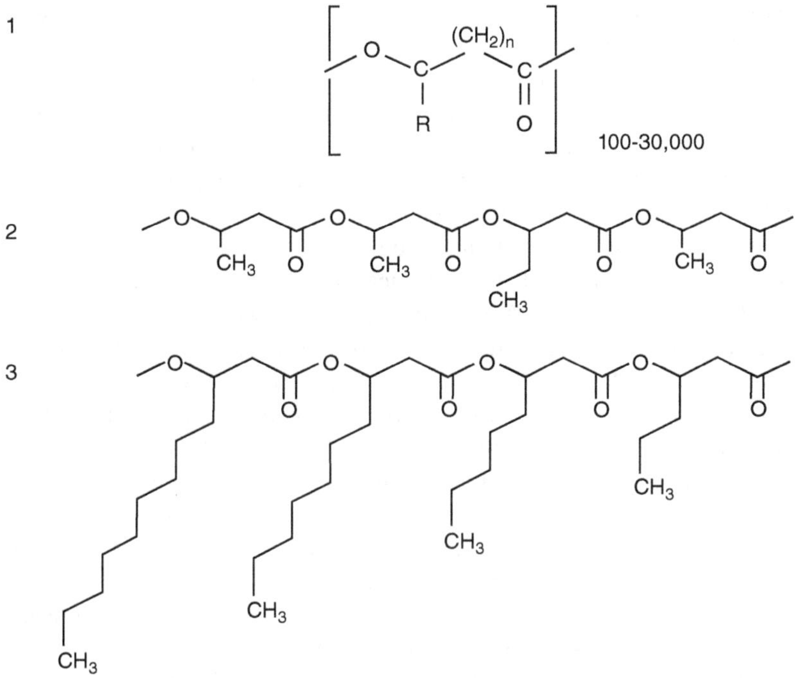

Figure 2.3 Chemical structures of PHAs. (1) General structural of PHAs. (2) The copolymer poly(3-hydroxybutyrate-*co*-3-hydroxyvalerate) P(3HB-*co*-3HV) as example of short chain length polymer (SCL). (3) P(HHX-*co*-HO-*co*-HD-*co*-HDD-*co*-....) as an example for PHA_{MCL}.

chemically [218]. Lütke-Eversloh et al. was the first to report on producing biopolymers with thioester linkages in the polymer backbone using *C. necator* in media containing 3-mercaptopropionate (3MP) or 3-mercaptobutyrate (3MB) in addition to 3-hydroxybutyrate as constituents [219–221]. Many factors affect the PHA's chemical composition like the microbial strain, the substrate, the cultivation condition, the extraction method, the number of *phaC*, *phaB* genes, the regulator *phaP* (phasin) and the presence of inhibitors. They inhibit different pathways, especially those which supply the synthases with different kinds of monomer or inhibit other pathways, which consume these monomers for their own or degrade it to shorter units like β oxidation pathway. In general, the PHA composition depends on the PHA synthases, the carbon source and the metabolic routes involved. The molecular weights of PHAs were established by light scattering, gel permeation chromatograph and sedimentation analysis. Their monomer composition was determined by gas chromatography (GC), mass spectroscopy (MS) and nuclear magnetic resonance (NMR) analysis [222]. PHAs show material properties that are similar to some common plastics such as polypropylene [223]. The bacterial origins of the PHAs make these polyesters a natural material, and many microorganisms have the ability to degrade these macromolecules [224]. The molecular mass of PHAs varies per PHA producer but is generally in the order of 50×10^3 to 1×10^6 Da. Inside the cell, P(3HB) exists in a fluid, amorphous state. However, after extraction from the cell with organic solvents, P(3HB) becomes highly crystalline [225] and in this state it is stiff but brittle material. Because of its brittleness, P(3HB) is not very stress resistant. The high melting temperature of P(3HB) (around 170 °C) is close to the temperature in which this polymer decomposes thermally and thus limits the ability to process the homopolymer. The incorporation of 3-hydroxyvalerate (3HV) into the P(3HB) resulted in P(3HB-*co*-3HV) copolymer that is less stiff and brittle than P(3HB), that can be used to prepare films that exert excellent water and gas barrier properties like polypropylene, and that can be processed at lower temperature while retaining most of the other excellent mechanical properties of P(3HB) [226]. (P(3HB-*co*-3HV)) has also low crystallinity and is more elastic than P(3HB) [227, 228]. The latex-like PHAs (PHA_{MCL}) display physical properties, which differ significantly from the PHA_{SCL}, such as P(3HB). Particularly with respect to the melting temperature and the extension to break value the two types of PHAs showed, mainly because of the lower crystallinity of PHA_{MCL}, striking differences. PHAs have been processed into fibers, which were then used to construct materials such as non-woven fabrics [229]. Moreover P(3HB) and P(3HB-*co*-3HV) were described as hot-melt adhesives [230]. They are considered for several applications in the packaging industry, medicine, pharmacy, agriculture and food industry or as raw materials for the synthesis of enantiomerically pure chemicals and the production of paints [231]. Possible applications of P(3HB) and copolymers are as packaging materials or agricultural foil [232]. As in the Biopol™ recovery process, the fermentor contents are heat treated to break down the nucleic acids, and proteases and detergents are added to solubilize the cells. Subsequent washing (i.e., removal of the solubilized cell material) and concentration of the resulting PHA latex is established by cross-flow microfiltration. To produce a coating, the PHA latex is sprayed onto a substrate such as paper. After evaporation of the water, the PHA latex particles readily coalesce into a film [218]. Due to the relatively high cost of PHA production, it is wise to apply PHAs for some cost-effective applications like medicinal instruments. PHAs were proved to be biocompatible in tissue engineering, implantations, etc. Many prokaryotic and eukaryotic organisms are able to produce LMW PHB molecules that are complexed with other biomolecules such as polyphosphates and that are present at low concentrations [233]. Over recent years, PHAs were used to develop many devices and material useful for clinical purposes such as suture fasteners, meniscus repair devices, rivets, tacks, staples, screws, (including interference screws), bone plates and bone plating systems, surgical mesh, repair patches, orthopedic pins (including bone filling augmentation material), adhesion barriers, stents, guided tissue repair regeneration devices, articular cartilage repair devices, nerve guides, tendon repair devices, pericardial patches, bulking and filling agents, vein valves, bone marrow scaffolds, meniscus regeneration devices, ligament, tendon grafts, ocular cell implants, spinal fusion cages, skin substitutes, dural substitutes, bone graft substitutes, bone dowels, wound dressing and hemostats [234–240]. Many biochemical engineering, molecular biology experiments and other tools were used to change the end products of the polyhydroxyalkanoate or to produce copolymers. To assess the biocompatibility of PHB, the structural organization of cellular molecules involved in adhesion was studied using osteoblastic and epithelial cell lines. On PHB, both cell lines revealed a rounded cell shape due to reduced spreading. The filamentous organization of the actin cytoskeleton was impaired. In double immunofluorescence, analyses the co-localization of the fibronectin with the fibril actin was demonstrated [241]. The investigated properties of PHB and PHB-co-PHV films proved to be fundamentally similar [242–246]. PHB-co-PHV film was chosen as a temporary substrate for growing retinal pigment epithelium cells as an organized monolayer before their subretinal transplantation. The surface of the PHB-co-PHV film was rendered hydrophilic by oxygen plasma treatment to increase the reattachment of D407 cells on the film surface. The cells were also grown to confluency as an organized monolayer suggesting PHB-co-PHV film as a potential temporary substrate for subretinal transplantation to replace diseased or damaged retinal pigment epithelium [247]. Tesema et al. and Malm et al. implanted PHB non-woven patches as transannular patches into the right ventricular outflow tract and pulmonary artery in 13 weanling sheep [248–250].

PHB non-woven patches can be used as a scaffold for tissue regeneration in low-pressure systems. The regenerated vessel had structural and biochemical qualities in common with the native pulmonary artery [250]. PHAs were used in tissue engineering, as antibiotic carriers, and many other medicinal applications [238, 251, 252]. Chen and Wu recently reported that PHAs possesses the biodegradability, biocompatibility and thermo-processibility for not only implant applications but also controlled drug release uses. PHAs show a promising future in pharmaceutical application such as drug delivery, which open a new approach. The many possibilities to tailor-make PHAs for medical implant applications have shown that this class of materials has a bright future as tissue engineering materials [253]. Different types of mutagenesis were applied for changing the substrate specificity, study the catalytic residues and to overproduce the PHAs [254–257].

2.6 Biopolymer Type Number 5: Polyisoprenoides

2.6.1 Natural Rubber

Natural rubber is a cis-1,4-polyisoprene-based biopolymer that has good resilience and damping behavior, but poor chemical resistance and processing capacity. It is collected from the milky secretion (latex) of individual trees, but the *Hevea brasiliensis* tree is the only important commercial source of natural rubber (sometimes called Pará rubber). Guayuleule is the only other plant under cultivation as a commercial source of rubber (*Parthenium argentatum*). Tyres, computer components, gloves, toys, shoe soles, elastic bands, flippers, erasers and athletic equipment are well-known uses of natural rubber. It is typically used for applications that need resistance to abrasion/wear; elastic resistance and properties that absorb damping or shock. In the production of synthetic rubber, oil is one of the necessary substituents. Natural rubber has enjoyed a rising market share due to the cost of oil and has become an attractive replacement for synthetic rubber. Since natural rubber has better properties compared to other synthetically manufactured rubber, rubber industries usually use it to enhance properties and extend applications of other rubber materials by blending. A very low level of adherence to other materials has also been documented. Natural rubber blended with virgin and recycled ethylene-propylene-diene monomer has been reported by Hayeemasae et al., the curing rate of natural rubber vulcanization was decreased. This was due to the incompatibility of these materials with natural rubber being cured. However, the maximum torque for the recycled material was increased with the addition of both virgin and recycled EPDM and was even higher. This was due to the higher density of cross-linking implemented by EPDM. Natural rubber is an economically important biopolymer with unparalleled performance characteristics, such as high elasticity, durability and efficient heat dispersion [258, 259]. Normal rubber is poly (cis-1,4-isoprene) 300 to 70000 isoperene molecules are coupled to form an irregular structure that cannot crystallize under normal conditions that mediate the amorphous rubber texture. Normal rubber is poly (cis-1,4-isoprene). Currently, the major rubber producing countries are Thailand, Indonesia, Malaysia, India and the People's Republic of China, which together account for 89% of the world's 9.33 million tons of the global rubber production.

2.7 Biopolymer Type Number 6: Inorganic Polyesters with Polyphosphate

The polyanhydride present in all living cells is inorganic polyphosphate. Commercial bacterial polyphosphate generation has not yet succceeded for economic reasons. Polyphosphates may have two cell regions as granules within the cytoplasm and related to the inner layer within the periplasmic space, the latter place being primarily related to the use of polyphosphate chemicals such as transport types. The chemical composition is that of a direct inorganic phosphate anhydride that changes in chain length from three to more than 10^3 units and consists more often than not of mixtures of distinct atomic sizes. Cations are attached to them. The polyphosphate increases particularly when a supplement lopsidedness occurs within the vicinity of phosphate overabundance. The Mg^{2+}-dependent polyphosphate kinase that moves the terminal phosphoryl group of ATP to polyphosphate is a second direct mechanism. The enzyme is found in various bacteria that are aerobic, anaerobic and facultative. Harland Wood and his colleagues studied *Propionibacterium shermanii* polyphosphate kinase and have shown that this monomeric enzyme (Mr 83000) catalyzes a strictly processive reaction [260]. Polyphosphate glucokinase, which makes the formation of glucose 6-phosphate without the intervention of ATP, has a fairly widespread distribution: (P)n + Glucose ~ (P) n-t + Glucose 6-P. The distribution of the enzyme is of taxonomic concern, being limited to a limited group of species (Actinomycetes, Propionibacteria, Micrococci, Brevibacteria and related species). Recently, Wood and colleagues made several major developments with the *P. shermanii* enzyme [260]. Polyphosphate glucokinase is four times more active in *P. shermanii* than the ATP glucokinase, illustrating the role of polyphosphate in this organism's metabolism. Some Acinetobacter species contained in sewage treatment plants using the active sludge process have the capacity, under suitable conditions, to accumulate up to 30% of their biomass of polyphosphate, that is, to be subjected to alternating anaerobiosis and aerobiosis cycles. This property has been put to practical use in the treatment of sewage, thereby allowing a biological method for the removal of phosphate from waste water. Owing to the over-enthusiastic use of fertilizers and detergents containing sodium tripolyphosphate, high phosphate levels in run-off water cause environmental issues with the production of familiar algal (cyanobacterial) blooms on lakes. In microbial cells, the excess phosphate in sewage is accumulated, which can then be removed along with the waste sludge from the process. It has been by van Steveninck and his group that polyphosphates located at the periphery of yeast cells are involved in the transport of sugars through the plasma membrane as energy donors. There are two hexose transport pathways in *Kluyveromyces marxianus* [261]. In the biosynthesis of cell wall mannoproteins, Kulaev et al. [262] discovered a new mechanism for the synthesis of high polymeric polyphosphate located outside the yeast cytoplasmic membrane and coupled with the conversion of dolichyl diphosphate to dolichyl monophosphate. The conversion of the terminal phosphate to polyphosphate is catalyzed by the special enzyme dolichyl diphosphate:polyphosphate phosphotransferase. There are known polyphosphatases which hydrolyze inorganic phosphate from long-chain polyphosphates. For nucleic acid and carbohydrate metabolism, transport processes and biosynthesis of cell wall polysaccharides, they constitute a pool of what Kulaev has called "activated phosphate" to be drawn on. They also play an essential role in controlling the intracellular concentrations of important metabolites containing phosphorus, including main molecules of the effector. Polyphosphate polyphosphate is an orthophosphate (Pi) residue polymer connected by P–O–P phosphoanhydride bonds. The majority of polyphosphates are stable even at high temperatures in neutral aqueous solutions, unlike long-chained polyphosphates, which are poorly soluble in water. Polyphosphates have a high negative charge density. The analogous structure of the RNA and other polyanions contributes to identical reactivity [263]. Polyphosphate is present in archaea, bacteria, algae, fungi, protists, plants, insects and mammals. Polyphosphate acts as a microbial phosphagen for a number of biochemical reactions, as a buffer against alkalis, as a storage of Ca^{2+} and as a metal-chelating agent due to its "high energy" bonds similar to those in ATP and its polyanion properties. In signaling and regulatory processes, cell viability and proliferation, pathogen virulence, as a structural component and chemical chaperone, and as a microbial stress reaction modulator, polyphosphate is essential. The majority of research on proteins involved in polyphosphate biosynthesis has focused on microorganisms, namely bacteria, including pathogenic and phosphate bacteria. Some orthologs were described in microorganisms of other taxonomic classes based on these findings. Other enzymes involved in the synthesis of polyphosphate polyphosphate phosphotransferase (EC 2.7.4.20) were associated with the synthesis of a small polyphosphate fraction associated with the vacuolar membrane of *Saccharomyces cerevisiae* [264]. Polyphosphate, similar to ATP, is composed of high-energy phosphate groups and is likely to be found in prebiotic soil. Polyphosphate is capable of stabilizing and preventing unfolding and aggregation of a wide range of proteins which maintain their competent conformations. Differentiated bacterial mutants with polyphosphate kinase show higher protein damage. Polyphosphate development by human gastrointestinal tract bacteria has been documented to protect the intestinal epithelium from oxidative stress [265]. Polyphosphate is associated with microorganisms in many physiological processes of vital importance, such as multilayer metabolic control, stress responses, resistance to pathogens, etc. Polyphosphate has also recently been documented to be involved in a variety of human health-related biological processes, such as cardiac ischaemia, blood coagulation, apoptosis, and cell death caused by stress [266, 267].

2.8 Biopolymer Type Number 7: Polyphenols

Phenolic compounds are defined chemically by the presence of at least one aromatic ring bearing one (phenols) or more (polyphenols) hydroxyl substituents, including their functional derivatives (such as esters and glycosides). Polyphenols can be roughly divided into LMW compounds and HMW Lignins [268]. Humic acids are important substances because they constitute the most ubiquitous source of non-living organic material that nature knows. Humic acids have important roles in soil fertility and stability. Humic acids also have industrial applications in the development of absorbents to be used at the source of metal-poisoning. Such properties extend their application to the fields of agriculture and medicine [269]. Tannin is another example of plyphenols. Polyphenols can be defined as secondary metabolites of relatively high molecular weight and diverse structural complexity that are synthesized by plants exclusively from the shikimate-derived phenylpropanoid and/or the polyketide pathway(s) in response to different types of stress (hydric or saline) or aggressive factors (bacteria, fungi, virus, ultraviolet radiation, etc.) [270–272]. Polyphenols are generally classified based on their chemical structures [273]. Four major classes of polyphenols are known and they are found in phenolic acids, flavonoids, lignans, and stilbenes. Phenolic acids are divided into hydroxybenzoic and hydroxycinnamic acids. The hydroxycinnamic acids are more common than hydroxybenzoic acids, and they include gallic acid, p-coumaric acid, caffeic and chlorogenic acids, and also ferulic and sinapic acids. These acids are rarely found in free form and are usually extracted as glycosylated derivatives or esters of quinic acid, shikimic acid and tartaric acid. Flavonoids are the most numerous of the phenolic compounds in plant products and are divided in several subclasses: flavonols, flavones, flavan-3-ols, flavanones, anthocyanidins and isoflavones, and other minor components of the diet such as coumarins or chalcones. Polyphenols are widely distributed in the higher plant kingdom; they are present in fruits, vegetables, herbs, spices, tea, and wine [274–276]. They show a great diversity of structures, ranging from rather simple molecules to polymers [277], with or without glycosylation and/or esterification. They may be classified in different groups as a function of the number of phenol rings that they contain and the structural elements that bind one ring to another [278]. Polyphenols have the ability to form a complex strongly with metal ions and macromolecules such as polysaccharides and proteins (Haslam 1998); hence, they present biological activities that make them attractive for nutraceutical and medicinal applications [279–283]. Particularly, polyphenols play a key role in the inhibition of enzymes related to cardiovascular and neurodegenerative diseases, as well as cancer and diabetes [272]. There is evidence that they prevent oxidation of LDL-lipoprotein [284–286], platelet aggregation [287, 288], and oxidative cell damage [289, 290]. Additionally, polyphenols act as antimutagens, anticarcinogens [291, 292], and antimicrobial agents [292–294]. Despite their demonstrated bioactive properties, the action of polyphenols on biological systems is complex and disputed because it is affected by bioavailability, doses, metabolism, and other biotransformations. Bioavailability is strictly related to their structures, like degree of glycosylation and conjugation with other polyphenols. For example, polymers such as proanthocyanidins may have direct effects on the stomach [292] and intestinal mucosa, protecting these tissues from oxidative stress or carcinogen action [277]. In addition, non-glycosylated phenolic compounds may be absorbed directly into the small intestine [277]. In turn, to become absorbed in the intestine, polyphenols present in the form of esters, glycosides, or polymers require hydrolyzation by enterocyte enzymes or through the action of colonic microbiote [295]. The antioxidant activity of polyphenols is arguable. On the one hand, studies in cell line cultures have shown that polyphenols are able to reduce oxidative stress and activate the antioxidant response of the cells. On the other hand, they could decrease cell viability and proliferation and induce cell apoptosis by acting as prooxidants and generating free radicals [296]. The effects of polyphenols in cell cultures (either protective antioxidant or prooxidant/cytotoxic) will depend on several factors such as their concentration, their ability to oxidize, their lipophilicity, the content of other antioxidants and metals, and the oxidative stress level of the cell culture [297–299]. Therefore, to safely use polyphenols as bioactive compounds it is necessary to carefully study aspects such as the specific dose, the delivery vehicle, the nutritional and health history, and the characteristics of the microbe.

2.9 Biopolymer Type Number 8: Polythioesters

In 2001, Lütke-Eversloh et al. published the first report on microbial polythioesters (PTEs). PTEs were synthesized by the same polymerase which synthesizes PHAs make PTEs unique biopolymers [220, 221, 300]. The first examples of PTEs were copolymers containing 3HB and 3-mercaptopropionate (3MP), which were obtained from the PHA accumulating bacterium Ralstonia eutropha strain H16 – a "model organism" in PHA research. Adding different precursor substrates is a tool for accelerating the polymerization process or for incorporating another monomric forma of PMAs. Polythioester production also includes incorporating monomeric formes of 3-hydroxybutyrate and 3-mercaptobutyrate, poly(3HB-co-3MB). The total polymer yield by *R. eutropha*, when 3-mercaptobutyric acid was fed as a carbon source in addition to gluconate contributed to up to 31% of the cellular dry weight. Mutants of *R. eutropha* with defective PHA synthase were not able to synthesize these copolymers. This demonstrated that the PHA synthase is responsible for the incorporation of 3MP and generally for biosynthesis of PTEs. If R. eutropha was cultivated in the presence of either 3-mercaptopropionic acid, 3,3' -thiodipropionic acid (TDP) or 3,3' -dithiodipropionic acid (DTDP), the copolymer poly- (3HB-co-3MP) was accumulated comprising molar fractions of 3MP of up to 54 mol-%. None of the sulfur-containing precursor substrates was utilized as sole carbon source for growth, thus, a second carbon source such as sodium gluconate was provided additionally to enable bacterial growth. The copolymer composition and polymer content referring to the cellular dry weight could be influenced by varying cultivation conditions and feeding regimes. When carbon sources, which are metabolized to acetyl coenzymeA(acetyl-CoA) were present in the culture medium of *R. eutropha*, the molar ratios of 3MP were usually less than 5 mol-%. It was observed that the total polymer yield decreased simultaneously to increasing 3MP content. However, this is not a strict rule, because other factors like the duration of fermentation also influenced the molecular weights of the accumulated polymers.

3-mercaptovalerate (3MV) were identified as constituents of PTE copolymers isolated from *R. eutropha*, extending the group of PTE constituents, which were referred to as 3-mercaptoalkanoates (3MA) [220, 221, 300].

2.10 Conclusion

The biopolymers contain eight groups based on their chemical structure. A ninth group could be added to include the monomers that are produced naturally and polymerized chemically such as polylactic acid. The polymeric forms of the inorganic structures are in need of more investigation. There are some unique criteria for biopolymers because they are derived from nature. They are degradable, bioavailable, compatible, renewable, safe, etc., so they are "green". They are the main income resources for a number of countries. Their applications are diverse. Some have not yet been commercialize due to petroleum oil-based synthetic polymers. Some species of the biopolymer have attracted attention in different times. For example, polyhydroxyalkanoates attract attention after the oil crises in 1973. Being produced by different types of biological cells their control through biochemical engineering or through different molecular tools enabled better management of their production in their mother wild type host cells or in foreigner cells. We should keep the different schools of biopolymer active. The ones which not being focused on today might be in great demand in the future.

Acknowledgement

The author acknowledges his mentors Professor Dr. Alexander Steinbüchel and Professor Dr. Bernd Rhem and the entire membership of the institute of Molecular Mikrobiologie und Biotechnologie, Mathematish-Naturwissenschaftlichen Fakultät der Westfälische Wilhelms-Universitüt Münster, Germany. Special thanks to the members of lab 06. The author acknowledges the DAAD for the grant provided as a PhD scholarship.

Conflict of Interest

The author declares that there is no any kind of conflict with any concerning this chapter

References

1 Ali, N.E., Kaddam, L.A., Alkarib, S.Y., Kaballo, B.G., Khalid, S.A., Higawee, A., AbdElhabib, A., AlaaAldeen, A., Phillips, A.O., and Saeed, A.M. (2020). Gum Arabic (Acacia Senegal) augmented total antioxidant capacity and reduced c-reactive protein among haemodialysis patients in phase II trial. *International Journal of Nephrology* 2020: 7214673.

2 Kaddam, L., Babiker, R., Ali, S., Satti, S., Ali, N., Elamin, M., Mukhtar, M., Elnimeiri, M., and Saeed, A. (2020). Potential role of Acacia Senegal (Gum Arabic) as immunomodulatory agent among newly diagnosed COVID 19 patients: a structured summary of a protocol for a randomised, controlled, clinical trial. *Trials* 21 (1): 766.

3 Kaddam, L.A. and Kaddam, A.S. (2020). Effect of Gum Arabic (Acacia senegal) on C-reactive protein level among sickle cell anemia patients. *BMC Research Notes* 13 (1): 162.

4 Amara, A.A. (2015). Kostenlose Viral Ghosts, Bacterial Ghosts, Microbial Ghosts and More (ed. A.A. Amara). Schüling Verlage Germany. ISBN: 978-3-86523-260-1.

5 Ahmed, A.S., Khalil, A., Ito, Y., van Loosdrecht, M.C.M., Santoro, D., Rosso, D., and Nakhla, G. (2021). Dynamic impact of cellulose and readily biodegradable substrate on oxygen transfer efficiency in sequencing batch reactors. *Water Research* 190: 116724.

6 Chaudhary, J., Thakur, S., Sharma, M., Gupta, V.K., and Thakur, V.K. (2020). Development of biodegradable agar-agar/gelatin-based superabsorbent hydrogel as an efficient moisture-retaining agent. *Biomolecules* 10 (6).

7 Dutta, S.D., Hexiu, J., Patel, D.K., Ganguly, K., and Lim, K.T. (2021). 3D-printed bioactive and biodegradable hydrogel scaffolds of alginate/gelatin/cellulose nanocrystals for tissue engineering. *International Journal of Biological Macromolecules* 167: 644–658.

8 Hai, L., Choi, E.S., Zhai, L., Panicker, P.S., and Kim, J. (2020). Green nanocomposite made with chitin and bamboo nanofibers and its mechanical, thermal and biodegradable properties for food packaging. *International Journal of Biological Macromolecules* 144: 491–499.

9 Ichimaru, H., Mizuno, Y., Chen, X., Nishiguchi, A., and Taguchi, T. (2020). Prevention of pulmonary air leaks using a biodegradable tissue-adhesive fiber sheet based on Alaska pollock gelatin modified with decanyl groups. *Biomaterials Science* 9: 861–883.

10 Li, M., Dong, Q., Xiao, Y., Du, Q., Huselsteind, C., Zhang, T., He, X., Tian, W., and Chen, Y. (2020). A biodegradable soy protein isolate-based waterborne polyurethane composite sponge for implantable tissue engineering. *Journal of Materials Science: Materials in Medicine* 31 (12): 120.

11 Olaiya, N.G., Nuryawan, A., Oke, P.K., Khalil, H., Rizal, S., Mogaji, P.B., Sadiku, E.R., Suprakas, S.R., Farayibi, P.K., Ojijo, V., and Paridah, M.T. (2020). The role of two-step blending in the properties of starch/chitin/polylactic acid biodegradable composites for biomedical applications. *Polymers (Basel)* 12 (3).

12 Wissamitanan, T., Dechwayukul, C., Kalkornsurapranee, E., and Thongruang, W. (2020). Proper blends of biodegradable polycaprolactone and natural rubber for 3D printing. *Polymers (Basel)* 12 (10): 2416.

13 Xu, J., Sagnelli, D., Faisal, M., Perzon, A., Taresco, V., Mais, M., Giosafatto, C.V.L., Hebelstrup, K.H., Ulvskov, P., Jorgensen, B., Chen, L., Howdle, S.M., and Blennow, A. (2021). Amylose/cellulose nanofiber composites for all-natural, fully biodegradable and flexible bioplastics. *Carbohydrate Polymers* 253: 117277.

14 Zhang, J., Xu, W.R., Zhang, Y.C., Han, X.D., Chen, C., and Chen, A. (2020). *In situ* generated silica reinforced polyvinyl alcohol/liquefied chitin biodegradable films for food packaging. *Carbohydrate Polymers* 238: 116182.

15 Bui, A.T., Williams, B.A., Hoedt, E.C., Morrison, M., Mikkelsen, D., and Gidley, M.J. (2020). High amylose wheat starch structures display unique fermentability

characteristics, microbial community shifts and enzyme degradation profiles. *Food and Function* 11 (6): 5635–5646.

16 Maevskaia, E.N., Shabunin, A.S., Dresvyanina, E.N., Dobrovol'skaya, I.P., Yudin, V.E., Paneyah, M.B., Fediuk, A.M., Sushchinskii, P.L., Smirnov, G.P., Zinoviev, E.V., and Morganti, P. (2020). Influence of the introduced chitin nanofibrils on biomedical properties of chitosan-based materials. *Nanomaterials (Basel)* 10 (5): 945.

17 Dashtban, M., Schraft, H., and Qin, W. (2009). Fungal bioconversion of lignocellulosic residues; opportunities & perspectives. *International Journal of Biological Sciences* 5 (6): 578–595.

18 Feeney, K.A., Wellner, N., Gilbert, S.M., Halford, N.G., Tatham, A.S., Shewry, P.R., and Belton, P.S. (2003). Molecular structures and interactions of repetitive peptides based on wheat glutenin subunits depend on chain length. *Biopolymers* 72 (2): 123–131.

19 Fritzsche, K., Lenz, R.W., and Fuller, R.C. (1990). Bacterial polyesters containing branched poly(beta-hydroxyalkanoate) units. *International Journal of Biological Macromolecules* 12 (2): 92–101.

20 Kim, Y.B., Lenz, R.W., and Fuller, R.C. (1992). Poly(β-hydroxyalkanoate) copolymers containing brominated repeating units produced by *Pseudomonas oleovorans*. *Macromolecules* 25 (7): 1852–1857.

21 Pourabedin, M., Xu, Z., Baurhoo, B., Chevaux, E., and Zhao, X. (2014). Effects of mannan oligosaccharide and virginiamycin on the cecal microbial community and intestinal morphology of chickens raised under suboptimal conditions. *Canadian Journal of Microbiology* 60 (5): 255–266.

22 Japelj, N., Suligoj, T., Zhang, W., Corte-Real, B., Messing, J., and Ciclitira, P.J. (2020). Natural variants of alpha-gliadin peptides with wheat proteins with reduced toxicity in coeliac disease. *British Journal of Nutrition* 123 (12): 1382–1389.

23 Pei, F., Sun, L., Fang, Y., Yang, W., Ma, G., Ma, N., and Hu, Q. (2020). Behavioral changes in glutenin macropolymer fermented by *Lactobacillus plantarum* LB-1 to promote the rheological and gas production properties of dough. *Journal of Agricultural and Food Chemistry* 68 (11): 3585–3593.

24 Sharma, I. and Kango, N. (2021). Production and characterization of keratinase by Ochrobactrum intermedium for feather keratin utilization. *International Journal of Biological Macromolecules* 166: 1046–1056.

25 Balitaan, J.N.I., Hsiao, C.D., Yeh, J.M., and Santiago, K.S. (2020). Innovation inspired by nature: biocompatible self-healing injectable hydrogels based on modified-beta-chitin for wound healing. *International Journal of Biological Macromolecules* 162: 723–736.

26 Celik, C., Ildiz, N., Sagiroglu, P., Atalay, M.A., Yazici, C., and Ocsoy, I. (2020). Preparation of nature inspired indicator based agar for detection and identification of MRSA and MRSE. *Talanta* 219: 121292.

27 Ferreira, L.M., Sari, M.H.M., Azambuja, J.H., da Silveira, E.F., Cervi, V.F., Marchiori, M.C.L., Maria-Engler, S.S., Wink, M.R., Azevedo, J.G., Nogueira, C.W., Braganhol, E., and Cruz, L. (2020). Xanthan gum-based hydrogel containing nanocapsules for cutaneous diphenyl diselenide delivery in melanoma therapy. *Investigational New Drugs* 38 (3): 662–674.

28 Hassan, M.A., Amara, A.A., Abuelhamd, A.T., and Haroun, B.M. (2010). Leucocytes show improvement growth on PHA polymer surface. *Pakistan Journal of Pharmaceutical Sciences* 23 (3): 332–336.

29 Hua, D., Gao, S., Zhang, M., Ma, W., and Huang, C. (2020). A novel xanthan gum-based conductive hydrogel with excellent mechanical, biocompatible, and self-healing performances. *Carbohydrate Polymers* 247: 116743.

30 Khan, M.U.A., Raza, M.A., Razak, S.I.A., Abdul Kadir, M.R., Haider, A., Shah, S.A., Mohd Yusof, A.H., Haider, S., Shakir, I., and Aftab, S. (2020). Novel functional antimicrobial and biocompatible arabinoxylan/guar gum hydrogel for skin wound dressing applications. *Journal of Tissue Engineering and Regenerative Medicine* 14 (10): 1488–1501.

31 Stark, M., DeBernardis, D., McDowell, C., Ford, E., and McMillan, S. (2020). Percutaneous skeletal fixation of painful subchondral bone marrow edema utilizing an injectable, synthetic, biocompatible hyaluronic acid-based bone graft substitute. *Arthroscopy Techniques* 9 (11): e1645–e1650.

32 Araujo, D., Alves, V.D., Lima, S.A.C., Reis, S., Freitas, F., and Reis, M.A.M. (2020). Novel hydrogels based on yeast chitin-glucan complex: characterization and safety assessment. *International Journal of Biological Macromolecules* 156: 1104–1111.

33 Johnson, W., Jr., Bergfeld, W.F., Belsito, D.V., Hill, R.A., Klaassen, C.D., Liebler, D.C., Marks, J.G., Jr., Shank, R.C., Slaga, T.J., Snyder, P.W., Gill, L.J., and Heldreth, B. (2020). Safety assessment of silk protein ingredients as used in cosmetics. *International Journal of Toxicology* 39 (3_suppl): 127S–144S.

34 Kasai, D. (2020). Poly(cis-1,4-isoprene)-cleavage enzymes from natural rubber-utilizing bacteria. *Bioscience, Biotechnology and Biochemistry* 84 (6): 1089–1097.

35 Amara, A. (2008). Polyhydroyalkanoates: from basic research and molecular biology to application. *IIUM Engineering Journal* 9 (1): 37–73.

36 Chaikaew, P., Adeyemi, O., Hamilton, A.O., and Clifford, O. (2020). Spatial characteristics and economic value of threatened species (Khaya ivorensis). *Scientific Reports* 10 (1): 6266.

37 Amara, A.A. and Moawad, H. (2011). PhaC synthases and PHA depolymerases: the enzymes that produce and degrade plastic. *IIUM Engineering Journal* 12 (4): 21–37.

38 Frausto de Silva, J.J.R. and Williams, R.J.P. (1993). The Biological Chemistry of the Elements: The Inorganic Chemistry of Life. Oxford: Clarendon Press.

39 Steinbüchel, A. (2001). Perspectives for biotechnological production and utilization of biopolymers: metabolic engineering of polyhydroxyalkanoate biosynthesis pathways as a successful example. *Macromolecular Bioscience* 1 (1): 1–24.

40 Asayama, S., Nogawa, M., Takei, Y., Akaike, T., and Maruyama, A. (1998). Synthesis of novel polyampholyte comb-type copolymers consisting of a Poly(L-lysine) backbone and hyaluronic acid side chains for a DNA carrier. *Bioconjugate Chemistry* 9 (4): 476–481.

41 Record, M.T. (1975). Effects of Na^+ and Mg^{++} ions on the helix-coil transition of DNA. *Biopolymers* 14 (10): 2137–2158.

42 Zheng, J., Zhu, G., Li, Y., Li, C., You, M., Chen, T., Song, E., Yang, R., and Tan, W. (2013). A spherical nucleic acid platform based on self-assembled DNA biopolymer for high-performance cancer therapy. *ACS Nano* 7 (8): 6545–6554.

43 Rau, I., Grote, J.G., Kajzar, F., and Pawlicka, A. (2012). DNA novel nanomaterial for applications in photonics and in electronics. *Comptes Rendus Physique* 13 (8): 853–864.

44 Smith, T.J. (1994). Calcium alginate hydrogel as a matrix for enteric delivery of nucleic acids. *BioPharm* 7 (3): 54–55.

45 Agostinacchio, F., Mu, X., Dire, S., Motta, A., and Kaplan, D.L. (2020). In Situ 3D printing: opportunities with silk inks. *Trends in Biotechnology* 39 (7): 19–30.

46 Xiao, C.Y., Zhu, Z.L., Zhang, C., Fu, M., Qiao, H.J., Wang, G., and Dang, E.L. (2020). Small interfering RNA targeting of keratin 17 reduces inflammation in imiquimod-induced psoriasis-like dermatitis. *Chinese Medical Journal (Engl)* 133 (24): 2910–2918.

47 Raychaudhuri, R., Naik, S., Shreya, A.B., Kandpal, N., Pandey, A., Kalthur, G., and Mutalik, S. (2020). Pullulan based stimuli responsive and sub cellular targeted nanoplatforms for biomedical application: synthesis, nanoformulations and toxicological perspective. *International Journal of Biological Macromolecules* 161: 1189–1205.

48 Jana, P., Ghosh, S., and Sarkar, K. (2020). Low molecular weight polyethyleneimine conjugated guar gum for targeted gene delivery to triple negative breast cancer. *International Journal of Biological Macromolecules* 161: 1149–1160.

49 Kraskouski, A., Hileuskaya, K., Kulikouskaya, V., Kabanava, V., Agabekov, V., Pinchuk, S., Vasilevich, I., Volotovski, I., Kuznetsova, T., and Lapitskaya, V. (2020). Polyvinyl alcohol and pectin blended films: preparation, characterization and mesenchymal stem cells attachment. *Journal of Biomedical Materials Research A* 109 (8): 1379–1392.

50 Randelli, F., Sartori, P., Carlomagno, C., Bedoni, M., Menon, A., Vezzoli, E., Sommariva, M., and Gagliano, N. (2020). The collagen-based medical device MD-tissue acts as a mechanical scaffold influencing morpho-functional properties of cultured human tenocytes. *Cells* 9 (12): 2641.

51 Augustine, R., Hasan, A., Dalvi, Y.B., Rehman, S.R.U., Varghese, R., Unni, R.N., Yalcin, H.C., Alfkey, R., Thomas, S., and Al Moustafa, A.E. (2021). Growth factor loaded in situ photocrosslinkable poly(3-hydroxybutyrate-*co*-3-hydroxyvalerate)/gelatin methacryloyl hybrid patch for diabetic wound healing. *Materials Science and Engineering C: Materials for Biological Applications* 118: 111519.

52 Djafari, J., Fernandez-Lodeiro, J., Santos, H.M., Lorenzo, J., Rodriguez-Calado, S., Bertolo, E., Capelo-Martinez, J.L., and Lodeiro, C. (2020). Study and preparation of Multifunctional Poly(L-Lysine)@Hyaluronic Acid Nanopolyplexes for the effective delivery of tumor suppressive MiR-34a into triple-negative breast cancer cells. *Materials (Basel)* 13 (23): 5309.

53 Alexakis, T., Boadi, D.K., Quong, D., Groboillot, A., O'Neill, I., Poncelet, D., and Neufeld, R.J. (1995). Microencapsulation of DNA within alginate microspheres and crosslinked chitosan membranes for in vivo application. *Applied Biochemistry and Biotechnology* 50 (1): 93–106.

54 Velusamy, V., Arshak, K., Korostynska, O., Oliwa, K., and Adley, C. (2010). An overview of foodborne pathogen detection: in the perspective of biosensors. *Biotechnology Advances* 28 (2): 232–254.

55 Doostan, M., Maleki, H., Khoshnevisan, K., Faridi-Majidi, R., and Arkan, E. (2021). Effective antibacterial electrospun cellulose acetate nanofibrous patches containing chitosan/erythromycin nanoparticles. *International Journal of Biological Macromolecules* 31 (168): 464–473.

56 Ridley, M. (2006). Genome. New York, NY: Harper Perennial.

57 Amara, A.A. (2013). Pharmaceutical and industrial protein engineering: where we are? *Pakistan Journal of Pharmaceutical Sciences* 26 (1): 217–232.

58 Bloise, N., Patrucco, A., Bruni, G., Montagna, G., Caringella, R., Fassina, L., Tonin, C., and Visai, L. (2020). *In vitro* production of calcified bone matrix onto wool keratin scaffolds *via* osteogenic factors and electromagnetic stimulus. *Materials (Basel)* 13 (14): 3052.

59 Barone, J., Schmidt, W., and Liebner, C. (2005). Compounding and molding of polyethylene composites reinforced with keratin feather fiber. *Composites Science and Technology* 65 (3–4): 683–692.

60 Gousterova, A., Braikova, D., Goshev, I., Christov, P., Tishinov, K., Vasileva-Tonkova, E., Haertle, T., and Nedkov, P. (2005). Degradation of keratin and collagen containing wastes by newly isolated thermoactinomycetes or by alkaline hydrolysis. *Letters in Applied Microbiology* 40 (5): 335–340.

61 Chon, C.H., Chung, S.Y., Ng, C.S., and Fuchs, G.J. (2005). Looped silk tie: surgical technique for management of the renal vein during laparoscopic live-donor nephrectomy. *Journal of Endourology* 19 (3): 401–405.

62 Elvin, C.M., Carr, A.G., Huson, M.G., Maxwell, J.M., Pearson, R.D., Vuocolo, T., Liyou, N.E., Wong, D.C., Merritt, D.J., and Dixon, N.E. (2005). Synthesis and properties of crosslinked recombinant pro-resilin. *Nature* 437 (7061): 999–1002.

63 Leknes, K.N., Roynstrand, I.T., and Selvig, K.A. (2005). Human gingival tissue reactions to silk and expanded polytetrafluoroethylene sutures. *Journal of Periodontology* 76 (1): 34–42.

64 Rising, A., Nimmervoll, H., Grip, S., Fernandez-Arias, A., Storckenfeldt, E., Knight, D.P., Vollrath, F., and Engstrom, W. (2005). Spider silk proteins–mechanical property and gene sequence. *Zoological Science* 22 (3): 273–281.

65 Mahabusarakam, W., Chairerk, P., and Taylor, W.C. (2005). Xanthones from *Garcinia cowa* Roxb. latex. *Phytochemistry* 66 (10): 1148–1153.

66 Rotondi, K.S. and Gierasch, L.M. (2005). Natural polypeptide scaffolds: beta-sheets, beta-turns, and beta-hairpins. *Biopolymers* 84 (1): 13–22.

67 Seebach, D., Hook, D.F., and Glattli, A. (2006). Helices and other secondary structures of beta- and gamma-peptides. *Biopolymers* 84 (1): 23–37.

68 Alemdar, A., Iridag, Y., and Kazanci, M. (2005). Flow behavior of regenerated wool-keratin proteins in different mediums. *International Journal of Biological Macromolecules* 35 (3–4): 151–153.

69 Alibardi, L. and Toni, M. (2004). Localization and characterization of specific cornification proteins in avian epidermis. *Cells Tissues Organs* 178 (4): 204–215.

70 Hamada, T., Kawano, Y., Szczecinska, W., Wozniak, K., Yasumoto, S., Kowalewski, C., and Hashimoto, T. (2005). Novel keratin 5 and 14 gene mutations in patients with epidermolysis bullosa simplex from Poland. *Archives of Dermatological Research* 296 (12): 577–579.

71 Abdolmaleki, K., Alizadeh, L., Hosseini, S.M., and Nayebzadeh, K. (2020). Concentrated O/W emulsions formulated by binary and ternary mixtures of sodium caseinate, xanthan and guar gums: rheological properties, microstructure, and stability. *Food Science and Biotechnology* 29 (12): 1685–1693.

72 Alizadeh-Sani, M., Rhim, J.W., Azizi-Lalabadi, M., Hemmati-Dinarvand, M., and Ehsani, A. (2020). Preparation and characterization of functional sodium caseinate/guar gum/

TiO_2/cumin essential oil composite film. *International Journal of Biological Macromolecules* 145: 835–844.

73 Antonov, Y.A. and Moldenaers, P. (2009). Inducing demixing of semidilute and highly compatible biopolymer mixtures in the presence of a strong polyelectrolyte. *Biomacromolecules* 10 (12): 3235–3245.

74 Antonov, Y.A., Van Puyvelde, P., and Moldenaers, P. (2004). Interfacial tension of aqueous biopolymer mixtures close to the critical point. *International Journal of Biological Macromolecules* 34 (1–2): 29–35.

75 Baracat, M.M., Nakagawa, A.M., Casagrande, R.B., Georgetti, S.R., Verri, W.A., and de Freitas, O. (2012). Preparation and characterization of microcapsules based on biodegradable polymers: pectin/casein complex for controlled drug release systems. *AAPS PharmSciTech* 13 (2): 364–372.

76 Nath, A., Dixit, M., Bandiya, A., Chavda, S., and Desai, A.J. (2008). Enhanced PHB production and scale up studies using cheese whey in fed batch culture of *Methylobacterium sp.* ZP24. *Bioresource Technology* 99 (13): 5749–5755.

77 Rollini, M., Musatti, A., Cavicchioli, D., Bussini, D., Farris, S., Rovera, C., Romano, D., De Benedetti, S., and Barbiroli, A. (2020). From cheese whey permeate to Sakacin-A/bacterial cellulose nanocrystal conjugates for antimicrobial food packaging applications: a circular economy case study. *Scientific Reports* 10 (1): 21358.

78 Singh, A.K. and Mallick, N. (2009). Exploitation of inexpensive substrates for production of a novel SCL-LCL-PHA *co*-polymer by Pseudomonas aeruginosa MTCC 7925. *Journal of Industrial Microbiology and Biotechnology* 36 (3): 347–354.

79 Surh, J., Vladisavljevi Cacute, G.T., Mun, S., and McClements, D.J. (2007). Preparation and characterization of water/oil and water/oil/water emulsions containing biopolymer-gelled water droplets. *Journal of Agricultural and Food Chemistry* 55 (1): 175–184.

80 Baran, R., Tosti, A., Hartmane, I., Altmeyer, P., Hercogova, J., Koudelkova, V., Ruzicka, T., Combemale, P., and Mikazans, I. (2009). An innovative water-soluble biopolymer improves efficacy of ciclopirox nail lacquer in the management of onychomycosis. *Journal of the European Academy of Dermatology and Venereology* 23 (7): 773–781.

81 Casaroto, A.R., da Silva, R.A., Salmeron, S., Rezende, M.L.R., Dionisio, T.J., Santos, C.F.D., Pinke, K.H., Klingbeil, M.F.G., Salomao, P.A., Lopes, M.M.R., and Lara, V.S. (2019). *Candida albicans*-cell interactions activate innate immune defense in human palate epithelial primary cells via Nitric Oxide (NO) and beta-Defensin 2 (hBD-2). *Cells* 8 (7): 707.

82 Bondarenko, L.B. (2004). Collagen transformation in organism: modern state of the problem. *Ukrainskii Biokhimicheskii Zhurnal* 76 (5): 5–15.

83 Brodsky, B. and Persikov, A.V. (2005). Molecular structure of the collagen triple helix. *Advances in Protein Chemistry* 70: 301–339.

84 Gonzalez-Masis, J., Cubero-Sesin, J.M., Guerrero, S., Gonzalez-Camacho, S., Corrales-Urena, Y.R., Redondo-Gomez, C., Vega-Baudrit, J.R., and Gonzalez-Paz, R.J. (2020). Self-assembly study of type I collagen extracted from male Wistar Hannover rat tail tendons. *Biomaterials Research* 24 (1): 19.

85 Kandan, P.V., Balupillai, A., Kanimozhi, G., Khan, H.A., Alhomida, A.S., and Prasad, N.R. (2020). Opuntiol prevents photoaging of mouse skin via blocking inflammatory responses and collagen degradation. *Oxidative Medicine and Cellular Longevity* 2020: 5275178.

86 Hamberg, L., Walkenstrom, P., and Hermansson, A.M. (2002). Shaping of gelling biopolymer drops in an elongation flow. *Journal of Colloid and Interface Science* 252 (2): 297–308.

87 Buescher, J.M. and Margaritis, A. (2007). Microbial biosynthesis of polyglutamic acid biopolymer and applications in the biopharmaceutical, biomedical and food industries. *Critical Reviews in Biotechnology* 27 (1): 1–19.

88 Dionisi, D., Majone, M., Miccheli, A., Puccetti, C., and Sinisi, C. (2004). Glutamic acid removal and PHB storage in the activated sludge process under dynamic conditions. *Biotechnology and Bioengineering* 86 (7): 842–851.

89 Inbaraj, B.S., Chiu, C.P., Ho, G.H., Yang, J., and Chen, B.H. (2008). Effects of temperature and pH on adsorption of basic brown 1 by the bacterial biopolymer poly(gamma-glutamic acid). *Bioresource Technology* 99 (5): 1026–1035.

90 Mooibroek, H., Oosterhuis, N., Giuseppin, M., Toonen, M., Franssen, H., Scott, E., Sanders, J., and Steinbuchel, A. (2007). Assessment of technological options and economical feasibility for cyanophycin biopolymer and high-value amino acid production. *Applied Microbiology and Biotechnology* 77 (2): 257–267.

91 Obst, M., Sallam, A., Luftmann, H., and Steinbuchel, A. (2004). Isolation and characterization of gram-positive cyanophycin-degrading bacteria-kinetic studies on cyanophycin depolymerase activity in aerobic bacteria. *Biomacromolecules* 5 (1): 153–161.

92 Reinecke, F. and Steinbuchel, A. (2009). Ralstonia eutropha strain H16 as model organism for PHA metabolism and for biotechnological production of technically interesting biopolymers. *Journal of Molecular Microbiology and Biotechnology* 16 (1–2): 91–108.

93 Stubbe, J., Tian, J., He, A., Sinskey, A.J., Lawrence, A.G., and Liu, P. (2005). Nontemplate-dependent polymerization processes: polyhydroxyalkanoate synthases as a paradigm. *Annual Review of Biochemistry* 74: 433–480.

94 Muirhead, H. and Perutz, M. (1963). Structure of hemoglobin. A three-dimensional fourier synthesis of reduced human hemoglobin at 5.5 A resolution. *Nature* 199 (4894): 633–638.

95 Kendrew, J., Bodo, G., Dintzis, H., Parrish, R., Wyckoff, H., and Phillips, D. (1958). A three-dimensional model of the myoglobin molecule obtained by x-ray analysis. *Nature* 181 (4610): 662–666.

96 Goodenough, P.W. (1995). A review of protein engineeringfor the food industry. *Molecular Biotechnology* 4: 151–166.

97 Crisman, R.L. and Randolph, T.W. (2009). Refolding of proteins from inclusion bodies is favored by a diminished hydrophobic effect at elevated pressures. *Biotechnology and Bioengineering* 102 (2): 483–492.

98 Amara, A.A. (2015). An overview of the molecular and cellular interactions of some bioactive compounds. In: Biotechnology of Bioactive Compounds (ed. V.K. Gupta and M.G. Tuohy, co-ed. A. O'Donovan and M. Lohani), 527–554. John Wiley & Sons, Ltd.

99 Runbingh, D.N. (1997). Protein Engineering from a bioindustrial point of view. *Current Opinion in Biotechnology* 8: 417–422.

100 Schäfer, T., Kirk, O., Borchert, T.V., Fuglsang, C.C., Pedersen, S., Salmon, S., Olsen, H.S., Deinhammer, R., and Lund, H. (2002). Enzymes for technical applications. In: Biopolymers Online (ed. R. Fahnestock and S.R. Steinbüchel), 377–437. Wiley VCH.

101 Rao, M.B., Tanksale, A.M., Ghatge, M.S., and Deshpande, V.V. (1998). Molecular and biotechnological aspects of microbial proteases. *Microbiology and Molecular Biology Reviews* 62 (3): 597–635.

102 Mozhaev, V.V. (1993). Mechanism-based strategies for protein thermostabilization. *Trends in Biotechnology* 11 (3): 88–95.

103 Krahe, M., Antranikian, G., and Mãrkl, H. (1996). Fermentation of extremophilic microorganisms. *FEMS Microbiology Reviews* 18 (2–3): 271–285.

104 Adams, M.W.W. and Kelly, R.M. (1998). Finding and using hyperthermophilic enzymes. *Trends in Biotechnology* 16 (8): 329–332.

105 Horikoshi, K. (1999). Alkaliphiles: some applications of their products for biotechnology. *Microbiology and Molecular Biology Reviews* 63 (4): 735–750.

106 Strausberg, S.L., Alexander, P.A., Gallagher, D.T., Gilliland, G.L., Barnett, B.L., and Bryan, P.N. (1995). Directed evolution of a subtilisin with calcium-independent stability. *Nature Biotechnology* 13 (7): 669–673.

107 Van Dyke, M.I., Lee, H., and Trevors, J.T. (1991). Applications of microbial surfactants. *Biotechnology Advances* 9 (2): 241–252.

108 Beer, H.D., Wohlfahrt, G., McCarthy, J.E.G., Schomburg, D., and Schmid, R.D. (1996). Analysis of the catalyic mechanism of a fungal lipase using computer-aided design and structural mutants. *Protein Engineering, Design and Selection* 9 (6): 507–517.

109 Brady, L., Brzozowski, A.M., Derewenda, Z.S., Dodson, E., Dodson, G., Tolley, S., Turkenburg, J.P., Christiansen, L., Huge-Jensen, B., Norskov, L., Thim, L., and Menge, U. (1990). A serine protease triad forms the catalytic centre of a triacylglycerol lipase. *Nature* 343 (6260): 767–770.

110 Martinelle, M., Holmquist, M., Clausen, I.G., Patkar, S., Svendsen, A., and Hult, K. (1996). The role of Glu87 and Trp89 in the lid of Humicola lanuginosa lipase. *Protein Engineering, Design and Selection* 9 (6): 519–524.

111 Rubingh, D.N. (1996). The influence of surfactants on enzyme activity. *Current Opinion in Colloid & Interface Science* 1 (5): 598–603.

112 Rubingh, D.N. (1997). Protein engineering from a bioindustrial point of view. *Current Opinion in Biotechnology* 8 (4): 417–422.

113 Shak, S., Capon, D.J., Hellmiss, R., Marsters, S.A., and Baker, C.L. (1990). Recombinant human DNase I reduces the viscosity of cystic fibrosis sputum. *Proceedings of the National Academy of Sciences* 87 (23): 9188–9192.

114 Breedveld, F.C. (2000). Therapeutic monoclonal antibodies. *The Lancet* 355 (9205): 735–740.

115 Jones, P.T., Dear, P.H., Foote, J., Neuberger, M.S., and Winter, G. (1986). Replacing the complementarity-determining regions in a human antibody with those from a mouse. *Nature* 321 (6069): 522–525.

116 Whittingham, J.L., Havelund, S., and Jonassen, I. (1997). Crystal structure of a prolonged-acting insulin with albumin-binding properties. *Biochemistry* 36 (10): 2826–2831.

117 Branningan, J.A. and Wilkinson, A.J. (2002). Protein engineering 20 years on. *Nature Reviews Molecular Cell Biology* 3: 964–970.

118 Baspinar, B. and Yardimci, H. (2020). Gluten-free casein-free diet for autism spectrum disorders: can it be effective in solving behavioural and gastrointestinal problems? *Eurasian Journal of Medicine* 52 (3): 292–297.

119 Jahromi, M., Niakousari, M., Golmakani, M.T., and Mohammadifar, M.A. (2020). Physicochemical and structural characterization of sodium caseinate based film-forming solutions and edible films as affected by high methoxyl pectin. *International Journal of Biological Macromolecules* 165 (Pt B): 1949–1959.

120 Pan, X., Mu, M., Hu, B., Yao, P., and Jiang, M. (2005). Micellization of casein-graft-dextran copolymer prepared through Maillard reaction. *Biopolymers* 81 (1): 29–38.

121 Nash, W., Pinder, D.N., Hemar, Y., and Singh, H. (2002). Dynamic light scattering investigation of sodium caseinate and xanthan mixtures. *International Journal of Biological Macromolecules* 30 (5): 269–271.

122 Polari, L., Alam, C.M., Nystrom, J.H., Heikkila, T., Tayyab, M., Baghestani, S., and Toivola, D.M. (2020). Keratin intermediate filaments in the colon: guardians of epithelial homeostasis. *International Journal of Biochemistry & Cell Biology* 129: 105878.

123 Di Foggia, M., Boga, C., Micheletti, G., Nocentini, B., and Taddei, P. (2021). Structural investigation on damaged hair keratin treated with alpha,beta-unsaturated Michael acceptors used as repairing agents. *International Journal of Biological Macromolecules* 167: 620–632.

124 Zahara, I., Arshad, M., Naeth, M.A., Siddique, T., and Ullah, A. (2021). Feather keratin derived sorbents for the treatment of wastewater produced during energy generation processes. *Chemosphere* 273: 128545.

125 Li, L., Yang, H., Li, X., Yan, S., Xu, A., You, R., and Zhang, Q. (2021). Natural silk nanofibrils as reinforcements for the preparation of chitosan-based bionanocomposites. *Carbohydrate Polymers* 253: 117214.

126 Battampara, P., Nimisha Sathish, T., Reddy, R., Guna, V., Nagananda, G.S., Reddy, N., Ramesha, B.S., Maharaddi, V.H., Rao, A.P., Ravikumar, H.N., Biradar, A., and Radhakrishna, P.G. (2020). Properties of chitin and chitosan extracted from silkworm pupae and egg shells. *International Journal of Biological Macromolecules* 161: 1296–1304.

127 Cabana, F. and Tay, C. (2020). The addition of soil and chitin into Sunda pangolin (Manis javanica) diets affect digestibility, faecal scoring, mean retention time and body weight. *Zoo Biology* 39 (1): 29–36.

128 Asapur, P., Mahapatra, S.K., and Banerjee, I. (2020). Secondary structural analysis of non-mulberry silk fibroin nanoparticles synthesized by using microwave and acetone method. *Journal of Biomolecular Structure and Dynamics* 4: 1–10.

129 Aparicio-Rojas, G.M., Medina-Vargas, G., and Diaz-Puentes, E. (2020). Thermal, structural and mechanical characterization of *Nephila clavipes* spider silk in southwest Colombia. *Heliyon* 6 (11): e05262.

130 Fernandez-Saiz, P., Lagaron, J.M., Hernandez-Muñoz, P., and Ocio, M.J. (2008). Characterization of antimicrobial properties on the growth of *S. aureus* of novel renewable blends of gliadins and chitosan of interest in food packaging and coating applications. *International Journal of Food Microbiology* 124 (1): 13–20.

131 Mariod, A.A. and Fadul, H. (2013). Extraction and characterization of gelatin from two edible Sudanese insects and its applications in ice cream making. *Food Science and Technology International* 21 (5): 380–391.

132 Selvakumar, G. and Lonchin, S. (2020). Fabrication and characterization of collagen-oxidized pullulan scaffold for biomedical applications. *International Journal of Biological Macromolecules* 164: 1592–1599.

133 Browning, A.C., Gray, T., and Amoaku, W.M. (2005). Isolation, culture, and characterisation of human macular inner choroidal microvascular endothelial cells. *British Journal of Ophthalmology* 89 (10): 1343–1347.

134 Claudepierre, T., Manglapus, M.K., Marengi, N., Radner, S., Champliaud, M.F., Tasanen, K., Bruckner-Tuderman, L., Hunter, D.D., and Brunken, W.J. (2005). Collagen XVII and BPAG1 expression in the retina: evidence for an anchoring complex in the central nervous system. *Journal of Comparative Neurology* 487 (2): 190–203.

135 Cosar, C.B., Ceyhan, N., Sevim, S., Sakaoglu, N., Sirvanci, S., San, T., Kurtkaya, O., and Acar, S. (2005). Corneal perforation with minor trauma: Ehlers-Danlos syndrome type VI. *Ophthalmic Surgery Lasers and Imaging* 36 (4): 350–351.

136 Diamantopoulos, C., Delliou, E., Kapranou, A., Balatsouras, D., and Elemenoglou, I. (2005). Collagenous spherulosis in epithelial-myoepithelial carcinoma of the submandibular gland: histologic and immunohistochemical study of a case. *Journal of Otolaryngology* 34 (4): 265–266.

137 Zhang, X., Wu, X., and Xia, K. (2013). Cellulose-wheat gluten bulk plastic materials produced from processing raw powders by severe shear deformation. *Carbohydrate Polymers* 92 (2): 2206–2211.

138 Markgren, J., Hedenqvist, M., Rasheed, F., Skepo, M., and Johansson, E. (2020). Glutenin and gliadin, a piece in the puzzle of their structural properties in the cell described through Monte Carlo simulations. *Biomolecules* 10 (8): 1095.

139 McMaster, T.J., Miles, M.J., Wannerberger, L., Eliasson, A.-C., Shewry, P.R., and Tatham, A.S. (1999). Identification of microphases in mixed alpha- and omega-gliadin protein films investigated by atomic force microscopy. *Journal of Agricultural and Food Chemistry* 47 (12): 5093–5099.

140 Fisichella, S., Amato, M.E., Lafiandra, D., Mantarro, D., Palermo, A., Savarino, A., and Scarlata, G. (2004). Structural studies of wheat flour glutenin polymers by CD spectroscopy. *Biopolymers* 74 (4): 287–301.

141 Altenbach, S.B., Chang, H.C., Rowe, M.H., Yu, X.B., Simon-Buss, A., Seabourn, B.W., Green, P.H., and Alaedini, A. (2020). Reducing the immunogenic potential of wheat flour: silencing of alpha gliadin genes in a U.S. wheat cultivar. *Frontiers of Plant Science* 11: 20.

142 Feng, Y., Zhang, H., Fu, B., Iftikhar, M., Liu, G., and Wang, J. (2020). Interactions between dietary fiber and ferulic acid change the aggregation of glutenin, gliadin and glutenin macropolymer in wheat flour system. *Journal of the Science of Food and Agriculture* 101 (5): 1979–1988.

143 Erdem, B.G. and Kaya, S. (2021). Production and application of freeze dried biocomposite coating powders from sunflower oil and soy protein or whey protein isolates. *Food Chemistry* 339: 127976.

144 Huang, T., Tu, Z., Shangguan, X., Wang, H., Zhang, L., and Bansal, N. (2021). Characteristics of fish gelatin-anionic polysaccharide complexes and their applications in yoghurt: rheology and tribology. *Food Chemistry* 343: 128413.

145 Di Bari, M., Cavatorta, F., Deriu, A., and Albanese, G. (2001). Mean square fluctuations of hydrogen atoms and water-biopolymer interactions in hydrated saccharides. *Biophysical Journal* 81 (2): 1190–1194.

146 Petersson, M., Loren, N., and Stading, M. (2005). Characterization of phase separation in film forming biopolymer mixtures. *Biomacromolecules* 6 (2): 932–941.

147 Chengyao, X., Yan, Q., Chaonan, D., Xiaopei, C., Yanxin, W., Ding, L., Xianfeng, Y., Jian, H., Yan, H., Zhongli, C., and Zhoukun, L. (2020). Enzymatic properties of an efficient glucan branching enzyme and its potential application in starch modification. *Protein Expression and Purification* 178: 105779.

148 Kamaldeen, O.S., Ariahu, C.C., and Yusufu, M.I. (2020). Application of soy protein isolate and cassava starch based film solutions as matrix for ionic encapsulation of carrot powders. *Journal of Food Science and Technology* 57 (11): 4171–4181.

149 Jabar, J.M., Alabi, K.A., and Lawal, A.K. (2020). Synthesis, characterization and application of novel 1, 3-bis[(furan-2-l)methylene]thiourea functional dye on wool and cotton fabrics. *SN Applied Sciences* 2 (11): 1850.

150 Jin, K., Tang, Y., Liu, J., Wang, J., and Ye, C. (2020). Nanofibrillated cellulose as coating agent for food packaging paper. *International Journal of Biological Macromolecules* 168: 331–338.

151 Kalia, S., Dufresne, A., Cherian, B.M., Kaith, B.S., Averous, L., Njuguna, J., and Nassiopoulos, E. (2011). Cellulose-based bio- and nanocomposites: a review. *International Journal of Polymer Science* 2011: 1–35.

152 Ferrari, R.R., Onuferko, T.M., Monckton, S.K., and Packer, L. (2020). The evolutionary history of the cellophane bee genus Colletes Latreille (Hymenoptera: Colletidae): molecular phylogeny, biogeography and implications for a global infrageneric classification. *Molecular Phylogenetics and Evolution* 146: 106750.

153 Abeer, M.M., Mohd Amin, M.C.I., and Martin, C. (2014). A review of bacterial cellulose-based drug delivery systems: their biochemistry, current approaches and future prospects. *Journal of Pharmacy and Pharmacology* 66 (8): 1047–1061.

154 Amin, M.C.I.M., Abadi, A.G., and Katas, H. (2014). Purification, characterization and comparative studies of spray-dried bacterial cellulose microparticles. *Carbohydrate Polymers* 99: 180–189.

155 Ciecholewska-Jusko, D., Zywicka, A., Junka, A., Drozd, R., Sobolewski, P., Migdal, P., Kowalska, U., Toporkiewicz, M., and Fijalkowski, K. (2020). Superabsorbent crosslinked bacterial cellulose biomaterials for chronic wound dressings. *Carbohydrate Polymers* 253: 117247.

156 Gao, G., Cao, Y., Zhang, Y., Wu, M., Ma, T., and Li, G. (2020). In situ production of bacterial cellulose/xanthan gum nanocomposites with enhanced productivity and properties using Enterobacter sp. FY-07. *Carbohydrate Polymers* 248: 116788.

157 Islam, S.U., Ul-Islam, M., Ahsan, H., Ahmed, M.B., Shehzad, A., Fatima, A., Sonn, J.K., and Lee, Y.S. (2020). Potential applications of bacterial cellulose and its composites for cancer treatment. *International Journal of Biological Macromolecules* 168: 301–330.

158 Phomrak, S., Nimpaiboon, A., Newby, B.Z., and Phisalaphong, M. (2020). Natural rubber latex foam reinforced with micro- and nanofibrillated cellulose via Dunlop method. *Polymers (Basel)* 12 (9): 1959.

159 Santos, T.A. and Spinace, M.A.S. (2021). Sandwich panel biocomposite of thermoplastic corn starch and bacterial cellulose. *International Journal of Biological Macromolecules* 167: 358–368.

160 Wang, F.P., Zhao, X.J., Wahid, F., Zhao, X.Q., Qin, X.T., Bai, H., Xie, Y.Y., Zhong, C., and Jia, S.R. (2021). Sustainable, superhydrophobic membranes based on bacterial cellulose for gravity-driven oil/water separation. *Carbohydrate Polymers* 253: 117220.

161 Kochumalayil, J.J., Zhou, Q., Kasai, W., and Berglund, L.A. (2013). Regioselective modification of a xyloglucan hemicellulose for high-performance biopolymer barrier films. *Carbohydrate Polymers* 93 (2): 466–472.

162 Butler, M.F., Clark, A.H., and Adams, S. (2006). Swelling and mechanical properties of biopolymer hydrogels containing chitosan and bovine serum albumin. *Biomacromolecules* 7 (11): 2961–2970.

163 Di Martino, A., Sittinger, M., and Risbud, M.V. (2005). Chitosan: a versatile biopolymer for orthopaedic tissue-engineering. *Biomaterials* 26 (30): 5983–5990.

164 Khan, R. and Dhayal, M. (2009). Chitosan/polyaniline hybrid conducting biopolymer base impedimetric immunosensor to detect Ochratoxin-A. *Biosensors and Bioelectronics* 24 (6): 1700–1705.

165 Xu, Z., Gao, N., Chen, H., and Dong, S. (2005). Biopolymer and carbon nanotubes interface prepared by self-assembly for studying the electrochemistry of microperoxidase-11. *Langmuir* 21 (23): 10808–10813.

166 Monteiro, O.A., Jr. and Airoldi, C. (1999). Some thermodynamic data on copper-chitin and copper-chitosan biopolymer interactions. *Journal of Colloid and Interface Science* 212 (2): 212–219.

167 Zhu, A., Zhang, M., and Shen, J. (2003). Covalent immobilization of O-butyrylchitosan with a photosensitive hetero-bifunctional crosslinking reagent on biopolymer substrate surface and bloodcompatibility characterization. *Journal of Biomaterials Science. Polymer Edition* 14 (5): 411–421.

168 Kim, D., Petrisor, G., and Yen, T.F. (2005). Evaluation of biopolymer-modified concrete systems for disposal of cathode ray tube glass. *Journal of the Air and Waste Management Association* 55 (7): 961–969.

169 Kim, D., Quinlan, M., and Yen, T.F. (2009). Encapsulation of lead from hazardous CRT glass wastes using biopolymer cross-linked concrete systems. *Waste Management* 29 (1): 321–328.

170 Gils, P.S., Ray, D., and Sahoo, P.K. (2009). Characteristics of xanthan gum-based biodegradable superporous hydrogel. *International Journal of Biological Macromolecules* 45 (4): 364–371.

171 Cortes, H., Caballero-Floran, I.H., Mendoza-Munoz, N., Escutia-Guadarrama, L., Figueroa-Gonzalez, G., Reyes-Hernandez, O.D., Gonzalez-Del Carmen, M., Varela-Cardoso, M., Gonzalez-Torres, M., Floran, B., Del Prado-Audelo, M.L., and Leyva-Gomez, G. (2020). Xanthan gum in drug release. *Cellular and Molecular Biology (Noisy-le-grand)* 66 (4): 199–207.

172 Alves, A., Miguel, S.P., Araujo, A., de Jesus Valle, M.J., Sanchez Navarro, A., Correia, I.J., Ribeiro, M.P., and Coutinho, P. (2020). Xanthan Gum-Konjac glucomannan blend hydrogel for wound healing. *Polymers (Basel)* 12 (1): 99.

173 Byram, P.K., Sunka, K.C., Barik, A., Kaushal, M., Dhara, S., and Chakravorty, N. (2020). Biomimetic silk fibroin and xanthan gum blended hydrogels for connective tissue regeneration. *International Journal of Biological Macromolecules* 165 (Pt A): 874–882.

174 Dzionek, A., Wojcieszynska, D., Adamczyk-Habrajska, M., Karczewski, J., Potocka, I., and Guzik, U. (2021). Xanthan gum as a carrier for bacterial cell entrapment: developing a novel immobilised biocatalyst. *Materials Science and Engineering C: Materials for Biological Applications* 118: 111474.

175 Castro, G.R., Panilaitis, B., and Kaplan, D.L. (2008). Emulsan, a tailorable biopolymer for controlled release. *Bioresource Technology* 99 (11): 4566–4571.

176 Ebert, K.H. and Schenk, G. (1968). Mechanisms of biopolymer growth: the formation of dextran and levan. *Advances in Enzymology and Related Areas of Molecular Biology* 30: 179–221.

177 van Oss, C.J. (1989). Energetics of cell-cell and cell-biopolymer interactions. *Cell Biophysics* 14 (1): 1–16.

178 Maciollek, A. and Ritter, H. (2014). One pot synthesis of silver nanoparticles using a cyclodextrin containing polymer as reductant and stabilizer. *Beilstein Journal of Nanotechnology* 5: 380–385.

179 Chen, L., Li, J., Ye, Z., Sun, B., Wang, L., Chen, Y., Han, J., Yu, M., Wang, Y., Zhou, Q., Seidler, U., Tian, D., and Xiao, F. (2020). Anti-high mobility group Box 1 neutralizing-antibody ameliorates dextran sodium sulfate colitis in mice. *Frontiers in Immunology* 11: 585094.

180 Bonnaud, M., Weiss, J., and McClements, D.J. (2010). Interaction of a food-grade cationic surfactant (lauric arginate) with food-grade biopolymers (pectin, carrageenan, xanthan, alginate, dextran, and chitosan). *Journal of Agricultural and Food Chemistry* 58 (17): 9770–9777.

181 Spyropoulos, F., Ding, P., Frith, W.J., Norton, I.T., Wolf, B., and Pacek, A.W. (2008). Interfacial tension in aqueous biopolymer-surfactant mixtures. *Journal of Colloid and Interface Science* 317 (2): 604–610.

182 Haghighatpanah, N., Mirzaee, H., Khodaiyan, F., Kennedy, J.F., Aghakhani, A., Hosseini, S.S., and Jahanbin, K. (2020). Optimization and characterization of pullulan produced by a newly identified strain of Aureobasidium pullulans. *International Journal of Biological Macromolecules* 152: 305–313.

183 Ganeshkumar, M., Ponrasu, T., Raja, M.D., Subamekala, M.K., and Suguna, L. (2014). Green synthesis of pullulan stabilized gold nanoparticles for cancer targeted drug delivery. *Spectrochimica Acta Part A: Molecular and Biomolecular Spectroscopy* 130: 64–71.

184 Shingel, K.I. (2004). Current knowledge on biosynthesis, biological activity, and chemical modification of the exopolysaccharide, pullulan. *Carbohydrate Research* 339 (3): 447–460.

185 Morris, A., Hillenbrand, M., Finkelman, M., George, M.P., Singh, V., Kessinger, C., Lucht, L., Busch, M., McMahon, D., Weinman, R., Steele, C., Norris, K.A., and Gingo, M.R. (2012). Serum (1>3)-β-D-glucan levels in HIV-infected individuals are associated with immunosuppression, inflammation, and cardiopulmonary function. *JAIDS Journal of Acquired Immune Deficiency Syndromes* 61 (4): 462–468.

186 Araujo, D., Alves, V.D., Marques, A.C., Fortunato, E., Reis, M.A.M., and Freitas, F. (2020). Low temperature dissolution of yeast chitin-glucan complex and characterization of the regenerated polymer. *Bioengineering (Basel)* 7 (1).

187 Wang, H., Chen, G., Li, X., Zheng, F., and Zeng, X. (2020). Yeast beta-glucan, a potential prebiotic, showed a similar probiotic activity to inulin. *Food and Function* 11 (12): 10386–10396.

188 Pornanek, P. and Phoemchalard, C. (2020). Dietary supplementation of beta-glucan-rich molasses yeast powder on antibody response to swine fever virus and hematology of starter-grower pigs. *Tropical Animal Health and Production* 53 (1): 43.

189 Huang, H., Liu, Y., and Liu, R. (2009). Sphingomonas sp.: an important microbial resource for biopolymer synthesis. *Wei Sheng Wu Xue Bao* 49 (5): 560–566.

190 Matyshevskaia, M.S., Gvozdiak, R.I., Maiko, I.I., Lipkes, M.I., and Dedusenko, G. (1979). Biopolymer produced by bacteria of the genus Xanthomonas and its use in the petroleum industry. *Mikrobiologicheskii Zhurnal* 41 (1): 88–92.

191 Tolstoguzov, V. (2002). Thermodynamic aspects of biopolymer functionality in biological systems, foods, and beverages. *Critical Reviews in Biotechnology* 22 (2): 89–174.

192 Zhong, J., Frases, S., Wang, H., Casadevall, A., and Stark, R.E. (2008). Following fungal melanin biosynthesis with solid-state NMR: biopolymer molecular structures and possible connections to cell-wall polysaccharides. *Biochemistry* 47 (16): 4701–4710.

193 Zhu, C., Lee, J.H., Raghavan, S.R., and Payne, G.F. (2006). Bioinspired vesicle restraint and mobilization using a biopolymer scaffold. *Langmuir* 22 (7): 2951–2955.

194 Recuenco, F.C., Kobayashi, K., Ishiwa, A., Enomoto-Rogers, Y., Fundador, N.G.V., Sugi, T., Takemae, H., Iwanaga, T., Murakoshi, F., Gong, H., Inomata, A., Horimoto, T., Iwata, T., and Kato, K. (2014). Gellan sulfate inhibits *Plasmodium falciparum* growth and invasion of red blood cells in vitro. *Scientific Reports* 4 (1).

195 Ding, Y., Jiang, F., Chen, L., Lyu, W., Chi, Z., and Liu, C. (2020). An alternative hard capsule prepared with the high molecular weight pullulan and gellan: processing, characterization, and in vitro drug release. *Carbohydrate Polymers* 237: 116172.

196 Gadhave, D., Rasal, N., Sonawane, R., Sekar, M., and Kokare, C. (2020). Nose-to-brain delivery of teriflunomide-loaded lipid-based carbopol-gellan gum nanogel for glioma: pharmacological and in vitro cytotoxicity studies. *International Journal of Biological Macromolecules* 16: 906–920.

197 Kozlowska, J., Prus-Walendziak, W., Stachowiak, N., Bajek, A., Kazmierski, L., and Tylkowski, B. (2020). Modification of collagen/gelatin/hydroxyethyl cellulose-based materials by addition of herbal extract-loaded microspheres made from gellan gum and xanthan gum. *Materials (Basel)* 13 (16): 3507.

198 Park, A., Choi, J.H., Lee, S., Been, S., Song, J.E., and Khang, G. (2020). Application of double network of gellan gum and pullulan for bone marrow stem cells differentiation towards chondrogenesis by controlling viscous substrates. *Journal of Tissue Engineering and Regenerative Medicine* 14 (11): 1592–1603.

199 Park, H., Kim, H., Kim, G.Y., Lee, M.Y., Kim, Y., and Kang, S. (2020). Enhanced biodegradation of hydrocarbons by *Pseudomonas aeruginosa*-encapsulated alginate/gellan gum microbeads. *Journal of Hazardous Materials* 406: 124752.

200 Rukmanikrishnan, B., Ismail, F.R.M., Manoharan, R.K., Kim, S.S., and Lee, J. (2020). Blends of gellan gum/xanthan gum/zinc oxide based nanocomposites for packaging application: rheological and antimicrobial properties. *International Journal of Biological Macromolecules* 148: 1182–1189.

201 Sriamornsak, P. (2011). Application of pectin in oral drug delivery. *Expert Opinion on Drug Delivery* 8 (8): 1009–1023.

202 Manjunath, M., Gowda, D.V., Kumar, P., Srivastava, A., Osmani, R.A., and Shinde, C. (2016). Guar gum and its pharmaceutical and biomedical applications. *Advanced Science, Engineering and Medicine* 8 (8): 589–602.

203 Patel, S. and Goyal, A. (2015). Applications of natural polymer Gum Arabic: a review. *International Journal of Food Properties* 18 (5): 986–998.

204 Balasubramaniam, S., Lee, H.C., Lazan, H., Othman, R., and Ali, Z.M. (2005). Purification and properties of a beta-galactosidase from carambola fruit with significant activity towards cell wall polysaccharides. *Phytochemistry* 66 (2): 153–163.

205 Teleman, A., Nordström, M., Tenkanen, M., Jacobs, A., and Dahlman, O. (2003). Isolation and characterization of O-acetylated glucomannans from aspen and birch wood. *Carbohydrate Research* 338 (6): 525–534.

206 Pasale, S.K., Cerroni, B., Ghugare, S.V., and Paradossi, G. (2014). Multiresponsive Hyaluronan-p(NiPAAm) "Click"-linked hydrogels. *Macromolecular Bioscience* 14 (7): 1025–1038.

207 Enrione, J., Díaz-Calderón, P., Weinstein-Oppenheimer, C.R., Sánchez, E., Fuentes, M.A., Brown, D.I., Herrera, H., and Acevedo, C.A. (2013). Designing a gelatin/chitosan/hyaluronic acid biopolymer using a thermophysical approach for use in tissue engineering. *Bioprocess and Biosystems Engineering* 36 (12): 1947–1956.

208 Esmonde-White, K.A., Le Clair, S.V., Roessler, B.J., and Morris, M.D. (2008). Effect of conformation and drop properties on surface-enhanced Raman spectroscopy of dried biopolymer drops. *Applied Spectroscopy* 62 (5): 503–511.

209 Figallo, E., Flaibani, M., Zavan, B., Abatangelo, G., and Elvassore, N. (2007). Micropatterned biopolymer 3D scaffold for static and dynamic culture of human fibroblasts. *Biotechnology Progress* 23 (1): 210–216.

210 Geissler, E., Hecht, A.M., and Horkay, F. (2007). Scaling equations for a biopolymer in salt solution. *Physical Review Letters* 99 (26): 267801.

211 Naskar, B., Ghosh, S., Nagadome, S., Sugihara, G., and Moulik, S.P. (2011). Behavior of the amphiphile CHAPS alone and in combination with the biopolymer inulin in water and isopropanol water media. *Langmuir* 27 (15): 9148–9159.

212 Dhanapal, V. and Subramanian, K. (2014). Recycling of textile dye using double network polymer from sodium alginate and superabsorbent polymer. *Carbohydrate Polymers* 108: 65–74.

213 Sukhlaaied, W. and Riyajan, S.-A. (2013). Synthesis and properties of carrageenan grafted copolymer with poly(vinyl alcohol). *Carbohydrate Polymers* 98 (1): 677–685.

214 Abd alFattah Amara, A. (2008). Polyhydroyalkanoates: from basic research and molecular biology to application. *IIUM Engineering Journal* 9 (1): 37–73.

215 Williams, S.F.a. and Martin, D.P. (2002). Applications of PHAs in medicine and pharmacy. In: Biopolymers, Vol. 4, Polyesters III; Applications and Commercial Products (ed. Y. Doi and A. Steinbüchel), 91–127. Germany: Wiley-VCH.

216 Anderson, A.J. and Dawes, E.A. (1990). Occurrence, metabolism, metabolic role, and industrial uses of bacterial polyhydroxyalkanoates. *Microbiological Reviews* 54 (4): 450–472.

217 Lageveen, R.G., Huisman, G.W., Preusting, H., Ketelaar, P., Eggink, G., and Witholt, B. (1988). Formation of polyesters by *Pseudomonas oleovorans*: effect of substrates on formation and composition of poly-(R)-3-hydroxyalkanoates and poly-(*R*)-3-hydroxyalkenoates. *Applied and Environmental Microbiology* 54 (12): 2924–2932.

218 de Koning, G.J.M., Kellerhals, M., van Meurs, C., and Witholt, B. (1997). A process for the recovery of poly(hydroxyalkanoates) from pseudomonads part 2: process development and economic evaluation. *Bioprocess Engineering* 17 (1): 15.

219 Lutke-Eversloh, T., Bergander, K., Luftmann, H., and Steinbuchel, A. (2001). Biosynthesis of poly(3-hydroxybutyrate-*co*-3-mercaptobutyrate) as a sulfur analogue to poly(3-hydroxybutyrate) (PHB). *Biomacromolecules* 2 (3): 1061–1065.

220 Lütke-Eversloh, T., Fischer, A., Remminghorst, U., Kawada, J., Marchessault, R.H., Bögershausen, A., Kalwei, M., Eckert, H., Reichelt, R., Liu, S.-J., and Steinbüchel, A. (2002). Biosynthesis of novel thermoplastic polythioesters by engineered *Escherichia coli*. *Nature Materials* 1 (4): 236–240.

221 Lütke-Eversloh, T. and Steinbüchel, A. (2003). Novel precursor substrates for polythioesters (PTE) and limits of PTE biosynthesis inRalstonia eutropha. *FEMS Microbiology Letters* 221 (2): 191–196.

222 Findlay, R.H. and White, D.C. (1983). Polymeric beta-hydroxyalkanoates from environmental samples and bacillus megaterium. *Applied and Environmental Microbiology* 45 (1): 71–78.

223 Byrom, D. (1987). Polymer synthesis by microorganisms: technology and economics. *Trends in Biotechnology* 5 (9): 246–250.

224 Jendrossek, D., Schirmer, A., and Schlegel, H.G. (1996). Biodegradation of polyhydroxyalkanoic acids. *Applied Microbiology and Biotechnology* 46 (5–6): 451–463.

225 Doi, Y. (1995). Microbial synthesis, physical properties, and biodegradability of polyhydroxyalkanoates. *Macromolecular Symposia* 98 (1): 585–599.

226 Marchessault, R.H. (1996). Tender morsels for bacteria: recent developments in microbial polyesters. *Trends in Polymer Science* 4: 163–168.

227 Gross, R.A., DeMello, C., Lenz, R.W., Brandl, H., and Fuller, R.C. (1989). The biosynthesis and characterization of poly(β-hydroxyalkanoates) produced by *Pseudomonas oleovorans*. *Macromolecules* 22 (3): 1106–1115.

228 Preusting, H., Nijenhuis, A., and Witholt, B. (1990). Physical characteristics of poly(3-hydroxyalkanoates) and poly(3-hydroxyalkenoates) produced by *Pseudomonas oleovorans* grown on aliphatic hydrocarbons. *Macromolecules* 23 (19): 4220–4224.

229 Son, H., Park, G., and Lee, S. (1996). Growth-associated production of poly-beta-hydroxybutyrate from glucose or alcoholic distillery wastewater by *Actinobacillus sp.* EL-9. *Biotechnology Letters* 18 (11): 1229–1234.

230 Kauffman, G.B. and Seymour, R.B. (1990). Elastomers: i. Natural rubber. *Journal of Chemical Education* 67 (5): 422.

231 Hocking, P.J. and Marchessault, R.H. (1994). Biopolyesters. In: Chemistry and Technology of Biodegradable Polymers (ed. G. Griffin), 48–96. Springer Netherlands.

232 Holmes, P.A. (1985). Applications of PHB – a microbially produced biodegradable thermoplastic. *Physics in Technology* 16 (1): 32–36.

233 Addison, C.J., Chu, S.H., and Reusch, R.N. (2004). Polyhydroxybutyrate-enhanced transformation of log-phase *Escherichia coli*. *Biotechniques* 37 (3): 376–378, 380, 382.

234 Knowles, J.C. (1993). Development of a natural degradable polymer for orthopaedic use. *Journal of Medical Engineering and Technology* 17 (4): 129–137.

235 Kunze, C., Freier, T., Kramer, S., and Schmitz, K.P. (2002). Anti-inflammatory prodrugs as plasticizers for biodegradable implant materials based on poly(3-hydroxybutyrate). *Journal of Materials Science: Materials in Medicine* 13 (11): 1051–1055.

236 Cheng, S., Chen, G.Q., Leski, M., Zou, B., Wang, Y., and Wu, Q. (2006). The effect of D,L-betahydroxybutyric acid on cell death and proliferation in L929 cells. *Biomaterials* 27 (20): 3758–3765.

237 Misra, S.K., Valappil, S.P., Roy, I., and Boccaccini, A.R. (2006). Polyhydroxyalkanoate (PHA)/inorganic phase composites for tissue engineering applications. *Biomacromolecules* 7 (8): 2249–2258.

238 Valappil, S.P., Misra, S.K., Boccaccini, A.R., and Roy, I. (2006). Biomedical applications of polyhydroxyalkanoates: an overview of animal testing and *in vivo* responses. *Expert Review of Medical Devices* 3 (6): 853–868.

239 van der Walle, G.A., de Koning, G.J., Weusthuis, R.A., and Eggink, G. (2001). Properties, modifications and applications of biopolyesters. *Advances in Biochemical Engineering/Biotechnology* 71: 263–291.

240 Wang, Y.W., Yang, F., Wu, Q., Cheng, Y.C., Yu, P.H., Chen, J., and Chen, G.Q. (2005). Effect of composition of poly(3-hydroxybutyrate-*co*-3-hydroxyhexanoate) on growth of fibroblast and osteoblast. *Biomaterials* 26 (7): 755–761.

241 Nebe, B., Forster, C., Pommerenke, H., Fulda, G., Behrend, D., Bernewski, U., Schmitz, K.P., and Rychly, J. (2001). Structural alterations of adhesion mediating components in cells cultured on poly β-hydroxy butyric acid. *Biomaterials* 22: 2425–2434.

242 Shishatskaya, I. and Volova, T.G. (2004). A comparative investigation of biodegradable polyhydroxyalkanoate films as matrices for in vitro cell cultures. *Journal of Materials Science: Materials in Medicine* 15 (8): 915–923.

243 Shishatskaya, I., Volova, T.G., Puzyr, A.P., Mogilnaya, O.A., and Efremov, S.N. (2004). Tissue response to the implantation of biodegradable polyhydroxyalkanoate sutures. *Journal of Materials Science: Materials in Medicine* 15 (6): 719–728.

244 Shishatskaya, I., Volova, T.G., Gordeev, S.A., and Puzyr, A.P. (2005). Degradation of P(3HB) and P(3HB-co-3HV) in biological media. *Journal of Biomaterials Science. Polymer Edition* 16 (5): 643–657.

245 Shishatskaya, I., Khlusov, I.A., and Volova, T.G. (2006). A hybrid PHB-hydroxyapatite composite for biomedical application: production, *in vitro* and *in vivo* investigation. *Journal of Biomaterials Science. Polymer Edition* 17 (5): 481–498.

246 Shishatskaya, I., Voinova, O.N., Goreva, A.V., Mogilnaya, O.A., and Volova, T.G. (2008). Biocompatibility of polyhydroxybutyrate microspheres: *in vitro* and *in vivo* evaluation. *Journal of Materials Science: Materials in Medicine* 19 (6): 2493–2502.

247 Tezcaner, A., Bugra, K., and Hasirci, V. (2003). Retinal pigment epithelium cell culture on surface modified poly(hydroxybutyrate-co-hydroxyvalerate) thin films. *Biomaterials* 24: 4573–4583.

248 Malm, T., Bowald, S., Karacagil, S., Bylock, A., and Busch, C. (1992). A new biodegradable patch for closure of atrial septal defect. *Scandinavian Journal of Thoracic and Cardiovascular Surgery* 26 (1): 9–14.

249 Malm, T., Bowald, S., Bylock, A., Saldeen, T., and Busch, C. (1992). Regeneration of pericardial tissue on absorbable polymer patches implanted into the pericardial sac. An immunohistochemical, ultrastructural and biochemical study in the sheep. *Scandinavian Journal of Thoracic and Cardiovascular Surgery* 26 (1): 15–21.

250 Malm, T., Bowald, S., Bylock, A., Busch, C., and Saldeen, T. (1994). Enlargement of the right ventricular outflow tract and the pulmonary artery with a new biodegradable patch in transannular position. *European Surgical Research* 26 (5): 298–308.

251 Cai, Z., Wang, L., Hou, X., and Cheng, G. (2002). Application of biodegradable polyhydroxybutyrate in medicine and tissue engineering. *Sheng Wu Yi Xue Gong Cheng Xue Za Zhi* 19 (2): 306–309.

252 Kassab, A.C., Xu, K., Denkbas, E.B., Dou, Y., Zhao, S., and Piskin, E. (1997). Rifampicin carrying polyhydroxybutyrate microspheres as a potential chemoembolization agent. *Journal of Biomaterials Science. Polymer Edition* 8 (12): 947–961.

253 Chen, G.Q. and Wu, Q. (2005). The application of polyhydroxyalkanoates as tissue engineering materials. *Biomaterials* 26 (33): 6565–6578.

254 Amara, A.A., Steinbuchel, A., and Rehm, B.H.A. (2002). *In vivo* evolution of the *Aeromonas punctata* polyhydroxyalkanoate (PHA) synthase: isolation and characterization of modified PHA synthases with enhanced activity. *Applied Microbiology and Biotechnology* 59: 477–482.

255 Amara, A.A. and Rehm, B.H. (2003). Replacement of the catalytic nucleophile cysteine-296 by serine in class II polyhydroxyalkanoate synthase from *Pseudomonas aeruginosa*-mediated synthesis of a new polyester: identification of catalytic residues. *Biochemical Journal* 374 (Pt 2): 413–421.

256 Taguchi, S., Maehara, A., Takase, K., Nakahara, M., Nakamura, H., and Doi, Y. (2001). Analysis of mutational effects of a polyhydroxybutyrate (PHB) polymerase on bacterial PHB accumulation using an in vivo assay system. *FEMS Microbiology Letters* 198 (1): 65–71.

257 Taguchi, S., Nakamura, H., Hiraishi, T., Yamato, I., and Doi, Y. (2002). In vitro evolution of a polyhydroxybutyrate synthase by intragenic suppression-type mutagenesis. *Journal of Biochemistry* 131 (6): 801–806.

258 van Beilen, J.B. and Poirier, Y. (2007). Guayule and Russian Dandelion as alternative sources of natural rubber. *Critical Reviews in Biotechnology* 27 (4): 217–231.

259 van Beilen, J.B. and Poirier, Y. (2007). Prospects for biopolymer production in plants. In: Green Gene Technology (ed. A. Ficheter and C. Sautter), 133–151. Berlin Heidelberg: Springer.

260 Clark, J.E., Beegen, H., and Wood, H.G. (1986). Isolation of intact chains of polyphosphate from "*Propionibacterium shermanii*" grown on glucose or lactate. *Journal of Bacteriology* 168 (3): 1212–1219.

261 Van den Broek, P.J.A., De Bruijne, A.W., and Van Steveninck, J. (1987). The role of ATP in the control of H+-galactoside symport in the yeast Kluyveromyces marxianus. *Biochemical Journal* 242 (3): 729–734.

262 Tsiomenko, A.B., Lupashin, V.V., and Kulaev, I.S. (1987). Export of enzymes into culture medium by yeasts of saccharomyces genus. In: Extracellular Enzymes of Microorganisms (ed. J. Chaloupka), 205–208. US: Springer.

263 Babes, V. (1895). Beobachtungen über die metachromatischen Körperchen, Sporenbildung, Verzweigung, Kolben-und Kapselbildung pathogener Bakterien. *Zeitschrift fur Hygiene und Infectionskrankheiten* 20 (1): 412–437.

264 Schomburg, D. and Stephan, D.R. (1997). Dolichyl-diphosphate-polyphosphate phosphotransferase. In: Enzyme Handbook (ed. D. Schomburg and D.R. Stephan), 417–419. Berlin Heidelberg: Springer.

265 Segawa, S., Fujiya, M., Konishi, H., Ueno, N., Kobayashi, N., Shigyo, T., and Kohgo, Y. (2011). Probiotic-derived polyphosphate enhances the epithelial barrier function and maintains intestinal homeostasis through integrin-p38 MAPK pathway. *PLoS ONE* 6 (8): e23278.

266 Dedkova, E.N. and Blatter, L.A. (2014). Role of beta-hydroxybutyrate, its polymer poly-beta- hydroxybutyrate and inorganic polyphosphate in mammalian health and disease. *Frontiers in Physiology* 5: 260.

267 Seidlmayer, L., Blatter, L.A., Pavlov, E., and Dedkova, E.N. (2012). Inorganic polyphosphate is a potent activator of the mitochondrial permeability transition pore in cardiac myocytes. *Heart* 97 (24): e8–e8.

268 Hattenschwilerr, S. and Vitousek, P.M. (2000). The role of polyphenols in terrestrial ecosystem nutrient cycling. *Trends in Ecology & Evolution* 15 (6): 238–243.

269 Pena-Mendez, E.M., Havel, J., and Patocka, J. (2005). Humic substances – compounds of still unknown structure: applications in agriculture, industry, environment, and biomedicine. *Journal of Applied Biomedicine* 3 (1): 13–24.

270 Hemingway, R.W. (1998). Practical polyphenolics: from structure to molecular recognition and physiological action (Edited by Haslam E. University of Sheffield). New York, NY: Cambridge University Press. *Journal of Natural Products* 61 (11): 1454–1455.

271 Serrano, J., Puupponen-Pimiã, R., Dauer, A., Aura, A.-M., and Saura-Calixto, F. (2009). Tannins: current knowledge of food sources, intake, bioavailability and biological effects. *Molecular Nutrition & Food Research* 53 (S2): S310–S329.

272 Quideau, S.P., Deffieux, D., Douat-Casassus, C.L., and Pouységu, L. (2011). Plant polyphenols: chemical properties, biological activities, and synthesis. *Angewandte Chemie International Edition* 50 (3): 586–621.

273 Crozier, A., Jaganath, I.B., and Clifford, M.N. (2009). Dietary phenolics: chemistry, bioavailability and effects on health. *Natural Product Reports* 26 (8): 1001.

274 Schieber, A., Stintzing, F.C., and Carle, R. (2001). By-products of plant food processing as a source of functional compounds—recent developments. *Trends in Food Science & Technology* 12 (11): 401–413.

275 Gruber, J.V., Lamoureux, B.R., Joshi, N., and Moral, L. (2000). Influence of cationic polysaccharides on polydimethylsiloxane (PDMS) deposition onto keratin surfaces from a surfactant emulsified system. *Colloids and Surfaces B: Biointerfaces* 19 (2): 127–135.

276 Djilas, S., Canadanovic-Brunet, J., and Cetkovic, G. (2009). By-products of fruits processing as a source of phytochemicals. *Chemical Industry and Chemical Engineering Quarterly* 15 (4): 191–202.

277 Manach, C., Scalbert, A., Morand, C., Rémésy, C., and Jiménez, L. (2004). Polyphenols: food sources and bioavailability. *The American Journal of Clinical Nutrition* 79 (5): 727–747.

278 Neveu, V., Perez-Jimenez, J., Vos, F., Crespy, V., du Chaffaut, L., Mennen, L., Knox, C., Eisner, R., Cruz, J., Wishart, D., and Scalbert, A. (2010). Phenol-Explorer: an online comprehensive database on polyphenol contents in foods. *Database* 2010 (0): bap024–bap024.

279 Collins, A.R. (2005). Assays for oxidative stress and antioxidant status: applications to research into the biological effectiveness of polyphenols. *The American Journal of Clinical Nutrition* 81 (1): 261S–267S.

280 Manach, C., Williamson, G., Morand, C., Scalbert, A., and Rémésy, C. (2005). Bioavailability and bioefficacy of polyphenols in humans. I. Review of 97 bioavailability studies. *The American Journal of Clinical Nutrition* 81 (1): 230S–242S.

281 Ramassamy, C. (2006). Emerging role of polyphenolic compounds in the treatment of neurodegenerative diseases: a review of their intracellular targets. *European Journal of Pharmacology* 545 (1): 51–64.

282 Hooper, L.V., Beranek, M.C., Manzella, S.M., and Baenziger, J.U. (1995). Differential expression of GalNAc-4-sulfotransferase and GalNAc-transferase results in distinct glycoforms of carbonic anhydrase VI in parotid and submaxillary glands. *Journal of Biological Chemistry* 270 (11): 5985–5993.

283 Jensen, G.S., Wu, X., Patterson, K.M., Barnes, J., Carter, S.G., Scherwitz, L., Beaman, R., Endres, J.R., and Schauss, A.G. (2008). *In vitro* and *in vivo* antioxidant and anti-inflammatory capacities of an antioxidant-rich fruit and berry juice blend. Results of a pilot and randomized, double-blinded, placebo-controlled, crossover study. *Journal of Agricultural and Food Chemistry* 56 (18): 8326–8333.

284 Frankel, E.N., German, J.B., Kinsella, J.E., Parks, E., and Kanner, J. (1993). Inhibition of oxidation of human low-density lipoprotein by phenolic substances in red wine. *The Lancet* 341 (8843): 454–457.

285 Fuhrman, B., Buch, S., Vaya, J., Belinky, P.A., Coleman, R., Hayek, T., and Aviram, M. (1997). Licorice extract and its major polyphenol glabridin protect low-density lipoprotein against lipid peroxidation: *in vitro* and *ex vivo* studies in humans and in atherosclerotic apolipoprotein E-deficient mice. *The American Journal of Clinical Nutrition* 66 (2): 267–275.

286 Landbo, A.-K. and Meyer, A.S. (2008). Ascorbic acid improves the antioxidant activity of European grape juices by improving the juicesâ€™ ability to inhibit lipid peroxidation of human LDL in vitro. *International Journal of Food Science & Technology* 36 (7): 727–735.

287 Guerrero, J.A., Navarro-Nuñez, L., Lozano, M.L., Martínez, C., Vicente, V., Gibbins, J.M., and Rivera, J. (2007). Flavonoids inhibit the platelet TxA2signalling pathway and antagonize TxA2receptors (TP) in platelets and smooth muscle cells. *British Journal of Clinical Pharmacology* 64 (2): 133–144.

288 Nardini, M., Natella, F., and Scaccini, C. (2007). Role of dietary polyphenols in platelet aggregation. A Review of the Supplementation Studies. *Platelets* 18 (3): 224–243.

289 Spormann, T.M., Albert, F.W., Rath, T., Dietrich, H., Will, F., Stockis, J.P., Eisenbrand, G., and Janzowski, C. (2008). Anthocyanin/polyphenolic-rich fruit juice reduces oxidative cell damage in an intervention study with patients on hemodialysis. *Cancer Epidemiology Biomarkers & Prevention* 17 (12): 3372–3380.

290 Paiva-Martins, F.t., Fernandes, J.o., Rocha, S., Nascimento, H., Vitorino, R., Amado, F., Borges, F., Belo, L.s., and Santos-Silva, A. (2009). Effects of olive oil polyphenols on erythrocyte oxidative damage. *Molecular Nutrition & Food Research* 53 (5): 609–616.

291 Kuroda, Y. and Hara, Y. (1999). Antimutagenic and anticarcinogenic activity of tea polyphenols. *Mutation Research/Reviews in Mutation Research* 436 (1): 69–97.

292 Cardador-Martinez, A., Castano-Tostado, E., and Loarca-Pina, G. (2002). Antimutagenic activity of natural phenolic compounds present in the common bean (*Phaseolus vulgaris*) against aflatoxin B 1. *Food Additives and Contaminants* 19 (1): 62–69.

293 Yoda, Y., Hu, Z.-Q., Shimamura, T., and Zhao, W.-H. (2004). Different susceptibilities of *Staphylococcus* and Gram-negative rods to epigallocatechin gallate. *Journal of Infection and Chemotherapy* 10 (1): 55–58.

294 Chung, K.-T., Wei, C.-I., and Johnson, M.G. (1998). Are tannins a double-edged sword in biology and health? *Trends in Food Science & Technology* 9 (4): 168–175.

295 Carrasco-Castilla, J., Hernández-Álvarez, A.J., Jiménez-Martínez, C., Gutiérrez-López, G.F., and Dávila-Ortiz, G. (2012). Use of proteomics and peptidomics methods in food bioactive peptide science and engineering. *Food Engineering Reviews* 4 (4): 224–243.

296 Xia, E.-Q., Deng, G.-F., Guo, Y.-J., and Li, H.-B. (2010). Biological activities of polyphenols from grapes. *International Journal of Molecular Sciences* 11: 622–646.

297 Wei, H., Bowen, R., Cai, Q., Barnes, S., and Wang, Y. (1995). Antioxidant and antipromotional effects of the soybean isoflavone genistein. *Experimental Biology and Medicine* 208 (1): 124–130.

298 Surh, Y.-J., Hurh, Y.-J., Kang, J.-Y., Lee, E., Kong, G., and Lee, S.J. (1999). Resveratrol, an antioxidant present in red wine, induces apoptosis in human promyelocytic leukemia (HL-60) cells. *Cancer Letters* 140 (1–2): 1–10.

299 Hou, D.-X., Fujii, M., Terahara, N., and Yoshimoto, M. (2004). Molecular mechanisms behind the chemopreventive effects of anthocyanidins. *Journal of Biomedicine and Biotechnology* 2004 (5): 321–325.

300 Lutke-Eversloh, T., Bergander, K., Luftmann, H., and Steinbuchel, A. (2001). Identification of a new class of biopolymer: bacterial synthesis of a sulfur-containing polymer with thioester linkages. *Microbiology (Reading)* 147 (Pt 1): 11–19.

3

Mushroom Pigments and Their Applications

Maura Téllez-Téllez[1] *and Gerardo Díaz-Godínez*[2]

[1]*Biological Research Center, Autonomous University of the State of Morelos, Morelos, México.*
[2]*Research Center for Biological Sciences, Autonomous University of Tlaxcala, Tlaxcala, México.*

3.1 Introduction

Color has been appreciated by humans since ancient times, pigments of different colors were used on ceremonial ornaments, as signs of mourning or hostility and adornment on their bodues. Each tribe and ethnic group in the world, manufactures exquisitely colored artifacts, such as masks, banners, and spiritual symbols, as part of their cultural heritage (Nordlund et al. 1989). Currently, there is renewed interest in using natural colorants in different areas such as food, cosmetics, textiles and pharmaceutical products, in addition, environmental awareness is being promoted, which is why the use of natural products has increased, with the purpose of causing less environmental damage. The use of natural colorants is an alternative to synthetic pigments, since they are not a health risk, do not generate polluting waste and are obtained from renewable resources. It should be mentioned that pigments can be classified according to their nature into synthetic pigments (they are obtained by laboratories), inorganic pigments (they are naturally found in minerals) and natural pigments that are produced by plants, microorganisms, animals and fungi (Bauernfeind 1981). Table 3.1 shows some dyes of natural origin and the chemical group to which they belong.

In the textile industry, dyes are a serious water pollution problem, mainly because they affect photosynthesis and aquatic life due to the toxic amount of metal ions and chlorine (Holkar et al. 2016), in addition, some people who are in direct or indirect contact with these waters may present allergies (dermatitis, difficulty in breathing) and genotoxic, carcinogenic and mutagenic effects (Tang et al. 2018). Within the food industry, colorants are added to maintain, intensify or add color to food; synthetic colorants have been used, but some have had harmful effects, so they are also trying to find alternatives such as natural food pigments, which are safer, have better pharmacological and health functions, hypotoxicity or non-toxicity, etc. Some of the main pigments of microorganisms that are used as food colorants are canthaxanthin, carotenoids, lycopene, melanin, phycocyanin, prodigiosin, riboflavin and violacein (Guo et al. 2020).

Some mushrooms are an alternative for obtaining natural dyes, there are reports that indicate that they have been used for this purpose in Europe, North Africa, Asia and America (Cardon 2007). Among the pigments that mushrooms present are betalains, carotenoids and other terpenoids. Besides the quinones or similar conjugated structures that are classified mainly according to the biosynthetic pathways that originate them (Figure 3.1): (1) derived

Biomolecules from Natural Sources: Advances and Applications, First Edition. Edited by Vijai Kumar Gupta, Satyajit D. Sarker, Minaxi Sharma, María Elida Pirovani, Zeba Usmani, and Chelliah Jayabaskaran.

Table 3.1 Chemical grouping of natural dyes.

Chemical group	Dye	Color
Tetraterpenes	Carotenoids	Yellow, orange and red
Flavonoids	Flavones	White–cream
	Flavonols	Yellow, white
	Chalcones	Yellow
	Aurone and Anthocyanins	Red, blue
Indigoids and indoles	Indigo	Blue, pink
	Betaleins	Yellow, red
Substituted pyrimidines	Pterin	White, yellow
	Flavin	Yellow
	Pheoxazine	Yellow, red
	Phenazine	Yellow, purple
Pyrano-quinone structures	Azaphilones	Yellow, orange and red
Polycyclic aromatic hydrocarbon	Quinones (naph- thoquinones benozoquinones)	Pale yellow, orange, red, purple, blue, brown
Polycyclic aromatic hydrocarbon	Quinones (anthraquinones)	Yellow, orange and red
Nitrogen pigments from quinones	Melanins	Dark brown
Tetrapyrroles (linear or cyclic)	Phycobilins	Blue, green Yellow, red
Xanthones	Xanthones	Yellow

(Lock Sing de Ugaz 1997; Urista et al. 2016)

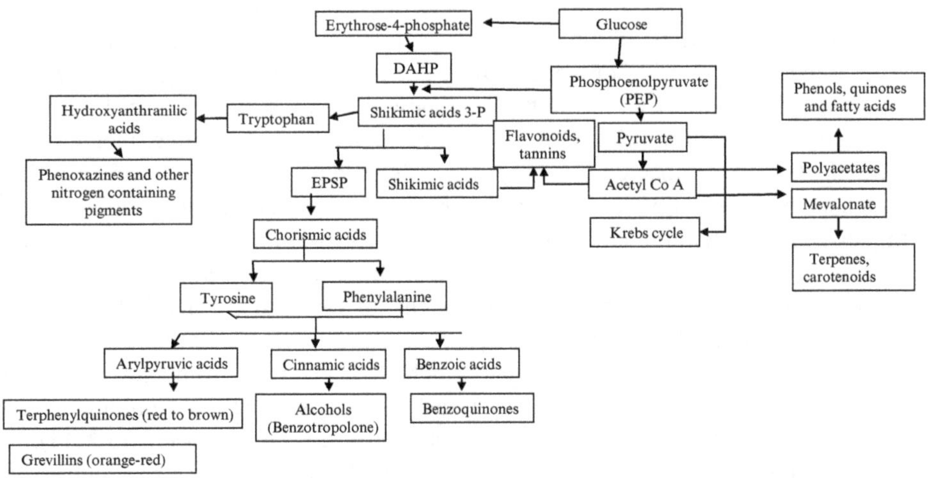

Figure 3.1 Biosynthetic pathways of quinones and similar conjugated structures.

from the shikimate pathway (chorismate), (2) the acetate-malonate pathway (polyketide), (3) via mevalonate (terpenoid) and (4) nitrogen-containing pigments. Diarylcyclopentenones are responsible for the color change in some fungi when they are damaged, it is formed by terphenylquinone ring contractions, while the oxidative cleavage of the hydroxyquinone ring in terphenylquinones, as well as tyrosine, hydroxylated to 3,4-dihydroxyphenylalanine (DOPA), is the precursor of red–violet betacyanins and yellow–orange betaxanthins. Both amino acids and their transformation products can be converted via quinones into melanins (heterogeneous dark pigments) by enzymatic browning reactions (oxidations and polymerizations) (Velíšek and Cejpek 2011).

3.2 Mushroom Pigments

Among the first works to describe pigments obtained from the fruiting body of fungi was that of Stahlschmidt and Thörner (1877) in which they reported terphenylquinones derived from L-tyrosine (atromentin) and L-phenylalanine (polyphoric acid) were reported which are quinoid pigments (Stadler and Hoffmeister 2015).

Many species of fungi (Basidiomycetes and Ascomycetes) have intense colors in their fruiting bodies: pink, red, orange, yellow, purple, olive green and gray, although the colors may fade due to the sun, water, and the age of the fungus, among others (Isaac 1994). Table 3.2 lists some pigments obtained from fruiting bodies of ascomycetes and basidiomycetes. Pigments are considered secondary metabolites that are produced by the mycelium when the supply of nutrients decreases and environmental conditions are not optimal, pigments can be produced to protect the fungus from sunlight, ultraviolet radiation, desiccation, bacterial and/or insect attack (Isaac 1994; Velíšek and Cejpek 2011). Among the most common fungal pigments are melanins (dark brown pigments), carotenoids (orange–red), lycopene (dark red) and xanthophylls, their presence being significant to improve fungal survival and spore resistance (Isaac 1994). Although carotenoids are colored pigments, their presence does not particularly affect the color of the fruiting bodies, some species of fungi present β-carotene in the fruiting body, however it is not reflected in its color (Ribeiro et al. 2011). The pigmentation of fungi can vary with age and play a biological role in the fruiting body, they can help to attract insects, or protect against bacterial attack and ultraviolet ray damage. For example, melanin is a pigment of great importance due to its role in resistance to adverse environmental factors, such as extreme temperatures, protection against desiccation and radioprotective (agent administered before or after exposure to radiation to minimize radiation toxicity) (Pagano and Dhar 2015). It is has been associated with protection against light, because it can counteract free radicals and protect against metal toxicity due to its chelating properties. For example, in the cultivated fungus *Agaricus bisporus* and others, melanin is formed starting from a benzoquinone, which is converted from the precursor γ-glutaminyl-4-hydroxybenzene (GHB) by tyrosinase, GHB is the main phenolic compound in the fruiting bodies and in the spores of *Agaricus bisporus* (Solano 2014). In Basidiomycota, tyrosine is the precursor of pigments (betalains), present in the genera *Amanita* and *Hygrocybe* (Musso 1979).

Table 3.2 Pigments found in some ascomycetes and basidiomycetes.

*Mushroom	Group	Ecology	Pigment	Fruiting body color	Reference
Albatrellus confluens (Alb. and Schwein.) Kotl. and Pouzar	Basidiomycota	Ectomycorrhizal	Albatrellin (meroterpenoid)	Pale orange or pink–orange	Yang et al. 2008
Albatrellus flettii Morse ex Pouzar	Basidiomycota	Ectomycorrhizal	Grifolin, neogrifolin, confluentin	Blue, gray	Yaqoob et al. 2020
Boletus pseudocalopus Hongo	Basidiomycota	Ectomycorrhizal	Grifolin derivatives 1–3 (Phenolic compounds)	Yellow	Song et al. 2009
Cantharellus cinnabarinus (Schwein.) Schwein	Basidiomycota	Ectomycorrhizal	Canthaxanthin, Beta-carotene, lycopene, alpha-carotene, and two other carotenes (delta and gamma isomers)	Pink- flamingo	Haxo 1950
Cantharellus lutescens Fr.	Basidiomycota	Ectomycorrhizal	Carotenoid (Favorhodin), lycopene and gamma-carotene	Pale yellow or cream	Haxo 1950
Cantharellus infundibuliformis (Scop.) Fr.	Basidiomycota	Ectomycorrhizal	Carotenoid (Favorhodin), lycopene and gamma-carotene	Brown, yellow	Haxo 1950
Chlorociboria aeruginosa (Oeder) Seaver ex C.S. Ramamurthi, Korf and L.R. Batra	Ascomycota	Saprobic	Xylindein	Green	Hinsch and Robinson 2016
Cordyceps pseudomilitaris Hywel-Jones and Sivichai	Ascomycota	Entomopathogenic	Mevalonic lactone, Phenol, 1,1,3-Trimethyl-3-phenylindan, 1,2,4-Triazolidine-3,5-dione, Pyrrolo (1,2-a) pyrazine-1,4-dione, hexahydro-3-(2- methylpropyl), 1,4-diaza-2,5-dioxo-3-isobutyl bicycle (4.3.0) nonane, N-Methyl-2-Propyl-5- Butylpiperidine and Phthalic acid	Orange to red	Sutthisa and Sanoamuang 2014
Cordyceps farinosa (Holmsk.) Kepler, B. Shrestha and Spatafora	Ascomycota	Entomopathogenic	Chromophore of the hydroxyanthraquinone type	–	Caro et al. 2012

(*Continued*)

Table 3.2 (Continued)

*Mushroom	Group	Ecology	Pigment	Fruiting body color	Reference
Cordyceps unilateralis (Tul. and C. Tul.) Sacc.	Ascomycota	Entomopathogenic	Six extracellular red naphthoquinone pigments, with erythrostominone as the major one, followed by 4-O-methyl erythrostominone, deoxyerythrostominol, deoxyerythrostominone, epierythrostominol, and in a smaller proportion 3,5,8-trihydroxy-6-methoxy-2-(5-oxohexa-1,3-dienyl)-1,4-naphthoquinone (shortened to 3,5,8-TMON)	–	Unagul et al. 2005
Cortinarius semisanguineus (Fr.) Gillet	Basidiomycota	Ectomycorrhizal	Physcion, dermoglausin, endocrocin, dermolutein, 5-chlorodermolutein, 5-chlorodermorubin, and erythroglausin	Blood-red, but turn cinnamon-brown on aging	Räisänen et al. 2020
Cortinarius sinapicolor Cleland	Basidiomycota	Ectomycorrhizal	Anthraquinones and naphthoquinones	Yellow	Durán et al. 2002
Cortinarius vitiosus (M.M. Moser) Niskanen, Kytöv., Liimat. & S. Laine	Basidiomycota	Ectomycorrhizal	Emodin, dermolutein, 5-chlorodermolutein and 5-chlorodermorubin	Red that turns brown over time	Räisänen 2019
Cortinarius violaceus (L.) Gray	Basidiomycota	Ectomycorrhizal	®-3′,4′-dihydroxy-β-phenylalanine [®-β- dopa]	Dark violet	von Nussbaum et al. 1998
Daldinia concentrica (Bolton) Ces. & De Not.	Ascomycota	Saprobic	BNT (1,1′–Binaphthalene–4,4′–5,5′–tetrol) (yellow), daldinol, 8–methoxy–1–napthol, 2–hydroxy–5–methylchromone, daldinal A–C (yellow), daldinin A–C (green-olivaceous-isabelline).	Reddish brown to blackish when young, then charcoal black	Caro et al. 2017
Daldinia eschscholtzii (Ehrenb.) Rehm	Ascomycota	Saprobic	BNT (1,1′–Binaphthalene–4,4′–5,5′–tetrol) (yellow), daldiol (dark brown), 8–methoxy–1–napthol, 2–hydroxy–5–methylchromone, daldinal A–C (yellow)	–	Caro et al. 2017

Dermocybe cardinalis E. Horak	Basidiomycota	Ectomycorrhizal	Anthraquinones and naphthoquinones	Crimson red, wine red or purple	Durán et al. 2002
Echinodontium tinctorium (Ellis & Everh.) Ellis & Everh.	Basidiomycota	Pathogen	Echinotinctone (orange)	Red brick	Ye et al. 1996
Entonaema splendens (Berk. & M.A. Curtis) Lloyd	Ascomycota	Saprobic	Mitorubrins variants such as entonaemins, rubiginosins, or hypomiltin	_	Caro et al. 2017
Fomes fomentarius (L.) Fr.	Basidiomycota	Saprobic	Melanin (black)	Matte, greyish surface, sometimes with brown or ocher colored bands	Tudor et al. 2013
Hypoxylon cohaerens (Pers.) Fr.	Ascomycota	Saprobic	Azaphilones, named cohaerins A and B	brown to blackish brown	Quang et al. 2005
Hypoxylon rickii Y.M. Ju & J.D. Rogers	Ascomycota	Saprobic	Rickenyl B and D (red and brown)	shiny to dark colored	Kuhnert et al. 2015
Isaria arinose (Holmsk.) Fr.	Ascomycota	Entomopathogenic	Main colored compound as anthraquinone	Yellow or cream	Velmurugan et al. 2010
Inonotus obliquus (Fr.) Pilát	Basidiomycota	Pathogen	Massive amounts of melanin	Black-brownish	Lin and Xu 2020
Lactarius indigo (Schwein.) Fr.	Basidiomycota	Ectomycorrhizal	Lactaroviolin (wine red) and stearoyldeterrol (blue)	Blue	Daniewski and Vidari 1999
Mycena rosea Sacc.	Basidiomycota	Saprobic	Red pyrroloquinoline alkaloids	Rose	Peters and Spiteller 2007a

(*Continued*)

Table 3.2 (Continued)

*Mushroom	Group	Ecology	Pigment	Fruiting body color	Reference
Pisolithus arhizus (Scop.) Rauschert	Basidiomycota	Ectomycorrhizal	Subvellerolactone B, Subvellerolactone D, Subvellerolactone E (Sesquiterpene hydroxylactones)	White	Kim et al. 2010
Polyporus australiensis Wakef.	Basidiomycota	Saprobic	Piptoporic acid	Orange or chestnut	Gill 1982
Pycnoporus sanguineus (L.) Murrill	Basidiomycota	Saprobic	Cinabarine (3-phenoxazine), o-acetyl -cinabarin, cinnabarinic acid, tramesanguine, 3-1 phenoxazine, 2-amino-phenoxazin -3-one, phenoxazine pycnosanguin ether	Bright reddish-orange colour	Achenbach and Blümm 1991
Pyrofomes albomarginatus (Zipp. Ex Lév.) Ryvarden	Basidiomycota	Saprobic	Echinotinctone	–	Ye et al. 1996
Tapinella atrotomentosa (Batsch) Šutara	Basidiomycota	Saprobic	Atromentin, leucomentin, spiromentin	Brown-ocher to dark-brown	Räisänen et al. 2020
Xylaria polymorpha (Pers.) Grev.	Ascomycota	Saprobic	Melanin (black)	Often black or brown	Tudor et al. 2013

*The names were taken from Index Fungorum.

It has been proposed that carotenoids (tetraterpenoids) in fungi, which serve as protection against lethal photooxidations, are attributed an antioxidant activity and inhibition of mutagenesis, among others (Durán et al. 2002), are pigments of light yellow to orange color and bright red, they are found in nature as cis and trans isomers, and can undergo isomerization due to their conjugated double bonds (responsible for the color). Carotenoids are divided into carotenes and xanthophylls that have their respective function as provitamin A and antioxidants. Carotenes include β-carotene, α-carotene, γ-carotene, lutein and compounds such as cryptoxanthin and canthaxanthin belong to the xanthophylls (Ribeiro et al. 2011), carotenoids are synthesized in two ways: one is the biosynthetic pathway of mevalonate and the other is the isoprene biosynthetic pathway.

Pigments are important within the taxonomy and systematics of fungi, Besl and Bresinsky (1997) reported that they can present some of the following compounds:

1) Terphenylquinones, are derived from tyrosine and are produced by a double condensation of two activated esters of the corresponding phenylpyruvic acid. Atromentin is important in the biosynthesis of pigments such as hydroxypulvinic acids and cyclopentenones. Simple oxidation products under conservation of the central quinone ring include cyclovariegatin and lephoric acid. All fungal terphenylquinones appear only sporadically and represent phylogenetic relics, therefore, they have a basal position of these pigments in the biosynthetic pathway.
2) Cyclopentenones, are formed by ring contractions of terphenylquinones, within these are chamonixin, involutin and the most oxidized gyroporin. Cyclopentenones can be detected in fungi at all levels of organization (from corticioid to gasteroid fungi).
3) Derivatives of pulvinic acid, Hydroxypulvinic acids, are formed through the formation of lactone, after the quinone ring of atromentin has been oxidized and opened. Derivatives of pulvinic acid include: gomphidic acid (which replaces the isomeric variegatic acid in *Gomphidius*), methyl bovinate (in *Suillus bovinus*) and methyl variegatate (Suillus collinitus). Xerocomic and variegatic acids play an important role in Boletales, these compounds represent a type of indicator pigment of this order.
4) Grevilins, during the biosynthesis of atromentin in a benzoquinone ring is produced forming two C–C bonds. However, if only a C–C bond is formed, the resulting lactonization produces grevilins, this type of pigment is apparently only found in the genus *Suillus*.
5) 1.2.4-Trihydroxybenzene is colorless, but is easily oxidized, it is responsible for the red and black coloration in species of the genus *Gomphidius*, the oxidation products are red gomphylactone and the corresponding biphenyls.
6) Prenylated phenols, the aromatic ring found in these compounds, carries an additional farnesyl or geranylgeranyl side chain, certain colorless phenols are stabilized by O-acetylation of the easily oxidizable hydroquinone system. The oxidation products of these chromogens produce shades of gray to brown in the upper half of the thin-layer chromatogram, together with benzoquinones, they are called lipophilic pigments, these form the characteristic encrustation in fasciculate cystidia that are colored by KOH.
7) Prenylated benzoquinones, the benzoquinones corresponding to the mentioned phenols appear more commonly as boviquinones, diboviquinone is a dimeric compound. Tridentoquinone and rhizopogone are quite interesting and special because the outer

end of the geranylgeranyl side chain joins the quinone system to form an intramolecular ring closure. This type of macrocyclic compound-loop is unique in nature.

8) Cavipetins, are five antifungal diterpene esters that were isolated from *Boletinus cavipes*, they are esters of 16-hydroxygeranylgeraniol and mesaconic or fumaric acid, derivatives of farnesol and geranylgeraniol also occur in chemotypes of the same species.
9) The presence of carotenoids has been reported in certain species of fungi.

3.3 Saprophytic Fungi Pigments

Many fungi that degrade wood produce certain types of pigmentation which are produced in response to antagonistic reactions between fungi, these compounds are secondary metabolites that act as physical and chemical barriers in the wood substrate. The ascomycete *Xylaria polymorpha* synthesizes pentaketide melanin from the polyketide monomer 1,8-dihydroxynaphthalene, however, some basidiomycetes can produce melanin from phenols and catechol precursors, characterized by a dark brown color, as is the case of *Trametes versicolor*. In a study, the effect of pH on the formation of pigments from *Xylaria polymorpha* and *Trametes versicolor* grown on malt extract agar added to wood was analyzed, both fungi showed pigmentation of the mycelium between pH 4 and 5.5, so this factor is important in the formation of fungal melanin, although it is necessary to analyze more factors (environmental and chemical) that can improve the production of these pigments (Tudor et al. 2013). The white-rot fungus *Echinodontium tinctorium* causes damage to living trees of *Abies grandis* and other conifers. It produces hoof-shaped conks attached to the tree, easily identifiable as greyish "teeth" that extend from the lower surface of the fruiting body. It has been reported that the Indians from Admirality Islands (Alaska) obtained from dehydrated fruiting bodies, a rust-red powder that was used to prepare dyes and paints. From the methanol extraction, an orange–brown solution was obtained that when adding ammonium changes the color to a purple–violet called echinotinctone (2,6-dihydroxy-1,8-dimethyl-3H-xanthen-3-one), it is a single fluorone chromophore (Ye et al. 1996).

The fungus *Laetiporus sulphureus* is a wood degrader that grows on oak stumps, produces fruiting bodies in the form of an orange–pink shelf, except for the fleshy margin that is bright yellow, the main pigment of the fruiting bodies was laetiporic acid (non-carotenoid; bright orange) which is a polyene of non-isoprenoid origin, presents a cis double bond at C-19 and was produced as a mixture of cis–trans isomers at C-7 in a ratio of 6:4, found at a concentration of 250 $\mu g\ g^{-1}$ dry weight. It resembles piptoporic acid, which is the pigment (bright orange) isolated from *Polyporus australiensis* (Table 3.2), both pigments share the terminal group (1-methyl-2-oxo-1-propylidene group conjugated with a polyene chain) represented by seven double bonds and laetiporic acid presents an unprecedented system as part of its chromophore (Weber et al. 2004). Davoli et al. (2005) grew three strains of *Laetiporus sulphureus* in liquid culture with a production of 0.1–6.7 $mg\ g^{-1}$ dry weight of the main pigment (polyene laetiporic acid A), another pigment was 2-dehydro-3-deoxylaetiporic acid A, both pigments were stable to oxygen and light, laetiporic acids B and C (low amounts) were also found. This mushroom has been used to obtain yellow dyes, and because it is an edible mushroom, it could also be used as a source of GRAS (generally recognized as safe) grade food colorants. The characteristic red or orange pigments of the fruiting bodies of the genus *Pycnoporus* are compounds derived from cinabarin (cinabarinic acid and tramesanguine) (Eggert et al. 1996). It has been reported that *Pycnoporus cinnabarinus* produces a red pigment (cinabarinic acid) by oxidative dimerization of the precursor 3-hydroxyanthranilic acid in cinabarinic acid (Table 3.2), this reaction is catalyzed by laccase enzymes, 3-hydroxyanthranilic acid (it is important in the shikimic acid pathway) a member of o-aminophenol that is a metabolite of the kynurenine pathway and is a precursor of cinnabarinic acid, in addition anthranilic acid is an intermediate in the synthesis of tryptophan and 3-HAA is produced by hydroxylation of anthranilic acid in the shikimic acid pathway (Li et al. 2001), these pigments can be found both in the fruiting body and in the culture medium (Eggert 1997).

The culture broth and biomass of *Tapinella atrotomentosa* (saprobium that decomposes wood, growing on coniferous stumps and decomposing coniferous trunks) grown for two months, were mixed with ether and xerocomic acid was obtained as the main pigment, followed by atromentic acid, in the fresh fruiting body none of the tetronic acids was detected. It has been suggested that atromentin is a precursor of atromentic acid, therefore, the vegetative phase of this species has a more complete biosynthetic capacity than the fruiting body (Gaylord et al. 1970). Besl et al. (1989) reported that in the fruiting bodies of *Tapinella atrotomentosa* and *Tapinella panuoides* they found as the main components a yellow–orange dye (flavomentines) and a purple dye (spiromentines), with a terphenylquinone structure (diphenylbenzenes or triphenyls) that has only been reported in fungi. *Tapinella panuoides* can be found in piles of sawdust or in wood from mines, cellars, etc., as it is an active wood destroyer. Gaylord and Brady (1971) characterized the pigment distribution pattern in the fruiting body and culture medium of *Tapinella atrotomentosus* and *Tapinella panuoides*, indicating that it was identical, except for the presence of lephoric acid in carpophores of *Tapinella atrotomentosus*, atromentin was isolated and identified from the fruiting bodies of both species, atromentic and xerocomic acids were found in the culture mixture. Apparently atromentin (4,4-dihydroxy polyphoric acid analog) is responsible for the reddish-brown color of the external parts of *Tapinella atrotomentosa*, also, leucomentins were determined that are colorless precursors of atromentin, in addition, this species of fungus biosynthesizes various lactone-like compounds, including osmundalactone and bis-osmundalactone (Béni et al. 2018).

The Hypoxyloideae subfamily includes several genera, their stromata show bright colors. The stromata of many Xylariaceae are rich in characteristic pigments that also serve as chemotaxonomic marker molecules (Stadler and Fournier 2006); it has been reported to contain large amounts of easy-to-extract pigments, such as lenormandins, mitorubrinol derivatives, hypoxyvermelhotins, or cohaerins. Recently, five compounds of *Hypoxylon rickii* (Table 3.2) belonging to ρ-substituted terphenyls were isolated, most of the fungal terphenyls have been found in fruiting bodies of basidiomycetes and represent an important class of pigments, but they have also been found in the fermentation broths (Kuhnert et al. 2015). The *Daldinia* genus present a wide range of shades (yellow to violet), they produce pigments in their stromal structures but also during mycelial growth, yellow BNT (1,1'-Binaphthalene-4,4',5,5'-tetrol) and daldinol in their stromata has been found in the following species: *Daldinia bambusicola, Daldinia caldariorum,*

Daldinia childiae, Daldinia clavata, Daldinia concentrica, Daldinia eschscholzii, Daldinia grandis, Daldinia lloydi, Daldinia loculata, Daldinia petriniae, Daldinia singularis. The yellowish daldinal pigment A, the daldinal B and the daldinal C were produced by *Daldinia concentrica* and *Daldinia eschscholzii*, colored 8-methoxy-1-naphthol and 2-hydroxy-5-methylchromone were found in the liquid cultures of all strains. The yellow azafilones and benzophenones found in *Daldinia childiae* did not show purple stromal colors. Most of the *Daldinia* spp. produce derivatives of naphthalene and chromane, and stromata produce mitorubrin but generally contain binaphthyls (Caro et al. 2017).

It has been reported that the nutritional type of saprotrophs and biotrophs are shared among all species within the fungal genera, thus nutritional modes are often assigned at the genus level (Nguyen et al. 2016). It is currently known that some ectomycorrhizal fungi are involved in the decomposition of organic material for the acquisition of nitrogen and phosphorus. Through genomic and transcriptome studies, it was determined that most ericoid mycorrhizal fungi have genes involved in decomposition and are versatile saprotrophs, as well as symbionts (Thoen et al. 2020). Also, various wood-decomposing fungi can form mycelial sheets (mantles) around the root tip and intercellular hyphal networks similar to the Hartig network, which are structures formed by ectomycorrhizal fungi (Smith et al. 2017). Such interactions can represent positions in a continuous saprotrophy–biotrophy system or, alternatively, a high ecological versatility, as is the case of the genus *Mycena* some species of which are known to have great ecological importance as decomposers. However, species associated with the root of *Pinus sylvestris* have been found, which is why it is suggested that they may be latent saprotrophs, living asymptomatically as root endophytes before clear cut, which would give them an advantage to access carbon as the roots age (Thoen et al. 2020). Peters and Spiteller (2007a) characterized two previously unknown red alkaloid pigments, mycenarubin A, and a dimer thereof, mycenarubin B, which were isolated from the fruiting bodies of *Mycena rosea* (Table 3.2), and are biosynthetically derived from tryptophan and S-adenosylmethionine. *Mycena sanguinolenta* is a recognized species due to the presence of a characteristic red latex, which is exuded if the fruiting bodies are cut or bruised, of which the structures of two blue alkaloids (sanguinone A and sanguinone B) and a blood-red alkaloid were described. It could be that the ecological role of the three pyrroloquinoline alkaloids, is not only to contribute to the color of the fruiting body, but also to protect against predators, however, due to instability and the available quantity being small, it is not was possible to analyze the application of these pigments. The presence of sanguinones, mycenarubins and hematopodins in *Mycena* species confirms that the presence of pyrroloquinoline alkaloids is not limited to marine sources, but that pigments are structurally related to damirones, isolated from a marine sponge (Peters and Spiteller 2007b).

3.4 Symbiotic Fungi Pigments

The *Suillus* species contain grevilins, the color of the fruiting bodies is orange to dark blue, however, many of the species that contain variegatic and xerocomic acids do not turn blue, which may be due to the lack of oxidase enzymes that are needed to produce the methides of blue quinone (Nelsen 2010). The genus *Cortinarius* is one of the largest groups of agaric basidiomycetes, it is distributed in both hemispheres as a component of boreal and sub-arctic soil communities. There are species that have been used for the purpose of obtaining pigments, as is the case of the two European species, *Cortinarius sanguineus* and *Cortinarius vitiosus*, which contain anthraquinones that produce colors that vary from yellow to violet or bluish-red due to their relatively short conjugated chromophores. The anthraquinone (carbonyl dye) can confer colors from yellow, red and blue, its conjugated chromophores are short, when these bind to donor and acceptor substituents they cause bathochromic changes that give red and even blue color, they have been used because they are intense dyes and they do not fade. Recently, it was reported that *Cortinarius sanguineus* yield 6% of dye from the fresh fruiting body, the most abundant compounds were: dermocybin, dermocybin glycoside, emodin, emodin glycoside and dermorubin. There were differences in the composition compared to *Cortinarius vitiosus* that has red dermocybin and dermocybin glycoside as main colorants, therefore, the pigment of this fungus is characterized by a stronger red color, since it contains fewer yellow compounds, in addition, dermocybin glycoside, dermoglausin, dermorubin and two new compounds (endocrocin and 5,7-dichloroendocrocin) among others (Table 3.2) (Räisänen 2019). The coloration presented by *Cantharellus cinnabarinus* has been attributed to the presence of carotenoids, the color is due to the presence of long conjugated double bonds, they absorb light in the 400–500 nm region of the spectrum and this gives rise to the colors yellow, orange and red (Ribeiro et al. 2011).

Cortinarius violaceus is a fungus that has attracted attention due to its characteristic dark blue–violet color and its smell of cedar wood, it is found in North America, Asia and Europe. The pigment is vacuolar, it is very unstable and polar; von Nussbaum et al. (1998) reported that the intensity of the violet color correlates with the iron concentration, therefore, the main dye is a 1:2 ligand–iron complex, but they could not determine if it is mono- or binuclear. The iron content of the fungus is 4.5–7.5 mg g^{-1} (dry weight), which is 100 times higher than that reported in other basidiomycetes (0.06 mg g^{-1}), the UV/Vis spectrum of the pigment at pH 6.2 (physiological fungus) presented maximum absorption at 571 nm, which is typical for iron(III)-catechol complex. Spiteller et al. (2000) reported that tyrosine is the precursor of ß-dopa, they confirmed this by the conversion of the precursor [3'-^{13}C] tyrosine to [3'-^{13}C] ß-dopa, which proceeds through tyrosine-2 which is then hydroxylated to -Dopa 3.

The classification of *Lactarius* (Basidiomycetes) species has historically been problematic, as mycologists often use different morphological characters for infrageneric classification; it is ectomycorrhizal, is characterized by the presence of latex, is distributed in all terrestrial ecosystems and includes several species as edible in different parts of the world (Boa 2004). Sesquiterpenes have been found in the vast majority of *Lactarius* species, which are responsible for their milky juice which is pungent and bitter and for the changes in the color of the latex when exposed to air (Yang et al. 2006). Colorless latex with an intense pungent taste has been reported to be due to the presence of unsaturated dialdehyde sesquiterpenes with marasmane, lactarane and secolactaran skeletons. Pungent unsaturated dialdehydes are formed enzymatically from a precursor sesquiterpenoid (fatty acid ester of a marasmane sesquiterpene) in response to injury to fruiting bodies, the formation and transformation of sesquiterpenes in pungent-tasting *Lactarius* species (*Lactarius vellereus* and *Lactarius torminosus*) appear to be part of its chemical defense system.

On the other hand, the latex of the fruiting body of *Lactarius deliciosus* and *Lactarius deterrimus* is initially orange in color, after some time (minutes) it darkens and finally turns green, which is due to the formation of purple (lactaroviolin), blue (deterrol) and yellow compounds (an alcohol, fatty acid esters and delicial), lactaroviolin and deterrol (sesquiterpenes) are slowly formed and accumulate in fungal tissue during the first hours after injury, however, it was not possible to isolate the delicial compound because it is very unstable, when exposed to light it polymerized rapidly. In previous studies it had been reported that in European specimens of *Lactarius deliciosus* guaiane sesquiterpenes, lactaroviolin, free dihydroazulene alcohol, stearic acid ester and lactarazulene were found, while lactarofulvene was isolated from Californian specimens and from Indian specimens of *Lactarius deterrimus* an aldehyde was isolated (Bergendorff and Sterner 1988). Yang et al. (2006) reported that they isolated two new red colored azulene pigments, 7-(1,2-dihydroxy-1-methylethyl)-4-methylazulene-1-carbaldehyde (purple–red color) and 7-acetyl-4-methylazulene-1-carbaldehyde (reddish-brown color) from the fruiting bodies of *Lactarius deliciosus* from the Yunnan province, China.

The fungi of the genus *Boletus* are species of variable morphology, an important characteristic for the identification of these fungi is the blue staining that the fungus suffers when it is collected due to oxidation, although the intensity of the coloration may vary according to the species (Bessette et al. 2000). This is caused by enzymatic oxidation of a derivative of pulvinic acid more highly hydroxylated than the dihydroxylated compound (variegatic acid). Several compounds arise from the enzymatic dimerization of ρ-hydroxyphenylpyruvic acid. This is derived by oxidation of the amino acid tyrosine, followed by the formation of CO bonds to produce grevilins (grevilin A) or the formation of CC bonds to produce hydroxylated terphenylquinone derivatives (atromentin) and products of subsequent reactions, including cleavage of the oxidative bond and rearrangement of atromentin with dehydration to dihydroxypulvinic acid or other oxidation reactions (Nelsen 2010).

3.5 Application of Fungal Pigments

Pigments have a wide range of applications in the pharmaceutical, cosmetic, food and textile industries. Mushroom pigments are extracellular metabolites, which can be used in many products. Color is very important in the appearance of foods, it is added to improve color, replace color lost during processing, and to minimize batch-to-batch variations (Mortensen 2006). Mushrooms produce safe and functional pigments to be used in food processing, for example, carotenoids act as a protector from intense sunlight to maintain food quality, in addition, they have antioxidant activity, so they can also prevent the incidence of many diseases such as cancer and heart disease, which is of great importance in the formulation of nutraceutical foods (Chattopadhyay et al. 2008). By understanding the characteristics of each pigment, favorable conditions for its use can be determined, for example: caraotenoids are more stable than lycopene, and carotenoids are soluble in lipids, so they cannot be used directly in water-based products (Mortensen 2006). *Cordyceps unilateralis* fungus presents a deep red pigment (3,5,8-TMON) with very low cytotoxic properties compared to the six naphthoquinones obtained from that fungus (Table 3.2), which makes it promising for food and cosmetic applications. An advantage is that from erythrostominone it is chemically converted to 3,5,8-TMON by heating the fermentation broth (100 ºC) under acidic conditions (pH 4) (Unagul et al. 2005). Caro et al. (2017) reported that the *Cordyceps farinose* pigment is excreted into the fermentation broth and extracted with a mixture of water and ethanol (1:1), it was resistant to heat, pH and temperature, making it an option to use as a colorant alimentary.

It is estimated that the international market revenue of the cosmetics industry will increase to 430 billion dollars by the year 2022, with an annual growth rate of 4.3% during the period 2016–2022 (World Cosmetics Market). Fungi present numerous biologically active compounds, many species of Basidiomycetes and Ascomycetes contain compounds that can be used in cosmetics, such as: surfactants, vitamins, antioxidants, enzymes, peptides and pigments. In addition, some pigments have antifungal and antimicrobial activities. Melanin shows resistance to ultraviolet light by absorbing a wide range of electromagnetic spectra and also offers protection from photoinduced damage, which is why it is widely used in cosmetics, eyewear and sunscreens. Carotenoids, lycopene and phenolic compounds can be used in cosmetics, sun lotions, anti-aging facial products, etc. (Barros et al. 2007), however, it is recommended that the clinical tests required be carried out to determine the effects and possible allergic, dermatological and endocrine responses.

In the pharmaceutical or nutraceutical industry, carotenoids are a great option since they play a very important role in human health. Because they serve as precursors of vitamin A, they can help in the prevention of age-related diseases (cataracts and degeneration muscle), reduce the incidence of coronary heart disease and carcinomas (lung, breast, prostate and colorectal cancer) (Fraser and Bramley 2004). It has been reported that carotenoids show antioxidant activity and produce free radical scavengers, lycopene also has health benefits including improving cardiovascular health, and protection against sunburn and certain types of cancers. On the other hand, cinnabarinic acid (Table 3.2) shows inhibitory effects towards several Gram-positive bacteria of the genus *Streptococcus*, due to the fact that it shares structural homology with actinomycins (antibiotic actinomycin D) produced by *Streptomyces* spp., making them an alternative in the pharmaceutical industry (Lin and Xu 2020).

Within the textile industry, it is required that dyed garments are color fast to rubbing and washing, it seeks to extend the useful life and value of textile articles, since the fading of fabrics generates contaminants. There are studies where the fastness of the color to washing and rubbing of fabrics dyed with fungal pigments is compared to the fastness of the color of commercial dyes. Cedano et al. (2001) reported that most of the 14 species of Aphyllophorales, presented brown, yellow and green tones, these colors are due to the presence of cinnamic acid and terphenylquinones present in polyphoric acid. The species *Albatrellus cristatus*, *Ganoderma applanatum*, *Ganoderma resinaceum*, *Inonotus radiatus* y *Phaeolus schweinitzii* required less biomass to perform the wool dyeing tests (1:1), most of the species presented pigments that were effectively fixed to the fibers of wool without mordants, probably due to the tannin content. However, Phaeolus schweinitzii offered great potential to be used as a dye, since five colors were obtained (olive brown, olive gray, dark brown, brownish gray and brown), all with excellent intensity. These colors could be obtained with less than the 1:1 ratio of wool fiber and mushrooms, in addition, the mushroom dye bath could be reused two or three more times. In another work, xylindein was tested, which is a blue–green pigment of the *Chlorociboria aeruginosa* fungus, it was demonstrated to be a very resistant dye, with a high affinity to adhere to cotton and silk fibers, either using the natural extract or in the presence of mordants. On the other hand, the draconin red of the fungus *Scytalidium cuboideum* produced inconsistent results. However, these fungal pigments outperformed commercial dyes in terms of color fastness to washing and wet and dry rubbing, and both pigments share properties such as stability to light, poor solubility and excellent bioadhesion (Hinsch and Robinson 2016). Gutierrez et al. (2018) reported that the red–pink pigment of *Scytalidium cuboideum* forms two crystals of different colors (red and orange), they are stable naphthoquinone crystals, which were produced using acetone and precipitation with liquid nitrogen. The crystalline nature of this pigment seems to make it a very stable colorant, since the crystals are stable due to the strong intermolecular bonding. The pigment in crystal form (lumpy powder) mainly adheres to itself, which means that the pigment is easy to add to oils, paints and other solvents as desired. It therefore has great potential to be of interest in different industries that use highly stable, durable natural colorants that can compete within the synthetic market, as well as within of the solar energy industry. Another widely tested fungus for textile coloring is *Dermocybe sanguinea*, the fastness of the pigments of this fungus on polyester was very good to excellent when tested for exposure to light, rubbing and washing. In the tests on polyamide and wool, there was color change when washing (Räisänen 2009).

3.6 Conclusion

Mushrooms are available as an alternative source to obtain natural pigments (fruiting body, culture medium and mycelium). They offer certain benefits, since many fungi can be grown all year round, they produce pigments with different color tones, they are not toxic, several fungi are also edible and/or medicinal, in addition, pigments obtained from fungi not only serve as a source of natural colorants, but also enrich the products where they are used, since as has been widely documented, they present molecules that can have effects as antioxidants, apoptosis inducers, antimicrobials, etc. Therefore, it is important to continue investigating alternative sources of fungal pigments and to continue working on obtaining a wider variety of shades, using pigments with health benefits, increasing the useful life of products and reducing production costs.

References

Achenbach, H. and Blümm, E. (1991). Investigation of the pigments of *Pycnoporus sanguineus* pycnosanguin and new phenoxazin-3-ones. *Archiv der Pharmazie* 324 (1): 3–6.

Barros, L., Ferreira, M.J., Queiros, B., Ferreira, I.C., and Baptista, P. (2007). Total phenols, ascorbic acid, β-carotene and lycopene in Portuguese wild edible mushrooms and their antioxidant activities. *Food Chemistry* 103 (2): 413–419.

Bauernfeind, J.C. (1981). Natural food colors. In: Carotenoids as Colorants and Vitamin A Precursors, Technological and Nutritional Applications (ed. J.C. Bauernfeind), 1–45. New York: Academic Press.

Béni, Z., Dékány, M., Kovács, B., Csupor-Löffler, B., Zomborszki, Z.P., Kerekes, E., Urbán, E., Hohmann, J., and Ványolós, A. (2018). Bioactivity-guided isolation of antimicrobial and antioxidant metabolites from the mushroom. *Tapinella atrotomentosa. Molecules* 23 (5): 1082.

Bergendorff, O. and Sterner, O. (1988). The sesquiterpenes of *Lactarius deliciosus* and *Lactarius deterrimus. Phytochemistry* 27 (1): 97–100.

Besl, H., Bresinsky, A., Geigenmüller, G., Herrmann, R., Kilpert, C., and Steglich, W. (1989). Pilzfarbstoffe, 61 Flavomentine und Spiromentine, neue Terphenylchinon-Derivate aus *Paxillus atrotomentosus* und *P. panuoides* (Boletales). *Justus Liebigs Annalen der Chemie* 1989 (8): 803–810.

Besl, H. and Bresinsky, A. (1997). Chemosystematics of Suillaceae and Gomphidiaceae (suborder Suillineae). *Plant Systematics and Evolution* 206 (1–4): 223–242.

Bessette, A.E., Roody, W.C., and Bessette, A.R. (2000). North American Boletes: A Color Guide to the Fleshy Pored Mushrooms. Syracuse, NY: *Syracuse University Press.*

Boa, E. (2004). Wild Edible Fungi-a Global Overview of Their Use and Importance to People. Non-wood Forest Products. Vol. 17. Rome: Food and Agriculture Organization of the United Nations.

Cardon, D. (2007). Natural Dyes: Sources, Tradition, Technology and Science. London: Archetype Publications.

Caro, Y., Anamale, L., Fouillaud, M., Laurent, P., Petit, T., and Dufossé, L. (2012). Natural hydroxyan-thraquinoid pigments as potent food grade colorants: an overview. *Natural Products and Bioprospecting* 2: 174–193.

Caro, Y., Venkatachalam, M., Lebeau, J., Fouillaud, M., and Dufossé, L. (2017). Pigments and colorants from filamentous fungi. In: Fungal Metabolites (ed. J.M. Merillon and K.G. Ramawat), 499–568. Cham, Switzerland: Springer International Publishing.

Cedano, M., Villaseñor, L., and Guzmán-Dávalos, L. (2001). Some Aphyllophorales tested for organic dyes. *Mycologist* 2 (15): 81–85.

Chattopadhyay, P., Chatterjee, S., and Sen, S.K. (2008). Biotechnological potential of natural food grade biocolorants. *African Journal of Biotechnology* 7 (17).2972–2985.

Daniewski, W.M. and Vidari, G. (1999). Constituents of *Lactarius* (mushrooms). In: fortschritte der chemie organischer naturstoffe/Progress in the Chemistry of Organic Natural Products, 69–171. Vienna: Springer Nature.

Davoli, P., Mucci, A., Schenetti, L., and Weber, R.W. (2005). Laetiporic acids, a family of non-carotenoid polyene pigments from fruit-bodies and liquid cultures of *Laetiporus sulphureus* (Polyporales, Fungi). *Phytochemistry* 66 (7): 817–823.

de Ugaz, O.L.S. (1997). Colorantes naturales. Perú: Pontificia Universidad Católica del Perú y Fondo Editorial.

Durán, N., Teixeira, M.F., De Conti, R., and Esposito, E. (2002). Ecological-friendly pigments from fungi. *Critical Reviews in Food Science and Nutrition* 42 (1): 53–66.

Eggert, C. (1997). Laccase-catalyzed formation of cinnabarinic acid is responsible for antibacterial activity of *Pycnoporus cinnabarinus. Microbiological Research* 152 (3): 315–318.

Eggert, C., Temp, U., and Eriksson, K.E. (1996). The ligninolytic system of the white rot fungus *Pycnoporus cinnabarinus*: purification and characterization of the laccase. *Applied and Environmental Microbiology* 62 (4): 1151–1158.

Fraser, P.D. and Bramley, P.M. (2004). The biosynthesis and nutritional uses of carotenoids. *Progress in Lipid Research* 43: 228–265.

Gaylord, M.C. and Brady, L.R. (1971). Comparison of pigments in carpophores and saprophytic cultures of *Paxillus panuoides* and *Paxillus atrotomentosus*. *Journal of Pharmaceutical Sciences* 60 (10): 1503–1508.

Gaylord, M.C., Benedict, R.G., Hatfield, G.M., and Brady, L.R. (1970). Isolation of diphenyl-substituted tetronic acids from cultures of *Paxillus atrotomentosus*. *Journal of Pharmaceutical Sciences* 59 (10): 1420–1423.

Gill, M. (1982). Polyolefinic 18-methyl-19-oxoicosenoic acid pigments from the fungus *Piptoporus australiensis* (Wakefield) Cunningham. *Journal of the Chemical Society, Perkin Transactions* 1: 449–1453.

Guo, L., Kong, D., Yao, K., Li, J., Li, H., Lan, N., and Hua, Y. (2020). Optimization and characterization of pigment production from *Boletus edulis* Bull.: Fr. by ultrasonic-assisted extraction. *Journal of Food Processing and Preservation* 44 (10) e14534.

Gutierrez, S.M.V., Hazell, K.K., Simonsen, J., and Robinson, S.C. (2018). Description of a naphthoquinonic crystal produced by the fungus *Scytalidium cuboideum*. *Molecules* 23 (8): 1905.

Haxo, F. (1950). Carotenoids of the mushroom *Cantharellus cinnabarinus*. *Botanical Gazette* 112 (2): 228–232.

Hinsch, E.M. and Robinson, S.C. (2016). Mechanical color reading of wood-staining fungal pigment textile dyes: an alternative method for determining colorfastness. *Coatings* 6 (3): 25.

Holkar, C.R., Jadhav, A.J., Pinjari, D.V., Mahamuni, N.M., and Pandit, A.B. (2016). A critical review on textile wastewater treatments: possible approaches. *Journal of Environmental Management* 182: 351–366.

Isaac, S. (1994). Mycology answers. Many fungi are brightly coloured; does pigmentation provide any advantage to those species? *Mycologist* 8 (4): 178–179.

Kim, K.H., Noh, H.J., Choi, S.U., Park, K.M., Seok, S.J., and Lee, K.R. (2010). Lactarane sesquiterpenoids from *Lactarius subvellereus* and their cytotoxicity. *Bioorganic & Medicinal Chemistry Letters* 20 (18): 5385–5388.

Kuhnert, E., Surup, F., Herrmann, J., Huch, V., Müller, R., and Stadler, M. (2015). Rickenyls A-E, antioxidative terphenyls from the fungus *Hypoxylon rickii* (Xylariaceae, Ascomycota). *Phytochemistry* 118: 68–73.

Li, K., Horanyi, P.S., Collins, R., Phillips, R.S., and Eriksson, K.E.L. (2001). Investigation of the role of 3-hydroxyanthranilic acid in the degradation of lignin by white-rot fungus *Pycnoporus cinnabarinus*. *Enzyme and Microbial Technology* 28 (4–5): 301–307.

Lin, L. and Xu, J. (2020). Fungal Pigments and their roles associated with human health. *Journal of Fungi* 6 (4): 280.

Mortensen, A. (2006). Carotenoids and other pigments as natural colorants. *Pure and Applied Chemistry* 78 (8): 1477–1491.

Musso, H. (1979). The pigments of fly agaric, *Amanita muscaria*. *Tetrahedron* 35 (24): 2843–2853.

Nelsen, S.F. (2010). Bluing components and other pigments of boletes. *Fungi* 3 (4): 11–14.

Nguyen, N.H., Song, Z., Bates, S.T., Branco, S., Tedersoo, L., Menke, J., ... Kennedy, P.G. (2016). FUNGuild: an open annotation tool for parsing fungal community datasets by ecological guild. *Fungal Ecology* 20: 241–248.

Nordlund, J.J., Abdel-Malek, Z.A., Boissy, R.E., and Rheins, L.A. (1989). Pigment cell biology: an historical review. *The Journal of Investigative Dermatology* 92 (s 4): 53–60.

Pagano, M.C. and Dhar, P.P. (2015). Fungal pigments: an overview. In: Fungal Biomolecules: Sources, Applications and Recent Developments (ed. V.K. Gupta, R.L. Mach, and S. Sreenivasaprasad), 173–181. UK: John Wiley and Sons.

Peters, S. and Spiteller, P. (2007a). Mycenarubins A and B, red pyrroloquinoline alkaloids from the mushroom *Mycena rosea*. *European Journal of Organic Chemistry* 2007 (10): 1571–1576.

Peters, S. and Spiteller, P. (2007b). Sanguinones A and B, blue pyrroloquinoline alkaloids from the fruiting bodies of the mushroom *Mycena sanguinolenta*. *Journal of Natural Products* 70 (8): 1274–1277.

Quang, D.N., Hashimoto, T., Nomura, Y., Wollweber, H., Hellwig, V., Fournier, J., Stadler, M., and Asakawa, Y. (2005). Cohaerins A and B, azaphilones from the fungus *Hypoxylon cohaerens*, and comparison of HPLC-based metabolite profiles in *Hypoxylon* sect. Annulata. *Phytochemistry* 66 (7): 797–809.

Räisänen, R. (2009). Anthraquinones from the fungus Dermocybe sanguinea as Textile Dyes. Home Economics and Craft Science. Dark Oy: University of Helsinki.

Räisänen, R. (2019). Fungal colorants in applications–focus on *Cortinarius* species. *Coloration Technology* 135 (1): 22–31.

Räisänen, R., Primetta, A., Nikunen, S., Honkalampi, U., Nygren, H., Pihlava, J.M., Berghe, I.V., and vson Wright, A. (2020). Examining safety of biocolourants from fungal and plant sources-examples from *Cortinarius* and *Tapinella, Salix* and *Tanacetum* spp. and dyed woollen fabrics. *Antibiotics* 9 (5): 266.

Ribeiro, B.P., Pinho, G.D., Andrade, P.B., Oliveira, C., Ferreira, A.C.S., Baptista, P., and Valentão, P. (2011). Do bioactive carotenoids contribute to the color of edible mushrooms? *The Open Chemical and Biomedical Methods Journal* 4: 14–18.

Smith, G.R., Finlay, R.D., Stenlid, J., Vasaitis, R., and Menkis, A. (2017). Growing evidence for facultative biotrophy in saprotrophic fungi: data from microcosm tests with 201 species of wood-decay basidiomycetes. *The New Phytologist* 215 (2): 747–755.

Solano, F. (2014). Melanins: skin pigments and much more-types, structural models, biological functions, and formation routes. *New Journal of Science*. https://doi.org/10.1155/2014/498276.

Song, J., Manir, M.M., and Moon, S.S. (2009). Cytotoxic grifolin derivatives isolated from the wild mushroom *Boletus pseudocalopus* (Basidiomycetes). *Chemistry and Biodiversity* 6 (9): 1435–1442.

Spiteller, P., Rüth, M., von Nussbaum, F., and Steglich, W. (2000). Detection of a 2, 3-aminomutase in the mushroom *Cortinarius violaceus*. *Angewandte Chemie International Edition* 39 (15): 2754–2756.

Stadler, M. and Hoffmeister, D. (2015). Fungal natural products-the mushroom perspective. *Frontiers in Microbiology* 6: 127.

Stadler, M. and Fournier, J. (2006). Pigment chemistry, taxonomy and phylogeny of the Hypoxyloideae (Xylariaceae). *Revista Iberoamericana de Micologia* 23 (3): 160–170.

Sutthisa, W. and Sanoamuang, N. (2014). Study of chemical compositions of *Cordyceps pseudomilitaris* pigments by Gas Chromatography-Mass Spectrometry (GC-MS). *Journal of Agricultural Technology* 10 (3): 583–593.

Tang, A.Y., Lo, C.K., and Kan, C.W. (2018). Textile dyes and human health: a systematic and citation network analysis review. *Coloration Technology* 134 (4): 245–257.

Thoen, E., Harder, C.B., Kauserud, H., Botnen, S.S., Vik, U., Taylor, A.F., Menkis, A., and Skrede, I. (2020). In vitro evidence of root colonization suggests ecological versatility in the genus *Mycena*. *The New Phytologist* 227 (2): 601–612.

Tudor, D., Robinson, S.C., and Cooper, P.A. (2013). The influence of pH on pigment formation by lignicolous fungi. *International Biodeterioration & Biodegradation* 80: 22–28.

Unagul, P., Wongsa, P., Kittakoop, P., Intamas, S., Srikitikulchai, P., and Tanticharoen, M. (2005). Production of red pigments by the insect pathogenic fungus *Cordyceps unilateralis* BCC 1869. *Journal of Industrial Microbiology and Biotechnology* 32 (4): 135–140.

Urista, C.M., Rodríguez, J.G., Corona, A.A., Cuenca, A.A., and Jurado, A.T. (2016). Pigments from fungi, an opportunity of production for diverse applications. *Biologia* 71 (10): 1067–1079.

Velíšek, J. and Cejpek, K. (2011). Pigments of higher fungi-a review. *Czech Journal of Food Sciences* 29 (2): 87–102.

Velmurugan, P., Lee, Y.H., Nanthakumar, K., Kamala-Kannan, S., Dufossé, L., Mapari, S.A., and Oh, B.T. (2010). Water-soluble red pigments from *Isaria farinosa* and structural characterization of the main-colored component. *Journal of Basic Microbiology* 50 (6): 581–590.

von Nussbaum, F., Spiteller, P., Rüth, M., Steglich, W., Wanner, G., Gamblin, B., Stievano, L., and Wagner, F.E. (1998). An iron (III)–catechol complex as a mushroom pigment. *Angewandte Chemie International Edition* 37 (23): 3292–3295.

Weber, R.W., Mucci, A., and Davoli, P. (2004). Laetiporic acid, a new polyene pigment from the wood-rotting basidiomycete *Laetiporus sulphureus* (Polyporales, Fungi). *Tetrahedron Letters* 45 (5): 1075–1078.

World Cosmetics Market-Opportunities and Forecasts, 2014–2022, Research and Markets. (2016). https://www.researchandmarkets.com/reports/3275915/world-cosmetics-market-opportunities-and (accessed 27 November 2020).

Yang, X.L., Qin, C., Wang, F., Dong, Z.J., and Liu, J.K. (2008). A new meroterpenoid pigment from the basidiomycete *Albatrellus confluens*. *Chemistry and Biodiversity* 5 (3): 484–489.

Yang, X.L., Luo, D.Q., Dong, Z.J., and Liu, J.K. (2006). Two new pigments from the fruiting bodies of the basidiomycete *Lactarius deliciosus*. *Helvetica chimica acta* 89 (5): 988–990.

Yaqoob, A., Li, W.M., Liu, V., Wang, C., Mackedenski, S., Tackaberry, L.E., Massicotte, H.B., Egger, K.N., Teimer, K., and Lee, C.H. (2020). Grifolin, neogrifolin and confluentin from the terricolous polypore *Albatrellus flettii* suppress KRAS expression in human colon cancer cells. *PloS One* 15 (5): e0231948.

Ye, Y., Josten, I., Arnold, N., Steffan, B., and Steglich, W. (1996). Isolation of a fluorone pigment from the Indian paint fungus *Echinodontium tinctorium* and *Pyrofomes albomarginatus*. *Tetrahedron* 52 (16): 5793–5798.

4

Pharmacological Potential of Pigments

M. C. Pagano[1], E. J. A. Corrêa[2], N. F. Duarte[3], and B. K. Yelikbayev[4]

[1]Federal University of Minas Gerais, Belo Horizonte, Minas Gerais, Brazil
[2]Empresa de Pesquisa Agropecuária de Minas Gerais EPAMIG-URECO, Pitangui, MG
[3]Instituto Federal de Minas Gerais
[4]Kazakh National Agrarian University, Almaty, Kazakhstan.
*Corresponding author: marpagano@gmail.com

4.1 Introduction

Most microorganisms on the planet have potential for bioproduction of biomolecules. Thus, research focusing on production of pigments is rising as the recognition of their functions on organisms and particularly for human welfare, such as anticancer, antileishmanial, antibacterial ability, and antioxidant potential.

Pigments are considered secondary metabolites produced by mycelium when the supply of essential nutrients decreases and environmental constraints are present. They may protect the host-fungi against the detrimental effect of sunlight and ultraviolet radiation, bacterial attack and also against insect attack (Velíšek and Cejpek 2011). The pigmentation can vary with age or when they are damaged or treated with alkali (Hanson 2008), this is one of the essential characters used in their identification (Velíšek and Cejpek 2011).

The most common fungal pigments are melanins (dark brown pigment), carotenoids (orange–red), lycopene (dark red) and xanthophylls; their presence being significant to improve fungal survival and spore resistance (Isaac 1994). Melanin is a special pigment with varied functions found in all biological kingdoms as a vital radioprotector (Eisenman and Casadevall 2012; Kunwar et al. 2012).

Fungi provide several natural substances that may also have the potential for use in medicine (Dufossé et al. 2014). Endophytic fungi (non-pathogenic fungi within plant tissues) can transform certain compounds into more potential analogs. This chapter examines recent information on microorganisms' potential for biotechnological processes focusing on pigments concerning the world's research results. First, we describe the fungal pigments showing their significance. We then review recent reports focusing on their potential use and importance for human welfare. Finally, we demonstrate the importance of research on microbial pigments both for anthropic and natural systems.

Biomolecules from Natural Sources: Advances and Applications, First Edition. Edited by Vijai Kumar Gupta, Satyajit D. Sarker, Minaxi Sharma, María Elida Pirovani, Zeba Usmani, and Chelliah Jayabaskaran.

4.2 Bacterial Pigments

It is known that 600 different carotenoids can be produced by plants, algae, bacteria, and fungi, some with nutraceutical properties for additional health benefits.

Extensive literature has been reviewed on bacterial pigments' biomedical applications (Chandi and Gill 2011). Below, we will discuss some aspects of this topic.

4.3 Fungal Pigments

Fungal pigments are produced as secondary metabolites by part of the mycelium when the supply of essential nutrients becomes depleted. It is also known that one fungal species can harbor a mixture of several different pigments (Isaac 1994), for example, carotenoids: β-carotene (orange), γ-carotene and lycopene (dark red). Additionally, xanthophylls can predominate. Carotenoids can protect the fungal mycelia from harmful sunlight and ultraviolet light. The colored basidioma (Basidiomycetes) and asci of Ascomycetes may be due to these pigments, part of the hyphal wall, cytoplasm, or oil droplets. Other pigments such as melanin and sporopollenin are found in spores and hyphal walls providing protection from radiation and drought, thus improving survival.

Moreover, the quantity and type of pigmented spores, which are more resistant against radiation, are dependent on altitude, latitude, season and other local factors (Isaac 1994).

Initial studies showed that the synthetic and natural melanins of *Tuber melanosporum* have the same physical and chemical properties; however, there are significant variations in the time to discoloration of the pigment (Harki et al. 1997).

Some reviews of fungal pigments (Velíšek and Cejpek 2011; Zhou and Liu 2010) focused on the structure of Basidiomycota's pigments.

Melanin is a special pigment with numerous functions found in the biological kingdom. Melanin increases fungi's ability to survive in severe environments (providing protection against environmental stresses and playing a role in fungal pathogenesis) (Eisenman and Casadevall 2012). Among abiotic stresses, melanin helps to cope with ultraviolet light, oxidizing agents and ionizing radiation.

Melanin prolonged survival and mitigated the effects of exposure to radiation in BALB/C mice (Kunwar et al. 2012). Moreover, melanin's protective effect minimizes radiation toxicity, as shown by the inhibition of radiation-induced hematopoietic damages, prevention of apoptosis in splenic tissue, and reduction of the oxidative stress hepatic tissue.

The occurrence of pigments together with colony characters can help the fungal identification. In this respect, malt extract agar (MEA), PDA and oatmeal agar (Gams et al. 2007) is commonly used. Plates are incubated at specific temperatures (e.g., for 14 days). Then, colony colors can be determined according to the NBS/IBCC Colour System or taken from the ISCC-NBS color charts (Kelly 1964). Colonies may vary in color, e.g., from yellowish-white (4.5Y 9.2/1.2) to pale yellow (4.7Y 9.0/3.8), and can show the same color for all media or not. For example, colonies isolated from dark septate endophytes of *Salix caprea* produced yellow pigmentation on PDA and MEA plates (Likar and Revgear 2013).

Table 4.1 Database survey conducted in 2021 (SCOPUS) for journal articles dealing with pigments.

Keywords
Pigments

*AMF = arbuscular mycorrhizal fungi.

Despite its importance and ubiquity, there are many fundamental concerns about the function of fungal pigments, e.g., the details of its chemical structure, increasing reports will reveal those questions (Table 4.1).

Photo-sensitive pigments (rhodopsins, light oxygen voltage proteins, phytochromes and cryptochromes) are sensitive to light variation and permit the organism to adjust to the light environmental conditions. They activate cellular signaling processes that control the physiology, development, and behavior, allowing organisms to perceive space and time (Heinzen 2012).

4.4 Pigments for the Food Industry

It is notable that with healthier lifestyles, and the growing market for natural food colorants worldwide, filamentous fungi have shown chemically diverse colorings. Additionally, from raw product detection, the potential of fungi from the food industry is immense. Dufossé et al. (2014) reviewed recent biotechnological processes for natural colorants for industrial food production. Filamentous fungi can produce a variety of pigments with different chemical structures (carotenoids, melanins, flavins, phenazines, quinones, monascins, violacein or indigo). It can also provide novel pigments of numerous color types and atypical chemical structures (Dufossé et al. 2014).

As natural pigments from plant and animal origins have numerous disadvantages (instability and low water solubility, less availability throughout the year), microbial pigments are of prime interest due to their higher stability and solubility (Gunasekaran and Poorniammal 2008).

Méndez et al. (2011) showed that the red pigment production and cellular growth of *Penicillium purpurogenum* in a submerged culture depended on the medium temperature and pH. For example, using a minimal medium maximal red pigment production was achieved using pH = 5 and 24 °C. Thus, they suggested an indirect relationship between fungal biomass and pigment production.

4.5 Pigments for Other Human Uses

It is known that endophytes constitute many sources of bioactive compounds (Pagano and Dhar 2013; Pimentel et al. 2011). The endophytic microorganisms provide protection and subsistence to the host plants through substances that may also be available in industry, agriculture, and medicine (Strobel and Daisy 2003; Strobel et al. 2004).

Endophytes are of great interest for the production of anticancer, antimicrobial, and antioxidant compounds for human consumption. Following are a few examples that illustrate the potential of fungal endophytes for human use.

The study of important medicinal plants such as *Taxus brevifolia* trees and their high predation rate by humans has required a search for other sources for the production of taxol (diterpenoid anticancer agent for the treatment of advanced breast, lung, and refractory ovarian cancer) which was isolated from the bark and has become expensive and scarce. However, the isolation of the same compound from a fungal endophyte (*Taxomyces andreanae*) isolated from *Taxus brevifolia* was a cheaper alternative and a more available product. The fungus produces taxol in a semi-synthetic liquid medium (Stierle et al. 1993). Then, taxol was found subsequently in several fungal endophytes associated or not to this plant species (see Pimentel et al. 2011).

Another example is the antifungal bioactive compound obtained from the fungal endophyte *Xylaria* sp. YX-28, isolated from *Ginkgo biloba* and with inhibitory activity against several foodborne and food spoilage microorganisms, and, thus, suggested being used as a natural preservative in food (Liu et al. 2008).

Furthermore, Simanjuntak et al. (2010) studied endophytic fungi for the ability to biotransform the pigment curcumin. Biotransformation is a promising method to produce chemical changes on bioactive compounds that are not their natural substrates as microorganisms show no limit to adaption to new environments (Borges et al. 2009).

An example of a natural yellow pigment isolated from *Curcuma* sp. is curcumin. An expensive component is used as an innovative therapeutic agent due to its pharmacological properties antioxidant, anti-inflammatory and anti-carcinogenic (Lin and Lin-Shiau 2001). Additionally, Lin and Lin-Shiau (2001) found that many analogs can exhibit more potent pharmacological activities. Lin and Lin-Shiau (2001) used fungal endophytes isolated from the rhizome of *Curcuma longa* L. selected from a collection in Indonesia to transform curcumin into a biotransformed compound (hexahydrocurcumin). They showed the use of endophytic fungi isolated from the same plant as the substrate and the need to investigate this analog's metabolism in mammals.

4.6 Pigments and Fungal Infection

It is known that melanins contribute to the survival of fungal spores by protecting against light damage and this is an important virulent factor. Inadequacy in the formation of appressoria due to melanin's changes was reported in the reduction of virulence.

Lower melanin production in *Alternaria alternata* was correlated with thinner and more U.V. light-sensitive conidia. Experiments with albino spore mutants of *Cochliobolus heterotrophs* (maize pathogen) showed melanin's need for fungus survival. Disruption of genes encoding melanin production enzymes can form pigmentless conidial phenotypes and significantly reduce fungal infection. In *Aspergillus fumigatus* (source of invasive aspergillosis), spore pigment is a virulent factor (Calvo et al. 2002). For more than 40 years, researchers have been identifying fungi that produce melanins and their biosynthesis pathways (identification of the genes and corresponding enzymes of the pathways were achieved). In contrast, recent investigations have linked melanin to virulence in some human pathogenic and phytopathogenic fungi. Researchers correlated the absence of melanin in humans and planted pathogenic fungi with a decrease in fungi virulence (Langfelder et al. 2003).

4.7 Pigment Production

Fungal laccases are enzymes related to pigment biosynthesis, fruit body formation, and plant pathogenesis (Alcalde 2007), broadly distributed in fungi. More than 100 laccases from Basidiomycetes and Ascomycetes have been characterized due to the increasing interest in their molecular regulation mechanisms and factors influencing their production. Several potential biotechnological interests include de-lignification of lignocellulosic complexes, biopulping and biobleaching, change of colorants in the textile industry, wastewater treatment, and the degradation of explosives and pesticides have been found in laccases (see Rivera-Hoyos et al. 2013).

Lopes et al. (2013) sustain the readily available alternative source of natural pigments produced by fungi as an eco-friendly alternative. They pointed out isolates of *Penicillium chrysogenum*, *Fusarium graminearum*, *Monascus purpureus* capable of growing and producing water-soluble pigment's agro-industrial residues. Their cultures can be used in the textile industry; nevertheless, they remarked that additional purification is required for food and pharmaceutical industrial purposes.

The effects of synthetic dyes used in food, pharmaceutical, textile, and cosmetic industries and the search for new sources of natural pigments have increased, mainly because of synthetic pigments' toxic effects. The advantages of producing pigments from microorganisms are: independence from weather conditions, colors of different shades, and growth on inexpensive substrates (Lopes et al. 2013).

Furthermore, Robinson et al. (2013) showed that additional fungal colonization could provide natural pigments for commercially available wood. It is known that pigments (pink or yellow) are produced by certain fungi and penetrate wood (Robinson 2012). Certain pigments are more light stable than other natural pigments of a similar type, such as the pink stain produced by *Acer negundo* L. (Robinson et al. 2013). For commercially available wood, additional colors can create colored zones being perfect for woodworkers and the industry of stained wood (Robinson et al. 2013).

Finally, a few studies have concluded a fungal production of biochar or chemically similar compounds suggesting that 9% of biochar in soil is formed biologically (Glaser and Knorr 2008). *Aspergillus niger* was broadly found in dark soils and biochar amended into soils (Brodowsky et al. 2005). These fungal species produce a black pigment named aspergillin; however, their contribution to biochar formation in soils needs more research (Wiedner and Glaser 2013).

4.8 Fungal Pigments and Plant Endophytes

As sessile organisms, plants are exposed to natural climatic or edaphic stresses and environmental modifications from human activities, consequently interaction with above-ground (Zheng and Dicke 2008) and below-ground (Kuyper and Goede 2005) microorganisms, such as a wide diversity of fungi (Vandenkoornhuyse et al. 2002). The identity and ecological function of the majority of them are unknown. Microfungal ascomycetes are commonly found in roots (having regularly septate hyphae with hyaline to quite darkly pigmented walls) named dematiaceous or dark septate endophytes – DSE (Addy et al. 2005).

Endophytes can provide a wide variety of bioactive secondary metabolites (alkaloids, benzopyranones, flavonoids, phenolic acids, quinones, steroids, terpenoids, tetralones, xanthones) (Tan and Zou 2001), with wide-ranging applications as agrochemicals, antibiotics, immunosuppressants, antiparasitics, antioxidants, and anticancer agents (Strobel 2003).

Evidence points to the bioactive compounds of medicinal plants as products of the plant itself or endophytes living inside the plant (Miller et al. 2012; Pagano and Dhar 2013).

In this sense, endophytic taxa with pigmented hyphae can be a weak pathogen, a saprotroph on senescent root tissues, or a root mutualist (Addy et al. 2005). However, a few studies focused on interactions between identified taxa of microfungal endophytes and host plants, which is crucial to clarify this symbiosis and for the precise identification of isolates of fungal endophytes for use in experiments (Addy et al. 2005).

Abundant reports on endophytes have been published; however, pigments are focussed on in fewer papers. One of which deals with foliar and four papers are about root fungal endophytes, respectively.

Several studies have exposed the great potential of endophytes for the production of bio-compounds with promising medicinal or agricultural applications (Aly et al. 2011). For this purpose, bio-prospection of plant-fungal endophytes has converted this investigation into a "hot spot" of interest to botany, ecology, natural medicine and pharmacology, and other disciplines (Sun et al. 2013).

Concerning foliar endophytes, Fernandes et al. (2011) allied them to the study of global changes, showing that hail storms can strongly affect fungi associated with leaves of an endemic threatened plant species (*Coccoloba cereifera*) in Brazil. Plant photosynthetic pigments increased in the leaves. The endophyte richness decreased considerably after the event, showing the importance of measuring plant species' susceptibility and survival with constrained distribution patterns.

Suryanarayanan et al. (2004) reported that the foliar endophyte *Phyllosticta capitalists* produce a black pigment considered to be melanin. As this fungus is widespread, they related melanin's presence with improved fitness in different habitats, such as mangroves, dry deciduous forest, moist deciduous and semi-evergreen forest, in both temperate and tropical biomes. That is to say, and the authors suggested that melanin content in the hyphae determines the success of this endophyte since melanin enhances the survival capability of fungi in stressful environments.

Concerning root endophytes, inoculation can increase plant fitness. For example, a filamentous fungus in plant roots, *Piriformospora indica*, can be inoculated in rice seedlings under high salt stress. The photosynthetic plant pigment content increased in inoculated rice under salt stress conditions compared with non-inoculated treatment, showing that endophytes can counteract decreases in plant growth throughout salt stress (Jogawat et al. 2013).

Addy et al. (2005) compiled a detailed review of some microfungal endophytes' descriptions from roots.

4.9 Pigments, Mycorrhizas and Endophytes

Most experiments investigating the effects of ascomycetous endophytes on host plants belong to fungi with pigmented septate hyphae, e.g., DSE, due to their easy visualization. They showed the effects of these microfungal endophytes on host plants ranging from negative to positive (see Addy et al. 2005)

Microfungal DSE was also observed in several vascular plants worldwide (Fernández et al. 2012; Mandyam and Jumpponen, 2005; Marins et al. 2010; Pagano 2012; Pagano et al. 2012; Urcelay 2012; Urcelay et al. 2011). DSE was also usually found in agronomic plants (Barrow 2003; Barrow and Aaltonen 2001; Chaudhry et al. 2005; Likar et al. 2008). Usually, they are in coexistence with AMF (Pagano et al. 2012).

Due to this facility, most researchers on AMF have begun to register DSE presence in the same analyzed root samples. However, whether DSE should be considered mycorrhizal or not remains controversial (Smith and Read 2008) as little evidence was reported. Finally, AMF has not been confirmed to associate functionally with Bryophytes. While fungal endophytes can improve their tolerance to pH and better growth, more experimentation is needed (Davey and Currah 2006). Despite their rich

Table 4.2 Selected books, book chapters or reviews dealing with microbial pigments.

Ecosystem type/ focus	References
Fungal biomolecules	Gupta et al. (2015)
Bacterial prodigiosins	Darshan and Manonmani (2015)
Microbial pigments as natural color sources	Tuli et al. (2015)
Natural Blue Colorants	Newsome et al. (2014)
Natural colorants	Shahid et al. (2013)
Bacterial pigments	Venil et al. (2013)
Higher fungi	Velíšek and Cejpek (2011)
Microbial Pigments	Babitha (2009)
Neotropic endophytic fungi.	Rosa (2021)

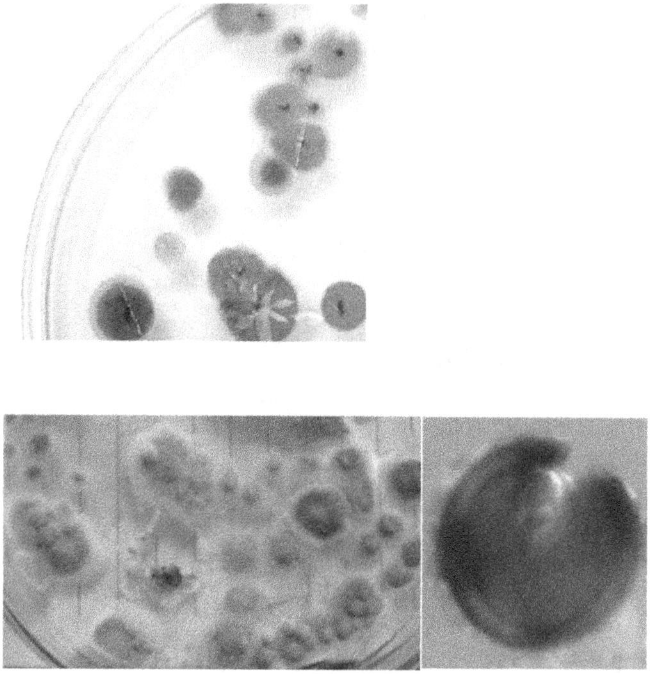

Figure 4.1 Fungi obtained compared against soil showing pigments. Clockwise, from upper left; cultivated fungi and brown spore of mycorrhizal fungi isolated from soil (Photo-credit: M. Pagano).

diversity, the endophytic fungi associated with plants of the Neotropical region remains mostly unknown (Rosa 2021).

4.10 Conclusion

Attention to the potential of microorganisms for the production of pigments has grown. These studies are significant both for humans and for knowledge of the biodiversity of the planet. Fungal pigments are investigated for multiple purposes improving plant defenses and natural production of pigments (industry, agriculture and medicine). We have demonstrated some melanized fungi commonly found in soil (Figure 4.1).

Finally, these studies related to ecology, natural medicine and biochemistry, and the effects of global change are of great interest to human health, but further research and experimentation are needed to cope with these challenges.

Acknowledgments

M. Pagano is grateful to the co-authors and editors.

References

Addy, H.D., Piercey, M.M., and Currah, R.S. (2005). Microfungal endophytes in roots. *Canadian Journal of Botany* 83 (1): 1–13.

Alcalde, M. (2007). Laccases: biological functions, molecular structure and industrial applications. In: Industrial Enzymes (ed. J. Polaona and A.P. MacCabe), 461–476. Springer Heidelberg.

Aly, A.H., Debbab, A., and Proksch, P. (2011). Fungal endophytes: unique plant inhabitants with great promises. *Applied Microbiology and Biotechnology* 90: 1829–1845.

Barrow, J.R. (2003). Atypical morphology of dark septate fungal root endophytes of *Bouteloua* in arid southwestern USA rangelands. *Mycorrhiza* 13: 239–247.

Barrow, J.R. and Aaltonen, R.E. (2001). A method of evaluating internal colonization of *Atriplex canescens* (Pursh) Nutt. roots by dark septate fungi and how they are influenced by host physiological activity. *Mycorrhiza* 11: 199–205.

Borges, K.B., Borges, W.D.S., Durán-Patrón, R., Pupo, M.T., Bonato, P.S., and Collado, I.G. (2009). Stereoselective biotransformations using fungi as biocatalysts. *Tetrahedron, Asymmetry* 20 (4): 385–397.

Brodowsky, S., Amelung, W., Haumaier, L., Abetz, C., and Zech, W. (2005). Morphological and chemical properties of black carbon in physical soil fractions as revealed by scanning electron microscopy and energy-dispersive X-ray spectroscopy. *Geoderma* 128: 116–129.

Calvo, A.M., Wilson, R.A., Bok, J.W., and Keller, N.P. (2002). Relationship between secondary metabolism and fungal development. *Microbiology and Molecular Biology Reviews* 66: 447–459.

Chandi, G.K. and Gill, B. (2011). Production and characterization of microbial carotenoids as an alternative to synthetic colors: a review. *International Journal of Food Properties* 14 (3): 503-513.

Chaudhry, M.S., Batoo, Z., and Khan, A.G. (2005). Preliminary assessment of plant community structure and arbuscular mycorrhizas in rangeland habitats of Cholistan desert, Pakistan. *Mycorrhiza* 15: 606–611.

Davey, M.L. and Currah, R.S. (2006). Interactions between mosses (Bryophyta) and fungi. *Canadian Journal of Botany* 84 (10): 1509–1519.

Dufossé, L., Fouillaud, M., Caro, Y., Mapari, S.A.S., and Sutthiwong, N. (2014). Filamentous fungi are large-scale producers of pigments and colorants for the food industry. *Current Opinion in Biotechnology* 26: 56–61.

Eisenman, H.C. and Casadevall, A. (2012). Synthesis and assembly of fungal melanin. *Applied Microbiology and Biotechnology* 93: 931–940.

Fernandes, G.W., Oki, Y., Sanchez-Azofeifa, A., Faction, G., and Amaro-Arruda, H.C. (2011). Hail impact on leaves and endophytes of the endemic threatened *Coccoloba cereifera* (Polygonaceae). *Plant Ecology* 212: 1687–1697.

Fernández, N., Fontenla, S., and Messuti, M.I. (2012). Co-occurrence of arbuscular mycorrhizas and dark septate endophytes in pteridophytes from a Valdivian temperate rainforest in Patagonia, Argentina. In: Mycorrhiza: Occurrence and Role in Natural and Restored Environments (ed. M.C. Pagano), 99–125. Hauppauge, NY: Nova Science Publishers.

Gams, W., Verkley, G.J.M., and Crous, P.W. (eds.) (2007). CBS Course of Mycology. Utrecht: Central bureau voor Schimmel cultures.

Glaser, B. and Knorr, K.H. (2008). Isotopic evidence for condensed aromatics from non-pyrogenic sources in soils – implications for current methods for quantifying soil black carbon. *Rapid Communications in Mass Spectrometry* 22: 935–942.

Gunasekaran, S. and Poorniammal, R. (2008). Optimization of fermentation conditions for red pigment production from *Penicillium* sp. under submerged cultivation. *African Journal of Biotechnology* 7 (12): 1894–1898.

Hanson, J.R. (2008). The Chemistry of Fungi. Cambridge: RSC Publishing, 222.

Harki, E., Talou, T., and Dargen, R. (1997). Purification, characterisation and analysis of melanin extracted from *Tuber mehnosporum* Vitt. *Food Chemistry* 58 (1–2): 69–73.

Heinzen, C. (2012). Plant and fungal photopigments. *WIREs: Membrane Transport and Signaling* 1: 411–432.

Jogawat, A., Saha, S., Bakshi, M., Dalaman, V., Kumar, M., Dua, M., Varma, A., Oelmüller, R., Tuteja, N., and Johri, A.K. (2013). *Piriformospora indica* rescues growth diminution of rice seedlings during high salt stress. *Plant Signaling & Behavior* 8 (10): e26891-1-6.

Isaac, S. (1994). Mycology answers.Many fungi are brightly coloured; does pigmentation provide any advantage to those species? *Mycologist* 8 (4): 178-179.

Kelly, K.L. (1964). Inter-Society Color Council National Bureau of Standards Color Name Charts Illustrated with Centroid Colors. Washington, DC: US Government Printing Office.

Kunwar, A., Adhikary, B., Jayakumar, S., Barik, A., Chattopadhyay, S., Raghukumar, S., and Priyadarsini, K.I. (2012). Melanin, a promising radioprotector: mechanisms of actions in a mice model. *Toxicology and Applied Pharmacology* 264: 202–211.

Kuyper, T.W. and Goede, R.G.M. (2005). Interaction between higher plants and soil-dwelling organisms. In: Vegetation Ecology (ed. E. der Maarel), 286–308. UK: Blackwell.

Langfelder, K., Streibel, M., Jahn, B., Haase, G., and Brakhage, A.A. (2003). Biosynthesis of fungal melanins and their importance for human pathogenic fungi. *Fungal Genetics and Biology* 38: 143–158.

Likar, M., Bukovnik, U., Kreft, I., Chrungoo, N.K., and Revgear, M. (2008). Mycorrhizal status and diversity of fungal endophytes in roots of common buckwheat (*Fagopyrum esculentum*) and Tartary buckwheat (*F. tataricum*). *Mycorrhiza* 18 (6–7): 309–315.

Likar, M. and Revgear, M. (2013). Isolates of dark septate endophytes reduce metal uptake and improve physiology of *Salix caprea* L. *Plant and Soil* 370: 593–604.

Lin, J.K. and Lin-Shiau, S.Y. (2001). Mechanisms of cancer chemoprevention by curcumin. *Proceedings of the National Science Council, Republic of China. Part B* 25: 59–66.

Liu, X., Dong, M., Chen, X., Jiang, M., Lv, X., and Zhou, J. (2008). Antimicrobial activity of an endophytic *Xylaria* sp.YX-28 and identification of its antimicrobial compound 7-amino- 4-methyl coumarin. *Applied Microbiology and Biotechnology* 78 (2): 241–247.

Mandyam, K. and Jumpponen, A. (2005). Seeking the elusive function of the root-colonising dark septate endophytic fungi. *Studies in Mycology* 53.

Marins, J.F., Carrenho, R., and Thomaz, S.M. (2010). Occurrence and coexistence of arbuscular mycorrhizal fungi and dark septate fungi in aquatic macrophytes in a tropical river-floodplain system. *Aquatic Botany* 91: 13–19.

Méndez, A., Pérez, C., Montañéz, J.C., Martínez, G., and Aguilar, C.N. (2011). Red pigment production by Penicillium purpurogenum GH2 is influenced by pH and temperature. *Journal of Zhejiang University Science B: Biomedicine & Biotechnology* 12 (12): 961–968.

Miller, K.I., Qing, C., Sze, D.M.-Y., Roufogalis, B.D., and Neilan, B.A. (2012). Culturable endophytes of medicinal plants and the genetic basis for their bioactivity. *Microbial Ecology* 64: 431–449.

Pagano, M.C. (ed.) (2012). Mycorrhiza: Occurrence and Role in Natural and Restored Environments. Hauppauge, NY: Nova Science Publishers, 327.

Pagano, M.C. and Dhar, P.P. (2013). Arbuscular mycorrhizal fungi association and bioactive compound in plants. In: Biology of Bioactive Compounds (ed. V.K. Gupta, A. O'Donovan, A. Pandey and M. Lohani), 225-243. UK: Wiley-Blackwell.

Pagano, M.C., Lugo, M., Araújo, F., Ferrero, M., Menoyo, E., and Steinaker, D. (2012). Native species for restoration and conservation of biodiversity in South America. In: Native Species: Identification, Conservation and Restoration (ed. L. Marín and D. Kovač), 1–55. New York: Nova Science Publishers.

Pimentel, M.R., Molina, G., Dionísio, A.P., Maróstica Junior, M.R., and Pastore, G.M. (2011). The use of endophytes to obtain bioactive compounds and their application in biotransformation process. *Biotechnology Research International* Article ID 576286 11.

Rivera-Hoyos, C.M., Morales- Alvarez, E.D., Poutou-Pinales, R.A., Pedroza-Rodriguez, A.M., Rodriguez-Vazquez, R., and Delgado-Boada, J.M. (2013). Fungal laccases. *Fungal Biology Reviews* 27 (3/4): 67-82.

Robinson, S.C. (2012). Developing fungal pigments for 'painting' vascular plants. *Applied Microbiology and Biotechnology* 93: 1389–1394.

Robinson, S.C., Tudor, D., MacDonald, G., Mansourian, Y., and Cooper, P.A. (2013). Repurposing mountain pine beetle blue wood for art through additional fungal colonization. *International Biodeterioration & Biodegradation* 85: 372–374.

Rosa, L.H. (ed.) (2021). Neotropical Endophytic Fungi. Diversity, Ecology, and Biotechnological Applications. Springer.

Simanjuntak, P., Prana, T.K., Wulandari, D., Dharmawan, A., Sumitro, E., and Hendriyanto, M.R. (2010). Chemical studies on a curcumin analogue produced by endophytic fungal transformation. *Asian Journal of Applied Sciences* 3: 60–66.

Smith, S.E. and Read, D.J. (2008). Mycorrhizal Symbiosis. New York: Elsevier.

Stierle, A., Strobel, G., and Stierle, D. (1993). Taxol and taxane production by *Taxomyces andreanae*, an endophytic fungus of Pacific yew. *Science* 260 (5105): 214–216.

Strobel, G. and Daisy, B. (2003). Bioprospecting for microbial endophytes and their natural products. *Microbiology and Molecular Biology Reviews* 67 (4): 491–502.

Strobel, G., Daisy, B., Castillo, U., and Harper, J. (2004). Natural products from endophytic microorganisms,". *Journal of Natural Products* 67 (2): 257–268.

Strobel, G.A. (2003). Endophytes as sources of bioactive products. *Microbes and Infection* 5: 535–544.

Sun, J., Xia, F., Cui, L., Liang, J., Wang, Z., and Wei, Y. (2013). Characteristics of foliar fungal endophyte assemblages and host effective components in *Salvia miltiorrhiza* Bunge. *Applied Microbiology and Biotechnology*. doi: 10.1007/s00253-013-5300-4.

Suryanarayanan, T.S., Ravishankar, J.P., Venkatesan, G., and Murali, T.S. (2004). Characterization of the melanin pigment of a cosmopolitan fungal endophyte. *Mycological Research* 108 (8): 974–978.

Tan, R.X. and Zou, W.X. (2001). Endophytes: a rich source of functional metabolites. *Natural Product Reports* 18: 448–459.

Tuli, H.S., Chaudhary, P., Beniwal, V., and Sharma, A.K. (2015). Microbial pigments as natural color sources: current trends and future perspectives. *Journal of Food Science and Technology* 52: 4669–4678.

Urcelay, C. (2012). Co-occurrence of three fungal root symbionts in *Gaultheria poeppiggi* DC in Central Argentina. *Mycorrhiza* 12: 89–92.

Urcelay, C., Acho, J., and Joffre, R. (2011). Fungal root symbionts and their relationship with fine root proportion in native plants from the Bolivian Andean highlands above 3,700 m elevation. *Mycorrhiza* 21: 323–330.

Vandenkoornhuyse, P., Baldauf, S.L., Leyval, C., Straczek, J., and Young, J.P.W. (2002). Extensive fungal diversity in plant roots. *Science* 295: 2051.

Velíšek, J. and Cejpek, K. (2011). Pigments of higher fungi: a review. *Czech Journal of Food Sciences* 29 (2): 87–102.

Wiedner, K. and Glaser, B. (2013). Biochar-Fungi interactions in soil. In: Biochar and Soil Biota (ed. N. Ladygina and F. Rineau), 69–99. Boca Raton, USA: CRC Press, Taylor and Francis Group.

Zheng, Si.-J. and Dicke, M. (2008). Ecological genomics of plant-insect interactions: from gene to community. *Plant Physiology* 146: 812–817.

Zhou, Z. and Liu, J. (2010). Pigments of fungi (macromycetes). *Natural Product Reports – The Royal Society of Chemistry*. doi: 10.1039/C004593D. Published on 09 August 2010 on http://pubs.rsc.org.

5

Bioactive Compounds

Encapsulation, Delivery, and Applications Using Albumins as Carriers

Flavia F. Visentini[a,b], *Adrian A. Perez*[a,b], *Joana B. Ferrado*[a,b], *María Laura Deseta*[a,b], *and Liliana G. Santiago*[a]

[a]*Área de Biocoloides y Nanotecnología, Instituto de Tecnología de Alimentos, Facultad de Ingeniería Química, Universidad Nacional del Litoral, 1 de Mayo 3250, Santa Fe, 3000, Argentina*
[b]*Consejo Nacional de Investigaciones Científicas y Técnicas (CONICET), Argentina*
**Corresponding author: lsanti@fiq.unl.edu.ar*

5.1 Introduction

A bioactive compound (BC) is a composite with the ability to interact with one or more components of living tissues and to produce a wide range of possible health benefits. Examples of relevant BCs are phenolic, vitamins and lipids molecules. All these compounds have different chemical structures, the hydrophobic ones are of particular interest due to their low water solubility and sensitiveness to injurious environmental factors, such as light and oxygen. It has been proven that their intake has numerous beneficial effects on human health; however, they are found in very small quantities in foods (Bender 2003; Guaadaoui et al. 2014; Manach et al. 2004). Therefore, different systems have been developed to use them in the field of pharmaceutical and food applications and to overcome problems linked with solubility and stability. In this sense, nanoencapsulation using albumins, appear to be a promising strategy. Albumins constitute a family of globular proteins, the most common one being the serum albumin, which is abundant in blood plasma. This specific albumin includes human serum albumin (HSA) and bovine serum albumin (BSA). Other albumin types comprise the storage proteins such as ovalbumin (OVA) in egg white, α-lactalbumin (ALA) in milk and different storage albumins present in some plant seeds, e.g., legumes, amaranth, and soybean seeds, etc. In general, BSA and HSA have been extensively investigated for pharmaceutical applications due to their immuno-compatibility, whereas OVA and ALA are widely used to improve the solubility and stability of different BCs.

Based on that, this chapter comments on some structural and conformational aspects of albumins, such as BSA, HSA, OVA, and ALA, which were chosen in terms of their abundance, importance, technological implications, and particularly, as protein systems

Biomolecules from Natural Sources: Advances and Applications, First Edition. Edited by Vijai Kumar Gupta, Satyajit D. Sarker, Minaxi Sharma, María Elida Pirovani, Zeba Usmani, and Chelliah Jayabaskaran.

for vehiculization and delivery of BCs. Subsequently, albumin binding sites and the BCs that could be transported are considered. In addition, methods and techniques to produce albumin nanocarriers for different applications in biomedical, biocontrol, and bioprocess fields and/or to enhance bioavailability of BCs in food purposes, are summarized. In this sense, it is very important to note that nanosupplements production using albumins as potential carriers of BCs have emerged as a vacancy area.

5.2 Bioactive Compounds

A bioactive compound is a composite with the capacity and ability to interact with one or more components of living tissues and to produce a wide range of possible health effects. The origin of these substances can be natural, terrestrial, or aquatic (vegetable, animal, or other source; e.g., microorganisms) or synthetic: partially or totally (Guaadaoui et al. 2014). As can be deduced from this definition, there is a huge number of BCs, in addition they could be divided in different groups: vitamins and pro-vitamins (water and fat-soluble), lipids (saponifiable and unsaponifiable), phenolic compounds (such as flavonoids, tannins, phenolic acids among others), etc. Within this range of compounds with actions on human health, the hydrophobic ones, which are often sensitive to different environmental factors such as light and oxygen, are of particular interest. Furthermore, their encapsulation becomes a necessity to both solubilize and to protect them.

Vitamins A, D, E, and K and carotenoids are fat-soluble compounds that are naturally present in food or are used as excipients in different industrial fields, such as pharmaceuticals or cosmetics. Although carotenoids do not belong to the usual classification of vitamins, they are studied as such, because 50 of the 600 carotenoids were identified as pro-vitamin A. Vitamins are critical for the correct performance of many biological processes, including vision, maintenance of calcium balance, different roles in cell signalling, regulation of gene expression, and cell differentiation, among others (Bender 2003). Fat-soluble vitamins are sensitive to oxidation, temperature, light, free radicals, and oxygen; therefore, encapsulation could constitute an appropriate method to preserve their properties during storage and to improve their physiological properties (Luo et al. 2012; Teng et al. 2013). Encapsulation can also promote great efficiency of administration, allowing smaller doses and thus reducing possible hypervitaminosis syndromes and side effects (Gonnet et al. 2010).

Polyphenols are secondary metabolites of plants generally involved in defense against ultraviolet radiation (UV) or the attack of pathogens. Its molecular structure consists of several phenolic rings. There are many molecules that have a polyphenolic structure and that can be classified according to the number of phenolic rings and the structural elements that join these rings together. This group includes phenolic acids, flavonoids (which include anthocyanins), tannins, stilbenes, and lignans. Regarding the beneficial effects on health, polyphenols would play an important role in the prevention of various diseases associated with oxidative stress, inflammatory processes, cancer and cardiovascular, and neurodegenerative diseases (Manach et al. 2004). Therefore, there is a considerable interest in the use of these BCs as nutraceutical ingredients in food and beverage products. However, the incorporation of these BCs in commercial products is currently limited by their low solubility in water, instability to changes in pH, sensitivity to UV radiation and low oral bioavailability. In this sense, different protein-based delivery systems were developed to protect them against harsh conditions (Gaber et al. 2019; Gholijani et al. 2020; Kalouta et al. 2020).

Lipids are divided in two large groups: saponifiable and unsaponifiable. Saponifiable lipids are classified as simple lipids, such as fatty acids, acylglycerides and waxes, and complex lipids such as phospholipids and glucolipids. Unsaponifiable lipids include terpenes, steroids, and prostaglandins (Dergal 2006). Although the structures of lipids are different and complexes, in general, present double bonds, (poly)-unsaturation or cyclic rings that make lipids very sensitive to lipolysis, oxidation, and isomerization (Dergal 2006). Lipids perform many different functions including energy reserve (such as triglycerides), structural (such as phospholipids in bilayers), and regulatory (such as steroid hormones). Nevertheless, lipids have very low water solubility and they are susceptible to oxidative deterioration, which decreases its nutritional value. To overcome these problems, it is necessary to protect/encapsulate them by using efficient delivery systems.

One way to protect BCs is through their encapsulation, which is defined as a process of entrapment of a substance (active agent) in another substance (wall material or carrier), producing particles on the nanometric (nanoencapsulation), micrometric (microencapsulation) or millimetric scale (Ray et al. 2016). In this sense, the transport of BCs through the use of nanoparticles has several advantages: (i) protection of nutrients against degradation by oxidation or other chemical or enzymatic reactions during production and/or storage, which would prevent the development of undesirable flavors and odours and loss of bioactivity; (ii) minimization of adverse effects on some sensory characteristics such as transparency in certain beverages and masking of unpleasant attributes (Robles-García et al. 2016; Zimet and Livney 2009); (iii) better bioavailability due to the properties of controlled release, solubility, and prolonged residence time at the gastrointestinal tract and efficient cellular absorption of nutrients (Zimet and Livney 2009); (iv) controlled release of the encapsulated BCs (Anandharamakrishnan 2014), (v) surface modifying (for example, by coatings) which may allow site-directed delivery of BCs (Yu et al. 2015), etc. Moreover, there has been a growing interest in encapsulating, protecting, and releasing hydrophobic BCs in order to develop delivery systems based on colloids, because they cannot be synthesized by humans and, therefore, must be incorporated into the diet (Matalanis et al. 2011). In this sense, different encapsulation systems have been developed including protein particles, microemulsions, liposomes, hydrogels, and systems based on biopolymers. Food proteins show excellent properties for encapsulation of BCs. Usually, these are GRAS (Generally Recognized as Safe), biodegradable, and biocompatible materials. These are important properties to promote nutraceutical and pharmaceutical applications. Many globular proteins have the intrinsic ability to bind lipophilic compounds (Ferrado et al. 2020; Visentini et al. 2017a, 2017b, 2020) and this property could be used as a strategy to protect and introduce BCs into food or pharmaceutical matrices.

5.3 Serum Albumins

5.3.1 Structural Properties

Serum albumins are globular proteins showing attractive intrinsic properties for the development of delivery systems for BCs and drugs. Albumin is the most abundant plasma protein (52–60%) produced in the liver, its main physiological function being the regulation of colloidal osmotic pressure and the distribution of fluids in the body. In addition, due to its high binding ability, this protein transports a wide variety of ligands, such as fatty acids, non-conjugated bilirubin, porphyrins, steroids, hemin, aminoacids, metals, and other metabolites. It is a small and elliptical shaped protein in comparison with other plasma proteins such as fibrinogen and immunoglobulin G. These natural features, in addition to its high abundance, have been exploited in the past decades for biomedical applications, among others.

The main albumins used for biopharmaceutical applications are bovine and human serum albumins (BSA and HSA). BSA and HSA produce innocuous degradation products resulting from *in vivo* metabolism, so they were approved for the US Food and Drug Administration (FDA) for pharmaceutical applications. Moreover, the presence of charged functional groups, such as carboxylic and amino groups, become suitable for surface modification allowing their targeting and functionalization (Bhushan et al. 2017).

BSA is a globular protein of 65.5 kDa, composed by 585 aminoacids, with a pI ~ 4.7. Its secondary structure is composed of 67% α-helix, 10% turns, and 23% extended chain at physiological conditions. BSA has a heart-shaped tridimensional structure including three homologous domains (I, II, and III); each one composed of ten helices packed in two subdomains (A and B) with six and four α-helices, respectively, connected by an extended loop. It also contains 35 Cys residues, which form 17 intramolecular disulfide bonds, and a free sulfhydryl residue (at Cys_{34}) (Barbosa et al. 2010; Carter and Ho 1994; Peters 1996). Moreover, domains I and II contain Trp residues, Trp_{134}, and Trp_{212}, respectively. Trp_{212} residue is surrounded by a hydrophobic environment forming a protein binding pocket in subdomain IIA, while Trp_{134} is close to the protein surface in a hydrophilic environment at domain I (Tayeh et al. 2009). BSA intrinsic fluorescence is mainly attributed to Trp_{212}, since Trp_{314} is located in a more hydrophilic environment, so its fluorescence emission is quenched by water molecules (Peters 1985). Papadopoulou et al. 2005 studied the interaction between BSA and the flavonoids quercetin, rutin, catechin, and epicatechin by using intrinsic fluorescence spectroscopy. It was highlighted that quercetin showed the strongest binding affinity, while epicatechin and catechin presented the lowest affinity, which would suggest that the molecular structure of flavonoids could determine the BSA binding properties.

BSA is widely used in BCs delivery applications because of their abundance, low cost, ease of purification, and versatile binding properties, which produce great acceptance in the pharmaceutical industry. BSA-based nanoparticles can be used to load several drugs and BCs (Elzoghby et al. 2012). Besides, some surface functional groups (i.e., carboxylic and amino groups) on BSA-based nanoparticles can allow the covalent attachment of cell-targeting agents, such as apolipoproteins, surfactants, PEG, folate, transferrin, monoclonal antibodies, cationic polymers, etc. (Hao et al. 2013). However, in order to avoid immunological responses, BSA is usually replaced by its human analogue HSA.

HSA shares 76% sequence identity with BSA and its tridimensional structure is considered similar to this one. Because of its endogen origin, HSA is very safe and well tolerated in humans (Lee et al. 2016). Regarding this, it is relevant to remark that BSA is widely used as a model protein for research, while HSA is the material employed in final products. Usually, HSA is produced by human plasma fractionation, which is generally a limited source. For pharmaceutical applications, mainly for parenteral formulations, the most important requirement is to avoid the potential contamination risk of HSA by blood-derived pathogens. In this regard, an existent alternative is recombinant DNA technology to produce pharmaceutical grade recombinant HSA (rHSA) (Fanali et al. 2011). HSA non-glycosylated feature has made it possible to produce it by using a wide range of host organisms including bacteria, such as Escherichia coli and Bacillus subtilis, yeasts such as Pichia pastoris and S. cerevisiae, and transgenic plants and animals. Nevertheless, yeast- and rice-based expression systems represent the main options for rHSA production (He et al. 2011; Kobayashi 2006).

Both, BSA and HAS, were extensively studied in terms of their structure, stability, binding, and conformational properties at different conditions of the aqueous medium such as pH, temperature, ionic strength, and presence of denaturant agents. It is known that serum albumins undergo conformational changes at different pH conditions (Barbosa et al. 2010; Bhattacharya et al. 2011; Dockal et al. 2000; Li et al. 2008; Murayama et al. 2004; Sen et al. 2008) giving rise to conformational isomers, showing differences in flexibility and binding properties. Changes upon thermal treatment has also been studied highlighting reversible and irreversible transitions in protein structure at above 60 °C. These conformational changes involve an increase in α-sheet structure and protein aggregation (Barone et al. 1992; Boye et al. 1996; Giancola et al. 1997; Michnik et al. 2006; Takeda et al. 1989).

Since the binding affinity affects the rate at which BCs and drugs are released at catabolism sites, binding studies assaying serum albumins are of great importance in pharmacology and pharmacodynamics. Hence, the ability of BSA and HSA bind a wide variety of natural and synthetic ligands had been studied. In this regard, serum albumins have two binding sites, namely site-I and site-II, which are located at subdomains IIA and IIIA, respectively. Although they differ in their relative affinities, they appear to be homologous in both BSA and HSA (Naveenraj et al. 2013). Site I is able to bind ligands by hydrophobic interactions; while site II seems to be smaller, and its binding ability is strongly affected by ligand molecular stereoselectivity and hydrophobic, hydrogen bond, and electrostatic interactions. The interaction with well-known small drugs, such as gemcitabine, 5-fluorouracil, diazepam, ibuprofen, cisplatin, and others, was studied and characterized by means of spectroscopic techniques (Kandagal et al. 2006; Sulkowska 2002; Sulkowska 2002). Also, there has been a special interest in the knowledge of serum albumins ability to bind natural BCs such as polyphenols, mainly flavonoids (Dufour and Dangles 2005; Zhang et al. 2009). Regarding this, fluorescence quenching measurements are often used to obtain binding affinity constants and thermodynamic parameters. HSA has just one Trp residue (Trp_{214}) located at the subdomain IIA. It was shown that the molecular microenvironment of Trp_{212} in BSA is similar to the one for Trp_{214} in HSA. Information concerning binding properties is important in terms of loading capacity of serum albumin-based structures used for delivery purposes.

5.3.2 Protein-based Delivery Systems

Nowadays, the design and development of protein structures from serum albumins has gained importance in the development of biopolymer-based delivery systems (Jahanshahi et al. 2008; Loureiro et al. 2016). The major developed systems are aimed to improve the therapeutic efficacy of well-known drugs in biomedical applications by reducing adverse effects and required doses. Some of them are intended to replace excipients or formulation compounds that manifest problems after human administration.

The first developed methods to obtain serum albumin-based delivery systems emerged in the 1970s. These methods were focused on the production of albumin microspheres (as support materials for pharmacological agents) by using multiple strategies, e.g., heat-denaturation, chemical modification, and emulsification. Then, they could be thermal or chemically stabilized by crosslinking with 2,3-butadiene or formaldehyde (Kramer et al. 1974; Scheffel et al. 1972). In the following years, the technology advanced towards the development of colloidal nanoparticles, in which the active compound is trapped, encapsulated or adsorbed onto the particle surface. The difficulty of removing the solvents constituted the main disadvantage of this method, so others gradually replaced it. As well, they allowed more stable nanoparticles to be obtained showing well-defined and controlled sizes.

Tables 5.1 and 5.2 show a summary of BSA and HSA nanosized structures reported in recent years for different purposes, respectively. The most frequently used method to prepare BSA- and HSA-based nanoparticle is desolvation. This process constitutes a simple method, moreover no specialized equipment or complex operating conditions are required (Jafari 2017). Briefly, BSA/HSA is dissolved into an aqueous solution (at pH 5.5) and ethanol is dropwise added under stirring, until the solution becomes turbid. Depending on its solubility, BCs or drugs are previously solubilize in ethanol or albumin solution. BSA/HSA solubility is reduced during the ethanol addition, and it precipitates giving rise to nanoparticles. However, they are not sufficiently stable, so tend to dissolve after aqueous dispersion. Therefore, a stabilization step is usually required. The most common strategy to stabilize albumin-based nanoparticles is crosslinking using glutaraldehyde, since its aldehyde group promotes a condensation reaction with amino moieties in Lys residues and Arg moieties in albumin guanidine of side chains (Weber et al. 2000). After crosslinking, nanoparticles are maintained under stirring to evaporate the ethanol and then are centrifuged to separate solid nanoparticles from the non-encapsulated drug and to remove the glutaraldehyde. Solid nanoparticles can be freeze-dried or maintained in aqueous suspension (de Oliveira et al. 2018). However, the use of glutaraldehyde is controversial due to its well-known toxic and carcinogenic properties, so its removal is extremely crucial. In addition, it was reported that it could potentially react with amino groups of encapsulated drugs, which could affect their biological properties (Langer et al. 2008; Merodio et al. 2001). Variables that should be controlled to optimize the desolvation method are protein solution pH, desolvation agent amount, rate of addition, agitation speed, protein concentration, and ionic strength (Weber et al. 2000). Another disadvantage of this method is the aggregation of the formed nanoparticles giving rise to microspheres. A common strategy to modulate the interparticle colloidal interactions (e.g., promoting repulsive forces) and avoid aggregation phenomena is to control the protein solution pH. Recently, several authors have published a modified-coacervation method with the aim of obtaining nanoparticles with different characteristics. Kayani et al. (2018) obtained doughnut-shaped BSA nanoparticles of 100–225 nm to encapsulate doxorubicin, using a desolvation method with the addition of dimethylsulfoxide (DMSO).

The emulsification method is not usual to obtain BSA nanoparticles because organic solvents are required and they must be removed after nanoparticle formation. Briefly, a BSA aqueous solution is prepared and emulsified into an organic phase, comprised by organic solvents and/or lipids. Hydrophobic drugs can be added in the organic phase and hydrophilic compounds are added to the BSA solution. The emulsification step can be realized by sonication or through a high-pressure homogenizer, or a combination of both. Then, organic solvents must be evaporated under vacuum, and the nanoparticles are centrifuged to separate the non-encapsulated drug from the solid nanoparticles. An advantage of this process is the absence of a crosslinking agent. Some authors have emphasized that when using high-pressure homogenization, the shear forces could induce the formation of new disulfide bonds, through oxidation and crosslinking of sulfhydryl groups in BSA (Zhao et al. 2015). In the emulsification method, the main process variables influencing mean size and size distribution are BSA concentration, ratio between aqueous and organic phases, and rate and time of homogenization, as well as the homogenization pressure and the number of cycles (Zu et al. 2013).

Other alternative methods that have emerged in this field to prepare albumin-based nanoparticles are self-assembly, thermal gelation, and nano-spray-drying. Self-assembly methods consist of covalent addition of a hydrocarbon chain to serum albumin, which then allows the formation of a core-shell structure or protein micelle. Gong et al. (2009) added an octyl group to BSA to obtain a micellar structure by chemical reaction which were loaded with paclitaxel. Following the same method, Xu et al. (2011) used HSA to prepare micellar structures with surface modifications in order to encapsulate doxorubicin. On the other hand, Ferrado et al. (2019) obtained BSA nanoparticles by protein self-aggregation induced by thermal treatment at different temperatures, pH and protein concentration in order to promote the vehiculization of the flavonoid chrysin, which antitumor properties are well known. Different populations of BSA nanoparticles were obtained, showing different diameters (13–28 nm), ζ potentials around − 10.0 mV, molecular weight (400–1,000 kDa), and spherical shape. Recently, the elaboration of a chrysin nanosupplement based on BSA nanoparticles (BSAnp) was reported (Ferrado et al. 2020). Their *in vitro* cytotoxic activity was assayed on MCF-7 and MDA-MB-231 cell lines (as models of cancer cell lines) revealing that the greatest cytotoxic effect was observed for the smallest sized chrysin-loaded BSA nanoparticles (obtained at pH 11.0 and 70 °C). Finally, this system was subjected to static *in vitro* gastrointestinal digestion using a standardized protocol, resulting in ~14% chrysin released from BSAnp after digestion. This evidence would suggest that smaller sized chrysin-loaded BSA nanoparticles could be a promising oral delivery system for potential cancer therapies.

Nano spray-drying is a recent encapsulation technique used for trapping active materials into an inert matrix (Arpagaus et al. 2018). Nano spray-dryer involve a vibrating mesh technology for producing particles of good quality and ultrafine size. The main advantage is that the particle formation occurs in a single step and the process is scalable. The process is based on the following steps: (i) atomization of the sample (protein + drug + surfactant); (ii) spray air contact (hot drying gas at enough temperature for solvent evaporation); (iii) spray drying (as the solvent is evaporated, the dried powder is formed); and (iv) separation of the dried powder from the drying air. In a simple protocol, a BSA solution containing the BC and a surfactant is prepared and spray dried. Parameters such as inlet temperature, pressure, gas flow rate, and mesh size must be previously defined. The dried powder is isolated from the particle collecting electrode using a particle scraper. Lee et al. (2011) obtained BSA nanoparticles by nano spray-drying and the particles presented 460 nm using the 4 μm spray mesh at 0.1% (w/v) BSA concentration, 0.05% (w/v) surfactant concentration, drying flow rate of 150 L min^{-1}, and inlet temperature of 120 °C. In general, BSA nanoparticle have sizes of < 300 nm, low PDI, spherical morphology, and negative α potential.

Finally, it is important to mention the research aimed at optimizing nanoparticle production processes in order to control parameters such as size, loading, and stability. In addition, surface modification with ligands to improve drug water solubility, to avoid immune response or to target action sites, is an important topic of research. Nowadays, the use of folate (Akbarian et al. 2020; Borah et al. 2020; Dong et al. 2019; Gawde et al. 2018; Lee 2019; Lei et al. 2017; Nosrati et al. 2018b; Qi et al. 2014; Song et al. 2016; Zhao et al. 2010) and polyethilenglycol (PEG; Geng et al. 2017; Saha et al. 2020; Xu et al. 2011) for targeting and avoiding immune response, respectively, is very common. From this analysis, it can be noted that the most frequently used preparation methods include desolvation, yielding EE values higher than 84% and 77% for HSA and BSA, respectively.

Encapsulation efficiency values (EE) reported for different BCs and drugs encapsulated by using BSA- and HSA-based nanoparticle systems are also shown in Tables 5.1 and 5.2. It is well known that EE depends on several factors, such as the nature and composition of nanoparticle systems (encapsulation matrix or carrier), drug chemical nature, loading method, and the medium conditions in which the nanocarrier is obtained. For this reason, different methods to determine EE were reported. The most common method for hydrophobic BC encapsulated into protein nanoparticles, is the indirect determination of the entrapped BC amount. Usually, BC-loaded nanoparticles are separated from unloaded free BC by ultrafiltration or centrifugation processes, and the free BC amount is extracted with a solvent from the supernatant, being its concentration determined by HPLC or UV-Vis spectroscopy. According to literature, EE values of albumin-based systems as carriers for BCs and drugs is generally higher than 60%. The BC quantification related to therapeutic proteins and monoclonal antibodies, is usually performed by a specific ELISA assay. In several cases, the direct determination is performed through a nanoparticle disruption method (using solvents or proteases), that release the whole quantity of the encapsulated BC followed by the HPLC quantification of released BC.

5.3.3 Antitumor Properties

From Tables 5.1 and 5.2 it can be deduced that the most important application of BSA and HSA nanoparticles is for biomedical purposes, specifically as delivery systems for BCs with cytotoxic or antitumor capacity. In this sense, in 2005, the FDA approved the commercialization of nanoparticles of albumin-bound paclitaxel (known as nanoparticle albumin-bound or nab-technology) under the brand name Abraxane®, for the treatment of metastatic breast cancer (Desai et al. 2006). This formulation allowed the administration of paclitaxel without cremophor EL, a formulation component that cause several adverse effects, such as allergic reactions and hypersensibility. Abraxane was the

Table 5.1 Encapsulation of BCs and drugs using BSA as a carrier.

Detailed application	Method	Size	Complementary material	Effect	Active material	Encapsulation efficiency (EE)	Reference
Antiviral	Coacervation	200–400 nm	None	Sustained release for treatment of cytomegalovirus	Ganciclovir	12.5 – 40.6 %	Merodio et al. 2001
Immunological	Desolvation	300–350 nm	None	Controlled release	Interferon gamma	85.01%	Segura et al. 2005
Antitumor	Desolvation	210 nm	None	Sustained release	5-fluorouracil	27.5–30.0%	Maghsoudi et al. 2008
Antitumor	Self-assembly	160–180 nm	Octaldehyde	Improved water solubility	Paclitaxel	90.5%	Gong et al. 2009
Antitumor	Conjugation	90 nm	Dextran	Decreased toxicity. Positive effect on the survivability of murine ascites hepatoma H22 tumor-bearing mice	Doxorubicin	74.4–97.2%	Deng et al. 2010
Antitumor	Desolvation	210 nm	Folate	Improved water solubility. Targeting cancer cell	Paclitaxel	95.3%	Zhao et al. 2010
Antioxidant	Modified desolvation	<10 nm	None	Improved antioxidant activity	Quercetin	Not reported	Fang et al. 2011
Antitumor	Desolvation	200 nm	None	Improved water solubility. Increased bioavailability and tissue targetability	Curcumin	86.42%	Jithan et al. 2011
Antioxidant Antitumor	Desolvation	186–300 nm	Chitosan or poly-Lys	Increased NPs stability in acid and basic pH. Increased permeability through Caco-2 cells	EGCG	32.3–35.4%	Li et al. 2014

(Continued)

Table 5.1 (Continued)

Detailed application	Method	Size	Complementary material	Effect	Active material	Encapsulation efficiency (EE)	Reference
Antitumor	Desolation Folate conjugation	195.3 nm	Folate Glutaraldehyde	Controlled and sustained release pattern. Enhanced cancer cell apoptotic effect and improved cellular internalization process for folate receptor positive MCF-7 cells.	Bexarotene	EE = 65.45% LC = 1.73%	Qi et al. 2014
Antioxidant Antitumor	Desolvation	45–48 nm	None	Improved thermostability, antioxidant and cytotoxic properties	Catechin Epicatechin	54.5–60.5%	Yadav et al. 2014
Antitumor	Nab-technology	250 nm	Paclitaxel (co-loaded)	Improved water solubility	Curcumin	88.6–96.5%,	Kim et al. 2016
Antioxidant	Desolvation	130 nm	None	Increased antioxidant activity	Quercetin	85%	Antônio et al. 2016
Antioxidant Antitumor	Desolvation	285–2942 nm	Chitosan	Reduced cellular oxidative damage in mice and expression of proapoptotic protein in mice	Green tea polyphenols	25.7–26.1%	Kumar et al. 2016
Stability Controlled released	Hydrogel	200–300 nm	Citrus peel pectin	Enhanced Vitamin C retention during storage and produced controlled released	Vitamin C	EE = 65.31%	Peng et al. 2016
Bioavailability	Conjugation Emulsification	158-255 nm	Dextran	Increase oral bioavailability in mice	Curcumin	98.4–99.5%	Wang et al. 2016
Antitumor	Desolvation	83 nm	None	Redox responsive delivery	Cisplatin	~100%	Catanzaro et al. 2017
Antioxidant	Coacervation	175 nm	None	Controlled scavenging activity	Resveratrol	60%	Fonseca et al. 2017

Antitumor	Micelles Core-shell NPs	<100 nm	2-deoxy-D-glucose (DG); amphiphilic starch; Folic Acid	NPs targeted folate receptor highly expressed on tumor cell *in vitro* and *in vivo*. *In vivo*, improved cytotoxicity with less toxicity	α-tocopheryl succinate	EE = 94.3% LC = 8.9%	Lei et al. 2017
Stability	Coacervation	204 nm	Poly-Lys Glutaraldehyde	Improved stability under storage	Curcumin	47–60%	Maldonado et al. 2017
Antioxidant Anti-inflammatory	Nab-technology	376 nm	None	Improved water solubility, antioxidant capacity and anti-inflammatory properties	Apigenin	82.6%	Pápay et al. 2017
Antitumor Anti-inflammatory	Desolvation	157 nm	None	Enhanced anticancer ability and anti-inflammatory behaviour	Ginsenoside compound K	-	Singh et al. 2017
Antitumor	Desolvation	180 nm	None	Increased stability and cytotoxicity over rhabdomyosarcoma cells	Citral	94%	White et al. 2017
Stability	Self-assembly	-	None	Improved stability of all BC. Protective effect after 26 days storage	Retinol Resveratrol EGCG	Not reported	Wu et al. 2017
Antiparasitic	Desolvation	180 nm	None	Anti-leishmanial efficacy (the same efficacy of commercial drug) and reduced toxicity	Amphotericin B	98%	Casa et al. 2018
Antioxidant	Antisolvent	-	None	Increased antioxidant activity	cis and trans resveratrol	Not reported	Cheng et al. 2018
Antioxidant	Self-assembly	<200 nm	Dextran	Improved stability and increased antioxidant activity	Curcumin	Not reported	Fan et al. 2018

(Continued)

Table 5.1 (Continued)

Detailed application	Method	Size	Complementary material	Effect	Active material	Encapsulation efficiency (EE)	Reference
Antitumor	Modified desolvation	146–198 nm	Folic acid	Increased cytotoxicity on tumor cells lines	Curcumin analog	78%	Gawde et al. 2018
Antitumor	Desolvation with addition of DSMO	100-225 nm	None	Improved water solubility, cytotoxicity and therapeutic effect on non-resistance and multi drug resistant cancer cells.	Doxorubicin	77–85%	Kayani et al. 2018
Gene delivery	Conjugation	250 nm	Oligochitosan	Improved transfection efficiency and reduced cytotoxicity	Plasmid EGFP-C3 (size 4.7 kb)	Not reported	Kumari et al. 2018
Antitumor	Desolvation and conjugation	127 nm	Folic Acid	Increased biocompatibility and cytotoxicity on tumor cells lines	Chrysin	Not reported	Nosrati et al. 2018a
Antitumor	Desolvation	97.5 nm	None	Sustained and controlled release of chrysene. Biocompatible and cytotoxic	Chrysin	Not reported	Nosrati et al. 2018b
Antitumor	Conjugation	120 nm	β-carotene (BC, crosslinking agent)	Improved tumor suppression effect when irradiated by 660 nm light	Chlorin e6	Ce6: 96.8% BC: 84.4%	Phuong et al. 2018
Antitumor	Coacervation	92 nm	None	Improved water solubility and cytotoxicity	Curcumin	78.12%	Salehiabar et al. 2018

Stability	Bilayer emulsion	221.27 nm	Arabic Gum Medium-chain triglyceride oil	Improved chemical stability under different environmental stresses (pH, thermal treatment, UV radiation, oxidants and storage time)	β-carotene	EE = 94.35%	Sheng et al. 2018
Antitumor	Desolvation Ultrasonication	106.9 nm	None	Improved cell uptake of paclitaxel, cytotoxicity, and anti-cancer efficacy in MCF-7/ADR tumor xenograft model	Vitamin E Paclitaxel	EE = 95.7% LC = 12.5%	Tang et al. 2018
Antioxidant Controlled released	solid lipid-polymer hybrid NPs	139–180 nm	Oxidized dextran	Solid lipid-polymer hybrid NPs enhanced antioxidant activity and showed a sustained release in simulated gastrointestinal fluids	Astaxanthin	EE = 70%	Wang et al. 2018
Antitumor	Desolvation and conjugation	201.4 nm	Folic Acid	Improved anticancer activity *in vitro* and *in vivo*	Ginsenoside Rg5	EE = 74% DL = 64 %	Dong et al. 2019
Antimicrobial	AgNPs encapsuled in BSA	19–31 nm	AgNPs	Improved thermal stability and antimicrobial efficacy against gram positive and negative bacterial strains	Ag	Not reported	Jose and Kuriakose 2019
Pharma (neural regeneration)	Desolvation and microfluidics	315.6 nm	Formaldehyde	Improved TQ thermal stability and decreased its toxicity. Enhanced head regeneration of planarians (model of neural regeneration)	Thymoquinone (TQ)	Not reported	Kazan et al. 2019

(Continued)

Table 5.1 (Continued)

Detailed application	Method	Size	Complementary material	Effect	Active material	Encapsulation efficiency (EE)	Reference
Pharma (eye delivery)	Complexation	~ 265 nm	Oleic acid (OA)	Improved intestinal absorption of FUC. After 15 days treatment, only FUC accumulation was discovered in eyes and the ocular antioxidant capability increased by 71.02%.	Fucoxanthin	EE = 83.7%	Liu et al. 2019
Antioxidant Stability	Desolvation	96 nm	None	Increased antioxidant properties Reduced photolytic degradation	Coumestrol	22.4%	Montero et al. 2019
Bioavailability	Desolvation	325.6 nm	Glutaraldehyde	Sustained release. Enhanced plasma concentration and prolonged drug residence time in systemic circulation. There was no evidence of toxicity in rats studies	EGCG	EE = 92%	Ramesh and Mandal 2019
Solubility Stability	Complexation	20 nm	None	Improved resveratrol solubility and stability in aqueous medium	Resveratrol	Not reported	Tabibiazar et al. 2019
Antifungal	Nanoemulsification	179.75 nm	Corn oil	Nanoeulsion showed high physical stability during 28 days' storage. Improved inhibitory activity in terms of mycelial growth and mycotoxin production.	Thyme oil	Not reported	Wan et al. 2019

Antioxidant	Complexation	-	None	Tri-ligand complexes maintain structural integrity and antioxidant property of all BCs simultaneously and to a greater degree than is observed for mono-ligand complexes	α-tocopherol Resveratrol EGCG	Not reported	Dong et al. 2020
Antitumor	Desolvation and cross-linking	BSA NP: 189 nm	None	Reduced cytotoxicity of 2,6-divanillylidenecyclohexanone in normal cell lines, and increased the antitumor activity without necrosis.	2,6-divanillylidenecyclohexanone	EE = 40.20% DL = 0.40–0.64 %	Fattahian Kalhor 2020
Antitumor	Thermal treatment	9–25 nm	None	Improved water solubility and cytotoxicity	Chrysin	44–84%	Ferrado et al. 2020
Pharma	Desolvation	148 nm	Glutaraldehyde	Reduced rheumatoid arthritis signs and release of nitric oxide and IL-17 inflammatory cytokine	Carvacrol	EE = 67.7% LC = 26.6%	Gholijani et al. 2020
Pharma	Complexation	99 nm	Palmitic acid	Targeted activated macrophages. Improved rheumatoid arthritis symptoms with fewer toxic effects.	Celastrol	EE = 94.4% LC = 2.6%	Gong et al. 2020
Antitumor	Core-shell nanogels	BSA-FA-AP: 89.5 nm BSA-AP: 105.58 nm	Folic acid Amylopectin	Nanogels were resilient to *in vitro* physiological oral-gastrointestinal digestion. Showed augmented cellular uptake and specific retention in folate receptor-positive HT29 cells, inducing apoptosis.	Curcumin	EE = 100%	Borah et al. 2020

(Continued)

Table 5.1 (Continued)

Detailed application	Method	Size	Complementary material	Effect	Active material	Encapsulation efficiency (EE)	Reference
Bioavailability	Self-assembly	78–128 nm	Oleic acid	Improved bioavailability	Astaxanthin	78.8–97.5%	Liu et al. 2020a
Controlled release	Desolvation and cross-linking	BSA-C NPs: 100 nm BSA-LD NPs: 65 nm	Glutaraldehyde	Controlled release of both drugs	L-dopa (LD) Catechol (C)	LD; EE = 88.1%; LC = 93.4 C: EE = 83.6%; LC = 89.7%	Rafipour et al. 2020
Anti-inflammatory	Self-assembly	˂200 nm	PPG-PEG-PPG Triblock copolymer	Increased stability and solubility Sustained prolonged release Potent anti-inflammatory activity	Quercetin	39–89%	Saha et al. 2020
Biocompatibility	Self-assembling	-	Carbon nanotubes (CNTs)	Strong free radical scavenging activity. Reduced cytotoxicity of CNTs and reactive oxygen species in endothelial cells.	Rutin	Not reported	Tian et al. 2020
Antitumor Controlled released	PLGA NPs coated with BSA	216 nm	Poly (lactic-co-glycolic) (PLGA)	Improved release profile and cellular uptake and reduce cytotoxicity in normal cells. Improved therapeutic efficacy of quinacrine	Quinacrine	EE = 49%	Vaidya et al. 2020

Stability Bioavailability	Alkaline method and emulsification	220 nm	Dextran Chlorogenic acid Corn oil	Improved physicochemical stability under different conditions and long-term storage. Improved bioaccessibility of Lutein under gastrointestinal fluids	Lutein Vitamin E	Not reported	Yan et al. 2020
Antitumor	Anti-solvent precipitation	150 nm	Paclitaxel (co-loaded)	Efficiently internalized by cells and exerted synergistic combination efficacy of both drugs resulting in dramatically in vitro cytotoxicity	Resveratrol	EE = 98.24% LC = 1.54%	Zhao et al. 2020

EE: encapsulation efficiency; EGCG: Epigallocatechin-3-gallate; LC: loading capacity; NP: nanoparticle

Table 5.2 Encapsulation of BCs and drugs using HSA as a carrier.

Detailed application	Method	Size	Complementary material	Effect	Active material	Encapsulation efficiency EE%	Reference
Antitumor	Nab-technology	130 nm	None	Improved water solubility and bioavailability	Paclitaxel	Not reported	Desai et al. 2006
Pharma	Desolvation	200 nm	Apolipoprotein E	Improved bioavailability. Facilitated transport of drugs across the blood-brain barrier	Loperamide	92.7 μg/ mg HSA	Michaelis et al. 2006
Antitumor	Desolvation	150–500 nm	None	Improved water solubility. Increase anticancer effects	Doxorubicin	EE > 95%	Dreis et al. 2007
Antitumor	Nab-technology	130–150 nm	None	Enhanced water solubility, increased *in vivo* accumulation in tumors and binding and transport in vascular endothelial cells	Curcumin	LC = 3.1-14.1%	Kim et al. 2011
Antitumor	Self-assembling Conjugation	30 nm	Peptide RGD Polyethylene glycols	Targeting delivery. Improved uptake and retention in human melanoma cells	Doxorubicin	LC = 56.2–62.3%	Xu et al. 2011
Antibacterial Antitumor	Nanoprecipitation Conjugation	126–174 nm	poly(lactide-co-glycolide)	Controlled released. Improved antibacterial effects against *Staphylococcus aureus* and *epidermidis* and *Pseudomonas aeruginosa*. Improved cytotoxicity in Hep G2 cells.	18-β-glycyrrhetinic acid	EE = 89.4%	Darvishi et al. 2015

Antitumor	Desolvation	200–220 nm	None	Targeting delivery of plasmid	Plasmid	-	Look et al. 2015
Antitumor	Desolvation (D) and emulsion / desolvation (ED) methods	D: 132 nm ED: 230 nm	Glutaraldehyde	Carvacol was released more slowly from NPs that were prepared by the D method than by the DE method. NPs showed antigastric cancer activity	Carvacrol	D: EE = 48.4%; LC = 45.1% DE: EE = 32%; LC = 32.3%	Maryam et al. 2015
Controlled released	Desolvation Covalent amide bond formation	117–180 nm	Polyethylenimine	Sustained released of GA. Protecting effect of NPs from the toxic effects of high doses of GA over rat pheochromocytoma cell line.	Gallic acid (GA)	EE = 62.05–29.5 % LC = 13.3–41.1 %	Mohammad-Beigi et al. 2015
Stability	Complexation	-	None	Improved catechins stability at 25°C	Epicatechin Epigallocatechin Epicatechingallate EGCG	Not reported	Zinellu et al. 2015
Antioxidant Antitumor	Desolvation	220 nm	None	Increases solubility, selectivity and cytotoxicity	Fisetin	EE = 84%	Ghosh et al. 2016
Controlled released Antitumor	Emulsion/ desolvation method and conjugation	165.6 nm	Folate	Sustained release, increase retention time *in vivo*, inhibition of HT 29-derived tumors without significant reduction in mice body weight.	Curcumin	EE = 88.7% LC = 7.9%	Song et al. 2016

(Continued)

Table 5.2 (Continued)

Detailed application	Method	Size	Complementary material	Effect	Active material	Encapsulation efficiency EE%	Reference
Antitumor	Desolvation Cross-linking	120 nm	Peptide RGD Polyethylene glycol	Promoted efficient cell uptake, prolonged circulation time and showed cytotoxic effect by apoptosis, biocompatibility and no significant systemic toxicity *in vivo*.	Resveratrol	EE = 62.5%	Geng et al. 2017
Antitumor	Solvent-displacement	169 nm	Olive oil Glutaraldehyde	Displayed better cell-uptake performance and showed better cytotoxicity	Curcumin	Not reported	Galisteo-González et al. 2018
Pharma	Desolvation	310 nm	Glutaraldehyde	Improved biodistribution of Bevacizumab in eyes with no effects on epithelial cells	Bevacizumab	EE = 89%	Luis de Redín et al. 2018
Antitumor	Reduction	122-205 nm	Glutathione as reducing agent	Improved cellular uptake, cytotoxicity, water solubility and accelerated release in response to inside reducing environment of cells.	Curcumin	Not reported	Saleh et al. 2018
Controlled released	Desolvation	77.38 nm	Polysorbate 80	Controlled released and good stability of NPs	Irinotecano	EE = 79% LC = 9.6 %	Taneja et al. 2018

Antitumor	Desolvation + GT Self-assembled + DTT	<200 nm	Glutaraldehyde (GT) Dithiothreitol (DTT)	Self-assembled NPs showed more cumulative drug release than desolvated NPs, causing them to be more cytotoxic on MCF-7 cells.	Piperine	Self-assembled: EE = 76.8%; LC = 8.92 Desolvation: EE = 34%; LC = 1.7 %	Abolhassani and Shojaosadati 2019
Antitumor	Oil core- HSA shell NPs conjugated with APBA	172.8 nm	Exemestane Phosphatidyl cholin Lauroglycol 90 3Aminophenylboronic acid (APBA)	Improved storage stability. Enhanced internalization into MCF-7 breast cancer cells and cytotoxic activity. *In vivo*, displayed superior anti-cancer activity in tumor-bearing mice.	Hesperetin	EE = 95.2% LC = 1.98%	Gaber et al. 2019
Pharma	Nanoemulsification Complexation	78.5 nm	Solutol HS 15 Soybean oil	Showed superior therapeutic effects and significantly improved safety profile in rheumatoid arthritis in mice	Celastrol	EE = 94% LC = 2.4%	Gong et al. 2019
Antitumor	Conjugation by EDC/NHS chemistry	-	Folate	Enhanced cellular uptake and cytotoxicity	Transforming growth factor-β1 antibody Methotrexate	Not reported	Lee 2019
Antitumor	Nab-technology	138.6 nm	Oleic acid	Enhanced uptake by cancer cells line and cytotoxicity. In tumor-bearing mice, NPs demonstrated an increased and lasting tumor distribution.	Gefitinib	EE = 85% LC = 6%	Pang et al. 2019

(Continued)

Table 5.2 (Continued)

Detailed application	Method	Size	Complementary material	Effect	Active material	Encapsulation efficiency EE%	Reference
Pharma	Desolvation	241–261 nm	Chitosan	Improved interaction between chitosan and rabbit nasal mucosa promoting higher penetrating potential of NPs	Sulforhodamine B sodium salt	63%	Piazzini et al. 2019
Antitumor	Desolvation	APT-HSA-CUR: 281.7 nm	HER-2 Aptamer (APT)	Increased solubility and stability. Improved long term stability, sustained drug release and higher cytotoxicity	Curcumin	EE = 71% LC = 3.4%	Saleh et al. 2019
Pharma	Desolvation Nanoprecipitation	180 nm	None	Controlled released Rate of wound healing was increased	Vitamin C	EE = 52.1%	Shojania et al. 2019
Antioxidant Antitumor	Electrostatic	188.3 nm	None	Reduced lipopolysaccharide-induced oxidative stress, restore cell viability and elevate superoxide dismutase and catalase activity *in vitro* and *in vivo*	Silymarin	EE = 97%	Sohrabi et al. 2019
Antitumor	Desolvation Conjugation	Folate-ARM-HSA: 198 nm	Folate	Improved cytotoxicity and drug cellular uptake. Prolonged the retention time of ARM and reduce possible drug side effects on normal cells.	Artemether (ARM)	EE = 77% LC = 6%	Akbarian et al. 2020

Controlled released	Salting out Electrostatic deposition	HSA-IFN-α-PSS/Chit/PS: 825 nm	Poly(sodium-4-styrenesulphonate) (PSS) Chitosan (Chit)	Both *in vitro* and *in vivo* release kinetics experiments have proven slow active agent dissolution from the formulation.	Interferon-α (IFN-α)	EE = 49% LC = 76%	Kristó et al. 2020
Pharma	Oil-in-water single emulsion-solvent evaporation	220 nm	Tannic acid Genipin	NPs delayed CUR release in simulated gastric fluid, prolonged its colon adhesion and increased its uptake in Caco-2 cells. NPs oral administration alleviated colitis symptoms in mice.	Curcumin (CUR)	EE = 86% LC = (material supleentario)	Luo et al. 2020

EE: encapsulation efficiency; EGCG: Epigallocatechin-3-gallate; LC: loading capacity; NP: nanoparticle

first approved medication based on nanoparticles derived from biomacromolecules (previously liposomal drugs had been approved in the form of polymer microspheres), this fact promoted the increase of research in this field. For example, Kim et al. (2016) prepared BSA nanoparticles co-loaded with paclitaxel and curcumin, and Desai et al. (2006) developed HSA nanoparticles to encapsulate paclitaxel by this method. Other methods had been used to produced paclitaxel-BSA nanoparticles as desolvation (Zhao et al. 2010) and self-assembly (Gong et al. 2009).

Another BC extensively studied as an antitumor agent encapsulated in BSA or HSA nanoparticles is curcumin. Curcumin is a natural phenolic compound in the plant Curcuma longa, but their low water solubility and deterioration due to injurious conditions such as light and pH led to the development of different encapsulation systems. In this sense, Jithan et al. (2011) and Kim et al. (2016) produced BSA nanoparticles to encapsulate curcumin with EE > 85% and improved water solubility, bioavailability, and cytotoxicity. Other examples are the production of BSA nanoparticles (Gawde et al. 2018) and nanogels (Borah et al. 2020) coated with folic acid to encapsulate curcumin. These nanovehicles showed improved cytotoxicity, increased cellular uptake and specific retention in folate receptor-positive cancer cells line as human colon adenocarcinoma cell line HT29. Kim et al. (2011) and Song et al. (2016), develop HSA-based nanoparticles with size < 170 nm to encapsulate curcumin, improving its water solubility. Moreover, these nanoparticles increased retention time and *in vivo* accumulation in tumors and inhibited HT29 derived tumors in mice. HSA nanoparticles were produced by a desolvation method and decorated with HER-2 Aptamer via the formation of covalent amide bond between amine groups of aptamer and carboxyl groups of HSA nanoparticles (Saleh et al. 2019). The final nanoparticles had a size of 281.7 nm, increased solubility and stability of curcumin, produced a sustained drug release and higher cytotoxicity.

Chrysin is another BC with antitumor activity encapsulated in BSA nanoparticles. Nosrati et al. (2018a, 2018b) developed BSA nanoparticles and folate-BSA nanoparticles to encapsulate chrysin. These systems produced a sustained and controlled released, increased biocompatibility (tested in human fibroblast cell line) and cytotoxicity (tested on human breast cancer cell line MCF-7). In other research, BSA nanoparticles were developed by heat treatment at different temperatures with sizes of 9–25 nm (Ferrado et al. 2020). The nanoparticles improved chrysin water solubility and showed cytotoxic activity in hormone-dependent (MCF-7) and hormone-independent (MDA-MB-231) cancer cell lines. Other phenolic compounds with antitumor activity, such as green tea polyphenols (Kumar et al. 2016), resveratrol (Zhao et al. 2020), etc., were studied to be encapsulated in BSA nanoparticles. In the case of HSA, different phenolic compounds with antitumor activity were studied. HSA-carvacrol (Maryam et al. 2015), HSA-fisetin (Ghosh et al. 2016) and HSA-resveratrol (Geng et al. 2017) nanoparticles were produced by the desolvation method with or without glutaraldehyde crosslinking. These nanoparticles had sizes between 120 and 220 nm and EE higher than 40%. All these nanoparticles improved water solubility, stability, andcytotoxicity of BCs. Another phenolic compound encapsuled in HSA was silymarin (Sohrabi et al. 2019). The nanocomplex was produced by an electrostatic interaction between negatively charged silymarin solution (KOH, pH 13) and HSA solution with positive charge distribution (acetic acid, pH 2.95). Equal volumes of the silymarin were added dropwise into the HSA solutions with a stirring rate of

300 rpm. The resultant silymarin-HSA nanocomplex solution was then centrifuged and the pellet was re-suspended in deionized water (Sohrabi et al. 2019). These nanocomplexes reduced lipopolysaccharide-induced oxidative stress, restore cell viability and increased superoxide dismutase and catalase activity *in vitro* and *in vivo*.

Vitamin E and α-tocopherol were also studied in relation to their antitumor activity. Tang et al. (2018) developed 160.9 nm BSA nanoparticles by the desolvation method for vitamin E and paclitaxel co-delivery. Compared with paclitaxel nanoparticles, paclitaxel-vitamin E nanoparticles exhibited higher cellular uptake and stronger cytotoxic effect in the breast cancer cell lines MCF-7 and MCF-7/ADR. However, paclitaxel is the active antitumor drug; vitamin E acts as a P-glycoprotein inhibitor improving the intracellular drug uptake and the therapeutic efficiency on resistant tumors (Tang et al. 2018). Non-toxicity is the main advantage of vitamin E in comparison with other P-glycoprotein inhibitors such as verapamil and tariquidar. Lei et al. (2017) developed a nanoparticle which consisted of a core of cationic amphiphilic starch, α-tocopheryl succinate and 2-deoxyglucose coated with a BSA-folic acid complex. These nanoparticles targeted folate receptors highly expressed on tumor cells *in vitro* and *in vivo* assays. Furthermore, *in vivo* anti-tumor experiments showed that nanoparticles had an improved biological function with less toxicity. α-tocopheryl acts as an inhibitor of mitochondrial oxidative phosphorylation, a metabolic pathway used to satisfy energy and biomass synthesis requirement by tumors (Lei et al. 2017). This research showed the versatility of BCs, since they can be used because of their own cytotoxic activity or as part of a system that enhances the efficacy of a drug.

Terpenes are BCs that have gained interest as cytotoxic compounds. They are unsaponifiable lipids, formed by two or more isoprene units (2-methyl-1,3-butadiene). Terpenes can be linear or cyclic molecules and some of them contain structures of both types. The terpenoid citral was encapsulated in BSA nanoparticles prepared by the desolvation technique. Nanoparticles had a size of 180 nm and EE of 94% and increased stability and cytotoxicity (White et al. 2017). HSA was used to encapsulate triterpenoid 18-β-glycyrrhetinic acid (Darvishi et al. 2015) and sesquiterpene lactone artemether (Akbarian et al. 2020). Darvishi et al. 2015 produced nanoparticles based in modified HSA (by introduction of thiol group) conjugated to PLGA to encapsulate 18-β-glycyrrhetinic acid. The nanoparticles produced a controlled release profile and improved cytotoxicity in a human liver cancer cell line Hep G2. Moreover, improved antibacterial effects against Staphylococcus aureus and epidermidis and Pseudomonas aeruginosa. Akbarian et al. 2020 produced HSA nanoparticles by the desolvation method to encapsulate artemether and folate using coupling agents NHS/EDC to activate the carboxyl groups of the folic acid and to form an amide bond with amine groups of HSA with a size of 198 nm. The nanoparticles improved cytotoxicity and BC cellular uptake in a breast cancer cell line (MDA-MB-231), prolonged the retention time and reduced possible artemether side effects on normal cells. In a recent study, two HSA nanoparticles were developed to encapsulate piperine, an alkaloid from Piper longum (Abolhassani and Shojaosadati 2019). The HSA nanoparticles were produced by two methods: desolvation followed by glutaraldehyde crosslinking and self-assembling by dithiothreitol reduction. Both nanoparticles sized less than 200 nm, but the piperine EE of the self-assembled HSA nanoparticles was superior (77 vs. 34%). Moreover, self-assembled nanoparticles showed more released piperine than the desolvation method, causing them to be more cytotoxic on breast cancer cell line MCF-7 (Abolhassani and Shojaosadati 2019).

5.3.4 Antioxidant Properties

Table 5.1 shows that antioxidant properties are important applications of BSA nanoparticles. Regarding to this, green tea polyphenols quercetin, epiogallocatechin-3-gallate (EGCG), catechin and epicatechin are extensively studied. Fang et al. (2011) and Antônio et al. (2016) produced BSA nanoparticles by the desolvation technique in order to encapsulate quercetin. The drug release profile of quercetin displayed that a biphasic pattern and 18% of quercetin was released after approximately 96 h. In addition, quercetin-BSA nanoparticles showed more effectiveness than free quercetin for scavenging radical ABTS · + and hypochlorous acid, after 96 h (Fang et al. 2011). Antônio et al. (2016) demonstrated that BSA nanoparticles have long-term antioxidant protection effects on the activity of quercetin in both, acidic and neutral conditions. Moreover, Kumar et al. 2016 developed HSA nanoparticles, by desolvation coated with chitosan, capable of reducing the loss of green tea polyphenols in a simulated gastric environment and subsequently resulted in higher biological therapeutic efficacy. Moreover, these green tea polyphenols-BSA nanoparticles reduced the ionizing radiation-induced oxidative damage to biological macromolecules including DNA by scavenging reactive oxygen species and by modulating the level of antioxidant machinery (superoxide dismutase, catalase, etc.).

Another phenolic compound studied due to its antioxidant activity is resveratrol. Cheng et al. (2018) produced nanocomplexes by mixing BSA with trans and cis resveratrol. BSA was able to protect resveratrol against irradiation, with only 5% reduction in scavenging capacity, after 240 min. A recent work developed a tri-ligand complex between BSA and three BCs: resveratrol and EGCG (phenolics compounds) and α-tocopherol (vitamin E) by mixing (Dong et al. 2020). Tri-ligand complexes maintain, simultaneously, the structural integrity and antioxidant property of α-tocopherol, resveratrol and EGCG, and to a greater degree than that observed for mono-ligand complexes with BSA.

In recent years, the terpene astaxanthin has attracted great interest. In this sense, Wang et al. 2018 developed a solid lipid-polymer hybrid nanoparticle, consisting of a combination of polymers and lipids. These nanoparticles were produced by an organic solvent-free technique that included the homogenization and sonication of BSA and oxidized dextran. Antioxidant activity of astaxanthin was dramatically enhanced in aqueous conditions, as well a sustained release was achieved in simulated gastrointestinal fluids.

5.3.5 Pharmacological and Nutraceutical Applications

Tables 5.1 and 5.2, also show the versatility of BSA and HSA that could be used to encapsulate different kinds of BCs which could be applied in different fields as pharmacology. For example, BSA nanoparticles were produced to encapsulate apigenin (Pápay et al. 2017), ginsenoside compound K (Singh et al. 2017) and quercetin (Saha et al. 2020) with an anti-inflammatory purpose. BSA and HSA nanoparticles encapsulating carvacrol and celastrol were used in rheumatoid arthritis (Gholijani et al. 2020; Gong et al. 2019, 2020). BSA and HSA nanoparticles encapsuling celastrol improved the symptoms of rheumatoid arthritis and reduced undesirable side effects in mice (Gong et al. 2019, 2020). Kazan et al. (2019) synthetized continuously BSA-thymoquinone nanoparticles by the desolvation method using a microfluidic platform. BSA and thymoquinone solutions were fed through tefzel tubing for both streams using syringe pumps and mixed inside a T-junction. As the water and ethanol streams met inside the T-junction, albumin was phase separated due to its diminished water solubility. The formed particles were stabilized by albumin crosslinking with formaldehyde. Nanoparticles (316 nm) improved thermal stability of thymoquinone and enhanced head regeneration of planarians (used as a model of neural regeneration) (Kazan et al. 2019). Wound healing is another example of pharmacology application. HSA nanoparticles to encapsulate vitamin C were prepared by the desolvation method with a size of 180 nm. When mice were treated with vitamin C-HSA nanoparticles, the rate of wound healing was increased. The wounds healed faster and completely in comparison with control and treated groups with free vitamin C or unloaded HSA nanoparticles (Shojania et al. 2019). Another field of application is against microorganisms. In this sense, thyme oil nanoemulsion stabilized by BSA, quillaja saponins or soy lysolecithin with antifungal (Fusarium graminearum) activity were produced (Wan et al. 2019). The mycotoxin inhibitory efficacy of thyme oil was enhanced considerably in nanoemulsion form as compared to pure oil. Among the three natural emulsifiers, BSA stabilized thyme oil nanoemulsion exhibited the highest mycotoxin inhibitory activity and mycelial growth.

Finally, BSA was successfully used to improve resveratrol solubility and stability in an aqueous medium. This approach could increase the daily intake of unchanged resveratrol by consumers of fortified beverages. Resveratrol, in its native form, has low aqueous stability and solubility restricting its use to fortify food products. Tabibiazar et al. (2019) studied the influence of BSA-trans-resveratrol complex on the solubility, photochemical stability and antioxidant activity. This study was undertaken to assess the potential of BSA-resveratrol complex formation as a primary step for designing a delivery system in the functional food field. The results have demonstrated that resveratrol solubility increases as function of BSA concentration up to 2.3:1 M ratio. Colloidal nanodispersion of BSA-resveratrol complex (20 nm), was stable for more than 30 days. Complex formation was confirmed using fluorescence spectroscopy, and FTIR. Complexation altered the BSA secondary structure. Formation of the complex significantly protected resveratrol from UV induced isomerization and oxidation.

In comparison with other nanoparticles (polymeric and lipid), the studies with protein-based nanoparticles for food/nutraceutical applications are incipient and limited to *in vitro* studies. A possible reason may be the instability of BSA nanoparticles in the intestinal medium due to its cleavage by proteases. The study of Kilic et al. (2019) brings an important discussion about the gastrointestinal behavior of BSA microparticles. The authors explain that pepsin prefers to cleavage peptide amide bonds of Phe, Trp, or Tyr, but BSA contains only a small percentage of these hydrophobic amino acids. These considerations could justify why BSA nanoparticles are stable in simulated gastric fluid containing pepsin and are able to protect the BCs against premature release at the stomach. However, trypsin is able to cleavage peptide bonds of Arg and Lys, and these amino acids comprise about 18% of BSA. Thus, since BSA is susceptible to proteases of pancreatin, when the particles reach the intestine, the drug could be released. Some authors do not consider this a problem, since after the release at the intestine, the compound would be absorbed into the bloodstream and distributed to tissues to exert their actions. However, the simple release at the intestine may not be enough for the compound to achieve its pharmacological effect. Intestinal barriers, including mucus, can be limiting factors to some compounds, depending on their physicochemical properties. Nanoparticles are specialized systems used not only to protect the drug from degradation but also to improve pharmacokinetic parameters. In this sense, we found in the literature some efforts to improve the oral applicability of BSA nanoparticles. The modification of the surfaces of BSA nanoparticle is the main strategy, by including more resistant and stable molecules as coatings (such as chitosan, pectin or dextran) (Li et al. 2014; Peng et al. 2016; Wang et al. 2016, 2018; Yan et al. 2020). Results showed the increase in the stability of BSA nanoparticles at gastrointestinal ambient, protecting the loaded drug from premature release at the stomach and intestine. Thus, nanoparticles *per se* are available to be absorbed and reach the blood and then be distributed releasing the compound at the tissues, securing its full activity. Another strategy was based on the fact that lipid-based systems promote the intestinal absorption of liposoluble BC. In fact, during gastrointestinal digestion, colloidal structures formed by the interaction of dietary lipids with intestinal bile salts are capable of solubilizing liposoluble BCs and transporting them to epithelium cells, where they are absorbed. Based on this, Liu et al. (2019) developed a complex between BSA and oleic acid to encapsulate fucoxanthin and discovered that the oleic acid–protein delivery system dramatically improved the absorption of fucoxanthin. Moreover, after administration of fucoxanthin–oleic acid–BSA complexes for 15 days in mice, only fucoxanthinol accumulation was discovered in eyes and the ocular antioxidant capability increased by 71%.

Therefore, considering the analyzed bibliography, it could be concluded that the application of BSA or HSA nanoparticles in food-related research, especially in nutraceutical delivery, opens a new avenue for the food industry. It should be noted that BCs encapsulated in BSA nanoparticles for food/nutraceutical applications are mainly phenolic compounds (such as EGCG and curcumin) and carotenoids (such as fucoxanthin and astaxanthin) leaving a vacancy to study other groups such as lipids or fat-soluble vitamins.

On the other hand, BSA and HSA were employed to prepare nanoparticles to encapsulate a wide variety of non-natural molecules as can be noted in Tables 5.1 and 5.2. Merodio et al. (2001) reported the preparation of spherical shaped BSA nanoparticles by desolvation in order to deliver ganciclovir, a drug against cytomegalovirus (CMV), yielding sizes in the 200–400 nm range. Segura et al. (2005) obtained BSA nanoparticles of 300–400 nm for controlled release of α-interferon. Other authors have obtained BSA and HSA nanoparticles to encapsulate drugs for cancer therapies such as 5-fluorouracil (Maghsoudi et al. 2008), doxorubicin (Deng et al. 2010; Dreis et al. 2007; Kayani et al. 2018; Xu et al. 2001), cisplatin (Catanzaro et al. 2017), methotrexate (Lee 2019), among others. Jose and Kuriakose (2019) encapsulated Ag nanoparticles in BSA by dissolving Ag nanoparticles in hexane and adding it to an aqueous solution of BSA. The study revealed that the Ag nanoparticles–BSA aggregates displayed appreciable thermal stability and better antimicrobial efficacy against gram-positive and gram-negative bacterial strains: Streptococcus haemolyticus, Corynebacterium diphtheriae, Klebsiella ornithinolytica and Salmonella paratyphi.

Ligands of different natures were encapsulated in HSA or BSA nanoparticles in order to improve their bioavailability or reduce adverse effects owing their administration. Among them, monoclonal antibodies (Lee 2019; Luis de Redín et al. 2018), lipoproteins (Michaelis et al. 2006) and plasmids (Look et al. 2015; Kumari et al. 2018) are found.

Other clinical applications for serum albumin is the delivery of engineered proteins and peptides, such as insulin and glucagon-like peptide 1 (GLP-1) for diabetes treatment, and the use of ^{99m}Tc-aggregated albumin for diagnosis purposes (Elsadek and Kratz 2012). Drug delivery technologies employing serum albumin as a drug nanocarrier to improve pharmacokinetics and to overcome difficulties in administration of therapeutic proteins and peptides are exhaustively discussed by Kratz (2008), Kratz and Elsadek (2012).

Albumin-based systems for intravenous administration of different drugs has been developed in recent years, and they include covalent coupling of drugs to albumin, fusion of albumin to bioactive proteins by genetic engineering, peptides and fatty acid bonded to albumin, in addition to encapsulation of drugs into albumin nanoparticles. In this field, serum albumin's special features are exploited, which improve intravenous delivery and preferential accumulation in inflamed and tumor tissues due to the EPR effect. In this field, the trend is directed towards a prodrug concept, which is based on the intravenous administration of drug derivatives able to specifically bind at Cys_{34} of circulating endogenous albumin (Kratz 2014).

5.4 Alpha-Lactalbumin

5.4.1 Structural Properties

The α-lactalbumin (ALA) is known as a small milk protein (14.2 KDa), acidic (with a pI ~ 4.5), and typical Ca^{2+} binding protein. In most mammals, ALA consists of 123 amino acid residues (Nitta and Sugai 1989), with the exception of its analogue in rat, which contains 17 additional C-terminal residues. Native protein is characterized by two main domains: a large α-helical and a small β-sheet domain, which are connected by the Ca^{2+} binding loop (Hiraoka et al. 1980; Permyakov et al. 1981).

ALA shows very interesting features from several points of view. Firstly, it is important to consider that ALA performs an important function in mammary secretory cells. It is one of the two components of lactose synthase, catalyzing the final step in lactose biosynthesis at the mammary gland (Hill and Brew 1975). However, at the moment, this biological function has not inspired technological applications.

One of the most relevant properties of ALA is its ability to bind metal cations, especially the Ca^{2+} ion. ALA has a single strong Ca^{2+} binding site, which is formed by oxygen bonds from carboxylic groups of three Asp residues (82, 87 and 88) and two carbonyl groups from the peptide backbone (79 and 84) at a loop between two α-helices (Hiraoka et al. 1980; Permyakov et al. 1981). Hence, it is very convenient to study calcium-binding effects on the interactions of ALA with proteins, peptides, membranes and low molecular weight organic compounds, which frequently manifest physiological actions. The Ca^{2+} binding to ALA promotes pronounced changes in their structure and function, mostly at the tertiary, but not the secondary structure (Permyakov et al. 1981). ALA is also able to bind Zn^{2+} (Permyakov et al. 1991a), and other physiologically significant cations such as Mg^{2+}, Mn^{2+}, Na^{+} and K^{+}, which may compete with Ca^{2+} at the same binding site (Anderson et al. 1997; Permyakov et al. 1985). Usually, cation binding increases the ALA thermal stability (Griko et al. 1994; Veprintsev et al. 1997). However, the binding of Zn^{2+} ions to Ca^{2+}-loaded ALA decreases thermal stability, promoting aggregation and increasing its susceptibility to protease action (Hirai et al. 1992; Permyakov et al. 1991a). Moreover, the Ca^{2+} binding stabilizes ALA under high pressure treatment (Dzwolak et al. 1999). Therefore, it is very important to remark that ALA denaturation is strongly dependent on temperature, pressure and metal ion concentration (especially the Ca^{2+} ion).

5.4.2 Protein-based Delivery Systems

ALA is able to interact with various low molecular weight organic compounds and peptides, but these interactions are modulated by cation binding (Permyakov et al. 1991a, 1991b; Permyakov and Kreimer 1986). ALA owns several classes of fatty acid binding sites (Cawthern et al. 1997). Regarding this, around 1995 it was discovered that a fraction of human milk was able to kill tumor cells by an apoptosis mechanism and besides, many different types of tumor cells were susceptible to this effect, while healthy cells were resistant. This activity was attributed to a complex formed between the human ALA and oleic acid, which is the most abundant fatty acid in human milk (Hiraoka et al. 1980). The complex was named HAMLET (human ALA made lethal to tumor cells) and it has been exhaustively characterized in order to determine the structural basis and mechanism of the antitumor activity (Gustafsson et al. 2004; Svensson et al. 2000). To become tumoricidal, ALA must undergo partial unfolding and bind the oleic acid in physiological conditions. HAMLET exemplifies how a change in three-dimensional structure enables a protein to alter its function in response to environmental signals. In addition, it is of potential interest as a model of unfolded protein cytotoxicity, particularly in view of its relative selectivity to tumor cells. The mechanism is not fully understood, it is well known that it requires both the ALA unfolding and the presence of the fatty acid to activate the uptake of HAMLET by tumor cells. Subsequently, many mechanisms have been proposed in order to explain the HAMLET cytotoxicity (Deretic 2006; Duringer et al. 2003; Hallgren et al. 2006; Kohler et al. 2001).

Table 5.3 summarizes applications of different ALA-BCs nanostructures. Similar to HAMLET preparation, complexation between ALA and different BCs are extensively used. In complexation, ALA stock solution is preparing by dissolving it in deionized water or buffer (such as PBS or Tris buffer) with stirring. Usually, the pH of ALA dispersion is adjusted to some value and is then mixed with the BC (mostly in ethanol or some organic solvent). Then, the nanoparticles ALA-BC could be lyophilized or spray dried. This technique was used to encapsulate phenolic compounds (Al-Hanish et al. 2016; Liu et al. 2018a), vitamin D3 (Delavari et al. 2015) and oleate (Kehoe and Brodkorb 2014).

Table 5.3 Encapsulation of BCs and drugs using ALA as a carrier.

Detailed application	Method	Size	Complementary material	Effect	Active material	Encapsulation efficiency EE%	Reference
Antimicrobial	Nanocomplexes	-	None	Reduction of Ag toxicity and maintenance of antimicrobial activity against against *E. coli* and *S. aureus*	Ag	Not reported	Zhang et al. 2011
Antitumor	Complexation	4.5 nm	None	ALA can bind at least a 20-fold molar excess of oleate, most likely in a non-specific manner.	Sodium oleate	Not reported	Kehoe and Brodkorb 2014
Nanosupplements	Complexation	125 nm	None	ALA contains one binding site for vitamin D3. Complex formation decreases the heat stability of the protein.	Vitamin D3	Not reported	Delavari et al. 2015
Antitumor	Complexation	227–292 nm	None	Improved cytotoxicity	Oleic acid	Not reported	Delgado et al. 2015
Stability Antioxidant	Nanoemulsification	~ 200 nm	EGCG Chlorogenic acid	Improved physical stability and inhibited degradation.	β-carotene	Not reported	Wang et al. 2015
Stability	Nanoemulsification	199–319 nm	Beet pectin	Protect citral from oxidation during the storage at 25°.	Citral	Not reported	Xiang et al. 2015

(Continued)

Table 5.3 (Continued)

Detailed application	Method	Size	Complementary material	Effect	Active material	Encapsulation efficiency EE%	Reference
Nanosupplements	Complexation	No	None	Improved uptake by monocytes.	Epigalocatechin-3-gallate	Not reported	Al-Hanish et al. 2016
Stability Antioxidant	Free radical method Nanoemulsification	162.7 nm	Catechin Corn oil	Improved antioxidant activity, retention and protects BC during storage	β-carotene	Not reported	Yi et al. 2016a
Antioxidant	Glycosylated ALA based nanocomplex	68 nm	Dextran	Improved DPPH-scavenging activity	Curcumin	EE = 98.8% LC = 1.98%	Yi et al. 2016b
Stability	Nanotubes (Mn^{2+})	Outer diameter: 20 nm; Inner diameter: 7 nm	None	Highly stable during a freeze-drying process but not after grinding.	Caffeine	EE = 96.96–97.16% LC = 9.21–12.3%	Fuciños et al. 2017
Stability	Micelles self-assembled of partial hydrolysed ALA	19–23 nm	None	Increased aqueous solubility and stability under 60°C heating or UV light irradiation of encapsulated β-carotene. Enhanced cellular absorption, transport efficiency and anti-oxidant activity	β-carotene	187 mg/g (β-carotene / ALA)	Du et al. 2018

Stability Antioxidant	Micelles self-assembled of partial hydrolysed ALA	ALA + CUR: 23 nm ALA + CUR + ANT: 21 nm	None	Improved stability at 60ºC heating an UV light exposure. Controlled released in simulated gastrointestinal fluid. Improved cellular uptake, transmembrane permeability and antioxidant activity in CaCO2 cell line models.	Anthocyanins (ANT) Curcumin (CUR)	Not reported	Jiang et al. 2018
Stability Antioxidant	Complexation	ALA: 87.8 nm ALA-RES: 95.3 nm	None	Increased water solubility and chemical stability of RES under pH 8.0 and high temperature 80ºC. Improved *in vitro* antioxidant activity than free RES.	Resveratrol (res)	EE = 99.6% – 37.4%. The highest LC was 1.4%.	Liu et al. 2018a
Bioavailability	Co-administration of ALA and INO	-	None	Co-administration in a single dose (healthy volunteers), significantly increased the plasma concentrations of INO. *In vitro*, INO absorption in Caco-2 cells was improved in presence of digested ALA.	Myo-Inositol	Not reported	Monastra et al. 2018
Stability	Micelles self-assembled of partial hydrolysed ALA	ALA-BCs between 24 and 26 nm	None	Improved solubility and UV stability.	Curcumin, Quercetin, Astaxanthin, Coenzyme Q10	Not reported	Hu et al. 2019

(Continued)

Table 5.3 (Continued)

Detailed application	Method	Size	Complementary material	Effect	Active material	Encapsulation efficiency EE%	Reference
Antitumor	Nanospheres (NS) self-assembled of partial hydrolysed ALA	NS-CUR: 29.72 nm RGD-NS-CUR-siPlk1: 41.5 nm	siRNA targeting human polo-like kinases (siPlk1) Tumor-penetrating peptides (RGD)	Controlled released. Improved tumor suppression, cellular uptake and tumor-targeting abilities. Prolonged circulation time and release of CUR and siPlk1 in cancer cells rather than in normal cells.	Curcumin	NS-CUR: LC = 17.9 % RGD-NS-CUR-siPlk1: LC = 16.3%	Li et al. 2019
Antitumor	Desolvation method	55.2 nm	Lysozyme Folate (FA) Glutaraldehyde	Reduced cell viability of different cancer cells line. Improved cytotoxicity in FA + cancer cell lines and showed pH-responsive drug release.	Curcumin	EE = 62.12% LC = 248.8 mg/g	Somu et al. 2019
Pharma	Electrospinning (nanofibers, NF)	NIC-NF: 692 nm NF: 646 nm	Polyethylene oxide	Efficient permeate through *ex vivo* porcine buccal tissue. Biocompatible, site-specific and fast-releasing drug delivery system.	Nicotine	Not reported	Kalouta et al. 2020
Stability Antioxidant	Electrostatic interactions	RES-ALA-CHI: 211.0 nm	Chitosan (CHI)	Improved UV light and heat stability, *in vitro* bioaccessibility, and antioxidant activity	Resveratrol (RES)	EE = 58.86% LC = 196.2 μg/mg	Liu et al. 2020b

EE: encapsulation efficiency; EGCG: Epigallocatechin-3-gallate; LC: loading capacity

One of the most used methods for ALA-based nanoparticle preparation is nanoemulsification. In general, ALA solution is prepared in buffer (such as phosphate or PBS) with a specific pH (usually 7–7.5) and then mixed with an oil phase (like corn oil, medium chain triacylglycerol oil, etc.) to form a coarse emulsion. Usually the emulsification is carried out in two steps; in the first, a coarse emulsion is obtained by using a blender or Ultra-Turrax and, in the second, a nanoemulsion is obtained using a homogenizer or a high-pressure microfluidizer. In general, the hydrophobic BC is pre-dissolved in the oil phase (Wang et al. 2015; Yi et al. 2016a). Moreover, it is possible to form a polysaccharide core upon the primary emulsion in order to improve the stability and protection of BC encapsulated (Xiang et al. 2015).

Another extensively used method is the development of micelles or nanospheres from hydrolyzed ALA. ALA and Bacillus licheniformis protease (BLP to ALA weight ratio, 1:70) are dissolved in a Tris-HCl buffer, pH 7.5 and heated at 50°C for 30 min, allowing the partial hydrolysis of ALA. The amphiphilic peptides obtained are self-assembled into micelles by gentle rotation on a mixer or by sonication. Then, the hydrophobic BC (in ethanol or similar) is mixed with ALA micelles for several hours, promoting BC incorporation into the micelles core. Usually, dialysis is then performed to remove unabsorbed BC and loaded micelles powder is obtained by a freeze-drying process (Hu et al. 2019; Jiang et al. 2018). In this method, amphiphilic peptides are obtained after the partial hydrolysis of ALA via specific cleavage of Glu and Asp bonds by Bacillus licheniformis protease. The micelles are formed via the self-assembly of those amphiphilic peptides under the optimum enzymatically hydrolysis condition. BC can be loaded into the core of the micelle through a hydrophobic interaction during the self-assembled process or after the micelle formation (Du et al. 2018). One advantage is that BC incorporation into the micelle can significantly improve its aqueous solubility because the ALA micelle possesses a hydrophobic core inside and hydrophilic corona outside. Moreover, micelle sizes are small around 19–25 nm and are present as monodispersed spherical nanospheres. An alternative to this process, is the use of a reduction agent (such as dithiothreitol, DTT) to form intermolecular disulphide bonds between peptides. Disulphide bonds cross-linked ALA nanostructures can respond to the high level of L-glutathione in tumor cells and release the BC in cancer cells rather than normal cells (Li et al. 2019)

Another documented feature of hydrolyzed ALA is the ability to form nanotubes. The BLP action on ALA results in the formation of large peptides depending on the hydrolysis degree, ALA initial concentration (~30 g L^{-1},), and Ca^{2+} presence (> 1.5 mol Ca^{2+} mol^{-1} ALA). Other metal ions have also been shown to have an effect on the formation of ALA nanotubes, but with major differences in structure (Graveland-Bikker and de Kruif 2006). In general, nanotubes are built from a ten-start right-handed helix with an outer diameter of ~ 20 nm, an inner diameter of ~ 8 nm, a pitch of ~ 110 nm, and they could reach lengths of several μm (Graveland-Bikker and de Kruif 2006). On the other hand, Geng et al. (2016) showed that building fragments for self-assembly could form nanotubes and gels over a wide pH range (7.5–4.0). Potential applications of ALA nanotubes include their use as a viscosifier, as a gelling agent for producing reversible and transparent gels, as an encapsulation matrix, etc. (Graveland-Bikker and de Kruif 2006). They could be very useful as thickener agents with the linear alignment being highly efficient to improve viscosity. In addition, their open ends can provide a more efficient delivery and controlled release of BCs. Encapsulated BCs could be released from nanotubes in a controlled manner, taking advantage of the ability of ALA nanotubes to disassemble at pH values below three, which could be exploited for the delivery of BCs on the gastrointestinal tract. Regarding this, ALA nanotubes were assayed as carriers for caffeine encapsulation being highly effective and achieving values of ~ 100% EE. Moreover, it was demonstrated that the structure of the nanotubes did not appear to be altered by caffeine encapsulation, so that they could maintain their functionality as a thickener agent (Fuciños et al. 2017).

ALA nanostructures based on enzymatic hydrolysis was also examined by other authors. Matsumura et al. (2000) studied the ALA polymerization caused by the incubation with transglutaminases from guinea pig liver (GTGase) and from Streptoverticillium mobaraense (MTGase) under the presence of dithiothreitol (DTT). Saricay et al. (2013) produced nanoparticles induced by horseradish peroxidase cross-linking, suggesting that the particle architecture (open and dilute, or dense and compact) could be controlled, finding an interesting topic for future research. Subsequently, the same authors elucidated the effect of heat treatment on the physical properties of peroxidase-cross-linked ALA aggregates, finding that the dilute dispersions of aggregates were stable against heat over a wide range of pH values and salt concentrations; while dense dispersions formed physical hydrogels. Furthermore, Dhayal et al. (2014) used horseradish peroxidase to produce nanoparticles with controlled size. Moreover, the speed of particle formation increased with the increase in ionic strength, but its meso-scale structure remained similar. These nanoparticles seemed to be similar in structure to most of the heat-induced protein particles made at pH around 7.0. However, in contrast to heat-induced protein aggregates, the peroxidase cross-linked protein nanoparticles are not much affected by the aqueous ionic strength. In another study, Balandrán-Quintana et al. (2013) ALA hydrolysis was examined using Staphylococcus aureus Glu-C V8 protease under Ca^{2+} presence. They obtained disk-like nanostructures with regular shapes and sizes around 430 nm, indicating an ordered self-assembly of peptides. ALA nanodisks were suggested as a promising tool for the delivery of sensitive substances to several environment conditions.

In order to produce interesting nanostructures, heat treatment can also influence on ALA physicochemical properties. In this sense, Wang et al. (2011) reported the heating of ALA solutions at pH 2.2 and 55 °C. Authors suggested that the resulting fibrils could be used to modulate the key properties of food products, such as protein beverages and yogurt. More recently, Monteiro et al. (2016) produced heat-induced nanoparticles by heating of ALA and lysozyme (Lys) mixed solutions at various pH values, temperatures and heating times. The minimum size (61 nm) was obtained at pH 11.0 for a thermal treatment at 75 °C, over 15 min, suggesting that these conditions could be critical for the formation of supramolecular structures. Their formation offers new opportunities to modify or improve functional properties of commercial products, such as foods and pharmaceuticals; and as a smart vehicle for BCs delivery. Heating and acidifying or alkalizing conditions can be optimized to control the albumin's properties and to improve its performance for tailor-made applications.

On the other hand, solvent desolvation processes were examined to produce ALA nanostructures. Arroyo-Maya et al. (2012) obtained glutaraldehyde cross-linked ALA nanoparticles by solvent desolvation using acetone, ethanol, or isopropanol and heat or a high-pressure treatment step before the desolvation process. The use of different desolvation agents and pretreatments resulted in the formation of ALA nanoparticles (100–200 nm) with a wide variety of properties. They also observed that gastrointestinal proteases easily degraded all types of ALA nanoparticles. In a further study, the same authors obtained glutaraldehyde cross-linked ALA nanoparticles (100–160 nm) in the presence of acetone (Arroyo-Maya et al. 2014). Moreover, in an interesting study Mahanta and Paul (2015) obtained glutaraldehyde cross-linked ALA nanoparticles (300 nm) by ethanol desolvation, which were conjugated with folic acid. Self-assembly was found to be highly stable against thermal, pH, and proteases stress. In the same way, Somu et al. (2019) obtained supramolecular nanoassemblies of lysozyme and ALA by the desolvation technique and glutaraldehyde cross-linking to encapsulate curcumin. Moreover, these nanoassemblies were conjugated to folate. The authors demonstrated that its administration produced strong anticancer activity in cancer cells including breast, cervical, and osteosarcoma, primarily caused by ROS mediated oxidative stress. In addition, folate-based targeting enhanced its cytotoxicity to cancer cells (Somu et al. 2019).

The electrospinning technique has been widely studied during the last few years due to the possibility to create fibres in the nano-scale for a variety of biomedical applications. In this sense, Guo et al. (2020) developed nanofibers of ALA electrospin with polycaprolactone with average diameters of 183–344 nm. Briefly, the polycaprolactone and ALA were dissolved in hexafluoro-isopropanol and the resulting solutions were loaded in a 5.0 mL syringe, which was installed into an electrospinning machine. The high electrical voltage (9–18 kV) was applied across a fixed distance of 15 cm between the needle tip and the collector and conditions were controlled during the process (25 °C and 40% humidity). The nanofibers were then vacuum dried for 48 h to completely remove the solvent (Guo et al. 2020). Their wettability and mechanical properties could be modulated by adjusting the mass ratios of ALA and polycaprolactone. The nanofibers exhibited negligible cytotoxicity and satisfactory adhesion to fibroblasts and promoted the proliferation of the fibroblasts. As compared to pure polycaprolactone-based nanofibers, the composite scaffolds could accelerate the wound healing process. Moreover, nanofibers could increase the synthesis of type I collagen and decrease the expression of α-smooth muscle actin, reducing the formation of scars. This study demonstrated that ALA is a promising biological material and could promote the regeneration of skin burns with reduced scar formation, when being loaded on ultrafine fibrous scaffolds, mimicking the structure of the natural extra cellular matrix (Guo et al. 2020). Moreover, nanofibers of high protein content were successfully produced via waterborne electrospinning of ALA with a minimum amount of poly(ethylene oxide) as a co-spinning polymer (Stie et al. 2020). These nanofibers displayed excellent biocompatibility *in vitro* with a diameter of 390–600 nm. ALA-based nanofibers were loaded with up to 6% ampicillin and they were able to maintain the activity of the antibiotic after electrospinning and cross-linking and inhibited the growth of Escherichia coli and Burkholderia thailandensis *in vitro*. The main advantage of these nanofibers is the possibility to electrospin ALA without the use of any organic or toxic agents and with a minimum amount of copolymer. Moreover, size, mechanical properties, and wettability of ALA nanofibers could indeed be rationally tuned by controlling the protein concentration and/or via a non-destructive, simple, and sustainable postproduction treatment (Stie et al. 2020). Kalouta et al. (2020)

developed a polyethylene oxide-ALA electrospin nanofibers as an efficient oromucosal delivery system for nicotine delivery. Nanofibers had a diameter of 646 nm and when loaded with nicotine the diameter increased to 692 nm. The nanofibers showed efficient permeation through *ex vivo* porcine buccal mucosa, are biocompatible, site-specific and a fast-releasing delivery system (Kalouta et al. 2020).

The different ALA nanostructures have found different applications, but are extensively used to protect, solubilize and increased the stability of BCs that could be applicated in pharmaceutical and food areas (Table 5.3). In this regard, Wang et al. (2015) and Yi et al. (2016a) produced nanoemulsions to encapsulate β-carotene with sizes of 160–200 nm. First, ALA was covalently conjugated with EGCG, catechin or chlorogenic acid and then a nanoemulsion with β-carotene (dissolved in corn oil or medium chain triacylglycerol oil) was produced. In general, the presence of ALA-EGCG or ALA-catechin to stabilize the nanoemulsion, enhanced the physical stability of β-carotene and retention, and inhibited its degradation during storage (Wang et al. 2015; Yi et al. 2016a). Another method to encapsulate β-carotene was to use micelles of partial hydrolysed ALA to incorporate the BC in the core via hydrophobic interactions with a final size of 23 nm (loaded) (Du et al. 2018). In the micelles, aqueous solubility and stability under 60 °C heating or UV light irradiation of encapsulated β-carotene improved significantly compared with free β-carotene. Moreover, it had an increased cellular uptake (three times) and transmembrane permeability in a Caco-2 cell monolayer model. Encapsulated β-carotene also exhibited an enhanced cellular anti-oxidant activity compared with free BC. In general, different structures were used to encapsulate, protect and improved the stability of a variety of BC such as curcumin (Hu et al. 2019), resveratrol (Liu et al. 2018a), quercetin (Hu et al. 2019), astaxanthin, citral (Xiang et al. 2015) among others. In general, results showed an increase in aqueous solubility and stability of BC during long storage or against injurious conditions (such as high temperatures, UV light irradiation, pH changes, industrial process such as freeze drying, etc.). Micelles of partial hydrolyzed ALA were developed to co-deliver anthocyanins and curcumin (Jiang et al. 2018). Both anthocyanins and curcumin loaded micelles showed improved stability (against heat and UV irradiation) and also exhibited an intestinal pH-responsive release property in simulated gastrointestinal fluid. Both compounds showed a slower release rate in the stomach acidic environment than in the intestinal environment, which is favourable for preventing the early release and damage of encapsulated ingredients under stomach conditions. As well, the micelles exhibited that an advanced cellular uptake and transmembrane permeability based on the Caco-2 cell monolayer model due to ALA micelles could have an active cellular uptake, which is more efficient than the passive diffusion of the free anthocyanins and curcumin. Finally, cellular anti-oxidant activity for anthocyanins and curcumin was markedly improved (Jiang et al. 2018). Liu et al. (2020b) used a similar strategy as used in BSA nanoparticles with surface modification to improve oral bioavailability. A 211 nm nanoparticle was fabricated by electrostatic interactions between oppositely charged ALA and chitosan at pH 6.5 for resveratrol encapsulation. Compared to free resveratrol, the UV light and heat stability, *in vitro* bioaccessibility, and antioxidant activity of resveratrol in nanoparticles were pronouncedly enhanced (Liu et al. 2020b).

5.4.3 Antitumor and Antioxidant Properties

BCs encapsulated in ALA nanostructures are extensively used with antioxidant and antitumor applications (Table 5.3). For example, Li et al. (2019) developed complex nanospheres to encapsulate curcumin. Firstly, ALA was partially hydrolysed by BLP and the obtained amphiphilic peptides were self-assembled into nanospheres. Curcumin was loaded in the hydrophobic core during the self-assembly process. Disulphide bonds were formed between the endogenous cysteines on ALA and two ends decorated sulfhydryl-siPlk1 (siRNA targeting human polo-like kinases). The disulphide bonds cross-linking of siPlk1 on ALA not only stabilized the siPlk1 against degradation but also enabled glutathione-triggered drug release. RGD peptide (tumor-penetrating peptides) was conjugated to the nanosphere's surface. The active targeting and positive charges enabled RGD/ALA/Cur/siPlk1 to be internalized actively in tumor cells. The positive-charged nanospheres induced a proton-sponge effect and rapidly escaped from the lysosome into cytosol, where disulphide bonds were cleaved by the intracellular glutathione and sequentially the loaded drugs were released (Li et al. 2019). These nanospheres improved tumor suppression and cellular uptake, prolonged circulation time and release of curcumin and siPlk1 in cancer cells rather than in normal cells. On the other hand, BCs as resveratrol, β-carotene and curcumin, among others were encapsulated in ALA nanostructures in order to improved their stability and water solubility increasing their antioxidant activity (Jiang et al. 2018; Liu et al. 2018a, 2020b; Wang et al. 2015; Yi et al. 2016a, 2016b).

Considering all of the above-mentioned, it is important to note the great potentiality of several kinds of ALA nanostructures, including micelles, nanocomplexes, nanotubes and nanoemulsions, for delivering a wide variety of drugs, nutraceutical compounds and metallic nanoparticles with applications in the pharmaceutical or food industry.

5.5 Ovalbumin

5.5.1 Structural Properties

Another storage albumin is ovalbumin (OVA), being the most abundant protein in egg white. In chicken eggs, OVA constitutes 54% of the total proteins of egg white, and 40% in turkey and duck. It is a monomeric phosphor-glycoprotein (Giosafatto et al. 2012; Weijers et al. 2002) and it is mainly responsible for the functional properties of egg white proteins, showing gelling capacity, foaming, emulsifying and water retention properties (Weijers et al. 2002). This is due not only to the fact that OVA is the major protein of the egg white, but also to its exceptional thermal and surface properties (Croguennec et al. 2007). OVA is available in large quantities and its use is widespread not only as a food ingredient but also as a standard in the studies of protein properties (Huntington and Stein 2001).

In terms of its structural features, OVA is included as a member of serpins superfamily (serine proteinase inhibitor), which includes more than 300 homologous members sharing a common tertiary structure, although OVA has no effect as a protease inhibitor

(Shinohara et al. 2005; Yamasaki et al. 2003). It is a globular protein constituted of 385 amino acid residues, being the half hydrophobic and one third charged, mostly acidic, giving a pI around 4.5. It has a molecular weight of 43 kDa and a mean hydrodynamic diameter of 6.4 nm. In addition, it has four sulfhydryl groups (SH) occluded into the globular structure and a disulphide bond (SS) per macromolecule (Croguennec et al. 2007). Almost all of the polypeptide chain is involved in defined secondary structure, with three β-sheets, nine α-helices and three short helical segments of three to four residues (Stein et al. 1991). In addition, OVA is composed of a mixture of three isoforms called A1, A2 and A3, according to the number of bound phosphate residues (two, one and zero, respectively) (Shinohara et al. 2005; Weijers et al. 2002). The phosphorylation sites are Ser_{68} and Ser_{344}. The protein also has a simple hydrocarbon chain that represents 3% of the weight and it is covalently bound to the amido nitrogen of Asn_{292} (Giosafatto et al. 2012).

Different terminologies are used to describe OVA forms, the native form (N) can be converted into stable form (S-) through the formation of an intermediary form (I-). All these species are able to self-aggregate. In addition, the N species can be formed by a fraction N1 and another N2. The last one can be denatured but not self-aggregated. The amount of S- form will depend on the storage time and the egg's pH (Weijers et al. 2003). In terms of its characteristics, OVA finds applications in different fields, such as food, biomedicine, industrial applications, etc. (Table 5.4). Currently, numerous systems that include it, either individually or in the company of other molecules such as polysaccharides, proteins, metallic ions etc., have been developed for encapsulation, transport and protection of different BCs. There is a wide range of scientific works in which the ligand binding properties of OVA were explored (Chen et al. 2018; Somchue et al. 2009; Sponton et al. 2015; Visentini et al. 2017a; Xie et al. 2017).

5.5.2 Protein-based Delivery Systems

On the other hand, many methods have been used to produce OVA nanostructures such as self-assembly (Liu et al. 2017, 2018a), thermal aggregation (Sponton et al. 2017a; Visentini et al. 2019, 2020), emulsification (Feng et al. 2018; Niu et al. 2016; Vicente et al. 2017), gelation (Chaux-Gutiérrez et al. 2020; Feng et al. 2019), and nanofibers (Song et al. 2013; Wongsasulak et al. 2010), among others. The nanoemulsification technique to obtain OVA nanoparticles is the same as that which was described to obtain ALA and BSA nanoparticles. In general, once a nanoemulsion is obtained a crosslinking process and/or freeze-dried step are realized (Vicente et al. 2017, 2018; Xie et al. 2019, 2020). Thermal aggregation is based in the fact that nanoparticles can be formed from globular proteins simply by heating their solutions. These proteins unfold above their denaturation temperature exposing the hydrophobic groups normally occluded. As a result, proteins begin to aggregate through hydrophobic interactions and SS formation. The aggregation process is irreversible and can be stopped by rapidly reducing the temperature. Ionic strength, pH and temperature strongly influence the denaturation and aggregation of globular proteins and, therefore, the size of the particles produced (Davidov-Pardo et al. 2015). The other method used is gelation, which consists of heating OVA solution during certain time. The principal difference with thermal aggregation is

Table 5.4 Encapsulation of BCs and drugs using OVA as a carrier.

Detailed application	Method	Size	Complementary material	Effect	Active material	Encapsulation efficiency (EE)	Reference
Food packaging	Nanofibers	265 ± 48 nm	Cellulose acetate	Edible polysaccharide–protein composite	-	Not reported	Wongsasulak et al. 2010
Controlled release	Microcapsules layer by layer adsorption	$D_{3,2}$:5-6 μm	High methoxyl pectin	Controlled released (an increasing in the number of layers delay release of limonene)	Limonene	Not reported	Humblet-Hua et al. 2011
Antimicrobial Antitumor	Nanofibers	200-400 nm diameter	Polyacrylonitrile Silver ions	Improved cytotoxicity on C2C12 cells and antimicrobial activities against *E. coil* and *S. aureus*	Ag	Not reported	Song et al. 2013
Bioavailability	Maillard reaction and self- assembling or crosslinked	285–339 nm	Dextran	Enhanced the apparent permeability coefficient (Papp) of EGCG on Caco2 cells monolayers.	EGCG	Self-assembling: EE = 23.4%; LC = 19.6%. Crosslinked: EE = 30.0%; LC = 20.9%	Li and Gu 2014
Pharma	Films	-	1-ethyl-3-3-dimethyl aminopropylcar-bodiimide hydrochloride	Films presnted oxygen permeability, water uptake, non-toxicity and cytocompatibility.	-	Not reported	Shojaee et al. 2015

(Continued)

Table 5.4 (Continued)

Detailed application	Method	Size	Complementary material	Effect	Active material	Encapsulation efficiency (EE)	Reference
Bioavailability	Maillard reaction Nanogels (90ºC, 30 min)	150-300 nm	Dextran	Increase gastrointestinal stability and bioavailability (solubilisation effect)	Curcumin	EE = 82.86% LC = 7.17%	Feng et al. 2016
Antimicrobial effect	Emulsification	~ 300 nm	Arabic Gum	Prolonged effect in *E. Coli* growth	Thyme oil	EE = 90%	Niu et al. 2016
Antioxidant	Complexation	103.57 ± 4.22 nm	None	Improved solubility and antioxidant activity	Curcumin	Not reported	Liu et al. 2017
Stability	Inclusion nanocomplexes Thermal aggregation.	77 nm	High methoxyl pectin (HMP)	Increased chemical stability of LA respect to system without HMP	Linoleic acid (LA)	LA-OVA NPs: EE = 36.11%; LC = 27.09% LA-OVA NPs-HMP: EE = 93.58%; LC = 56.29%	Sponton et al. 2017a
Stability	Emulsification	-	Pectin Xanthan gum	Slow omega-3 release profile.	Omega 3 (Sacha Inchi oil)	EE = 86.6% – 96.6%	Vicente et al. 2017
Stability	Thermal aggregation	286–376 nm	High methoxyl pectin	Increase in photochemical stability of Retinol.	Retinol	Not reported	Visentini et al. 2017b
Pharma	Films	-	Silk Fibroin	Promoted cell viability and adhesion	-	Not reported	You et al. 2017

Antioxidant	Acid hydrolysation at 85 °C Emulsification	$D_{3,2} < 400$ nm	None	Improved emulsifying effect, emulsifying stability and antioxidant activity	Mulberry seed oil	Not reported	Chang et al. 2018
Emulsifier	Complexation Emulsification	~ 300 nm	Soybean oil	Increased emulsion stability	Tannic Acid	Not reported	Chen et al. 2018
Pharma	Complexation Electrostatic deposition	157.5 ± 9.86 nm	Chitosan	Enhance uptake in Dendritic Cells and promote it to maturate *in vitro*. *In vivo*, trigger antigen-specific immune responses.	Genipin	Not reported	Dong et al. 2018
Stability Antioxidant	Conjugation Emulsification	230–400 nm	Fish oil	Improve antioxidant capacity and stability against droplet aggregation or coalescence. Reduced lipid hydroperoxide and TBARS.	Catechin Epigallocatechin EGCG	Not reported	Feng et al. 2018
Pharma	Heat treatment (80 ºC, 20 min)	170-480 nm	None	Suppress secretion of pro-inflammatory mediators and promote production of anti-inflammatory interleukin-10. Therapeutic efficacy in alleviating ulcerative colitis in mouse model.	EGCG	EE = 80-99% LC = 19-109%	Gou et al. 2018

(Continued)

Table 5.4 (Continued)

Detailed application	Method	Size	Complementary material	Effect	Active material	Encapsulation efficiency (EE)	Reference
Stability	Self-assembly	-	None	Improve solubility and photostability	Curcumin	Not reported	Liu et al. 2018b
Stability	Conjugation by Maillard reaction Emulsification	Nanoemulsion (~200 nm)	Lactose	Improved emulsifying capacity and thermal stability. Improved AOPO emulsions stability against temperature, pH, ionic strength and storage.	Aged orange peel oil (AOPO)	Not reported	Liu et al. 2018c
Stability	Emulsification	-	Pectin (Pec) Xanthan Gum (XG)	Emulsion were stable at 25ºC and at least 14 days	Sacha Inchi Oil	Not reported	Vicente et al. 2018
Stability Bioaccessibility	Nanocomplexes NPs (90 ºC, 30 minutes)	207-305 nm	Carboxymethyl-cellulose	Improves stability under UV irradiation and bioaccessibility	Resveratrol	Nanocomplexes: EE ~ 26%; LC ~ 13% NPs: EE ~ 70%; LC ~ 35%	Xiong et al. 2018
Sensing system	Complexation	3.6 nm	Cupper ions and hydrazine	Sensing of vitamin B1 and deoxyciclyne with high sensitivity	-	Not reported	Yang et al. 2018

Stability Antioxidant	Maillard reaction and heat gelation	185 nm	Dextran Pectin	Improved chemical stability under degradation models and simulated gastrointestinal tract. Improved cellular uptake and permeability and restored viability of cells and activities of endogenous antioxidant enzymes	Cyanidin-3-O-glucoside	EE = 65.43 LC = 2.55%,	Feng et al. 2019
Antitumor	Emulsion-solvent evaporation	279.3 nm	Polymethyl methacrylate (PMMA); Graphene oxide (GO)	NPs do not affect cell viability of normal VERO cells. Improved cytotoxicity in CaCO2 cell line.	Doxorubicin	OVA-PMMA: EE = 60%; LC = 4.8 % OVA-PMMA-GO: EE = 80%; LC = 6.5%	Prabakaran et al. 2019
Stability	Emulsification Desolvation	397.2 nm	κ carrageenan	Improved stability under heat (60ºC) and natural light. Controlled released in simulated gastrointestinal fluids	Curcumin	EE = 98.1% LC = 2.53%	Xie et al. 2019

(Continued)

Table 5.4 (Continued)

Detailed application	Method	Size	Complementary material	Effect	Active material	Encapsulation efficiency (EE)	Reference
Pharma	Conjugation	3.22 nm	Tetrachloroauric acid	Stable adjuvant formulation with the ability to generate antigen-specific serum IgG subcutaneously administered. Improved cellular uptake and *in vivo* adjuvant activity.	Gold NPs	Not reported	Yang et al. 2019
Antitumor	Thermal aggregation	25-280 nm	None	Cytotoxic effect against HT-29 colorectal adenocarcinoma cell line	Conjugated linoleic acid	> 95%	Visentini et al. 2019, 2020
Controlled released	Heat gelation Electrostatic deposition	-	Low methoxyl-amidated pectin	Improved EE and controlled released	Betalain	EE = 88%–97%.	Chaux-Gutiérrez et al. 2020
Antitumor	Conjugation	-	None	Inhibited proliferation of H22 hepatoma cells, blocked cell cycle and accelerated apoptosis of solid tumors cells	Selenium	Not reported	Sun et al. 2020

Antibacterial	Heat gelation Minced	132.2 nm	None	Improved bactericide activity of free carvacrol against *B. cereus* and *Salmonella*	Carvacrol	EE = 51.37 %	Rao et al. 2020a
Antibacterial	Heat gelation Minced	64 –179 nm.	None	Improved bactericide activity of free carvacrol against *B. cereus* and *Salmonella*	Carvacrol	%EE = 35–91% %LC = 13–43%	Rao et al. 2020b
Stability Controlled released	Complexation Antisolvent precipitation	250 nm	high methoxyl pectin	Improved storage stability (4 °C and 31 days) Sustained release in the simulate gastrointestinal tract	Vitamin D3	EE = 96.37% LC = 1.93%	Xiang et al. 2020
Stability	Emulsification Crosslinking	200 nm	Tannic acid Gum Arabic	Protection of astaxanthin against UV light and heat.	Camelina oil extract	EE = 70%	Xie et al. 2020

EE: encapsulation efficiency; EGCG: Epigallocatechin-3-gallate; LC: loading capacity; NP: nanoparticle

OVA concentration. For example, Chaux-Gutiérrez et al. (2020) heat a 5% OVA solution at pH 8, 15 min and 85 °C to form gels.

The EE values reported for different BCs and drugs encapsulated using OVA-based nanoparticle systems are shown in Table 5.4. The indirect determination of the entrapped compound amount by HPLC or UV-Vis spectroscopy is the most common method to determine EE of BCs encapsulated into OVA nanoparticles, as in the case of BSA and HSA. As can be seen in the table, EE values of OVA-based systems as carriers are generally higher than 50%. In the case of OVA nanoparticles, it is not possible to draw conclusions about predominant preparation methods or to compare EE or loading capacity (LC) magnitudes because of the variety of formulations and other complementary materials used.

The analysis of the results shown in Table 5.4 suggests that OVA has been used for improved stability, solubility, bioavailability and antioxidant activity of BCs, with antitumor and antibacterial purposes, with pharmaceutical applications, etc. In addition, the versatility of this protein was demonstrated, not only in the area of application, but also in the diversity of methodologies and complementary materials used to produce OVA nanostructures. In this sense, an extensive characterization of the binding mechanism and interaction between OVA and curcumin was carried out (Liu et al. 2017, 2018a), revealing an improvement in water solubility and in antioxidant activity of the BC. Besides, the photolysis was < 30% for OVA-curcumin complex respect to free curcumin.

The production and characterization of OVA nanoparticles (OVAn) by thermal self-assembly was widely examined by Sponton et al (2016, 2017a, 2017b, 2018). In addition, they studied the binding with linoleic acid (as a model polyunsaturated fatty acid) through the formation of an inclusion complex, and designed a separation method for OVA-linoleic acid complex. An electrostatic deposition of high methoxyl pectin (HMP) was produced over OVAn-linoleic acid complex with EE values of 93.5%. While studying linoleic acid oxidative deterioration, the authors discovered that only 20% of linoleic acid deteriorated in HMP-OVAn-linoleic acid system, whereas 70 % deteriorated in OVAn-linoleic acid complex, after 13 days of storage (Sponton et al. 2017a). Using the same method, Visentini et al (2017b) developed biopolymer nanoparticles (BNPs) based on OVAn and a further electrostatic deposition of high methoxyl pectin onto its surface. BNPs, with a size of 286–376 nm, were tested to evaluate retinol stability against light radiation and oxygen, finding promising results. Vulnerability to retinol photodecomposition was 34%, whereas in the same conditions for retinol encapsulated in the BNPs was 14%.

In addition, from Table 5.4, conclusions that lipids are encapsulated principally by emulsification technique using OVA can be drawn. Chang et al. (2018) developed an emulsion between egg albumen and mulberry seed oil. In the first place, an acidic hydrolysis of egg albumen at pH 2.5 and different times was carried out and then an emulsion was produced with 10% mulberry seed oil. The size of the emulsion was < 400 nm. The acid hydrolysis improved the emulsifying effect, emulsifying stability and antioxidant activity (Chang et al. 2018). Moreover, an emulsion between OVA and sacha inchi oil was produced and then, xanthan gum or pectin was added (Vicente et al. 2018). The authors explored different concentrations of the components and concluded that it was possible to form oil-in-water emulsions composed of 8% sacha inchi oil–OVA

(1.0–2.0%) and pectin (1–3%), which are stable at 25 °C for at least 14 days. While the emulsions with 8% sacha inchi oil–OVA (1–2%) are stable with xanthan gum at concentrations of 0.5–1.0%, they are more viscous emulsions and can contain aggregates (Vicente et al. 2018).

On the other hand, phenolic compounds were used to produce oil in water emulsion as well. A conjugate OVA–lactose was developed by the Maillard reaction, and used to produce an oil in water emulsion with aged orange peel oil (AOPO, rich in poly-methoxy-flavones) with a size of 200 nm (Liu et al. 2018b). The lactose conjugation to OVA improved emulsifying capacity and thermal stability. Moreover, improved the stability of AOPO emulsions against temperature, pH, ionic strength and storage. Feng et al. (2018) confirmed that the OVA antioxidant capacity was appreciably improved via complexation with catechins enhancing the storage and oxidative stability of fish oils. Xie et al. (2020) used egg albumin and gum Arabic to emulsify camelina oil extract rich in astaxanthin. The resultant emulsions were crosslinked by tannic acid, a natural phenolic cross-linker. The entrapment efficiency of astaxanthin in emulsions was around 70%. After crosslinking by tannic acid, emulsions showed better protection of astaxanthin against UV light and heat (Xie et al. 2020).

OVA has been extensively used in combination with polysaccharides as pectin, xanthan gum, dextran, etc. to improve the oral bioavailability of BCs such as vitamin, lipids and phenolic compounds. Li and Gu (2014) obtained two nanoparticles, EGCG-OVA-dextran conjugated nanoparticle and EGCG-OVA-dextran conjugated nanoparticles by using glutaraldehyde as a cross-linking agent. The authors reported that cross-linked OVA nanoparticles promoted minor release of EGCG under *in vitro* gastrointestinal digestion in comparison with non-cross-linked nanoparticles. In addition, an experiment of EGCG absorption on Caco-2 cell line monolayers revealed that EGCG in EGCG-OVA-dextran conjugated nanoparticles showed higher apparent permeability coefficient on Caco-2 cell monolayers with respect to EGCG suggesting improvement on the EGCG gastrointestinal absorption. Feng et al. (2016) obtained nanogels based on OVA-dextran conjugates prepared by the Maillard reaction to encapsulate curcumin. After *in vitro* simulated gastrointestinal digestion, the curcumin bioaccessibility was higher for loaded nanogels in comparison with free curcumin, highlighting that nanogels protected curcumin from degradation in the gastrointestinal tract and improved its accessibility due to efficient solubilisation. Moreover, Feng et al. (2019) produced nanogels of OVA–dextran–pectin conjugates for cyanidin-3-O-glucoside encapsulation. Nanogel enhanced the chemical stability of encapsulated cyanidin-3-O-glucoside under accelerated degradation models and a simulated gastrointestinal tract. Cellular uptake and apparent permeability coefficients of encapsulated BC were also improved. Finally, encapsulated cyanidin-3-O-glucoside was more efficient in restoring the viability of cells and activities of endogenous antioxidant enzymes than free BC in oxidative models (Feng et al. 2019). Interestingly, Xiong et al. (2018) developed nanocomplexes and nanoparticles between OVA and carboxymethylcellulose (CMC) to encapsulate resveratrol. The electrostatic assembly between OVA and CMC formed nanocomplexes and the nanoparticles were fabricated through further heating the nanocomplexes at 90 °C for 30 min. The nanocomplexes EE were around 26% and increased to around 70% for nanoparticles. Both formulations improved the photostability of resveratrol when exposed to UV light, but the nanoparticles were more effective. Moreover, the bioaccessibility of resveratrol embedded in nanocomplexes increased to 60% and to 80% in nanoparticles compared to pure resveratrol (Xiong et al. 2018).

An antisolvent precipitation method was used to prepare vitamin D3-OVA complexes and then the complexes were added to a pectin solution (Xiang et al. 2020). When vitamin D3-OVA-pectin complexes were digested, a small amount of vitamin D3 is released from nanocomplexes in simulated gastric fluid, while a large amount of vitamin D3 is released in simulated intestinal fluid. Vitamin D3-OVA-pectin complexes were resistant to pepsin because pectin chains protect the cleavage site of pepsin on OVA, thereby inhibiting the rate and extent of proteolysis. Pepsin and pectin chains are negatively charged under gastric conditions (acidic pH) so the electrostatic repulsions lead to limited pepsin binding to complexes. However, the complexes were easily hydrolysed in intestinal conditions because the structure of vitamin D3-OVA-pectin complexes will be dissociated or swell under neutral conditions, thereby promoting intestinal fluid penetration. In addition, adsorbed polysaccharide chains can be replaced by the bile salts which are more surface-active thereby disrupting the conformation of the protein and accelerating its proteolysis (Xiang et al. 2020). Moreover, microcapsules formulated with OVA, sacha inchi oil and xanthan gum or pectin were produced (Vicente et al. 2017). The microcapsules effectively protected w-3 fatty acids after *in vitro* simulated gastrointestinal digestion, showing a very slow α-3 fatty acids release profile.

5.5.3 Biological Properties and Applications

As was said before, antitumor and antibacterial applications were also investigated for BCs encapsulated in OVA nanostructures (Table 5.4). Similar to Sponton et al. (2015, 2016), Visentini et al. (2019, 2020) developed and characterized an OVAn-conjugated linoleic acid (CLA) nanocomplex. Under a gastrointestinal digestion model, a small released of CLA was achieved in gastric conditions (around 5%) and a higher released was observed in intestinal conditions (20–30%). Moreover, CLA encapsulated in OVAn had the same cytotoxic effect on HT29 human adenocarcinoma cell line as free CLA. Therefore, OVAn was capable of encapsulating, solubilizing and protecting CLA without interfering in their cytotoxic effect *in vitro* (Visentini et al. 2020). Niu et al. (2016) formulated emulsions based on mixtures of OVA, gum Arabic and thyme oil at different pH, with sizes around 300 nm. Antimicrobial activity experiments showed that all the emulsions had prolonged effects on inhibition of E. coli growth in comparison with pure thyme oil. Rao et al (2020a, 2020b) developed nanogels to encapsulate carvacrol with antibacterial activity. Nanogels were obtained from egg white at pH 2, 5, 7 and 9 heating the solutions for 30 min at 90 °C; nevertheless, carvacrol was added prior to heat treatment. Both works evaluated extensively the psychochemical characteristics of nanogels obtained at different pH, highlighting that OVA–carvacrol nanogels inhibited bacterial growth of Bacillus cereus and Salmonella at lower concentrations than free carvacrol (Rao et al. 2020a, 2020b). Particularly, inhibition experiments revealed that the minimum inhibitory concentration of nanogels obtained at pH 2 was lower than that obtained at pH 7 and 9. Moreover, they possessed a better dense gel structure, higher stability and encapsulation rate (Rao et al. 2020b).

Table 5.4 shows that different OVA systems as films, nanofibers and nanoparticles were developed for biomedical applications as the fabrication of tissue scaffolds. In this sense, Shojaee et al. (2015) produced a thin film from egg white cross-linked with 1-ethyl-3-3-dimethyl aminopropyl carbodiimide hydrochloride (EDC) for wound dressing applications. The film showed a suitable oxygen permeability and water uptake, essential characteristics to prevent excessive dehydration. Cell viability analysis demonstrated no toxicity, highlighting potential biocompatibility. As well, You et al. (2017) developed an egg white–fibroin film (EW-SF) for biomedical use, showing an excellent biocompatibility. Interestingly, more human umbilical vein endothelial cells (HUVECs) were adhered to the EW and composite EW/SF films, suggesting the addition of EW provided a better cell adhesion. Other studies were focused on preparing functional OVA self-cross-linked nanoparticles with genipin (Dong et al. 2018). The authors developed a delivery platform based on genipin cross-linked OVA nanoparticles with the advantages of self-fluorescence, biocompatibility and low toxicity properties. Results indicated that the uptake of OVA in dendritic cells (DC) was enhanced and further, promoted DC activation and maturation *in vitro*. *In vivo*, genipin cross-linked OVA nanoparticles triggered an antigen-specific immune response. On the other hand, the self-fluorescence of OVA nanoparticles was used for the real-time tracking of cellular uptake *in vitro* and monitoring the migration of DC into the lymph nodes and sustained release of antigen *in vivo*. Another OVA biomedical application is the production of sensing systems. Yang et al. (2018) developed a system for the detection of vitamin B1 and doxycycline in biological samples such as urine. It is based on facile-green copper nanoclusters (CuNCs) derived from OVA (CuNCs-OVA) fabricated by a simple method. These sensing systems were tested with human urine samples and showed detection limits of the order of pM or nM, which converted it to a promising system for application in clinical diagnosis. A single-step conjugation between gold nanoparticles and OVA (Au-OVA) via sodium borohydride (NaBH4) reduction was developed by Yang et al. (2019) to be used as a vaccine adjuvant. These nanoparticles have a mean diameter of 3.22 nm determined by TEM image analysis. Results revealed that Au-OVA was a stable adjuvant formulation with the ability to generate antigen-specific serum IgG when administered subcutaneously. Besides, improved *in vitro* cellular uptake and *in vivo* adjuvant activity. Moreover, the dosage for aluminium adjuvant required to reach an equivalent OVA-specific antibody titer was almost five times higher than Au–OVA (Yang et al. 2019). Another OVA application was in ulcerative colitis therapy where a nanoparticle was developed by heat treatment at 80 °C for 15 min and then an EGCG solution was added (Gou et al. 2018). The nanoparticles sized 170–480 nm and EE values were 80–99%, depending on EGCG/OVA molar ratio. The nanoparticles possessed excellent biocompatibility and achieved a much higher cellular uptake rate of EGCG than pure EGCG. Furthermore, EGCG–OVA nanoparticles had remarkably stronger capacity to suppress the secretion of pro-inflammatory mediators (e.g., tumor necrosis factor α, interleukin-6, interleukin-12) and promote the production of anti-inflammatory factor (interleukin-10), in comparison with pure EGCG. Finally, *in vivo* experiments demonstrated the excellent therapeutic efficacy of EGCG–OVA nanoparticles in alleviating ulcerative colitis in a mouse model (Gou et al. 2018).

From all of the above mentioned, it can be concluded that many techniques were used to produce OVA nanostructures, either using it in native form, generating nanoparticles by different methods or by combining it with other materials such as oils and polysaccharides (Table 5.4). With respect to BCs, a diversity of them were encapsulated based on OVA nanostructures. In this sense, the principal phenolic compounds investigated in order to improve their stability, antioxidant activity and bioavailability were catechins, epigallocatechin and EGCG (the main components of green tea), curcumin and resveratrol. Different lipids were studied with the same purpose such as sacha inchi oil, linoleic acid and mulberry seed oil; while vitamin D3 and retinol were investigated to a lesser extent. With respect to antibacterial applications of BCs encapsulated in OVA nanostructures, carvacrol and thyme oil (the principal constituent is thymol) were the most studied. Other BCs were encapsulated in OVA nanoparticles but not extensively studied, for example, CLA with antitumor activity or EGCG for ulcerative colitis therapy.

On the other hand, not only BC have been used in combination with OVA with antibacterial, antitumor or food industry purpose. For example, Song et al. (2013) developed silver/ovalbumin (Ag-OVA) films by using layer-by-layer self-assembled polyacrylonitrile (PAN) nanofibers with antimicrobial activity. PAN nanofibers were obtained by electrospinning, and subsequent assay for apoptosis detection in C2C12 cells (myocyte) revealed that assembled Ag–OVA film had favourable biocompatibility, being this less cytotoxic than the pure PAN one. On the other hand, the films showed greater activity against E. coli and S. aureus in comparison with pure Ag or PAN, being behaviour independent of the Ag amount in the film. Prabakaran et al. (2019) produced 280 nm nanoparticles to encapsulate doxorubicin, being the first step an emulsion–solvent evaporation technique to produce OVA-polymethyl methacrylate (PMMA) nanocarriers and in a second step, they were coated with graphene oxide particles (GO). The time taken to release the drug from OVA–PMMA–GO carrier was greater than the one to release the drug from OVA–PMMA carrier due to the graphene scaffold. Moreover, nanoparticles do not affect cell viability of normal VERO cells and improved cytotoxicity in CaCO2 colon cancer cell line (Prabakaran et al. 2019). In the last year, a modified selenium-ovalbumin was produced with antitumor purpose (Sun et al. 2020). Briefly, the seleno-ovalbumin was prepared by dissolving OVA in seleninic acid solution with 6% hydrogen peroxide, and reacted at 40 °C for 5 h. Then ethanol was added into the reacted solution to precipitate modified ovalbumin. Finally, the precipitate was freeze-dried to obtain seleno-ovalbumin. The authors found that seleno-ovalbumin could reduce the damage of cancer to immune organs, improve the proliferation capacities of T and B lymphocytes, enhance the cytotoxicity of the natural killer cells and increase the phagocytic activity of peritoneal macrophages of H22 hepatoma cell-bearing mice. Besides, seleno-OVA could block the cell cycle of solid tumors cells in the G0/G1 phase and accelerate the apoptosis of solid tumor cells through mitochondrial pathway (Sun et al. 2020).

Finally, in the food industry, Wongsasulak et al. (2010), produced composite edible nanofibers from cellulose acetate and egg albumen obtaining diameters between 265 and 410 nm. The authors suggested the edible polysaccharide-protein complex could find applications in food packaging materials due to its potential to promote safety and quality of packed foods.

5.6 Conclusion

In summary, BSA is the more important albumin and it has been widely used to produce nanoparticles by different methods such as coacervation, self-assembly, thermal treatment, desolvation, nab-technology, etc. Its main application is as a nanocarrier of paclitaxel, curcumin, chrysin, resveratrol, quercetin, EGCG and doxorubicin for medical purposes as antitumor and antioxidant and to a lesser extent for anti-inflammatory application and to improve stability and bioavailability of BCs. Because of its endogen origin, HSA is very safe and well tolerated in humans and, in consequence, BSA is usually used as a model for research while HSA is the material present in the final products. For this reason, HSA nanostructures have been extensively used for antitumor applications. With regard to this, the desolvation technique and nab-technology are the main methods used to produced HSA nanostructures. Moreover, phenolic compounds and in particular curcumin were encapsulated in HSA nanostructures developed in order to improve cytotoxic effects, cellular uptake and targeting delivery of BCs encapsulated. With respect to ALA, it has been used to design nanoparticles to carry curcumin, β-carotene, oleic acid, etc. The main applications include enhanced stability and bioavailability, anti-tumor, and anti-oxidative activities, which could be relevant for designing nanosupplements to be incorporated in functional foods. Finally, nanocarriers using OVA have been produced by a diversity of methods like thermal aggregation, complex self-assembly, emulsification, nanogelation, etc. The applications are very diverse and the main uses included improved stability, bioavailability, antioxidant capacity and antibacterial activity of BCs encapsulated. Many BCs have been delivered including phenolic compounds (catechin, EGCG, curcumin, resveratrol, carvacrol, etc.), lipids (linoleic acid, CLA, sacha inchi oil) and vitamins (Vitamin D3 and retinol). From the analysis of the literature, it could be concluded that the main use of albumins is as nanocarriers of BCs and drugs for medical purposes. In this sense, the production of nanosupplements using albumins as carriers and as functional food ingredients has emerged as a vacancy area.

References

Abolhassani, H. and Shojaosadati, S.A. (2019). A comparative and systematic approach to desolvation and self-assembly methods for synthesis of piperine-loaded human serum albumin nanoparticles. *Colloids and Surfaces B: Biointerfaces* 184: 110534.

Akbarian, A., Ebtekara, M., Pakravan, N. et al. (2020). Folate receptor alpha targeted delivery of artemether to breast cancer cells with folate-decorated human serum albumin nanoparticles. *International Journal of Biological Macromolecules* 152: 90–101.

Al-Hanish, A., Stanic-Vucinic, D., Prodic, J.M.I. et al. (2016). Noncovalent interactions of bovine α-lactalbumin with green tea polyphenol, epigalocatechin-3-gallate. *Food Hydrocolloids* 61: 241–250.

Anandharamakrishnan, C. (2014). Chapter 1: nanoencapsulation of food bioactive compounds. In: *Techniques for Nanoencapsulation of Food Ingredients* (ed. R.W. Hartel), 1–6. New York: Springer.

Anderson, P.J., Brooks, C.L., and Berliner, L.J. (1997). Functional identification of calcium binding residues in bovine α-lactalbumin. *Biochemistry* 36: 11648–11654.

Antônio, E., Khalil, N.M., and Mainardes, R.M. (2016). Bovine serum albumin nanoparticles containing quercetin: characterization and antioxidant activity. *Journal of Nanoscience and Nanotechnology* 16 (2): 1346–1353.

Arpagaus, C., Collenberg, A., Rütti, D. et al. (2018). Nano spray drying for encapsulation of pharmaceuticals. *International Journal of Pharmaceutics* 546 (1-2): 194–214.

Arroyo-Maya, I.J., Rodiles-López, J.O., Cornejo-Mazón, M. et al. (2012). Effect of different treatments on the ability of α-lactalbumin to form nanoparticles. *Journal of Dairy Science* 95 (11): 6204–6214.

Arroyo-Maya, I.J., Hernández-Sánchez, H., Jiménez-Cruz, E. et al. (2014). α-Lactalbumin nanoparticles prepared by desolvation and cross-linking: structure and stability of the assembled protein. *Biophysical Chemistry* 193–194: 27–34.

Balandrán-Quintana, R.R., Valdéz-Covarrubias, M.A., Mendoza-Wilson, A.M. et al. (2013). α-Lactalbumin hydrolysate spontaneously produces disk-shaped nanoparticles. *International Dairy Journal* 32 (2): 133–135.

Barbosa, L.R.S., Ortore, M.G., Spinozzi, F. et al. (2010). The importance of protein-protein interactions on the pH-induced conformational changes of bovine serum albumin: a small-angle X-Ray scattering study. *Biophysical Journal* 98 (1): 147–157.

Barone, G., Giancola, C., and Verdoliva, A. (1992). DSC studies on the denaturation and aggregation of serum albumins. *Thermochimica Acta* 199: 197–205.

Bender, D.A. (2003). Chapter 1: the vitamins. In: *Nutritional Biochemistry of the Vitamins*, 1–29. Cambridge: Cambridge University Press.

Bhattacharya, M., Jain, N., Bhasne, K. et al. (2011). pH-induced conformational isomerization of bovine serum albumin studied by extrinsic and intrinsic protein fluorescence. *Journal of Fluorescence* 21: 1083–1090.

Bhushan, B., Khanadeev, V., Khlebtsov, B. et al. (2017). Impact of albumin based approaches in nanomedicine: imaging, targeting and drug delivery. *Advances in Colloid and Interface Science* 246: 13–39.

Borah, P.K., Das, A.S., Mukhopadhyay, R. et al. (2020). Macromolecular design of folic acid functionalized amylopectin–albumin core–shell nanogels for improved physiological stability and colon cancer cell targeted delivery of curcumin. *Journal of Colloid and Interface Science* 580: 561–572.

Boye, J.I., Alli, I., and Ismail, A.A. (1996). Interactions involved in the gelation of bovine serum albumin. *Journal of Agricultural and Food Chemistry* 44: 996–1004.

Carter, D.C. and Ho, J.X. (1994). Structure of serum albumin. *Advances in Protein Chemistry* 45: 153–203.

Casa, D.M., Scariot, D.B., Khalil, N.M. et al. (2018). Bovine serum albumin nanoparticles containing amphotericin B were effective in treating murine cutaneous leishmaniasis and reduced the drug toxicity. *Experimental Parasitology* 192: 12–18.

Catanzaro, G., Curcio, M., Cirillo, G. et al. (2017). Albumin nanoparticles for glutathione-responsive release of cisplatin: new opportunities for medulloblastoma. *International Journal of Pharmaceutics* 517: 168–174.

Cawthern, K.M., Narayan, M., Chaudhuri, D. et al. (1997). Interactions of α-lactalbumin with fatty acids and spin label analogs. *Journal of Biological Chemistry* 272: 30812–30816.

Chang, J., Kang, X., and Yuan, J.L. (2018). Enhancing emulsification and antioxidant ability of egg albumin by moderately acid hydrolysis: modulating an emulsion-based system for mulberry seed oil. *Food Research International* 109: 334–342.

Chaux-Gutiérrez, A.M., Pérez-Monterroza, E.J., Granda-Restrepo, D.M. et al. (2020). Cryogels from albumin and low methoxyl amidated pectin as a matrix for betalain encapsulation. *Journal of Food Processing and Preservation* 44 (11): e14843.

Chen, Y., Hu, J., Yi, X. et al. (2018). Interactions and emulsifying properties of ovalbumin with tannic acid. *LWT – Food Science and Technology* 95: 282–288.

Cheng, H., Fang, Z., Wusigale, A.M.B. et al. (2018). Complexation of trans- and cis-resveratrol with bovine serum albumin, β-lactoglobulin or α-lactalbumin. *Food Hydrocolloids* 81: 242–252.

Croguennec, T., Renault, A., Beaufils, S. et al. (2007). Interfacial properties of heat-treated ovalbumin. *Journal of Colloid and Interface Science* 315: 627–636.

Darvishi, B., Manoochehri, S., Esfandyari-Manesh, M. et al. (2015). Enhanced cellular cytotoxicity and antibacterial activity of 18-β-glycyrrhetinic acidby albumin-conjugated PLGA nanoparticles. *Drug Research* 65: 617–623.

Davidov-Pardo, G., Joye, I.J., and McClements, D.J. (2015). Food-grade protein-based nanoparticles and microparticles for bioactive delivery: fabrication, characterization, and utilization. In: *Advances in Protein Chemistry and Structural Biology* (ed. R. Donev), Vol. 98, 293–325. Academic Press.

Delavari, B., Saboury, A.A., Atri, M.S. et al. (2015). Alpha-lactalbumin: a new carrier for vitamin D3 food enrichment. *Food Hydrocolloids* 45: 124–131.

Delgado, Y., Morales-Cruza, M., Figueroa, C.M. et al. (2015). The cytotoxicity of BAMLET complexes is due to oleic acid and independent of the α-lactalbumin component. *FEBS Open Bio* 5: 397–404.

Deng, W., Li, J., Yao, P. et al. (2010). Green preparation process, characterization and antitumor effects of doxorubicin-BSA-dextran nanoparticles. *Macromolecular Bioscience* 10 (10): 1224–1234.

de Oliveira, J.K., Ronik, D.F.V., Ascari, J. et al. (2018). Nanoencapsulation of apocynin in bovine serum albumin nanoparticles: physicochemical characterization. *Nanoscience and Nanotechnology* 8 (1): 90–99.

Deretic, V. (2006). Autophagy as an immune defense mechanism. *Current Opinion in Immunology* 18: 375–382.

Dergal, S.B. (2006). Chapter 4: lipids. In: *Food Chemistry* (ed. E.Q. Duarte), 4e, 245–300. Mexico: Pearson Education.

Desai, N., Trieu, V., Yao, Z. et al. (2006). Increased antitumor activity, intratumor paclitaxel concentrations and endothelial cell transport of cremophor-free, albumin-bound paclitaxel, ABI-007, compared with cremophor-based paclitaxel. *Clinical Cancer Research* 12 (4): 1317–1324.

Dhayal, S.K., Gruppen, H., De Vries, R. et al. (2014). Controlled formation of protein nanoparticles by enzymatic cross-linking of α-lactalbumin with horseradish peroxidase. *Food Hydrocolloids* 36: 53–59.

Dockal, M., Carter, D.C., and Rúcker, F. (2000). Conformational transitions of the three recombinant domains of human serum albumin depending on pH. *The Journal of Biological Chemistry* 275: 3042–3050.

Dong, X., Sun, Z., Liang, J. et al. (2018). A visible fluorescent nanovaccine based on functional genipin crosslinked ovalbumin protein nanoparticles. *Nanomedicine: Nanotechnology, Biology, and Medicine* 14: 1087–1098.

Dong, Y., Fu, R., Yang, J. et al. (2019). Folic acid-modified ginsenoside Rg5-loaded bovine serum albumin nanoparticles for targeted cancer therapy in vitro and in vivo. *International Journal of Nanomedicine* 14: 6971–6988.

Dong, H., Yin, X., Wusigale, C.H. et al. (2020). Antioxidant activity and stability of α-tocopherol, resveratrol and epigallocatechin-3-gallate in mixture and complexation with bovine serum albumin. *International Journal of Food Science and Technology*. doi: 10.1111/ijfs.14804.

Dreis, S., Rothweiler, F., Michaelis, M. et al. (2007). Preparation, characterization and maintenance of drug efficacy of doxorubicin-loaded human serum albumin (HSA) nanoparticles. *International Journal of Pharmaceutics* 341: 207–214.

Du, Y., Bao, C., Huang, J. et al. (2018). Improved stability, epithelial permeability and cellular antioxidant activity of β-carotene via encapsulation by self-assembled α-lactalbumin micelles. *Food Chemistry* 271: 707–714.

Dufour, C. and Dangles, O. (2005). Flavonoid-serum albumin complexation: determination of binding constants and binding sites by fluorescence spectroscopy. *Biochimica et Biophysica Acta* 1721 (1-3): 164–173.

Duringer, C., Hamiche, A., Gustafsson, L. et al. (2003). Hamlet interacts with histones and chromatin in tumor cell nuclei. *Journal of Biological Chemistry* 278: 42131–42135.

Dzwolak, W., Kato, M., Shimizu, A. et al. (1999). Fourier-transform infrared spectroscopy study of the pressure-induced changes in the structure of the bovine α-lactalbumin: the stabilizing role of the calcium ion. *Biochimica et Biophysica Acta* 1433: 45–55.

Elsadek, B. and Kratz, F. (2012). Impact of albumin on drug delivery- New applications on the horizon. *Journal of Controlled Release* 157: 4–28.

Elzoghby, A.O., Samy, W.M., and Elgindy, N.A. (2012). Albumin-based nanoparticles as potential controlled release drug delivery systems. *Journal of Controlled Release* 157: 168–182.

Fan, Y., Yi, J., Zhang, Y. et al. (2018). Fabrication of curcumin-loaded bovine serum albumin (BSA)-dextran nanoparticles and the cellular antioxidant activity. *Food Chemistry* 239: 1210–1218.

Fanali, G., di Masi, A., Trezza, V. et al. (2011). Human serum albumin: from bench to bedside. *Molecular Aspects of Medicine* 33: 209–290.

Fang, R., Jing, H., Chai, Z. et al. (2011). Design and characterization of protein-quercetin bioactive nanoparticles. *Journal of Nanobiotechnology* 9: 19.

Fattahian Kalhor, N., Saeidifar, M., Ramshini, H. et al. (2020). Interaction, cytotoxicity and sustained release assessment of a novel anti-tumor agent using bovine serum albumin nanocarrier. *Journal of Biomolecular Structure & Dynamics* 38 (9): 2546–2558.

Feng, J., Wu, S., Wang, H. et al. (2016). Improved bioavailability of curcumin in ovalbumin-dextran nanogels prepared by Maillard reaction. *Journal of Functional Foods* 27: 55–68.

Feng, J., Cai, H., Wang, H. et al. (2018). Improved oxidative stability of fish oil emulsion by grafted ovalbumin catechin conjugates. *Food Chemistry* 241: 60–69.

Feng, J., Wu, Y., Zhang, L. et al. (2019). Enhanced chemical stability, intestinal absorption, and intracellular antioxidant activity of cyanidin-3-O-glucoside by composite nanogel encapsulation. *Journal of Agricultural and Food Chemistry* 67 (37): 10432–10447.

Ferrado, J.B., Pérez, A.A., Visentini, F.F. et al. (2019). Formation and characterization of self-assembled bovine serum albumin nanoparticles as chrysin delivery systems. *Colloids and Surfaces B: Biointerfaces* 173 (1): 43–51.

Ferrado, J.B., Perez, A.A., Ruiz, M.C. et al. (2020). Chrysin-loaded bovine serum albumin particles as bioactive nanosupplements. *Food & Function* 11: 6007–6019.

Fonseca, P.D., Khalil, N.M., and Mainardes, R.M. (2017). Bovine serum albumin-based nanoparticles containing resveratrol: characterization and antioxidant activity. *Journal of Drug Delivery Science and Technology* 39: 147–155.

Fuciños, C., Míguez, M., Fuciños, P. et al. (2017). Creating functional nanostructures: encapsulation of caffeine into α-lactalbumin nanotubes. *Innovative Food Science and Emerging Technologies* 40: 10–17.

Gaber, M., Hany, M., Mokhtar, S. et al. (2019). Boronic-targeted albumin-shell oily-core nanocapsules for synergistic aromatase inhibitor/herbal breast cancer therapy. *Materials Science and Engineering C* 105: 110099.

Galisteo-González, F., Molina-Bolívar, J.A., Navarro, S.A. et al. (2018). Albumin-covered lipid nanocapsules exhibit enhanced uptake performance by breast-tumor cells. *Colloids and Surfaces B: Biointerfaces* 165: 103–110.

Gawde, K.A., Kesharwani, P., Sau, S. et al. (2018). Synthesis and characterization of folate decorated albumin bio-conjugate nanoparticles loaded with a synthetic curcumin difluorinated analogue. *Journal of Colloid and Interface Science* 15 (496): 290–299.

Geng, X.L., Bjerrum, M.J., Arleth, L. et al. (2016). Formation of nanotubes and gels at a broad pH range upon partial hydrolysis of bovine α-lactalbumin. *International Dairy Journal* 52: 72–81.

Geng, T., Zhao, X., Ma, M. et al. (2017). Resveratrol-loaded albumin nanoparticles with prolonged blood circulation and improved biocompatibility for highly effective targeted pancreatic tumor therapy. *Nanoscale Research Letters* 12: 437–447.

Gholijani, N., Abolmaali, S., Kalantar, K. et al. (2020). Therapeutic effect of carvacrol-loaded albumin nanoparticles on arthritic rats. *Iranian Journal of Pharmaceutical Research* 19: 312–320.

Ghosh, P., Roy, A.S., Chaudhury, S. et al. (2016). Preparation of albumin-based nanoparticles for delivery of fisetin andevaluation of its cytotoxic activity. *International Journal of Biological Macromolecules* 86: 408–417.

Giancola, C., De Sena, C., Fessas, D. et al. (1997). DSC studies on bovine serum albumin denaturation. Effects of ionic strength and SDS concentration. *International Journal of Biological Macromolecules* 20: 193–204.

Giosafatto, C.V.L., Rigby, N.M., Wellner, N. et al. (2012). Microbial transglutaminase-mediated modification of ovalbumin. *Food Hydrocolloids* 26: 261–267.

Gong, J., Huo, M., Zhou, J. et al. (2009). Synthesis, characterization, drug-loading capacity and safety of novel octyl modified serum albumin micelles. *International Journal of Pharmaceutics* 376: 161–168.

Gong, T., Zhang, P., Deng, C. et al. (2019). An effective and safe treatment strategy for rheumatoid arthritis based on human serum albumin and Kolliphor® HS 15. *Nanomedicine* 14 (16): 2169–2187.

Gong, T., Tan, T., Zhang, P. et al. (2020). Palmitic acid-modified bovine serum albumin nanoparticles target scavenger receptor-A on activated macrophages to treat rheumatoid arthritis. *Biomaterials* 258: 120296.

Gonnet, M., Lethuaut, L., and Boury, F. (2010). New trends in encapsulation of liposoluble vitamins. *Journal of Controlled Release* 146 (3): 276–290.

Gou, S., Chen, Q., Liu, Y. et al. (2018). Green fabrication of ovalbumin nanoparticles as natural polyphenol carriers for ulcerative colitis therapy. *ACS Sustainable Chemistry and Engineering* 6 (10): 12658–12667.

Graveland-Bikker, J.F. and de Kruif, C.G. (2006). Unique milk protein based nanotubes: food and nanotechnology meet. *Trends Food Science Technology* 17: 196–203.

Griko, Y.V., Freire, E., and Privalov, P.L. (1994). Energetics of the alpha-lactalbumin states: a calorimetric and statistical thermodynamic study. *Biochemistry* 33: 1889–1899.

Guaadaoui, A., Benaicha, S., Elmajdoub, N. et al. (2014). What is a bioactive compound? A combined definition for a preliminary consensus. *International Journal of Food Sciences and Nutrition* 3 (3): 174–179.

Guo, X., Liu, Y., Bera, H. et al. (2020). α-lactalbumin-based nanofiber dressings improve burn wound healing and reduce scarring. *ACS Applied Materials & Interfaces* 12 (41): 45702–45713.

Gustafsson, L., Leijonhufvud, I., Aronsson, A. et al. (2004). Treatment of skin papillomas with topical α-lactalbumin-oleic acid. *The New England Journal of Medicine* 350: 2663–2672.

Hallgren, O., Gustafsson, L., Irjala, H. et al. (2006). Hamlet triggers apoptosis but tumor cell death is independent of caspases, Bcl-2 and p53. *Apoptosis* 11: 221–233.

Hao, H., Ma, Q., Huang, C. et al. (2013). Preparation, characterization, and in vivo evaluation of doxorubicin loaded BSA nanoparticles with folic acid modified dextran surface. *International Journal of Pharmaceutics* 444 (1–2): 77–84.

He, Y., Ning, T., Xie, T. et al. (2011). Large-scale production of functional human serum albumin from transgenic rice sedes. *PNAS* 108 (47): 19078–19083.

Hill, R.L. and Brew, K. (1975). Lactose synthetase. *Advances in Enzymology and Related Areas of Molecular Biology* 43: 411–490.

Hirai, Y., Permyakov, E.A., and Berliner, L.J. (1992). Proteolytic digestion of α-lactalbumin: physiological implications. *Journal of Protein Chemistry* 11: 51–57.

Hiraoka, Y., Segawa, T., Kuwajima, K. et al. (1980). α-Lactalbumin: a calcium metalloprotein. *Biochemical and Biophysical Research Communications* 95: 1098–1104.

Hu, Y., Bao, C., Li, D. et al. (2019). The construction of enzymolyzed α-lactalbumin based micellar nanoassemblies for encapsulating various kinds of hydrophobic bioactive compounds. *Food & Function* 10: 8263–8272.

Humblet-Hua, K.N.P., Scheltens, G., van der Linden, E. et al. (2011). Encapsulation systems based on ovalbumin fibrils and high methoxyl pectin. *Food Hydrocolloids* 25 (3): 307–314.

Huntington, J.A. and Stein, P.E. (2001). Structure and properties of ovalbumin. *Journal of Chromatography B* 756: 189–198.

Jafari, S.M. (2017). Chapter 1: an overview of nanoencapsulation techniques and their classification. In: *Nanoencapsulation Technologies for the Food and Nutraceutical Industries* (ed. S.M. Jafari), 1–34. Academic Press.

Jahanshahi, M. and Babaei, Z. (2008). Protein nanoparticle: a unique system as drug delivery vehicles. *African Journal of Biotechnology* 7 (25): 4926–4934.

Jiang, P., Huang, J., Bao, C. et al. (2018). Enzymatically partially hydrolyzed α-lactalbumin peptides for self-assembled micelle formation and their application for coencapsulation of multiple antioxidants. *Journal of Agricultural and Food Chemistry* 66 (49): 12921–12930.

Jithan, A.V., Madhavi, K., Madhavi, M. et al. (2011). Preparation and characterization of albumin nanoparticles encapsulating curcumin intended for the treatment of breast cancer. *International Journal of Pharmaceutical Investigation* 1 (2): 119–125.

Jose, L.M. and Kuriakose, S. (2019). Spectroscopic and thermal investigation of silver nanoparticle dispersed biopolymer matrix bovine serum albumin: a promising antimicrobial agent against the pathogenic bacterial strains. *Macromolecular Research* 27: 670–678.

Kalouta, K., Stie, M.B., Janfelt, C. et al. (2020). Electrospun α-lactalbumin nanofibers for site-specific and fast-onset delivery of nicotine in the oral cavity: an in vitro, ex vivo and tissue spatial distribution study. *Molecular Pharmaceutics* 17 (11): 4189–4200.

Kandagal, P.B., Ashoka, S., Seetharamappa, J. et al. (2006). Study of the interaction of an anticancer drug with human and bovine serum albumin: spectroscopic approach. *Journal of Pharmaceutical and Biomedical Analysis* 41: 393–399.

Kayani, Z., Firuzi, O., and Bordbar, A.K. (2018). Doughnut-shaped bovine serum albumin nanoparticles loaded with doxorubicin for overcoming multidrug-resistant in cancer cells. *International Journal of Biological Macromolecules* 107: 1835–1843.

Kazan, A., Yesil-Celiktas, O., and Zhang, Y.S. (2019). Fabrication of thymoquinone-loaded albumin nanoparticles by microfluidic particle synthesis and their effect on planarian regeneration. *Macromolecular Bioscience* 19 (11): 1900182.

Kehoe, J.J. and Brodkorb, A. (2014). Interactions between sodium oleate and α-lactalbumin: the effect of temperature and concentration on complex formation. *Food Hydrocolloids* 34: 217–226.

Kilic, E., Novoselova, M., Lim, S. et al. (2019). Formulation for oral delivery of lactoferrin based on bovine serum albumin and tannic acid multilayer microcapsules. *Scientific Reports* 7: 44159.

Kim, T.H., Jiang, H.H., Youn, Y.S. et al. (2011). Preparation and characterization of water-soluble albumin-bound curcumin nanoparticles with improved antitumor activity. *International Journal of Pharmaceutics* 403: 285–291.

Kim, B., Lee, C., Lee, E.S. et al. (2016). Paclitaxel and curcumin co-bound albumin nanoparticles having antitumor potential to pancreatic cancer. *Asian Journal of Pharmaceutical Sciences* 11: 708–714.

Kobayashi, K. (2006). Summary of recombinant human serum albumin development. *Biologicals* 34: 55–59.

Kohler, C., Gogvadze, V., Hakansson, A. et al. (2001). A folding variant of human α-lactalbumin induces mitochondrial permeability transition in isolated mitochondria. *European Journal of Biochemistry* 268: 186–191.

Kramer, P.A. (1974). Albumin microspheres as vehicles for achieving specificity in drug delivery. *Journal of Pharmaceutical Sciences* 63 (10): 1646–1647.

Kratz, F. (2008). Albumin as a drug carrier: design of prodrugs, drug conjugates and nanoparticles. *Journal of Controlled Release* 132: 171–183.

Kratz, F. (2014). A clinical update of using albumin as a drug vehicle — a commentary. *Journal of Controlled Release* 190: 331–336.

Kratz, F. and Elsadek, B. (2012). Clinical impact of serum proteins on drug delivery. *Journal of Controlled Release* 161: 429–445.

Kristó, K., Szekeres, M., Makai, Z., Márki, Á., Kelemen, A., Bali, L., Pallai, Z., Dékány, I., and Csóka, I. (2020). Preparation and investigation of core–shell nanoparticles containing human interferon-α. *International Journal of Pharmaceutics* 5 (573): 118825.

Kumar, S., Meena, R., and Rajamani, P. (2016). Fabrication of BSA-green tea polyphenols-chitosan nanoparticles and their role in radioprotection: a molecular and biochemical approach. *Journal of Agricultural and Food Chemistry* 64 (30): 6024–6034.

Kumari, M., Liu, C.H., and Wu, W.C. (2018). Efficient gene delivery by oligochitosan conjugated serum albumin: facile synthesis, polyplex stability, and transfection. *Carbohydrate Polymers* 183: 37–49.

Langer, K., Anhorn, M.G., Steinhauser, I. et al. (2008). Human serum albumin (HSA) nanoparticles: reproducibility of preparation process and kinetics of enzymatic degradation. *International Journal of Pharmaceutics* 347: 109–117.

Lee, S.H., Heng, D., Ng, W.K. et al. (2011). Nano spray drying: a novel method for preparing protein nanoparticles for protein therapy. *International Journal of Pharmaceutics* 403 (1-2): 192–200.

Lee, E.S. and Youn, Y.S. (2016). Albumin-based potential drugs: focus on half-life extension and nanoparticle preparation. *Journal of Pharmaceutical Investigation* 46: 305–315.

Lee, S. (2019). Human serum albumin: a nanomedicine platform targeting breast cancer Cells. *Journal of Drug Delivery Science and Technology* 52: 652–659.

Lei, X., Li, K., Liu, Y. et al. (2017). Co-delivery nanocarriers targeting folate receptor and encapsulating 2-deoxyglucose and α-tocopheryl succinate enhance anti-tumor effect in vivo. *International Journal of Nanomedicine* 12: 5701–5715.

Li, Y., Lee, J., Lal, J. et al. (2008). Effects of pH on the interactions and conformation of bovine serum albumin: comparison between chemical force microscopy and small-angle neutron scattering. *The Journal of Physical Chemistry B* 112: 3797–3806.

Li, Z. and Gu, L. (2014). Fabrication of self-assembled (−)-epigallocatechin gallate (EGCG) ovalbumin–dextran conjugate nanoparticles and their transport across monolayers of human intestinal epithelial Caco-2 cells. *Journal of Agricultural and Food Chemistry* 62: 1301–1309.

Li, Z., Ha, J., Zou, T. et al. (2014). Fabrication of coated bovine serum albumin (BSA)–epigallocatechin gallate (EGCG) nanoparticles and their transport across monolayers of human intestinal epithelial Caco-2 cells. *Food & Function* 5 (6): 1278–1285.

Li, D., Shi, M., Bao, C. et al. (2019). Synergistically enhanced anticancer effect of codelivered curcumin and siPlk1 by stimuli-responsive α-lactalbumin nanospheres. *Nanomedicine* 14 (5): 595–612.

Liu, Y., Ying, D., Cai, Y. et al. (2017). Improved antioxidant activity and physicochemical properties of curcumin by adding ovalbumin and its structural characterization. *Food Hydrocolloids* 72: 304–311.

Liu, Y., Fan, Y., Gao, L. et al. (2018a). Enhanced pH and thermal stability, solubility and antioxidant activity of resveratrol by nanocomplexation with α-lactalbumin. *Food & Function* 9: 4781–4790.

Liu, Y., Cai, Y., Ying, D. et al. X. (2018b). Ovalbumin as a carrier to significantly enhance the aqueous solubility and photostability of curcumin: interaction and binding mechanism study. *International Journal of Biological Macromolecules* 116: 893–900.

Liu, G., Yuan, D., Wang, Q. et al. (2018c). Maillard-reaction-functionalized egg ovalbumin stabilizes oil nanoemulsions. *Journal of Agricultural and Food Chemistry* 66 (16): 4251–4258.

Liu, Y., Qiao, Z., Liu, W. et al. (2019). Oleic acid as a protein ligand improving intestinal absorption and ocular benefit of fucoxanthin in water through protein-based encapsulation. *Food & Function* 10 (7): 4381–4395.

Liu, Y., Huang, L., Li, D. et al. (2020a). Re-assembled oleic acid-protein complexes as nano-vehicles for astaxanthin: multispectral analysis and molecular docking. *Food Hydrocolloids* 103: 105689.

Liu, Y., Gao, L., Yi, J. et al. (2020b). α-Lactalbumin and chitosan core–shell nanoparticles: resveratrol loading, protection, and antioxidant activity. *Food & Function* 11: 1525–1536.

Look, J., Wilhelm, N., von Briesen, H. et al. (2015). Ligand-modified human serum albumin nanoparticles for enhanced gene delivery. *Molecular Pharmaceutics* 12: 3202–3213.

Loureiro, A., Azoia, N.G., Gomes, A.C. et al. (2016). Albumin-based nanodevices as drug carriers. *Current Pharmaceutical Design* 22 (10): 1371–1390.

Luis de Redín, I., Boiero, C., Martínez-Ohárriz, M.C. et al. (2018). Human serum albumin nanoparticles for ocular delivery of bevacizumab. *International Journal of Pharmaceutics* 541: 214–223.

Luo, Y., Teng, Z., and Wang, Q. (2012). Development of zein nanoparticles coated with carboxymethyl chitosan for encapsulation and controlled release of vitamin D3. *Journal of Agricultural and Food Chemistry* 60 (3): 836–843.

Luo, R., Lin, M., Zhang, C. et al. (2020). Genipin-crosslinked human serum albumin coating using a tannic acid layer for enhanced oral administration of curcumin in the treatment of ulcerative colitis. *Food Chemistry* 330: 127241.

Maghsoudi, A., Shojaosadati, S.A., and Farahani, E.V. (2008). 5-Fluorouracil-loaded BSA nanoparticles: formulation optimization and in vitro release study. *AAPS PharmSciTech* 9 (4): 1092–1096.

Mahanta, S. and Paul, S. (2015). Stable self-assembly of bovine α-lactalbumin exhibits target-specific antiproliferative activity in multiple cancer cells. *ACS Applied Materials & Interfaces* 7 (51): 28177–28187.

Maldonado, L., Sadeghi, R., and Kokini, J. (2017). Nanoparticulation of bovine serum albumin and poly-d-lysine through complex coacervation and encapsulation of curcumin. *Colloids and Surfaces B: Biointerfaces* 159: 759–769.

Manach, C., Scalbert, A., Morand, C. et al. (2004). Polyphenols: food sources and bioavailability. *The American Journal of Clinical Nutrition* 79 (5): 727–747.

Maryam, K., Shakeri, S., and Kiani, K. (2015). Preparation and in vitro investigation of antigastric cancer activities of carvacrol-loaded human serum albumin nanoparticles. *IET Nanobiotechnology* 9 (5): 294–299.

Matalanis, A., Jones, O.G., and McClements, D.J. (2011). Structured biopolymer-based delivery systems for encapsulation, protection, and release of lipophilic compounds. *Food Hydrocolloids* 25 (8): 1865–1880.

Matsumura, Y., Lee, D.S., and Mori, T. (2000). Molecular weight distributions of α-lactalbumin polymers formed by mammalian and microbial transglutaminases. *Food Hydrocolloids* 14 (1): 49–59.

Merodio, M., Arnedo, A., Renedo, M.J. et al. (2001). Ganciclovir-loaded albumin nanoparticles: characterization and in vitro release properties. *European Journal of Pharmaceutical Sciences* 12: 251–259.

Michaelis, K., Hoffmann, M.M., Dreis, S. et al. (2006). Covalent linkage of apolipoprotein E to albumin nanoparticles strongly enhances drug transport into the brain. *The Journal of Pharmacology and Experimental Therapeutics* 317: 1246–1253.

Michnik, A., Michalik, K., Kluczewska, A. et al. (2006). Comparative DSC study of human and bovine serum albumin. *Journal of Thermal Analysis and Calorimetry* 84: 113–117.

Mohammad-Beigi, H., Shojaosadati, S.A., Morshedi, D. et al. (2015). Preparation and in vitro characterization of gallic acid-loaded human serum albumin nanoparticles. *Journal of Nanoparticle Research* 17 (4): 167.

Monastra, G., Sambuy, Y., Ferruzza, S. et al. (2018). Alpha-lactalbumin Effect on Myo-inositol Intestinal Absorption: In vivo and In vitro. *Current Drug Delivery* 15 (9): 1305–1311.

Monteiro, A.A., Monteiro, M.R., Pereira, R.N. et al. (2016). Design of bio-based supramolecular structures through self-assembly of α-lactalbumin and lysozyme. *Food Hydrocolloids* 58: 60–74.

Montero, G., Arriagada, F., Günther, G. et al. (2019). Phytoestrogen coumestrol: antioxidant capacity and its loading in albumin nanoparticles. *International Journal of Pharmaceutics* 1 (562): 86–95.

Murayama, K. and Tomida, M. (2004). Heat-induced secondary structure and conformation change of bovine serum albumin investigated by Fourier transform infrared spectroscopy. *Biochemistry* 43 (36): 11526–11532.

Naveenraj, S. and Anandan, S. (2013). Binding of serum albumins with bioactive substances-Nanoparticles to drugs. *Journal of Photochemistry and Photobiology C* 14: 53–71.

Nitta, K. and Sugai, S. (1989). The evolution of lysozyme and α-lactalbumin. *European Journal of Biochemistry* 182: 111–118.

Niu, F., Pan, W., Su, Y. et al. (2016). Physical and antimicrobial properties of thyme oil emulsions stabilized by ovalbumin and gum Arabic. *Food Chemistry* 212: 138–145.

Nosrati, H., Abbasi, R., Charmi, J. et al. (2018a). Folic acid conjugated bovine serum albumin: an efficient smart and tumor targeted biomacromolecule for inhibition folate receptor positive cancer cells. *International Journal of Biological Macromolecules* 117: 1125–1132.

Nosrati, H., Rakhshbahar, A., Salehiabar, M. et al. (2018b). Bovine serum albumin: an efficient biomacromolecule nanocarrier for improving the therapeutic efficacy of chrysin. *Journal of Molecular Liquids* 271: 639–646.

Pang, X., Yang, P., Wang, L. et al. (2019). Human serum albumin nanoparticulate system with encapsulation of gefitinib for enhanced anti-tumor effects in non-small cell lung cancer. *Journal of Drug Delivery Science and Technology* 52: 997–1007.

Papadopoulou, A., Green, R.J., and Frazier, R.A. (2005). Interaction of flavonoids with bovine serum albumin: a fluorescence quenching study. *Journal of Agricultural and Food Chemistry* 53 (1): 158–163.

Pápay, Z.E., Kósa, A., Böddi, B. et al. (2017). Study on the pulmonary delivery system of apigenin-loaded albumin nanocarriers with antioxidant activity. *Journal of Aerosol Medicine and Pulmonary Drug Delivery* 30: 1–15.

Peng, H., Chen, S., Luo, M. et al. (2016). Preparation and self-assembly mechanism of bovine serum albumin–citrus peel pectin conjugated hydrogel: a potential delivery system for vitamin C. *Journal of Agricultural and Food Chemistry* 64 (39): 7377–7384.

Permyakov, E.A., Yarmolenko, V.V., Kalinichenko, L.P. et al. (1981). Calcium binding to α-lactalbumin: structural rearrangement and association constant evaluation by means of intrinsic protein fluorescence changes. *Biochemical and Biophysical Research Communications* 100: 191–197.

Permyakov, E.A., Morozova, L.A., and Burstein, E.A. (1985). Cation binding effects on the pH, thermal and urea denaturation transitions in α-lactalbumin. *Biophysical Chemistry* 21: 21–31.

Permyakov, E.A. and Kreimer, D.I. (1986). Effects of pH, temperature and Ca2+ content on the conformation of alpha-lactalbumin in a medium modelling physiological condition. *General Physiology and Biophysics* 5: 377–390.

Permyakov, E.A., Shnyrov, V.L., Kalinichenko, L.P. et al. (1991a). Binding of Zn(II) ions to α-lactalbumin. *Journal of Protein Chemistry* 10: 577–584.

Permyakov, E.A., Grishchenko, V.M., Kalinichenko, L.P. et al. (1991b). Calcium-regulated interactions of human α-lactalbumin with bee venom melittin. *Biophysical Chemistry* 39: 111–117.

Peters, T., Jr. (1985). Serum albumin. *Advances in Protein Chemistry* 37: 161–245.

Peters, T., Jr. (1996). *All about Albumins: Biochemistry, Genetics and Medical Applications.* San Diego, CA: Academic Press.

Phuong, P.T.T., Lee, S., Lee, C. et al. (2018). Beta-carotene-bound albumin nanoparticles modified with chlorin e6 for breast tumor ablation based on photodynamic therapy. *Colloids and Surfaces B: Biointerfaces* 171: 123–133.

Piazzini, V., Landucci, E., D'Ambrosio, M. et al. (2019). Chitosan coated human serum albumin nanoparticles: a promising strategy for nose-to-brain drug delivery. *International Journal of Biological Macromolecules* 129: 267–280.

Prabakaran, S., Jeyaraj, M., Nagaraj, A. et al. (2019). Polymethyl methacrylate–ovalbumin @ graphene oxide drug carrier system for high anti-proliferative cancer drug delivery. *Applied Nanoscience* 9: 1487–1500.

Qi, L., Guo, Y., Luan, J. et al. (2014). Folate-modified bexarotene-loaded 774 bovine serum albumin nanoparticles as a promising tumor-targeting delivery system. 775 *Journal of Materials Chemistry B* 2: 8361–8371.

Rafipour, A., Yari, A., Farhadi, K. et al. (2020). Encapsulation of L-dopa and catechol in bovine serum albumin nanocarrier using the desolvation method and their in vitro release studies. *Journal of the Chinese Chemical Society* 67: 2082–2090.

Ramesh, N. and Mandal, A.K.A. (2019). Encapsulation of epigallocatechin-3-gallate into albumin nanoparticles improves pharmacokinetic and bioavailability in rat model. *3 Biotech* 9 (6): 238.

Rao, S., Xu, G., Lu, X. et al. (2020a). Characterization of ovalbumin-carvacrol inclusion complexes as delivery systems with antibacterial application. *Food Hydrocolloids* 105: 105753.

Rao, S., Xu, G., Zeng, H. et al. (2020b). Physicochemical and antibacterial properties of fabricated ovalbumin–carvacrol gel nanoparticles. *Food & Function* 11: 5133–5141.

Ray, S., Raychaudhuri, U., and Chakraborty, R. (2016). An overview of encapsulation of active compounds used in food products by drying technology. *Food Bioscience* 13: 76–83.

Robles-García, M.A., Rodríguez-Félix, F., Márquez-Ríos, E. et al. (2016). Applications of nanotechnology in the agriculture, food, and pharmaceuticals. *Journal of Nanoscience and Nanotechnology* 16 (8): 8188–8207.

Saha, S., Kundu, J., Verma, R.J. et al. (2020). Albumin coated polymer nanoparticles loaded with plant extract derived quercetin for modulation of inflammation. *Materialia* 9: 100605.

Saleh, T., Soudi, T., and Shojaosadati, S.A. (2018). Redox responsive curcumin-loaded human serum albumin nanoparticles: preparation, characterization and in vitro evaluation. *International Journal of Biological Macromolecules* 114: 759–766.

Saleh, T., Soudi, T., and Shojaosadati, S.A. (2019). Aptamer functionalized curcumin-loaded human serum albumin (HSA) nanoparticles for targeted delivery to HER-2 positive breast cancer cells. *International Journal of Biological Macromolecules* 130: 109–116.

Salehiabar, M., Nosrati, H., Javani, E. et al. (2018). Production of biological nanoparticles from bovine serum albumin as controlled release carrier for curcumin delivery. *International Journal of Biological Macromolecules* 115: 83–89.

Saricay, Y., Wierenga, P., and De Vries, R. (2013). Nanostructure development during peroxidase catalysed cross-linking of α-lactalbumin. *Food Hydrocolloids* 33 (2): 280–288.

Scheffel, U., Rhodes, B.A., Natarajan, T.K. et al. (1972). Albumin microspheres for study of the reticuloendothelial system. *Journal of Nuclear Medicine* 13: 498–503.

Segura, S., Espuelas, S., Renedo, M.J. et al. (2005). Potential of albumin nanoparticles as carriers for interferon gamma. *Drug Development and Industrial Pharmacy* 31: 271–280.

Sen, P., Ahmad, B., and Khan, R.H. (2008). Formation of a molten globule like state in bovine serum albumin at alkaline pH. *European Biophysics Journal* 37: 1303–1308.

Sheng, B., Li, L., Zhang, X. et al. (2018). Physicochemical properties and chemical stability of β-carotene bilayer emulsion coated with bovine serum albumin and Arabic gum compared to monolayer emulsions. *Molecules* 23 (2): 495.

Shinohara, H., Iwasaki, T., Miyazaki, Y. et al. (2005). Thermostabilized ovalbumin that occurs naturally during development accumulates in embryonic tissues. *Biochimica et Biophysica Acta* 1723: 106–113.

Shojaee, M., Navaee, F., Jalili–Firoozinezhad, S. et al. (2015). Fabrication and characterization of ovalbumin films for wound dressing applications. *Materials Science and Engineering C* 48: 158–164.

Shojania, H.R., Momeni-Moghaddam, M., Hossini, S.E. et al. (2019). MicroRNA 155 downregulation by vitamin C-loaded human serum albumin nanoparticles during cutaneous wound healing in mice. *The International Journal of Lower Extremity Wounds* 18: 143–152.

Singh, P., Singh, H., Castro-Aceituno, V. et al. (2017). Bovine serum albumin as a nanocarrier for the efficient delivery of ginsenoside compound K: preparation, physicochemical characterizations and in vitro biological studies. *RSC Advances* 7: 15397–15407.

Sohrabi, M.J., Dehpour, A.R., Attar, F. et al. (2019). Silymarin-albumin nanoplex: preparation and its potential application as an antioxidant in nervous system in vitro and in vivo. *International Journal of Pharmaceutics* 572: 118824.

Somchue, W., Sermsri, W., Shiowatana, J. et al. (2009). Encapsulation of α-tocopherol in protein-based delivery particles. *Food Research International* 42 (8): 909–914.

Somu, P. and Paul, S. (2019). Supramolecular nanoassembly of lysozyme and α-lactalbumin (apo α-LA) exhibits selective cytotoxicity and enhanced bioavailability of curcumin to cancer cells. *Colloids and Surfaces B: Biointerfaces* 178: 297–306.

Song, R., Yana, J., Xua, S. et al. (2013). Silver ions/ovalbumin films layer-by-layer self-assembled polyacrylonitrile nanofibrous mats and their antibacterial activity. *Colloids and Surfaces B: Biointerfaces* 108: 322–328.

Song, Z., Lu, Y., Zhang, X. et al. (2016). Novel curcumin-loaded human serum albumin nanoparticles surface functionalized with folate: characterization and in vitro/vivo evaluation. *Drug Design, Development and Therapy* 10: 2643–2649.

Sponton, O.E., Perez, A.A., Carrara, C.R. et al. (2015). Linoleic acid binding properties of ovalbumin nanoparticles. *Colloids and Surfaces B: Biointerfaces* 128: 219–226.

Sponton, O.E., Perez, A.A., Carrara, C.R. et al. (2016). Complexes between ovalbumin nanoparticles and linoleic acid: stoichiometric, kinetic and thermodynamic aspects. *Food Chemistry* 211: 819–826.

Sponton, O.E., Perez, A.A., and Santiago, L.G. (2017a). Protein-polysaccharide associative phase separation applied to obtain a linoleic acid dried ingredient. *Food Hydrocolloids* 71: 158–167.

Sponton, O.E., Perez, A.A., Ramel, J. et al. (2017b). Protein nanovehicles produced from egg white. Part 1: effect of pH and heat treatment time on particle size and binding capacity. *Food Hydrocolloids* 73: 67–73.

Sponton, O.E., Perez, A.A., Ramel, J. et al. (2018). Protein nanovehicles produced from egg white. Part 2: effect ofprotein concentration and spray drying on particle size and linoleic acid binding capacity. *Food Hydrocolloids* 77: 863–869.

Stein, P.E., Leslie, A.W.G., Finch, J.T. et al. (1991). Crystal structure of uncleaved ovalbumin at l.95 Å resolution. *Journal of Molecular Biology* 221: 941–959.

Stie, M.B., Corezzi, M., Juncos Bombin, A.D. et al. (2020). Waterborne electrospinning of α-lactalbumin generates tunable and biocompatible nanofibers for drug delivery. *ACS Applied Nano Materials* 3 (2): 1910–1921.

Sulkowska, A. (2002). Interaction of drugs with bovine and human serum albumin. *Journal of Molecular Structure* 614: 227–232.

Sun, N., Teng, A., Zhao, Y., Juncos Bombin, A.D. et al. (2020). Waterborne electrospinning of α-lactalbumin generates tunable and (2020). Immunological and anticancer activities of seleno-ovalbumin (Se-OVA) on H22-bearing mice. *International Journal of Biological Macromolecules* 163: 657–665.

Svensson, M., Hakansson, A., Mossberg, A.K. et al. (2000). Conversion of α-lactalbumin to a protein inducing apoptosis. *Proceedings of the National Academy of Sciences* 97: 4221–4226.

Tabibiazar, M., Mohammadifar, M.A., Roufegarinejad, L. et al. (2019). Improvement in dispersibility, stability and antioxidant activity of resveratrol using a colloidal nanodispersion of BSA-resveratrol. *Food Bioscience* 27: 46–53.

Takeda, K., Wada, A., Yamamoto, K. et al. (1989). Conformational change of bovine serum albumin by heat treatment. *Journal of Protein Chemistry* 8: 653–659.

Taneja, N. and Singh, K.K. (2018). Rational design of polysorbate 80 stabilized human serum albumin nanoparticles tailored for high drug loading and entrapment of irinotecan. *International Journal of Pharmaceutics* 536 (1): 82–94.

Tang, B., Qian, Y., Gou, Y. et al. (2018). VE-albumin core-shell nanoparticles for paclitaxel delivery to treat MDR breast cancer. *Molecules* 23: 2760.

Tayeh, N., Rungassamy, T., and Albani, J.R. (2009). Fluorescence spectral resolution of tryptophan residues in bovine and human serum albumins. *Journal of Pharmaceutical and Biomedical Analysis* 50: 107–116.

Teng, Z., Luo, Y., and Wang, Q. (2013). Carboxymethyl chitosan–soy protein complex nanoparticles for the encapsulation and controlled release of vitamin D3. *Food Chemistry* 141 (1): 524–532.

Tian, R., Long, X., Yang, Z. et al. (2020). Formation of a bovine serum albumin diligand complex with rutin and single-walled carbon nanotubes for the reduction of cytotoxicity. *Biophysical Chemistry* 256: 106268.

Vaidya, B., Kulkarni, N.S., Shukla, S.K. et al. (2020). Development of inhalable quinacrine loaded bovine serum albumin modified cationic nanoparticles: repurposing quinacrine for lung cancer therapeutics. *International Journal of Pharmaceutics* 577: 118995.

Veprintsev, D.B., Permyakov, S.E., Permyakov, E.A. et al. (1997). Cooperative thermal transitions of bovine and human apo-α-lactalbumins: evidence for a new intermediate state. *FEBS Letters* 412: 625–628.

Vicente, J., de Souza Cezarino, T., Pereira, L.J.B. et al. (2017). Microencapsulation of sacha inchi oil using emulsion-based delivery systems. *Food Research International* 99: 612–622.

Vicente, J., Barreto Pereira, L.J., Heckert Bastos, L.P. et al. (2018). Effect of xanthan gum or pectin addition on Sacha Inchi oil-inwater emulsions stabilized by ovalbumin or tween 80: droplet size distribution, rheological behavior and stability. *International Journal of Biological Macromolecules* 120 part A: 339–345.

Visentini, F.F., Sponton, O.E., Perez, A.A. et al. (2017a). Formation and colloidal stability of ovalbumin-retinol nanocomplexes. *Food Hydrocolloids* 67: 130–138.

Visentini, F.F., Sponton, O.E., Perez, A.A. et al. (2017b). Biopolymer nanoparticles for vehiculization and photochemical stability preservation of retinol. *Food Hydrocolloids* 70: 363–370.

Visentini, F.F., Ferrado, J.B., Perez, A.A. et al. (2019). Simulated gastrointestinal digestion of inclusion complexes based on ovalbumin nanoparticles and conjugated linoleic acid. *Food & Function* 10 (5): 2630–2641.

Visentini, F.F., Perez, A.A., Baravalle, M.E. et al. (2020). In vitro gastrointestinal digestion and cytotoxic effect of ovalbumin-conjugated linoleic acid nanocomplexes. *Food Research International* 137: 109381.

Wan, J., Zhong, S., Schwarz, P. et al. (2019). Enhancement of antifungal and mycotoxin inhibitory activities of food-grade thyme oil nanoemulsions with natural emulsifiers. *Food Control* 106: 106709.

Wang, S.S.S., Liu, K.N., Wen, W.S. et al. (2011). Fibril formation of bovine α-lactalbumin is inhibited by glutathione. *Food Biophysics* 6 (1): 138–151.

Wang, X., Liu, F., Liu, L. et al. (2015). Physicochemical characterisation of β-carotene emulsion stabilised by covalent complexes of α-lactalbumin with (−)-epigallocatechin gallate or chlorogenic acid. *Food Chemistry* 173: 564–568.

Wang, C., Liu, Z., Xu, G. et al. (2016). BSA-dextran emulsion for protection and oral delivery of curcumin. *Food Hydrocolloids* 61: 11–19.

Wang, T., Hu, Q., Lee, J.Y. et al. (2018). Solid lipid–polymer hybrid nanoparticles by in situ conjugation for oral delivery of astaxanthin. *Journal of Agricultural and Food Chemistry* 66: 9473–9480.

Weber, C., Kreuter, J., and Langer, K. (2000). Desolvation process and surface characteristics of HSA-nanoparticles. *International Journal of Pharmaceutics* 196: 197–200.

Weijers, M., Sagis, L.M.C., Veerman, C. et al. (2002). Rheology and structure of ovalbumin gels at low pH and low ionic strength. *Food Hydrocolloids* 16 (3): 269–276.

Weijers, M., Barneveld, P.A., Cohen Stuart, M.A. et al. (2003). Heat-induced denaturation and aggregation of ovalbumin at neutral pH described by irreversible first-order kinetics. *Protein Science* 12 (12): 2693–2703.

White, B., Evison, A., Dombi, E. et al. (2017). Improved delivery of the anticancer agent citral using BSA nanoparticles and polymeric wafers. *Nanotechnology, Science and Applications* 10: 163–175.

Wongsasulak, S., Patapeejumruswong, M., Weiss, J. et al. (2010). Electrospinning of food-grade nanofibers from cellulose acetate and egg albumen blends. *Journal of Food Engineering* 98: 370–376.

Wu, Y., Cheng, H., Chen, Y. et al. (2017). Formation of a multiligand complex of bovine serum albumin with retinol, resveratrol, and (−)-epigallocatechin-3-gallate for the protection of bioactive components. *Journal of Agricultural and Food Chemistry* 65 (14): 3019–3030.

Xiang, J., Liu, F., Fan, R. et al. (2015). Physicochemical stability of citral emulsions stabilized by milk proteins (lactoferrin, α-lactalbumin, β-lactoglobulin) and beet pectin. *Colloids and Surfaces A: Physicochemical and Engineering Aspects* 487: 104–112.

Xiang, C., Gao, J., Ye, H. et al. (2020). Development of ovalbumin-pectin nanocomplexes for vitamin D3 encapsulation: enhanced storage stability and sustained release in simulated gastrointestinal digestion. *Food Hydrocolloids* 106: 105926.

Xie, L., Wehling, R.L., Ciftci, O. et al. (2017). Formation of complexes between tannic acid with bovine serum albumin, egg ovalbumin and bovine beta-lactoglobulin. *Food Research International* 102: 195–202.

Xie, H., Xiang, C., Li, Y. et al. (2019). Fabrication of ovalbumin/κ-carrageenan complex nanoparticles as a novel carrier for curcumin delivery. *Food Hydrocolloids* 89: 111–121.

Xie, L., Ciftci, O., and Zhang, Y. (2020). Encapsulation of astaxanthin-enriched camelina oil extract in ovalbumin/gum arabic stabilized emulsion with/without crosslinking by tannic acid. *ES Food and Agroforestry* 1: 77–84.

Xiong, W., Ren, C., Li, J. et al. (2018). Characterization and interfacial rheological properties of nanoparticles prepared by heat treatment of ovalbumin-carboxymethylcellulose complexes. *Food Hydrocolloids* 82: 355–362.

Xu, R., Fisher, M., and Juliano, R.L. (2011). Targeted albumin-based nanoparticles for delivery of amphipathic drugs. *Bioconjugate Chemistry* 22: 870–878.

Yadav, R., Kumar, D., Kumari, A. et al. (2014). Encapsulation of catechin and epicatechin on BSA NPS improved their stability and antioxidant potential. *EXCLI Journal* 13: 331–346.

Yamasaki, M., Takahashi, N., and Hirose, M. (2003). Crystal structure of S-ovalbumin as a non-loop-inserted thermostabilized serpin form. *The Journal of Biological Chemistry* 278 (37): 35524–35530.

Yan, Y., Zhu, Q., Diao, C. et al. (2020). Enhanced physicochemical stability of lutein-enriched emulsions by polyphenol-protein-polysaccharide conjugates and fat-soluble antioxidant. *Food Hydrocolloids* 101: 105447.

Yang, K., Wang, Y., Lu, C. et al. (2018). Ovalbumin-directed synthesis of fluorescent copper nanoclusters for sensing both vitamin B1 and doxycycline. *Journal of Luminescence* 196: 181–186.

Yang, Y., Zhang, Y., Thakur, A. et al. (2019). One-pot synthesis and characterization of ovalbumin-conjugated gold nanoparticles: a comparative study of adjuvanticity against the physical mixture of ovalbumin and gold nanoparticles. *International Journal of Pharmaceutics* 571: 118704.

Yi, J., Fan, Y., Zhang, Y. et al. (2016a). Characterization of catechin-α-lactalbumin conjugates and the improvement in β-carotene retention in an oil-in-water nanoemulsion. *Food Chemistry* 205: 73–80.

Yi, J., Fan, Y., Zhang, Y. et al. (2016b). Glycosylated α-lactalbumin-based nanocomplex for curcumin: physicochemical stability and DPPH-scavenging activity. *Food Hydrocolloids* 61: 369–377.

You, R., Zhang, J., Gu, S. et al. (2017). Regenerated egg white/silk fibroin composite films for biomedical applications. *Materials Science and Engineering C* 79: 430–435.

Yu, F., Li, Y., Liu, C.S. et al. (2015). Enteric-coated capsules filled with mono-disperse micro-particles containing PLGA-lipid-PEG nanoparticles for oral delivery of insulin. *International Journal of Pharmaceutics* 484 (1-2): 181–191.

Zhao, D., Zhao, X., Zu, Y. et al. (2010). Preparation, characterization, and in vitro targeted delivery of folate-decorated paclitaxel-loaded bovine serum albumin nanoparticles. *International Journal of Nanomedicine* 5: 669–677.

Zhao, L., Zhou, Y., Gao, Y. et al. (2015). Bovine serum albumin nanoparticles for delivery of tacrolimus to reduce its kidney uptake and functional nephrotoxicity. *International Journal of Pharmaceutics* 483 (1-2): 180–187.

Zhao, Y., Cai, C., Liu, M. et al. (2020). Drug-binding albumins forming stabilized nanoparticles for co-delivery of paclitaxel and resveratrol: in vitro/in vivo evaluation and binding properties investigation. *International Journal of Biological Macromolecules* 153: 873–882.

Zhang, G., Chen, X., Guo, J. et al. (2009). Spectroscopic investigation of the interaction between chrysin and bovine serum albumin. *Journal of Molecular Structure* 921: 346–351.

Zhang, B., Luo, Y., and Wang, Q. (2011). Development of silver/α-lactalbumin nanocomposites: a new approach to reduce silver toxicity. *International Journal of Antimicrobial Agents* 38 (6): 502–509.

Zimet, P. and Livney, Y.D. (2009). Beta-lactoglobulin and its nanocomplexes with pectin as vehicles for ω-3 polyunsaturated fatty acids. *Food Hydrocolloids* 23 (4): 1120–1126.

Zinellu, A., Sotgia, S., Scanu, B. et al. (2015). Human serum albumin increases the stability of green tea catechins in aqueous physiological conditions. *PLoS One* 10 (7): e0134690.

Zu, Y., Meng, L., Zhao, X. et al. (2013). Preparation of 10-hydroxycamptothecin-loaded glycyrrhizic acid-conjugated bovine serum albumin nanoparticles for hepatocellular carcinoma-targeted drug delivery. *International Journal of Nanomedicine* 8: 1207–1222.

6

The Protein Structure, Function and Specificity

PhaC Synthases Type I, II, III and IV as a Model

Amro Abd Al Fattah Amara[1,2]

[1]Protein Research Department, Genetic Engineering and Biotechnology Research Institute, City of Scientific Research and Technological Applications, Alexandria, Egypt
[2]Head of the Scientific Publishing office, City of Scientific Research and Technological Applications, Alexandria, Egypt
*amroamara@web.de

6.1 Introduction

6.1.1 Basic Background of Polyhydroxyalkanoates

PHB synthesis, degradation and the structure of the polymer attracted the attention of scientists in Germany, the UK and the USA during the 1960s and then later Japan. In 1974 the heteropolymer containing 3-hydroxyacyl monomers other than 3-hydroxybutyrate was discovered (Wallen and Rohwedder 1974). Findlay and White (1983) reported that PHB is a general class of optically active polyesters (Findlay and White 1983). Witholt's group observed that intracellular granules accumulated by *Pseudomonas oleovorans* growing on *n*-octane were composed of 3-hydroxyoctanoate with a little 3-hydroxyhexanoate (de Smet et al. 1983). Holmes et al. (1985) at ICI had shown growing *Alcaligenes eutrophus* on glucose and propionic acid producing random relative concentrations of copolymer of 3-hydroxybutyrate (3HB) and 3-hydroxyvalerate (3HV) (Holmes 1985). Witholt's group report accumulating two or more 3-hydroxyacids from single *n*-alkanes (C8 to C12) (Lageveen et al. 1988), and polyesters containing unsaturated monomer units were produced from I-nonene and I-decene (Lageveen et al. 1988). Fuller, Lenz and their colleagues at Amherst (Brandl et al. 1989; Gross et al. 1989) showed that *P. oleovorans* accumulates PHA containing up to five monomer components from *n*-alkanoic acids (C6–C12), and branched-chain 3-hydroxyacids is incorporated when 5-, 6- or 7-methyloctanoate is provided as co-substrate with octanoic acid; unsaturated substrates·such as 3-hydroxyoctenoic acids are also incorporated (Fritzsche et al. 1990a, 1990b). Dawes and his group found substantial PHA formation from C8–C10 *n*-alcohols (Haywood et al. 1989a, 1989b) and reclaimed that carbon chain length of the substrate determines the range of monomer units incorporated (Haywood et al. 1989a, 1989b). *Rhodospirillum rubrum* displays a pattern of polymer accumulation differing from that of *Pseudomonas species*, producing PHA containing C4 to C6 3-hydroxyacids from a range (C2 to Cl0) of *n*-alkanoic acid substrates (Brandl et al. 1989). *P. aeruginosa*,

Biomolecules from Natural Sources: Advances and Applications, First Edition. Edited by Vijai Kumar Gupta, Satyajit D. Sarker, Minaxi Sharma, María Elida Pirovani, Zeba Usmani, and Chelliah Jayabaskaran.

P. putida, *P. fluorescens* and *P. testosteronii* accumulate polyesters from C6 to Cl0 straight-chain alkanes, alcohols and alkanoic acids as sole carbon sources (Haywood et al. 1989a, 1989b). Accumulating C6 to C12 3-hydroxyacids (but not 3-hydroxybutyrate) seems to be a characteristic of the fluorescent pseudomonads (Huisman et al. 1989,). Dawes' group and Steinbüchel's group, have shown that several Pseudomonads produce a PHA containing chiefly 3-hydroxydecanoate when grown with unrelated substrates and certain *Rhodococcus species* accumulate a copolymer of 3HB and 3HV from one substrate, glucose. *Ralstonia eutrophus* polymers are essentially limited to C4 and C5 monomer units (small amounts of 3-hydroxypropionate can also be incorporated). Doi and his colleagues discovered a novel polymer in which the backbone unit is extended from the C3 characteristic of PHB to C4 and C5 by incorporating 4-hydroxybutyrate and 5-hydroxyvalerate monomers into the 3-hydroxybutyrate structure (Doi 1987, 1988, 1990). The structure of PHB was first investigated by Marchessault and his colleagues in the 1960s (Alper et al. 1963). The early belief that PHB granules *in vivo* are crystalline (Alper et al. 1963; Ballard et al. 1987) was revised following application of a number of physical techniques by Barham's group and 13C-NMR spectroscopy by Barnard and Sanders (1988). The genes involved in the biosynthesis of PHB in *A. eutrophus* were cloned and expressed in *E. coli* independently by Schubert et al. (1988) and Slater et al. (1988) .The recombinant *E. coli* strains accumulated substantial amounts of PHB as intracellular granules. The structural genes for the three biosynthetic enzymes of *A. eutrophus* were sequenced by Peoples and Sinskey (1989) (Masamune et al. 1989a, 1989b; Peoples and Sinskey 1989a, 1989b). Transcription occurs in the sequence synthase, thiolase, reductase. The ability to transfer the biosynthetic genes between unrelated bacteria opens the way to produce the polymers commercially by organisms that are readily cultivated on an industrial scale with opportunities for using recombinant overproduced strains. The other progress and key discovery in PHAs will be discussed. Some important companies produced different PHAs are ICI, Chimie Linz (currently Biomer); Biogreen; Biocycle; Zeneca; Monsanto Metabolix; Procter and Gamble and Kaneka Co.

6.1.2 PHA nomenclature

There are many ways to abbreviate PHA species, however, there are two which are used most commonly: the first is based on the length of their carbon atom (SCL, MCL and LCL) and the second is based on describing the name of its monomeric constituents. The first classifies them to short, medium or long chain length. The second classifies them based on the total percentage of the various monomeric constituents in their backbone. One should not mix between the number of carbon atoms from the same monomeric constituents and those from various monomeric constituents. As a matter of fact, even this abbreviation system is based on the number of carbon atoms in the composition of their monomeric unit, at the same time it shows the various physical properties where SCL PHAs are solid or semisolid (Figure 6.1) while MCL PHAs are latex-like, rubber or elastomers. PHAs created from three to five carbon atoms are called short chain length (SCL), while PHAs with six to fourteen carbon atoms are called medium chain length (MCL) [48]. The PHAs, which contain more than fourteen carbon atoms, are called long chain length (LCL). Genes coding for

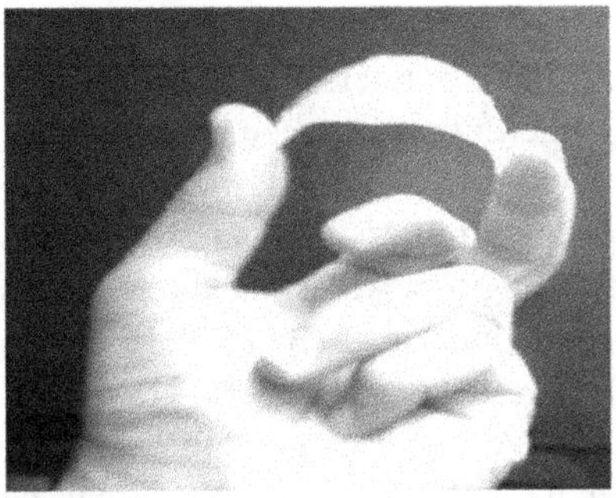

Figure 6.1 PHAs are a plastic-like and a biodegradable natural biopolymer.

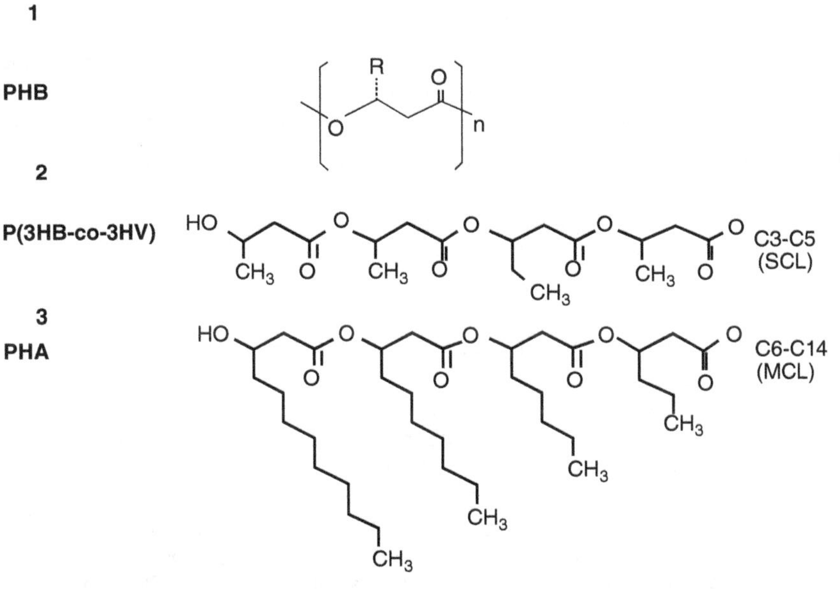

Figure 6.2 Chemical structure of PHAs. (1) PHAs are generally composed of (*R*)-β-hydroxy fatty acids, where the pendant group (*R*) varies from methyl (C1) to tridecyl (C13). Other fatty acids that were incorporated have the hydroxy group at the γ, δ or ε position, while the pendant group may be saturated or unsaturated or contain substituents. (2) P(3HB-*co*-3HV) is poly(3-hydroxybutyrate-*co*-3-hydroxyvalerate) copolymer (SCL is short chain length). (3) General formula for PHA_{MCL} (MCL is medium chain length).

proteins involved in the biosynthesis of PHA are referred to in alphabetical order as *phaA* (β ketothiolase), *phaB* (acetoacetyl-Co A reductase), *phaC* (PHA synthase), *phaG* (3hydroxyacyl-acyl carrier protein-Co A transacylase), and so on (Anderson and Dawes 1990; Rehm and Steinbuchel 2001) (Figure 6.2).

The names of the genes involved in PHA degradation are referred to in reverse alphabetical order such as *phaZ* for PHA depolymerases, *phaY*, *phaX*, *phaW* and so on. The genes for phasins and regulator proteins are referred to as *phaP* and *phaR*, respectively. The origin of the gene or protein can be also referred to using the first letter of the genus and the species as subscript, for example, phaCPa refers to *P. aeruginosa*. It is essential to note that using the numbering style to differentiate between the monomer structures is prefered when a copolymer is present such as P(3HB-*co*-4HB). If the polymer is well known, it is preferrable to use their abbreviation rather than using SCL/MCL style. If co-monomeric forms exist in the short chain length polymer backbone -co- is used. For example, polyhydroxybutyrate is usually abbreviated as PHB or P3(HB) while polyhydroxybutyrate-co-polyhydroxyvalirate is abbreviated as PHB-*co*-PHV or P (3HB-co-3HV). On the other hand, in the case of MCL PHAs and because of the big number of their monomeric composition PHA_{MCL} is usually used. Generally, PHAs can be abbreviated using any of the described styles depending on the situation. An example of the various abbreviation styles for polyhydroxybutyrate-*co*-polyhydroxyvalerate are PHB-*co*-PHV, P3 (HB-*co*-HV), P (3HB-*co*-3HV), P (3-βHB-*co*-3-βHV), P3β (HB-*co*-HV). A few authors are still using β in PHA abbreviation. PHAs will refer to various kinds of polyhydroxyalkanoates including SCL, MSL and LCL. From the nomenclature of the various types of the polyhydroxyalkanoates one could imagine the efforts of hundreds of active scientists throughout the world who are working hard to set up this science, to commercialize those types of natural, safe biopolymer which could keep our planet safe from many environmental problems. However, key scientists are reviewing the huge work through many communications particularly the International Symposium on Biological Polyesters (ISBP) meetings.

Historical timeline for polyhydroxyalkanoates and PhaC synthases

1888	Perhaps, the first report about lucent granules of PHAs in bacterial cells was made by Beijerinck in 1888 (Beijerinck 1988; Chowdhury 1963). Beijerinck observed granules in a microscope inside Rhizobium cells. Such granules were present in the "bacteroides" isolated from nodules and were described as being extremely refractile globule.
1923 to 1951	The French scientist Maurice Lemoigne, who worked at the Lille branch of the Pasteur Institute, France, had an interest in and characterized these inclusion bodies, which were found in *Bacillus spp*. Lemoigne noticed that, when cultures of Bacillus were followed by autolysis in distilled water, the pH value decreased because they formed an unknown acid. This acid was subsequently identified as a monomer of poly-β-hydroxybutyric acid. The work done by Lemoigne and coworkers was published in 27 publications from 1923 to 1951. Lemoigne was the first to report that the bacterial granule constituents (Lemoigne describes it also as reserved material) were not ether soluble, as in lipids, and later reported that polyhydroxybutanoate (PHB) is the major constituent of these granules. Using microscopic observations, saponification numbers, and autolysis whose solubility and melting point (T_m) variation were correlated with molecular size proposed a polyester structure with formula $(C_4H_6O_2)_n$. Lemoigne has also reported that PHB could cast into film like cellulose nitrate material (de la Riviére 1967; Lemoigne 1923, 1925, 1926, 1927). Stapp suggested that *Azotobacter chroococcum* inclusions could be easily extracted with chloroform, and identified this structure as poly-β-hydroxybutyrate (Stapp 1924).
	In the following years, interest in this new material was negligible.
1953	In 1953 Weibull had isolated the granules of *Bacillus megaterium* by dissolution of the cell wall with a lysozyme (Weibull 1953).

1958	The first report on functions of PHB appeared in 1958 by Wilkinson and Wilkinson who obtained morphologically intact granules from *Bacillus cereus* using alkaline hypochlorite, and Macrae and Wilkinson in 1958 who described poly-β hydroxybutyrate metabolism in washed suspensions of *B. cereus* and *B. megaterium* (Macrae and Wilkinson 1958; Weibull 1953; Williamson and Wilkinson 1958).
1959	Doudoroff and Stanier (1959) and Stanier et al. (1959) found that PHB can be produced by photosynthetic assimilation of organic compounds by phototrophic bacteria and they described that the reaction involved in the metabolic pathway is responsible for the biosynthesis of PHB from acetic acid (Doudoroff and Stanier 1959; Stanier et al. 1959).
1961	Merrick and Doudoroff (1961) isolated and described native PHB granules of chemoheterotrophic bacteria, *R. rubrum* and *B. megaterium* (Merrick and Doudoroff 1961).
1961	Schlegal at the University of Göttingen, Germany in 1961 detected that *Cupriavidus necator* (formally *Wautersia eutropha, R. eutropha* and *Alcaligenes eutrophus*), could accumulate large amount of PHB in a media with low nitrogen content (Schlegel et al. 1961).
1964, 1968	Merrick and Doudoroff (1964) were the first to identify the enzyme responsible for PHB polymerization followed by work done by Griebel et al. (1968) (Merrick and Doudoroff 1961, 1964). In 1966, Lusty and Doudoroff conducted a study on depolymerases which were able to hydrolyze PHB (Lusty and Doudoroff 1966).
1973	Schlegal working with *C. necator* and Dawes working with *Azotobacter berjerinckii* (University of Hull, England) in 1973 simultaneously were the first to isolate and characterize the enzymes, which catalyze the PHB monomer synthesis (Oeding and Schlegel 1973; Senior and Dawes 1973).
1974	Wallen and Rohwedder (1974) gave the first indication that the polymer discovered by Lemoigne may contain proportions of 3-hydroxyacids other than 3-hydroxybutyrate and in 1974, reported heteropolymers of 3-hydroxybutyrate and 3-hydroxyvalerate in chloroform extracts of activated sewage sludge as major constituents with C6 and possibly C7 3-hydroxyacids as minor constituents (Wallen and Rohwedder 1974).
1983	These heteropolymers (aqs above) had a lower melting point than PHB and, unlike the homopolymer, was soluble in hot ethanol. Findlay and White (1983) using GC analysis, detected at least 11 short chain 3-hydroxyacids, the principal ones being 3HB and 3HV (Findlay and White 1983). De Smet et al. characterized PHA_{MCL} in *Pseudomonas oleovorans* during growth on octane (de Smet et al. 1983).
	Huisman et al. confirmed that PHA_{MCL} accumulation was the common feature of fluorescent pseudomonads [46]. PHAs are also produced by gram-positive, gram-negative and phototrophic bacteria and also archaea (Anderson and Dawes 1990; Fernandez-Castillo et al. 1986; Huisman et al. 1989; Shabeb et al. 2005; Steinbüchel 1991; Steinbüchel et al. 1998; Steinbüchel and Hein 2001).

6.2 PHAs chemical properties

Like any polymer, PHAs are chains of repeated monomers either homopolymer or heteropolymer. Overall, PHAs known to be linear, are composed of 3-hydroxy alkanoic acid monomers units, each unit form an ester bond with the hydroxyl group of the other one. The hydroxyl-substituted carbon atom has *R* configuration in all characterized PHAs (Figure 6.2). At the C-3 atom (β position), an alkyl group, the length of which can vary from methyl (C1) to tridecyl (C13), is found (Anderson and Dawes 1990). This alkyl side chain is not necessarily saturated. Unsaturated, aromatic, epoxidized, halogenated, and branched monomers were reported as well (Abe et al. 1990; Choi and Yoon 1994; Curley et al. 1996; Doi and Abe 1990; Fritzsche et al. 1990a, 1990b; Kim et al. 1991, 1992; Lageveen et al. 1988; Song and Yoon 1996). Cross-linking of unsaturated bond substations in the side chains of PHAs can be added chemically (de Koning et al. 1997). Lütke-Eversloh et al. were the first to report the production of biopolymers with thioester linkages in the polymer backbone using *C. necator* in media containing 3-mercaptopropionate (3MP) or 3-mercaptobutyrate (3MB) besides 3-hydroxybutyrate as constituents (Lutke-Eversloh et al. 2001, 2002). Many factors affect the chemical composition of lPHAs, such as the microbial strain, the substrate, the cultivation condition, the extraction method, the number of *phaC*, *phaB* genes, the regulator *phaP* (phasin) and the presence of inhibitors. They inhibit various pathways especially those which supply the synthases with various kinds of monomer or inhibit other pathways, which consume these monomers for their own or degrade it to shorter units like the inhibition of β oxidation pathway. The PHA_{MCL} is synthesized with an average molar mass ranging from 5×10 to 2×10 g mol^{-1}, whereas PHB can be produced with an average molar mass of $> 1 \times 10$ g/mol. Generally, the PHA composition and molecular mass depends on the PHA synthases, the carbon source and the metabolic routes involved (Steinbuchel and Fuchtenbusch 1998). The molecular weights of the PHAs were established by light scattering, gel permeation chromatograph and sedimentation analysis. Their monomer composition was determined by gas chromatography (GC), mass spectroscopy (MS) and nuclear magnetic resonance (NMR) analysis (Findlay and White 1983; Lutke-Eversloh et al. 2001, 2002).

6.3 The Physical properties of PHAs

PHAs show material properties that are similar to some common plastics such as polypropylene (Byrom 1987). The bacterial origins of the PHAs make these polyesters a natural material, and many microorganisms have the ability to degrade these macromolecules (Jendrossek et al. 1996). In most cases the producer microbe has the gene of the degradation. That is because the microbe is accumulating these polymers for its own use as a carbon storage material. The molecular mass of PHAs varies for each producer but is generally in the order of 50×10^3 to 1×10^6 Da. In the entire cell, P(3HB) exists in a fluid, amorphous state. However, after extraction from the cell with organic solvents, P(3HB) becomes highly crystalline (Doi 1995) and in this state it is a stiff or even brittle material. Because of its brittleness, P(3HB) is not very stress resistant. The high melting temperature of P(3HB) (around 170 °C) is close to the temperature in which this polymer decomposes thermally and thus limits the ability to process the homopolymer. Incorporating 3-hydroxyvalerate (3HV) into the P(3HB) resulted in P(3HB-*co*-3HV), a copolymer that is less stiff and brittle than P(3HB), that can be used to prepare films that demonstrated excellent water and gas barrier properties like polypropylene, and that can be processed at lower temperature while retaining most of the other excellent mechanical properties of P(3HB) (Marchessault 1996). P(3HB-*co*-3HV) also has low crystallinity and is more elastic than P(3HB) (Gross et al. 1989; Preusting et al. 1990). The latex-like PHAs (PHA_{MCL}) have physical properties, which differ significantly from the PHA_{SCL}, such as P(3HB). Particularly with respect to the melting temperature and the extension to break

Table 6.1 Properties of some PHAs and polypropylene.

Parameter	P(3HB)	P(3HB-*co*-3HV)	P(3HB-*co*-4HB)	P(3HO-*co*-3HH)	PP
Tm (°C)	177	145	150	61	176
Tg (°C)	2	−1	−7	−36	−10
Crystallinity (%)	70	56	45	30	60
Extension to break (%)	5	50	444	300	400

value, the two types of PHAs showed striking differences chiefly becouse of the lower crystallinity of PHA_{MCL} (Table 6.1).

Table 6.1 represents the physical properties of some selected PHAs. T_m is melting temperature. T_g is glass transition temperature. P(3HB) is poly(3-hydroxybutyrate), P(3HB-*co*-3HV) is poly(3-hydroxybutyrate-*co*-3-hydroxyvalerate) containing 20% 3HV, P(3HB-*co*-4HB) is poly(3-hydroxybutyrate-*co*-4-hydroxybutyrate) containing 16% 4HB, P(3HO-*co*-3HH) is poly(3-hydroxyoctanoate-*co*-3-hydroxyhexanoate) containing 11% 3HH, and PP is polypropylene.

6.4 PHAs Biosynthesis

6.4.1 Metabolic Pathways and Monomer-supplied for PHA Synthesis

PhaC synthases are enzymes that produce their catalytic work mechanically. For that they polymerize the monomeric forms when the suitable conditions exist. Each of the PhaC synthases is able to produce a specific type of polymer through a specific pathway and nutritional conditions. However, natural isolates prove their ability to incorporate other monomeric forms to their essential backbone. One should differentiate between changing the substrate specificity and the ability of various PhaC synthases to incorporate other monomeric forms. Changing the substrate specificity means turning one type to another using the same PhaC and the unrelated substrate. For example, the production of PHA_{MCL} by PhaC synthase that produce PHB (PHA_{SCL}) using the substrates of the MCL polyhydroxyalkanoates. However, PhaC synthases show high ability to involving non-related monomeric forms in their backbone when growing in non-related substrate particularly through inhibiting the β-oxidation pathway or adding non-related substrate beside the related one.

6.4.2 PhaCSCL Synthases

PhaC synthases are the key enzymes for PHA biosynthesis. $PhaC_{SCL}$ synthases use coenzyme A thioesters of hydroxybyturic acids (HB) as substrates and catalyze polymerizing HAs into PHB with the concomitant release of CoA. In three different laboratories, 33 years ago, PHB synthase operon of *C. necator* was cloned (Peoples and Sinskey 1989; Schubert et al. 1988; Slater et al. 1988). PHB is the most extensively investigated PHA.

The PHB biosynthesis pathway consists of three distinct enzymatic reactions. Examination of the enzymes leading to form PHB has shown that regulating PHB synthesis is controlled by acetyl-co-enzymes A (CoA) acyl-transferase (Asada et al. 1999; Jackson and Dawes 1976). PHB is synthesized from acetyl coenzyme-A (acetyl-CoA) by a sequence of three enzymatic reactions. Synthesis of PHB is controlled by the first enzyme, 3-ketothiolase that is inhibited by coenzyme A (CoASH) (Mansfield et al. 1995). β-ketoacyl-CoA thiolase (encoded by phaA) is the key that starts the process. The first reaction comprises condensing of two acetyl coenzyme A molecules into acetoacetyl-CoA by β-ketoacyl-CoA thiolase (β-ketothiolase). NADPH-dependent acetoacetyl-CoA dehydrogenase (encoded by *phaB*) is the enzyme producing the monomeric form, where the second reaction is reducing acetoacetyl-CoA to (*R*)-3-hydroxybutyryl-CoA by an NADPH-dependent acetoacetyl-CoA dehydrogenase to form *R*)-3-hydroxybutyryl-CoA monomers. PHB synthases (encoded by *phaC*) are responsible for the polymerization steps where the (*R*)-3-hydroxybutyryl-CoA monomers are polymerized to PHB. Under balanced growth conditions, acetyl-CoA is fed into the tricarboxylic acid cycle and the resultant CoA inhibits acetyl-CoA acyltransferase and PHB synthesis. Under nutrient limitation (such as oxygen limitation) and carbon excess, NADH oxidase activity decreases, NADH increases and inhibits citrate synthase and isocitrate dehydrogenase, and acetyl-CoA acyltransferase by CoA is overcome. The ensuing condensation reaction forms acetoacetyl-CoA and initiates PHB synthesis. The high NADH/NAD ratio caused oxygen limitation to rapidly readjust as PHB synthesis starts, and PHB assumes the role of an alternative electron acceptor (Jackson and Dawes 1976). Degrading of PHB is initiated by PHB depolymerase to form D(-)hydroxybutyric acid. An NAD-specific dehydrogenase oxidizes the acid to acetoacetate, which is an intermediate common to the biosynthesis and degrading PHB (Doi et al. 1992).

6.4.3 PHAMCL Synthases

PHA_{MCL} is biosynthesized from related substrates where in 1983, De Smet described accumulating PHA, the PHA_{MCL}, in *P. oleovorans* grown on octane as the sole carbon source. The compositions of the PHAs formed by pseudomonads of the rRNA homology group I were directly interconnected to the structure of the alkane, alkene or fatty acid used as the carbon source. These substrates are referred to as "related substrates". When the carbon source consists of fatty acids with 6 to 12 carbon atoms, monomers of the PHA are of the same length as the carbon source or are shortened by 2, 4 or 6 carbon atoms (Brandl et al. 1988a; de Smet et al. 1983; Huisman et al. 1989). *Aeromonas punctata* produces a random copolymer of 3HB and 3HHX, when growing on even-numbered fatty acids or olive oil as the sole carbon source. PHA production in *A. punctata* proceeds from the β-oxidation pathway intermediate 3-hydroxyenoyl-CoA, which is hydroxylated by the enantiospecific enoyl-CoA hydratase. PHA_{MCL} is biosynthesis from unrelated substrates which are not interconnected and their growth substrates occur relatively rarely in nature (Steinbuchel and Fuchtenbusch 1998). These substrates, which do not resemble the monomers of the accumulated polymer, are named "unrelated substrates". The first indication about the ability of synthases to utilize products from other pathways in their monomeric composition was reported by Wallen and Rohwedder (1974) who reported heteropolymers of HB and 3-hydroxyvalerate in chloroform extracts of activated sewage sludge as major constituents with C6 and possibly C7 3-hydroxyacids as minor constituents (Wallen and Rohwedder 1974). Slater et al. (1998) identified that a second β-ketothiolase gene in *C. necator* is responsible for synthesis of 3-ketovaleryl-CoA (Slater et al. 1998). Pseudomonads of the rRNA homology group I like *P. putida* KT2442, accumulate PHAs that consist primarily of C10 and C8 monomers, when grown on sugars or gluconate (Huijberts et al. 1992; Timm and Steinbüchel 1990). These monomers are derived from intermediates of fatty acid biosynthesis and the composition of the PHAs is reflecting the pool of fatty acid biosynthetic intermediates.

6.5 Metabolic Engineering of PHAMCL

The primary aims of metabolic engineering of PHA include the management of various factors that determine physical properties, such as monomeric composition, molecular weight, copolymer microstructure, and optimizing yield. Efficient PHA production was reached using the fermentation technology of naturally occurring PHA-producing bacteria based on external substrate manipulation. One critical point that should be addressed is that PHB has a melting point close to its decomposition which is a bad aspect of this polymer during its industrial fabrication. However, incorporating various other co-monomeric forms in their backbone improve their mechanical properties. Additionally, one should not neglect the natural oxidation of the polymer which decreases its durability in the market for example in the necks of bottles which are subjected to natural stress or because of the load put on them during closing and opening. There is a real demand for new formla of PHAs which could satisfy the production processses and the marketing demand. Many technical applications were improved by changing the monomeric constituents of the PHAs either by using natural islets or mutants able to produce better polymers. The β-oxidation pathway inhibition, the fatty acid *de nevo* pathway etc. succeeded in adding various types of polymers with various mechanical properties. Subsequent fortification in recombinant gene technology supported the metabolic engineering of the PHAs. Enzyme engineering is becoming a new approach for PHA production. Innovation in the chemical synthesis of macromolecules with desirable properties was achieved by the development of prominent chemical catalysts through "catalyst evolution". Many tools have become available to produce various forms of PHAs using various strategies. One should address that heterologous expression of various *phaC* genes enables great opportunities either in overall cost reduction or in discovering new metabolic pathways. Some knock-out mutants enabled the identification of the structural differences between the various PhaC synthases and the opportunity to use one or another. Some important pathways are summarized in Figure 6.3. Use of inhibitors, particularly for the β-oxidation pathways, drive unrelated substrates to other pathways where PhaC synthase Class I incorporate the intermediates and form co-polymers. PhaG (R)-3-hydroxyacyl-ACP:CoA transacylase enable linking different pathways and transferring (R)-3-hydroxyacyl-ACP to (R)-3-hydroxyacyl-CoA for both of the PHA_{SCL} synthases and the PHA_{MCL} synthases to produce different PHA.

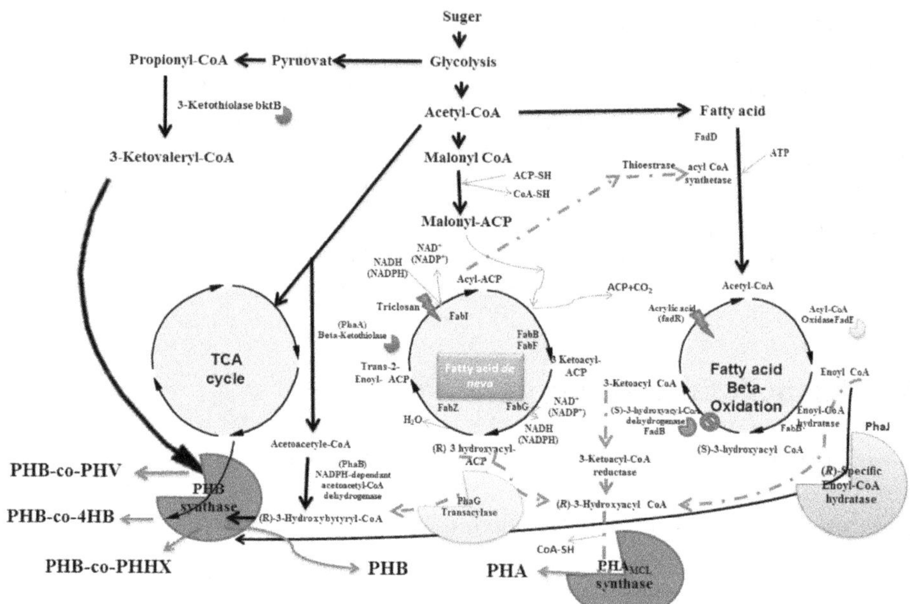

Figure 6.3 Different natural and engineered metabolic pathways (generated in this study) leading to biosynthesis of PHB, PHA, PHB-co-PHV, PHB-co-4HB and PHB-co-PHHX. Acrylic acid inhibits the fatty acid β-Oxidation pathway while triclosan inhibit the fatty acid *de nevo* biosynthesis pathway.

6.5.1 Linking Metabolic Pathways

From the time of its discovery by Dr. Maurice Lemoigne to the 3-hydroxybutyrate (3HB) unit, P(3HB) (de la Riviére 1967; Lemoigne 1923, 1925, 1926, 1927) various studies were conducted to understand the possibility of linking various metabolic pathways to the PHAs basic ones. The main aim is the production of PHAs from a simple, cheap, carbon source such as waste materials like whey lactose, coal and methane, and agriculture sources. It is essential to highlight that the abilities of microbes to utilize various carbon sources are variables. To recognize such differences PhaC synthase type I in *E: coli*, requires gluconate to produce PHB which is expensive, while in the case of using Bacillus, agricultural wastes could be used. That is due to the differences in the ability of different microbes to convert one substrate to another finally to supply the PHA pathways with the required monomeric constituents. Some interesting studies were in utilizing more than one Bacillus strain to be able to utilize a complicated substrate containing lipid, sugars and proteins instead of doing in-cell pathway linking. The point of view of a biotechnologist might be different from the point of view of a molecular biologist or a biochemist because biotechnology is more general while molecular biology and biochemistry are more specific. For example, an expert in biotechnology will not spend time improving a certain microbial strain alone but will also search for a naturally existing different wild type which could satisfy their demand in one move. For that reason many microbes are isolated in developing countries with high biodiversity, and so many technical applications are flourishing. Overexpression of the PhaC synthase will affect the amount of the polymer, while the cells will be exhausted. For example, a researcher might hope to put the PhaC synthase under phage T_7 promoter to over produce the PHA. The T7 promotor will change the microbe to PhaC synthase (protein) producer and the cell will overload with the enzyme and it will no longer be able to produce the PHA. The outer cells metabolic link by using more than one organism and optimizing the production conditions using an experimental design has proved to be an efficient strategy, in particular to produce biotechnological products beside PHB such as the technical enzymes (PHB, lipase and protease at one batch) (Amara 2011).

Steinbüchel in 2001 mentioned that in order to produce PHAs other than PHB from CO_2 or renewable resources it would be necessary to link the central metabolic pathway with PHA synthesis (Steinbüchel 2001). Rehm et al., determined that the gene product of phaG, (*R*)-3-hydroxyacyl-ACP:CoA transacylase, is responsible for the conversion of (*R*)-3-hydroxyacyl-ACP to (*R*)-3hydroxyacyl-CoA (Rehm et al. 1998). An example is drawing a metabolic intermediate from other pathways concerning the citric acid cycle for generating PHA precursors in recombinant *E. coli*; a polymer consisting of 4-hydroxybutyrate (4HB) was produced by drawing succinyl-CoA from the citric acid cycle, which is converted to 4hydroxybutyryl-CoA, a precursor for PHA biosynthesis (Valentin and Dennis 1997). Wild type *P. putida* is able to produce PHA_{MCL} from related and non-related substrates (Brandl et al. 1988; Huijberts et al. 1992; Huisman et al. 1989; Timm and Steinbüchel 1990). An isolated *P. putida* mutant can grow on any carbon source tested and was able to synthesize PHA_{MCL} from fatty acids even when not from gluconate or glucose (Rehm et al. 1998). Biochemical studies on the affected enzyme revealed that this enzyme catalyzes the transfer of (*R*)-3hydroxydecanoyl moieties from the acyl carrier protein: coenzyme A transferase. The respective gene, which complemented the mutant, was referred to as phaG, this was also identified and cloned from *P. aeruginosa* (Hoffmann et al. 2000) and *P. oleovorans* (Hoffmann et al. 2000; Valentin and Dennis 1997). Hoffmann and Amara, with cooperation from others under the supervision of Dr. Rhem succeded in 2002 carrying out a biochemical characterization of the *P. putida* 3-Hydroxyacyl ACP:CoA transacylase, which diverts intermediates of fatty acid *de novo* biosynthesis (Hoffmann et al. 2002).

6.6 PhaC: The Location, Structure and Function

6.6.1 Organizing PHA Synthase Genes in Selected Microbes

The PHA biosynthesis genes and genes for other proteins interconnected the metabolism of PHA are often clustered in the bacterial genomes. In *R. eutropha*, the genes for PHA synthase (*phaC*), β-ketothiolase (*phaA*) and the NADP-dependent acetoacetyl-CoA dehydrogenase (*phaB*) constitute the *phaCAB* operon (Peoples and Sinskey 1989; Schubert et al. 1988; Slater et al. 1998). Approximately 4 kbp downstream of this operon a second β-ketothiolase gene (*bktB*) was identified (Slater et al. 1998). *BktB* is, in contrast to *phaA*, able to synthesize 3-ketovaleryl-CoA. As well as *R. eutropha*, also *Alcaligenes latus* and *Burkholderia cepacia* seem to possess a *phaCAB* operon. In *Pseudomonas sp.* 61-3 and *Acinetobacter sp.* these genes are also clustered. In contrast, *phaC* seems to be separated from other *pha* genes in *Zoogloea ramigera, Methylobacterium extorquens, Sinorhizobium meliloti 41, Nocardia corallina, R. ruber, Paracoccus denitrificans, Rhodobacter sphaeroides, R. rubrum Ha, Rhodobacter capsulatus and A. punctata. PhaA*, *phaB* or other genes of PHA metabolism are not directly linked to the *phaC*s genes in these genomes. An exception is *A. punctata*, in this bacterium the gene encoding an enoyl-CoA hydratase is found downstream of *phaC*. *P. oleovorans, Pseudomonas sp.* 61-3 and *P. aeruginosa* posses two various *phaC* genes which are in the genome separated by the structural gene for an intracellular PHA depolymerase. *Pseudomonas sp.* 61-3 also contains these two *phaC* genes besides another *phaC* gene, which is co-localized with *phaB* and *phaA*. In all bacteria, which possess a two-constituent PHA synthase, *phaC* and *phaE* are directly linked in the genomes constituting most probably single operons. In *A. vinosum phaA* and *phaB* are found in the opposite direction in a gene cluster interconnected to PHA metabolism. Organizing the genes is most probably similar if not identical in *Thiocystis violacea* and *Thiocapsa pfennigii*, whereas in *Synechocystis sp.* PCC 6803 further *pha* genes definitely do not map close to the *phaEC* locus. In *R. ruber* and *A. vinosum* the structural genes for phasin proteins (ORF3 and ORF5, respectively) map close to the respective *phaC* loci (Rehm et al. 2001). Solaiman and Ashby (2005) suggested that the Class IV phaC sequences can be divided into at least two subgroups because of the extent of nucleic acid sequence matches (Solaiman and Ashby 2005). The *B. megaterium* subgroup has a pha locus which consists of the *phaR-phaB-phaC* operon and *phaP-phaQ* operon in the opposite direction. The *B. cereus* subgroup has a *pha* locus which consists of the *phaR-phaB-phaC* operon and *phaP-phaQ-phaJ* operon in the opposite direction.

6.6.2 PHA Classes

Some bacterial strains have been studied in more depth such as *R. eutropha*, *P. oleovorans*, and *A. vinosum* (Lenz and Marchessault 2005; Stubbe and Tian 2003) and *B. megaterium*. Based on the substrate specificity and subunit composition, PHA synthases were first grouped into three classes I, II and III (Rehm 2003). The PHA synthase of *B. megaterium* is a two-constituent enzyme, similar to a Class III synthase (PhaEC) (McCool and Cannon 1999). Unlike those in Class III, the 40 kDa PhaE was not required, and furthermore, the 22 kDa PhaRBm had no homology to PhaE. Class I and III PHA synthases have substrate specificities toward short chain length hydroxyacyl monomers (C3–C5), whereas Class II PHA synthases utilize medium chain length hydroxyacyl monomers (C6–C14) (McCool and Cannon 1999). Classes I and II have a single subunit (PhaC). The Class I PHA synthases (*R. eutropha*) are composed of one single type of polypeptide chain and use chiefly (*R*)-3-hydroxybutyryl-CoA, (*R*)-3-hydroxyvaleryl-CoA and other short-carbon chain length hydroxyalkanoic acid CoA thioesters including 3-mercaptoalkanoic acid CoA thioesters as substrates (Lutke-Eversloh et al. 2001). The Class III PHA synthases (*A. vino*sum) are composed of two various subunits each of about 40 kDa. The substrate specificity is similar to that of Class I synthases, although some medium chain length 3-hydroxyfatty acids are also incorporated (Liebergesell et al. 2000). Both classes of PHA synthases were purified, and *in vitro* activity has already been reached (Gerngross et al. 1994; Jossek et al. 1998; Liebergesell et al. 1994). The Class II enzymes and Class I PHA synthases, which are composed of only one type of subunit, are chiefly found in pseudomonads such as *P. aeruginosa*. One major difference between Class II and both Class I and III, PHA synthases is the substrate specificity. Class II PHA synthases incorporate preferentially 3-hydroxyfatty acids of medium-chain length (C6–C14) into PHAs, and the resulting product is a latex-like polymer (Qi et al. 1997, 1998; Timm and Steinbuchel 1992). These substrates are chiefly derived from intermediates of fatty acid *β*-oxidation or from fatty acid *de novo* biosynthesis (Hoffmann et al. 2000; Rehm et al. 1998, 2001) provided fatty acids or simple non-related carbon sources (Rehm et al. 2001). Class II PHA synthases were purified, and *in vitro* activity was reached (Qi et al. 2000; Rehm et al. 2001). Class I and II PHA synthases possess two thiol groups provided by conserved Cys-319, and in Class III PHA synthases the second thiol group might be provided by conserved Cys-130 of subunit PhaE from *A. vinosum*. Class II PHA synthases, have a catalytic triad (Cys-296–Asp- 452–His-453) with the originally proposed general base catalyst His-480 functionally replaced by His-453 (Amara and Rehm 2003). Figures 6.4 and 6.5 show the four classes where each class shows high similarity within its members and shows homology with the other synthases.

6.6.3 PhaC Structure

PHA synthases are the key enzymes of PHA biosynthesis and catalyze the conversion of 3-hydroxyacyl-CoA substrates to PHAs with the concomitant release of CoA. The multiple alignment of the primary structures of these PHA synthases showed an overall identity. PHA synthases were first assigned to three classes based on their substrate specificity and subunit composition. Immunoelectron microscopy studies of granules isolated from *Allochromatium. vinosum*, employing gold-labeled anti-PhaC antibodies clearly suggested a PHA synthase complex at the surface of the PHA granule (Liebergesell et al. 1994; Mayer et al. 1996). This homogenous population of particles measuring 11.2–12.8 nm in diameter and data derived from gel filtration chromatography suggest that this PHA synthase might be composed of ten subunits (Liebergesell et al. 1994; Mayer et al. 1996).

Class I PHA synthases are composed of one single type of polypeptide chain and use chiefly (*R*)3-hydroxybutyryl-CoA, (*R*)-3-hydoxyvaleryl-CoA and other short-carbon chain length hydroxyalkanoic acid CoA thioesters including 3-mercaptoalkanoic acid CoA thioesters as substrates (Lutke-Eversloh et al. 2002; Rehm and Steinbüchel 2002).

Müh et al. 1999 reported that PhaC synthase Class III from *Chromatium vinosum* which is able to catalyze the conversion of 3-hydroxybutyryl-CoA (HB-CoA) to polyhydroxybutyrate (PHB) and CoA. The synthase is composed of a ~1:1 mixture of two subunits, PhaC and PhaE. The holo-enzyme, digestion of [3H]-sT-synthase with trypsin and HPLC analysis resulted in isolation of three labeled peptides. The enzymatic lag phase is reduced and the specific activity increased in the presence of trimeric analogs (Müh et al. 1999; Wodzinska et al. 1996). A BLAST sequence homology search with the *A. vinosum* Class III PHA synthase and the *P. aeruginosa* Class II PHA synthase has shown homology to lipases, particularly to the lipase from *B. cepacia*, and the putative active

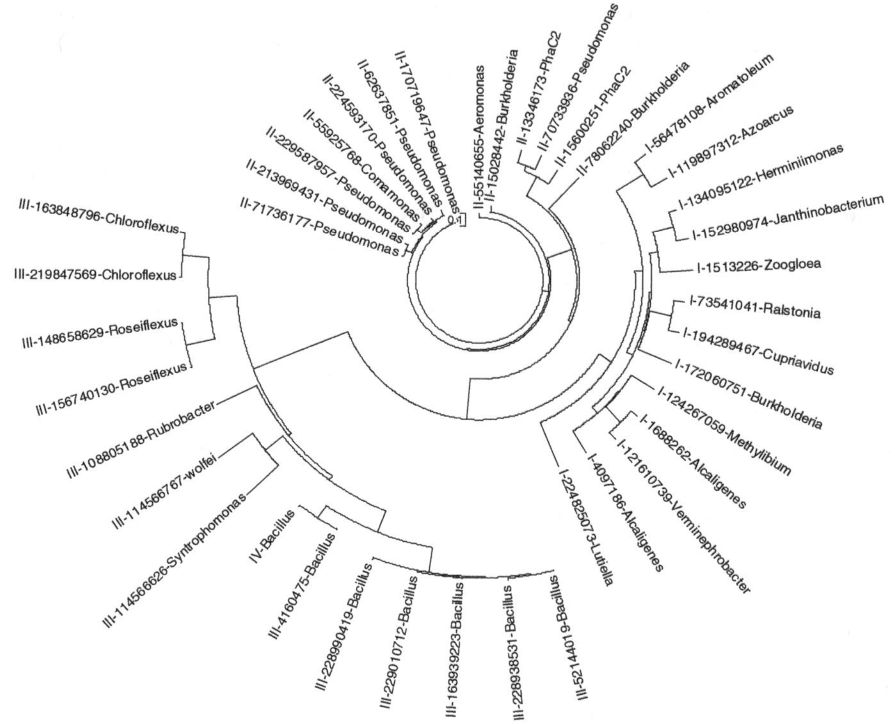

Figure 6.4 Phylogenetic tree of 40 PHA synthases (generated in this study) represented Class I, II, III and IV. The tree shows perfect grouping of each class. Only Class III is similar to Class IV and clusters with it. The tree was generated using the software UGene 33-win (Okonechnikov et al. 2012). The sequences used are: I-73541041-*Ralstonia eutropha* JMP134; I-194289467-*Cupriavidus taiwanensis*; I-172060751-*Burkholderia ambifaria* MC40-6; I-1688262-*Alcaligenes sp.*; I-124267059-*Methylibium petroleiphilum* PM1]; I-4097186-*Alcaligenes latus*; I-121610739-*Verminephrobacter eiseniae* EF01-2; I-134095122-*Herminiimonas arsenicoxydans*; I-152980974-*Janthinobacterium sp. Marseille*; I-1513226-*Zoogloea ramigera*; I-56478108-*Aromatoleum aromaticum* EbN1; I-119897312-*Azoarcus sp.* BH72; I-224825073-*Lutiella nitroferrum* 2002; II-170719647-*Pseudomonas putida* W619; II-55140655-*Aeromonas hydrophila*; II-15028442-*Burkholderia caryophylli*; II-224593170-*Pseudomonas chlororaphis*; II-229587957-*Pseudomonas fluorescens* SBW25; II-71736177-*Pseudomonas syringae pv. phaseolicola* 1448A; II-213969431-*Pseudomonas syringae pv. tomato* T1; II-62637851-*Pseudomonas corrugate*; II-15600251-PhaC2 *Pseudomonas aeruginosa* PAO1; II-13346173-PhaC2 *Pseudomonas nitroreducens*; II-78062240-*Burkholderia sp.* 383; II-70733936-*Pseudomonas fluorescens* Pf-5; III-52144019-*Bacillus cereus* E33L; III-228938531-*Bacillus thuringiensis serovar berliner* ATCC 10792; III-163939223-*Bacillus weihenstephanensis* KBAB4; III-229010712-*Bacillus mycoides* DSM 2048; III-228990419-*Bacillus pseudomycoides* DSM 12442; III-4160475-*Bacillus megaterium*; III-108805188-*Rubrobacter xylanophilus* DSM 9941; III-156740130-*Roseiflexus castenholzii* DSM 13941; III-114566626-*Syntrophomonas wolfei subsp.* wolfei str. Goettingen; III-148658629-*Roseiflexus sp.* RS-1; III-114566767-wolfei subsp.wolfei str. Goettingen; III-219847569-*Chloroflexus aggregans* DSM 9485; III-163848796-*Chloroflexus aurantiacus* J-10-fl; IV-*Bacillus megaterium*.

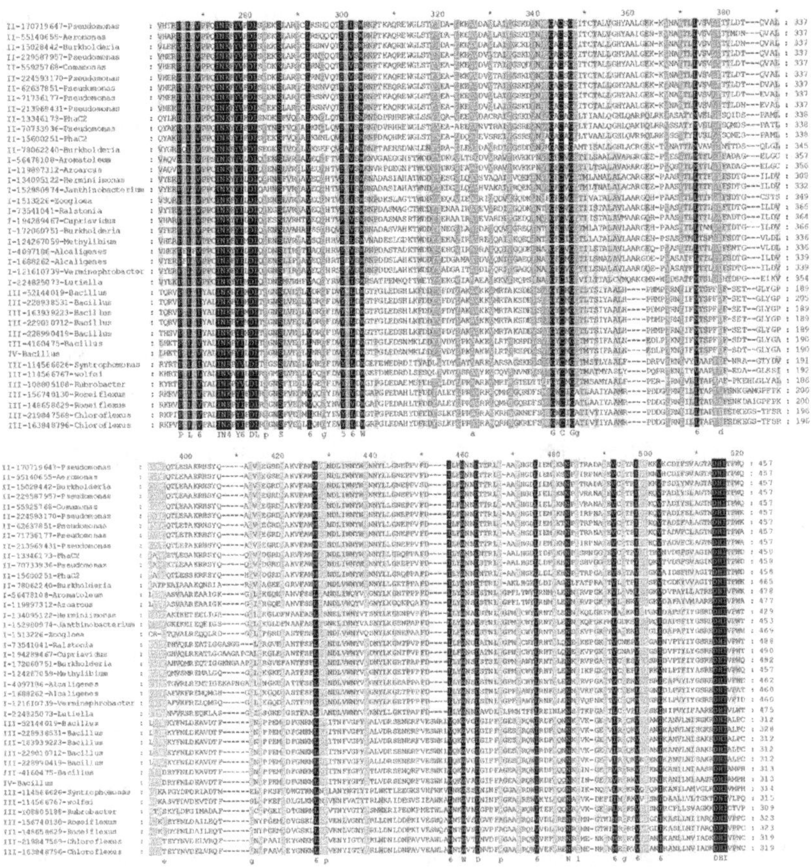

Figure 6.5 Multiple alignment of the primary structure of 40 amino acid sequences of different PhaC synthases (generated in this study) represent Class I, II, III and IV. The alignment shows perfect grouping of each class. Only Class III is similar to Class IV and aligned near to it. The alignment was generated by BioEdit software (Hall 1999) and visualized by Genedoc software (Nicholas et al. 1997) in this study.

site cysteine residue aligned with the active site serine of the lipase (Jia et al. 2000; Rehm and Steinbüchel 2002). A threading model of the Class II PHA synthase from *P. aeruginosa* was developed with the aid of the SAMT98 algorithm (Karplus et al. 1998) based on the homology to the structure of the lipase from *B. cepacia*. In the multiple alignment of PHA synthases only one cysteine residue (Cys-319) is highly conserved. The essential role of Cys-319 of the *R. eutropha* PHA synthase for the reaction mechanism was obtained from site-specific mutagenesis (Gerngross et al. 1994). The weakly conserved Cys-459 was supposed to be involved in the catalytic cycle, providing the second thiol group. However, site-specific mutagenesis (Gerngross et al. 1994) clearly indicated, that this amino acid residue is not essential for catalytic activity. To investigate post-translational modification by 4-phosphopantetheine, radiolabeling experiments were conducted and no evidence for covalent post-translational modification by 4-phosphopantetheine is considered (Hoppensack et al. 1999). Exchange of amino acid residue Ser-260 against alanine and threonine, respectively, abolished *in vivo* and *in vitro* activity of PHA synthase from *R. eutropha* (Hoppensack et al. 1999). Rehm et al. (2002) have developed a threading model of the *R. eutropha* PhaC synthase based on the homology to the *Burkholderia glumae* lipase structure (Noble et al. 1993). Part of Professor Sterinbucher's group under the supervision of Professor Rhem has conducted many mutation analyses using different tools for the different PhaC synthases. Some of those mutants will be described. Replacement of the two amino acid residues Y445 (F) and L446 (K) respectively result in reducing the activity. The E267K mutant has only slightly reduced activity with a surface-exposed mutation site. Four site-specific deletions were generated to evaluate the role of the C-terminus and variant amino acid sequence regions. Deleted regions were D281–D290, A372–C382, E578–A589 and V585–A589 and the respective PHA synthases showed no detectable activity. The fussion of the N-terminal part of the Class II PHA synthase ($PhaC_{Pa}$) from *P. aeruginosa* and the C-terminal part of the Class I PHA synthase ($PhaC_{Re}$) from *R. eutropha* resulting in three fusion proteins with no detectable *in vivo* activity. However, the fusion protein F1 ($PhaC_{Pa}$-1-265-$PhaC_{Re}$-289-589) showed 13% of wild type *in vitro* activity with the fusion point found at a surface-exposed loop region. For more details about those mutants refer to Rehm et al. (2002).

Also mutants created by other scientist were analysed by applying them on the PhaC synthase model generated by Rehm et al. (2002). Inspection of the proposed $PhaC_{Re}$ model showed that the identified catalytic residues C319, H508 and D480 are found adjacent to the core structure of the $PhaC_{Re}$ protein model with the catalytic C319 found at the nucleophile elbow. The active site C319 lies at the C-terminal corresponding to strand L5 of the lipases, in the strand-turn-helix motif. Mutational studies revealed that the *R. eutropha* PHA synthase possesses an essential catalytic dyad (C319–H508) in which the C319 is involved in covalent catalysis and H508 serves as a general base catalyst (Jia et al. 2001). A conserved D480 was shown to be not required for acylation of C319 by sT-CoA (a terminally saturated trimer of (*R*)-3-hydroxybutyryl-CoA) and is proposed to function as a general base catalyst to activate the hydroxyl of (*R*)-3-hydroxybutyryl-CoA for ester formation (Jia et al. 2001). W425 was found at the surface. Furthermore, the site-specific mutations Y445F and L446K resulted in reduced activity, and since Y445 and L446 are hydrophobic residues and are found adjacent to W425 at the surface of the protein model, these residues might exert a similar function in view of dimerization. The permissive double mutant GS3 (R386C, K139R) contained one mutation site R386C found in the lid region. Two site-specific mutants (A391T, T393A) were generated by Taguchi et al. (Taguchi et al. 2001), which resided in the lid region and still has PHA synthase activity. This model suggested that the surface-exposed lid-like structure is not essential for enzyme function. Respective number site-specific mutations were performed with Class I PHA synthase from *R. eutropha* (Gerngross et al. 1994;

Hoppensack et al. 1999; Jia et al. 2001; Schubert et al. 1991; Taguchi et al. 2001). Single gene shuffling with the *R. eutropha* PHA synthase gene resulted in isolating five mutants (double or triple mutants) with altered enzyme activity, whose mutation sites were chiefly found at the surface of the $PhaC_{Re}$ model. Only in the case of the triple mutant GS4 a rather conserved residue, the P230, was affected. However, P230 was found at the surface of the $PhaC_{Re}$ model.

Amara and Rehm (2003) developed the first threading model of a Class II PHA synthase, the PHA synthase I from *P. aeruginosa*, based on the homology to the epoxide hydrolase (1ek1) from mice, the structure of which was determined by X-ray analysis (Amara and Rehm 2003). Many of the features of the α/β -hydrolase fold are maintained in the protein model. The central β-sheet in the core of the molecule conforms to the α/β-hydrolase fold. Inspection of the proposed PhaC1 model showed that the proposed catalytic residues Cys-296, His-480 and Asp-452 are found adjacent to the core structure of PhaC1 protein model with the catalytic Cys- 296 found at the nucleophile elbow. The active site Cys-296 lies at the C-terminal end of a β-strand, corresponding to strand β 5 of the lipases, in the strand-turn-helix motif. Amara and Rehm (2003) generated for the first time site-specific mutants of a Class II PHA synthase (Amara and Rehm 2003). The highly conserved and proposed catalytic residues Cys-296, Asp-452 and His-480 were considered to constitute the catalytic triad, as was found for Class I and Class III PHA synthases (Amara and Rehm 2003). Surprisingly, replacement of the catalytic nucleophile Cys-296 with Ser only reduced *in vitro* enzyme activity to 14% of wild type enzyme activity, whereas the same replacement in Class I and Class III synthases strongly reduced enzyme activity to 0.001% and 0.1% of wild type enzyme activity, respectively (Jia et al. 2000, 2001).

The highly conserved Trp-425 in the *R. eutropha* Class I PHA synthase was replaced by alanine, which reduced the *in vivo* activity to 19% and *in vitro* activity to 0.003% of wild type activity. Replacement of the corresponding Trp-398 with Ala abolished *in vivo* and *in vitro* enzyme activity, suggesting the importance of this residue for enzyme activity presumably by enabling dimerization. Dimerization of PhaC1 was induced by addition of substrate (Rehm et al. 2001). The discrepancies detected for *in vivo* and *in vitro* activity of the respective mutants suggest that the enzymes were still capable of efficient acylation even so catalysis of polymerization reaction was impaired (Amara 2002; Rehm et al. 2002). Yuan et al. (Yuan et al. 2001) reported that 4-phosphopantetheinylation might play a role for enzyme activity of Class I and III PHA synthases. However, analysis of 4-phosphopantethenylation of the Class II PHA synthase did not provide evidence for such a post-translational modification. So the second nucleophilic group might be provided through dimerization of PhaC1 (Rehm et al. 2001). Amara and Rehm (2003) describe the first approach to investigate amino acid residues involved in catalysis of Class II PHA synthases, which suggested a catalytic triad (Cys-296–Asp- 452–His-453) with the originally proposed general base catalyst His-480 functionally replaced by His-453. Moreover, replacement of the catalytic nucleophile Cys-296 with Ser kept the enzyme highly active, which was not the case in the respective Class I and III PHA synthases. Interestingly, mutants C296S and H453Q showed a significantly changed substrate specificity, which has not been described for the

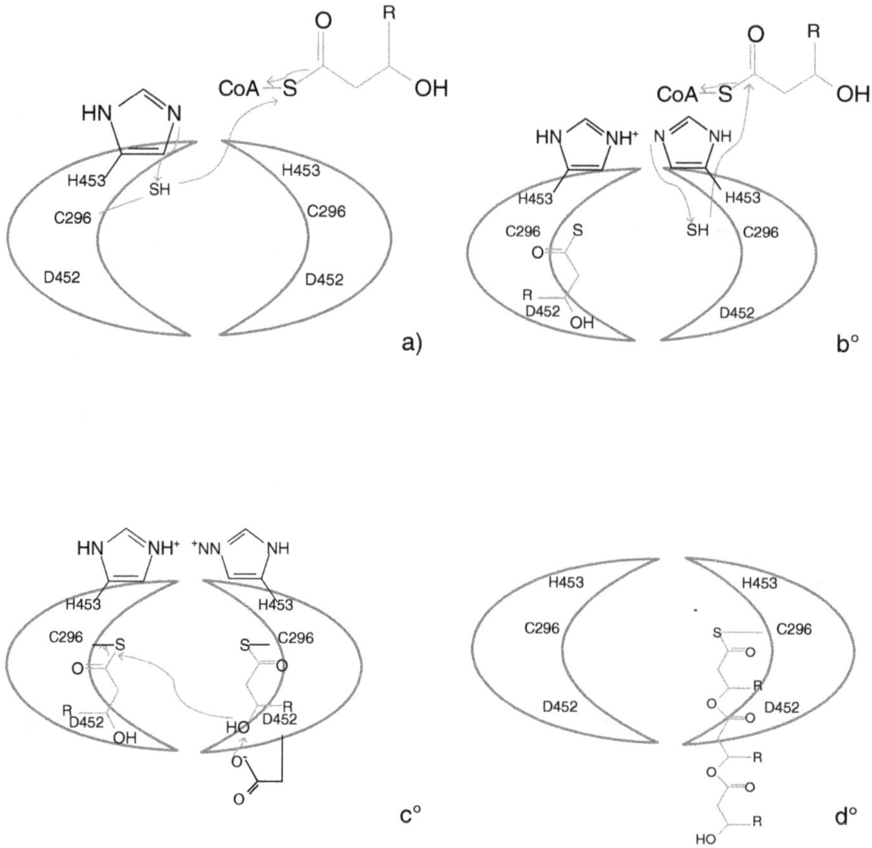

Figure 6.6 PHA_{MCL} polymerization in PhaC synsase Class II (*Pseudomonas aeruginosa*) in four steps (a)–(d).

respective Class I and Class III PHA synthases. This is an example where a single amino acid replacement exerted a strong impact on substrate specificity. The study concluded that besides differences in substrate specificity the Class II PHA synthases differ strikingly from Class I and III PHA synthases with respect to catalytic residues. Figure 6.6 show the polymerization of the PhaCpa synthase Class II following Amara and Rehm (2003). The PhaC synthases of different 3D structures of the different classes were represented as in Figure 6.7. As in Figure 6.7 and Table 6.2 Class I and II are bigger than Class III and IV. And Class III are bigger than Class IV. That might explain why Class III and Class IV need a complementary protein for optimum polymerization. Additionally the amino acids analysis as in Table 6.2 show the differences and the similarity between the classes. The comparative amino acids analysis shows also that the catalytic residue is represented by a lower number if compared with other amino acids such as the alanine.

Table 6.2 The average % of the amino acids profiles for different classes of PhaC synthases generated by the software Mega 6 (Tamura et al. 2013) for different sequences obtained from BLAST database (Madden 2002).

Class	Ala	Cys	Asp	Glu	Phe	Gly	His	Ile	Lys	Leu	Met	Asn	Pro	Gln	Arg	Ser	Thr	Val	Trp	Tyr	Total
I	11.42	0.71	5.12	5.14	3.98	7.36	1.68	4.18	4.92	10.00	2.26	4.80	5.87	5.00	4.51	5.70	5.26	6.16	2.78	3.15	572.93
II	9.19	0.87	5.50	4.78	3.56	6.61	2.69	3.71	4.82	11.27	2.06	6.13	5.44	4.19	5.23	6.44	5.56	6.49	2.67	2.79	559.85
III	6.68	0.73	6.18	5.93	3.88	6.51	1.68	6.01	6.42	10.22	3.13	4.71	5.55	3.01	5.48	4.76	5.12	6.70	2.39	4.90	379.36
IV	6.35	0.28	6.91	5.80	4.14	5.52	1.10	6.35	9.12	9.94	3.31	4.97	5.80	2.21	4.14	5.52	4.70	6.35	1.93	5.52	362.00

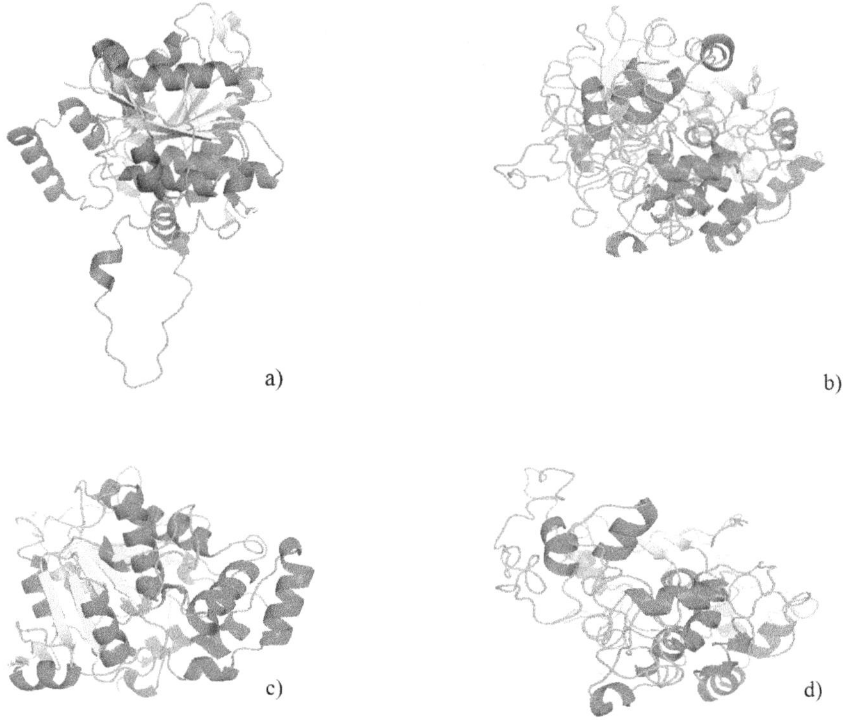

Figure 6.7 PhaC synthases 3D models representing (a) Class I PhaC synthase *Ralstonia eutropha*, (b) Class II PhaC sunthase *Pseudomonas-putida*, (c) Class I PhaC sunthase *Bacillus thrugenesis*, (d) Class I PhaC sunthase *Bacillus thrugenesis*. (a), (b), (c) generated by Modeller v9.8. (Gilbert 2000; Šali and Blundell 1993), (d) generated from the intracellular polyhydroxyalkanoate synthase structure of PhaC from *Chromobacterium sp.* through Swiss protein modeling web page (Arnold et al. 2006).

6.7 Mutagenesis Case Studies

6.7.1 Case Study for Type I Synthases *In Vitro* Random Mutagenesis

Kichise et al. performed successful *in vitro* molecular evolution experiments on PhaC from *A. caviae* by applying the *in vitro* evolutionary technique to a limited region of the *phaC* and coexpressing the mutants with the monomer-supplying enzyme genes, *phaAB* from *R. eutropha* and *phaJ* from *A. caviae*, to supply monomers from glucose or dodecanoate, respectively. Two evolvants have increased activity toward 3HB-CoA of 56% and 21%, respectively, compared to the wild type enzyme in *in vitro* assays. These mutations led to enhanced accumulation (up to 6.5-fold higher than the wild type PhaC) of P(3HB-co-3HHx) and increases in the mole fraction of 3HHx (16–18 mol.% compared to 10 mol.% of the wild type PHA synthase) in recombinant *E. coli* LS5218 strains grown on dodecanoate. Two single mutations, Asn149Ser (N149S) and Asp171Gly (D171G), are not highly conserved among PHA synthases. As an extended study, a combination of these two beneficial mutations (N149S/D171G) synergistically altered enzymatic properties, leading to synthesis of PHA copolymers with enhanced 3HA fraction and increased molecular weight in recombinant *R. eutropha* (Kichise et al. 2002; Taguchi et al. 2012; Tsuge et al. 2007).

6.7.2 Case Study for Type I Synthases *In Vivo* Random Mutagenesis

XL1-Red, an *E. coli* strain which is deficient in three of the primary DNA repair pathways, was used to mutates phaC synthases genes from *P. aeruginosa*, *R. eutropha* and *A. punctate*. The study took nearly three and a half years and the three genes were subjected to daily mutagenesis. The three deficient genes in XL1Red are The *mutS* (error-prone mismatch repair) (Glickman and Radman 1980), *mutD* (deficient in 3′-to 5′- exonuclease of DNA polymerase III) (Scheuermann et al. 1983) and *mutT* (unable to hydrolyze 8-oxodGTP) by Stratagene (LaJolla, USA). This strain grows extremely slowly in rich media (such as LB media), having a doubling time of ~90–120 minutes. The XL1-Red strain carries the Tn*10* transposon that comprises the tetracycline-resistance gene. However, becouse of the rapid mutator phenotype, the cells will frequently give rise to tetracycline-sensitive (Tc^S) variants. Therefore, the resulting transformants may or may not be tetracycline resistant, so it is not recommended that tetracycline be added to the growth media. Each gene was in a plasmid which collected after each cycle and retransformed in a stable recombinant *E. coli* XL1 Blue or XL1 Gold (with high transforming percentage). Figures 6.8 and 6.9 represent both of the random mutagenesis for screening mutants and that for changing the substrate specificity. Each gene then either transformed to an *E. coli* with PhaAB to complement the PhaC of the *R. eutropha* and *A. punctate* (plasmids with different markers are used to enable the correct selection for the cells contain the two plasmids) or a leaky *P. putida* (*P. putida* GPp104) in the case of the $phaC_{P.a}$. to search for the change in the substrate activity. Selected mutants which give more flurescence (under UV) were sequenced. Finally, the mutant of interest with overproduction capacity was published Amara 2001 (Amara 2001, Amara et al. 2002). Higher *in vitro* specific activity (up to fivefold) and higher PHA content (up to 126% of wild type) were detected for the overproducing mutants. Also, mutants synthesized PHAs with increased weight-average molecular weight with incorporation of the 3HHx are reported.

6.7.3 Case Study Comparative Analysis of PhaC Synthases Class I, II, III and IV

In a previous study, 301 PHA synthases protein sequences were used to analyse the differences between the four classes. The question addressed in this study is: why are similar enzymes classified to various subclasses? To answer this question the study introduced the hypothesis, that as well as the catalytic residue the overall amino acids are another crucial factor responsible for adjusting the structure/function/specificity. Basically, many previous studies used the protein alignment to analyse various PhaC synthases and to determined the amino acids that are responsible for or are involved in the catalysis. The differences between the classes in various PhaC synthases were investigated.

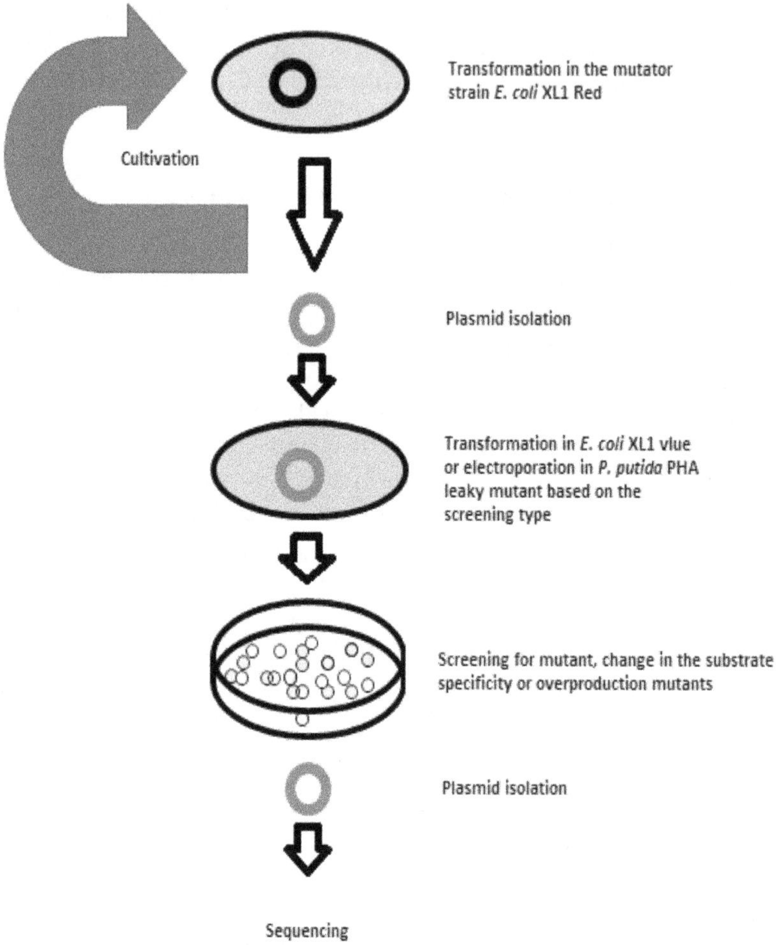

Figure 6.8 Strategy for random mutagenesis using XL1Red and Nile red. The strategy has been used to mutate plasmids harbouring *phaC* synthases different genes. The mutated genes then transferred to a stable recombinant *E. coli* such as XL1 Blue and complemented with the necessary genes (*phaAB*) in the case of the PHB producing synthases and transferred to *P. putuda* leaky mutant in case of phaC II. To search for changes in the substrate specificity each plasmid transferred to the opposite strain by changing the substrate in the media (Amara et al. 2001, 2002; Rehm 2002a; Spiekermann 1999).

The homology and phylogenic tree of some selected PhaC synthases of various strains (representing the four used classes) were demonstrated. Rehm et al. (2002) (Rehm et al. 2002), Amara et al. (2001, 2002) (Amara et al. 2001, 2002) and Taguchi et al. (2001, 2002) showing and proving that mutants in places other than the PhaC synthases (catalytic residues) active sites could influence their activity. They have used various protocols for conducting such mutations (Amara et al. 2001; Taguchi et al. 2001, 2002). This clearly

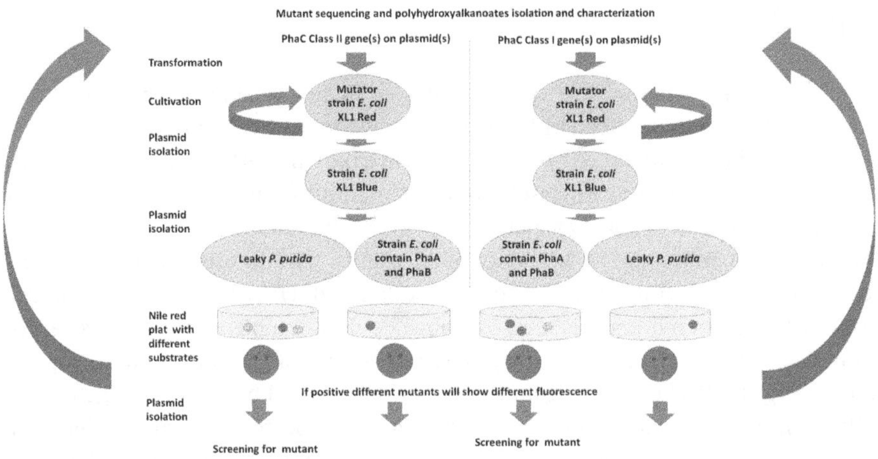

Figure 6.9 Random mutagenesis for screening mutants such as overproducer, leaky, negative, change in substrate specificity etc. (Amara et al. 2001; Amara et al. 2002; Rehm et al. 2002; Spiekermann 1999).

shows that amino acids other than the catalytic residue involved in the protein stability therefore effect its activity. In the EMBO conferences for comparative genomics of Eukaryotic microorganisms (2009) Amara et al. (2009, 2011) show a new trend in evaluating the various classes of one type of enzyme based on their constituents of amino acids (Amara et al. 2009, 2011). The percentage of amino acids in the same class is high, and comparing with other classes the number of differences is high. Only Class III and IV show high similarity. To investigate the differences in amino acids between PhaC synthases and various Class I, II, III and IV, 100 sequences from each of the first three classes were collected randomly from the BLAST protein database and one sequence from Class IV (Amara et al. (2012)).

Various PhaC synthases classes are even polymerases in their nature, have differences in their end products, structures and substrate specificity. Why are enzymes doing the same functions different? Various classes have different amino acids profiles even they have minimal differences in their catalytic residues as proved by Amara and Rehm (2003) (Amara and Rehm 2003; Amara et al. 2001, 2002; Rehm et al. 2002). To re-evaluate this finding alignment was produced for the four classes. The analysis of the four classes of amino acids give similer results to those found in 2009 and 2011. The extracted phylogenic tree put Class III and IV protein sequences in one group and Class I and II sequences each in separate groups. In addition a protein model for each of Class I, II and III is the same as in Amara et al. 2001. A new model for Class IV were generated in this study according to the Swiss model generation web site based on intracellular polyhydroxyalkanoate synthase Structure of PhaC from *Chromobacterium sp.* USM2. PhaC Class IV in its 3D structure appear slightly diferent from the other classes in that it is shorter. The Class III and IV are significantly shorter than Class I and II. That explains why Class III and IV need a complementary sequence.

6 Conclusion

Polyhydroxyalkanoates (PHAs) are various types of biodegradable polymer produced chiefly by microbes under reduced or imbalanced nutritional conditions. There are four known classes of PhaC synthase enzymes that polymerize various hydroxyalkanoic acids to the various PHAs. Their nomenclature as short/medium/length chain form is dominant. Their physical and chemical properties are various based on the differences in their monomeric constituents and the branched side chains. The interest in PHAs ranges overages from weak to strong base on the market demand. However, its beginning was like most newly discovered materials, in the lab. The first notable interest was after the petroleum oil crises at 1973. At that time, there was a tangible political plan to support commercializing the PHAs mostly from the countries affected by the rise in petroleum prices. Different studies were conducted to analyze the expected catalytic nucleophilic residues. New polymers were produced and overproducer mutants were identified. The site directed mutagenesis and the random mutagenesis (various experiments) succeeded in identifying the nucleophilic residues of the various PhaC synthases plus other essential amino acids involved in the catalysis. Overall, PHAs were subjected to five phases of research. Phase I was the first discovery and identification. Phase II was isolating and identifying various natural producers. Phase III started by cloning the various PhaC synthases (which were named polymerases at that time) and the other genes involved in their production pathways were cloned. Phase IV was concerned with the study of the functions of the PhaC synthases structures by conducting various types of mutagenesis. And Phase V was the production of new PHAs either biologically or chemically. PhaC I, II, III and IV show high homology, but even higher homology was shown in individual classes compared with the other classes. Lütke-Eversloh et al. was the first to report the production of biopolymers with thioester linkages in the polymer backbone using C. necator in media containing 3-mercaptopropionate (3MP) or 3-mercaptobutyrate (3MB) with 3-hydroxybutyrate as constituents. Each of the PhaC synthases is able to produce a specific type of polymer through a specific pathway and nutritional conditions. However, natural isolates prove their ability to incorporate other monomeric forms to their essential backbone. One should differentiate between changing the substrate specificity and the ability of various PhaC synthases to incorporate other monomeric forms. Changing the substrate specificity means moving from one type to another, such as production of PHAMCL by PhaC synthase that produce PHB (PHA_{SCL}) using the substrates of the MCL polyhydroxyalkanoates and the PhaC enzymes of the SCL PhaC synthases. Many research efforts, huge funding, publications, conferences, patents, and so on are providing PHA with a push and many products are being produced. However, apparently the time of PHA commercialization as a plastic has not yet come. However, alternative medical applications might save the research on it. PhaC synthases are a perfect example of enzymes which are highly similar but carrying out different specific functions due to specific differences in their backbone.

From the various small examples concerning the available data one could conclude that PhaC synthases are similar generally and more similar within each class. An exception is that Class III is highly similar to Class IV. Class IV is smaller than Class III and both are smaller than I and II. The PHAs are dependent on the type of the PhaC synthases, the producer microbes, the available substrate and the production conditions. In nature there are many powerful wild type microbes. Intermediates from the different non-related pathways could be involve if using different substrates or inhibiting the β-oxidation pathway, or through the fatty acids *de nevo* biosynthesis pathways. The catalytic residues of various PhaC synthases are identified. In general PhaC synthases is one type of enzyme that produces different PHAs. That proves the role of their amino acid compositions and number, their folding form and their nucleophilic residue (involved in the catalysis).

Acknowledgement

The author gratefully acknowledges his monitors Professor Dr. Alexander Steinbüchel and Professor Dr. Bernd Rhem also the entire membership of the Institute of Molecular Mikrobiologie und Biotechnologie, Mathematish-Naturwissenschaftlichen Fakultät der Westfälische Wilhelms-Universitüt Münster, Germany. Special thanks to the members of the lab 06 during the period 1999–2003. The author acknowledges the DAAD for the grant provided as a PhD scholarship.

Conflict of interest

The author declares that there is no conflict of interest concerning this chapter

References

Abe, C., Taima, Y., Nakamuraand, Y., and Doi, Y. (1990). New bacterial copolyesters of 3 hydroxyalkanoates and 3-hydroxy-ù-fluoroalkanoates produced by *Pseudomonas oleovorans*. *Polymer Communications* 31: 404–406.

Alper, R., Lundgren, D.G., Marchessault, R.H., and Cote, W.A. (1963). Properties of poly-?-hydroxybutyrate. I. General considerations concerning the naturally occurring polymer. *Biopolymers* 1 (6): 545–556.

Amara, A.A., Matar, M.S., and Serour, E. (2009). Comparative study on the amino acids constituents of PhaC synthases Class I, II, III and IV. EMBO –Conference- "Comparative genomics of Eukaryotic microorganisms, 59.

Amara, A.A., Matar, S.M., and Serour, E. Comparative amino acids studies on phac synthases and proteases as well as establishing a new trend in experimental design. *IIUM Engineering Journal* 13 (1).

Amara, A.A. and Rehm, B.H. (2003). Replacement of the catalytic nucleophile cysteine-296 by serine in class II polyhydroxyalkanoate synthase from Pseudomonas aeruginosa-mediated synthesis of a new polyester: identification of catalytic residues. *The Biochemical Journal* 374 (Pt 2): 413–421.

Amara, A.A., Rehm, B.H.A., and Steinbüchel, A. (2001). Biopolymer overproduction by new mutants using simple methods for selection. In DAAD-Bioforum-Berlin "Grenzenlos forschen". *DAAD, Biotechnologische Methoden*: 231–239.

Amara, A.A., Steinbuchel, A., and Rehm, B.H. (2002). In vivo evolution of the Aeromonas punctata polyhydroxyalkanoate (PHA) synthase: isolation and characterization of modified PHA synthases with enhanced activity. *Applied Microbiology and Biotechnology* 59 (4–5): 477–482.

Anderson, A.J. and Dawes, E.A. (1990). Occurrence, metabolism, metabolic role, and industrial uses of bacterial polyhydroxyalkanoates. *Microbiological Reviews* 54 (4): 450–472.

Arnold, K., Bordoli, L., Kopp, J., and Schwede, T. (2006). The SWISS-MODEL workspace: a web-based environment for protein structure homology modelling. *Bioinformatics* 22: 195–201.

Asada, Y., Miyake, M., Miyake, J., Kurane, R., and Tokiwa, Y. (1999). Photosynthetic accumulation of poly-(hydroxybutyrate) by cyanobacteria–the metabolism and potential for CO2 recycling. *International Journal of Biological Macromolecules* 25 (1–3): 37–42.

Ballard, D.G.H., Holmes, P.A., and Senior, P.J. (1987). In: Recent Advances in Mechanistic and Synthetic Aspects of Polymerization (ed. M. Fontanille and A. Guyot), 215: 293–314. Lancaster, UK: Reidel (Kluwer) Pub. Co.

Barnard, G.N. and Sanders, J.K.M. (1988). Observation of mobile poly(Î²-hydroxybutyrate) in the storage granules ofMethylobacteriumAM1 by in vivo13C-NMR spectroscopy. *FEBS Letters* 231 (1): 16–18.

Beijerinck, M.W. (1888). Cultur des Bacillus radicola aus den Knöllchen. *Bot. Ztg.* 46: 740–750.

Brandl, H., Gross, R.A., Lenz, R.W., and Fuller, R.C. (1988). Pseudomonas oleovorans as a source of poly(β-hydroxyalkanoates) for potential applications as biodegradable polyesters. *Applied and Environmental Microbiology* 54 (8): 1977–1982.

Brandl, H., Knee, E.J., Fuller, R.C., Gross, R.A., and Lenz, R.W. (1989). Ability of the phototrophic bacterium Rhodospirillum rubrum to produce various poly (Î²-hydroxyalkanoates): potential sources for biodegradable polyesters. *International Journal of Biological Macromolecules* 11 (1): 49–55.

Byrom, D. (1987). Polymer synthesis by microorganisms: technology and economics. *Trends in Biotechnology* 5 (9): 246–250.

Choi, M.H. and Yoon, S.C. (1994). Polyester biosynthesis characteristics of pseudomonas citronellolis grown on various carbon sources, including 3-methyl-branched substrates. *Applied and Environmental Microbiology* 60 (9): 3245–3254.

Chowdhury, A.A. (1963). Poly-β-hydroybuttersäure abbauende Bakterien und Exoenzyme. *Archiv für Mikrobiologie* 47 (2): 167–200.

Curley, M., Hazer, B., Lenz, R.W., and Fuller, R.C. (1996). Production of poly(3hydroxyalkanoates) containing aromatic substituents by Pseudomonas oleovorans. *Macromolecules* 29: 1762–1766.

de Koning, G.J.M., Kellerhals, M., van Meurs, C., and Witholt, B. (1997). A process for the recovery of poly(hydroxyalkanoates) from Pseudomonads Part 2: process development and economic evaluation. *Bioprocess Engineering* 17 (1): 15.

de la Riviére, R.D. (1967). Notice Ne´crologique sur M. Maurice Lemoigne (1883–1967). *C. R. Acad. des Sci.*, Paris, t: 264.

de Smet, M.J., Eggink, G., Witholt, B., Kingma, J., and Wynberg, H. (1983). Characterization of intracellular inclusions formed by *Pseudomonas oleovorans* during growth on octane. *Journal of Bacteriology* 154 (2): 870–878.

Doi, Y. (1995). Microbial synthesis, physical properties, and biodegradability of polyhydroxyalkanoates. *Macromolecular Symposia* 98 (1): 585–599.

Doi, Y. and Abe, C. (1990). Biosynthesis and characterization of a new bacterial copolyester of 3-hydroxyalkanoates and 3-hydroxy-.omega.-chloroalkanoates. *Macromolecules* 23 (15): 3705–3707.

Doi, Y., Kawaguchi, Y., Koyama, N., Nakamura, S., Hiramitsu, M., Yoshida, Y., and Kimura, H. (1992). Synthesis and degradation of polyhydroxyalkanoates inAlcaligenes eutrophus. *FEMS Microbiology Letters* 103 (2–4): 103–108.

Doi, Y., Kunioka, Y., Nakamura, Y., and Soga, K. (1988). *Macromolecules* 21: 2722–2727.

Doi, Y., Segawa, A., and Kunioka, M. (1990). *International Journal of Biological Macromolecules* 12: 106–111.

Doi, Y., Tamaki, A., Kunioka, M., and Soga, K. (1987). *Makromol. Chem., Rapid Commun* 8: 631–635.

Doudoroff, M. and Stanier, R.Y. (1959). Role of poly-β-hydroxybutyric acid in the assimilation of organic carbon by bacteria. *Nature* 183 (4673): 1440–1442.

Fernandez-Castillo, R., Rodriguez-Valera, F., Gonzalez-Ramos, J., and Ruiz-Berraquero, F. (1986). Accumulation of poly (beta-hydroxybutyrate) by halobacteria. *Applied and Environmental Microbiology* 51 (1): 214–216.

Findlay, R.H. and White, D.C. (1983). Polymeric beta-hydroxyalkanoates from environmental samples and *Bacillus megaterium*. *Applied and Environmental Microbiology* 45 (1): 71–78.

Fritzsche, K., Lenz, R.W., and Fuller, R.C. (1990a). Bacterial polyesters containing branched poly(beta-hydroxyalkanoate) units. *International Journal of Biological Macromolecules* 12 (2): 92–101.

Fritzsche, K., Lenz, R.W., and Fuller, R.C. (1990b). Production of unsaturated polyesters by Pseudomonas oleovorans. *International Journal of Biological Macromolecules* 12 (2): 85–91.

Gerngross, T.U., Snell, K.D., Peoples, O.P., Sinskey, A.J., Csuhai, E., Masamune, S., and Stubbe, J. (1994). Overexpression and purification of the soluble polyhydroxyalkanoate Synthase from *Alcaligenes eutrophus*: evidence for a required posttranslational modification for catalytic activity. *Biochemistry* 33 (31): 9311–9320.

Gilbert, H.F. (2000). Chapter 5 Expression of genetic information. In: Basic Concepts in Biochemistry a Student's Survival Guide. McGraw-Hill.

Glickman, B.W. and Radman, M. (1980). Escherichia coli mutator mutants deficient in methylation-instructed DNA mismatch correction. *Proceedings of the National Academy of Sciences* 77 (2): 1063–1067.

Gross, R.A., DeMello, C., Lenz, R.W., Brandl, H., and Fuller, R.C. (1989). The biosynthesis and characterization of poly(Î²-hydroxyalkanoates) produced by Pseudomonas oleovorans. *Macromolecules* 22 (3): 1106–1115.

Hall, T.A. (1999). BioEdit: a user-friendly biological sequence alignment editor and analysis program for Windows 95/98/NT. *Nucleic Acids Symposium Series* 41: 95–98.

Haywood, G.W., Anderson, A.J., and Dawes, E.A. (1989a). The importance of PHB-synthase substrate specificity in polyhydroxyalkanoate synthesis byAlcaligenes eutrophus. *FEMS Microbiology Letters* 57 (1): 1–6.

Haywood, G.W., Anderson, A.J., and Dawes, E.A. (1989b). A survey of the accumulation of novel polyhydroxyalkanoates by bacteria. *Biotechnology Letters* 11 (7): 471–476.

Hoffmann, N., Amara, A.A., Beermann, B.B., Qi, Q., Hinz, H., Jr, and Rehm, B.H.A. (2002). Biochemical characterization of the Pseudomonas putida3-hydroxyacyl ACP:CoA transacylase, which diverts intermediates of fatty acid de novo biosynthesis. *Journal of Biological Chemistry* 277 (45): 42926–42936.

Hoffmann, N., Steinbüchel, A., and Rehm, B.H.A. (2000). The *Pseudomonas aeruginosa* phaGgene product is involved in the synthesis of polyhydroxyalkanoic acid consisting of medium-chain-length constituents from non-related carbon sources. *FEMS Microbiology Letters* 184 (2): 253–259.

Holmes, P.A. (1985). *Phys. Technol* 16: 32–36.

Hoppensack, A., Rehm, B.H.A., and Steinbüchel, A. (1999). Analysis of 4-phosphopantetheinylation of polyhydroxybutyrate synthase from *Ralstonia eutropha*: generation of β-alanine auxotrophic Tn5 mutants and cloning of the panD gene region. *Journal of Bacteriology* 181 (5): 1429–1435.

Huijberts, G.N., Eggink, G., de Waard, P., Huisman, G.W., and Witholt, B. (1992). *Pseudomonas putida* KT2442 cultivated on glucose accumulates poly(3-hydroxyalkanoates) consisting of saturated and unsaturated monomers. *Applied and Environmental Microbiology* 58 (2): 536–544.

Huisman, G.W., de Leeuw, O., Eggink, G., and Witholt, B. (1989). Synthesis of poly-hydroxyalkanoates is a common feature of fluoresent pseudomonads. *Applied and Environmental Microbiology* 55: 1949–1954.

Jackson, F.A. and Dawes, E.A. (1976). Regulation of the tricarboxylic acid cycle and poly-beta-hydroxybutyrate metabolism in Azotobacter beijerinckii grown under nitrogen or oxygen limitation. *Journal of General Microbiology* 97 (2): 303–312.

Jendrossek, D., Schirmer, A., and Schlegel, H.G. (1996). Biodegradation of polyhydroxyalkanoic acids. *Applied Microbiology and Biotechnology* 46 (5–6): 451–463.

Jia, Y., Kappock, T.J., Frick, T., Sinskey, A.J., and Stubbe, J. (2000). Lipases provide a new mechanistic model for polyhydroxybutyrate (PHB) synthases: characterization of the functional residues in *Chromatium vinosum* PHB synthase. *Biochemistry* 39 (14): 3927–3936.

Jia, Y., Yuan, W., Wodzinska, J., Park, C., Sinskey, A.J., and Stubbe, J. (2001). Mechanistic studies on class I polyhydroxybutyrate (PHB) synthase from *Ralstonia eutropha*:: class I and III synthases share a similar catalytic mechanism. *Biochemistry* 40 (4): 1011–1019.

Jossek, R., Reichelt, R., and Steinbüchel, A. (1998). *In vitro* biosynthesis of poly(3-hydroxybutyric acid) by using purified poly(hydroxyalkanoic acid) synthase of *Chromatium vinosum*. *Applied Microbiology and Biotechnology* 49 (3): 258–266.

Karplus, K., Barrett, C., and Hughey, R. (1998). Hidden Markov models for detecting remote protein homologies. *Bioinformatics* 14 (10): 846–856.

Kichise, T., Taguchi, S., and Doi, Y. (2002). Enhanced accumulation and changed monomer composition in polyhydroxyalkanoate (PHA) copolyester by in vitro evolution of aeromonas caviae PHA synthase. *Applied and Environmental Microbiology* 68 (5): 2411–2419.

Kim, Y.B., Lenz, R.W., and Fuller, R.C. (1991). Preparation and characterization of poly(.beta.-hydroxyalkanoates) obtained from Pseudomonas oleovorans grown with mixtures of 5-phenylvaleric acid and n-alkanoic acids. *Macromolecules* 24 (19): 5256–5260.

Kim, Y.B., Lenz, R.W., and Fuller, R.C. (1992). Poly(β-hydroxyalkanoate) copolymers containing brominated repeating units produced by *Pseudomonas oleovorans*. *Macromolecules* 25 (7): 1852–1857.

Lageveen, R.G., Huisman, G.W., Preusting, H., Ketelaar, P., Eggink, G., and Witholt, B. (1988). Formation of Polyesters by Pseudomonas oleovorans: effect of Substrates on Formation and Composition of Poly-(R)-3-Hydroxyalkanoates and Poly-(R)-3-Hydroxyalkenoates. *Applied and Environmental Microbiology* 54 (12): 2924–2932.

Lemoigne, M. (1923). Production d'acide β -oxybutyrique par certaines bactéries du groupe du *Bacillus subtilis*. *C.R. Hebd. Seances Acad. Sci* 176: 1761.

Lemoigne, M. (1925). Etudes sur l'autolyse microbie acidification par formation d'acide β oxybutyrique. *Ann. Inst. Pasteur* 39: 144.

Lemoigne, M. (1926). Produits de l'hydration et de polymerisation de l'acide β-oxobutyrique. *Bull. Soc. Chem. Biol.* (Paris) 8: 770–782.

Lemoigne, M. (1927). Etudes sur l'autolyse microbienne. Origine de l'acide β -oxybutyrique form'e par autolyse. *Ann. Inst. Pasteur* (Paris) 39: 144.

Lenz, R.W. and Marchessault, R.H. (2005). Bacterial polyesters: biosynthesis, biodegradable plastics and biotechnology. *Biomacromolecules* 6 (1): 1–8.

Liebergesell, M., Rahalkar, S., and Steinbüchel, A. (2000). Analysis of the *Thiocapsa pfennigii* polyhydroxyalkanoate synthase: subcloning, molecular characterization and generation of hybrid synthases with the corresponding *Chromatium vinosum* enzyme. *Applied Microbiology and Biotechnology* 54 (2): 186–194.

Liebergesell, M., Sonomoto, K., Madkour, M., Mayer, F., and Steinbuchel, A. (1994). Purification and characterization of the poly(hydroxyalkanoic acid) synthase from *Chromatium vinosum* and localization of the enzyme at the surface of poly(hydroxyalkanoic acid) granules. *European Journal of Biochemistry* 226 (1): 71–80.

Lusty, C.J. and Doudoroff, M. (1966). Poly-β-hydroxybutyrate depolymerases of Pseudomonas lemoignei. *Biochemistry* 56: 960–965.

Lutke-Eversloh, T., Bergander, K., Luftmann, H., and Steinbuchel, A. (2001). Identification of a new class of biopolymer: bacterial synthesis of a sulfur-containing polymer with thioester linkages. *Microbiology* (Reading) 147 (Pt 1): 11–19.

Lutke-Eversloh, T., Kawada, J., Marchessault, R.H., and Steinbuchel, A. (2002). Characterization of microbial polythioesters: physical properties of novel copolymers synthesized by Ralstonia eutropha. *Biomacromolecules* 3 (1): 159–166.

Macrae, R.M. and Wilkinson, J.F. (1958). Poly-β -hyroxybutyrate metabolism in washed suspensions of *Bacillus cereus* and *Bacillus megaterium*. *Journal of General Microbiology* 19 (1): 210–222.

Madden, T. (2002). Oct 9 (Updated 2003 Aug 13). The BLAST Sequence analysis tool. In: The NCBI Handbook [Internet] (ed. J. McEntyre and J. Ostell). Bethesda, MD: National Center for Biotechnology Information (US) Chapter 16.

Mansfield, D.A., Anderson, A.J., and Naylor, L.A. (1995). Regulation of PHB metabolism inAlcaligenes eutrophus. *Canadian Journal of Microbiology* 41 (13): 44–49.

Marchessault, R.H. (1996). Tender morsels for bacteria: recent developments in microbial polyesters. *Trends in Polymer Science* 4: 163–168.

Masamune, S., Palmer, M.A.J., Gamboni, R., Thompson, S., Davis, J.T., Williams, S.F., Peoples, O.P., Sinskey, A.J., and Walsh, C.T. (1989a). Bio-Claisen condensation catalyzed by thiolase from Zoogloea ramigera. Active site cysteine residues. *Journal of the American Chemical Society* 111 (5): 1879–1881.

Masamune, S., Walsh, C.T., Sinskey, A.J., and Peoples, O.P. (1989b). Poly-(R)-3-hydroxybutyrate (PHB) biosynthesis: mechanistic studies on the biological Claisen condensation catalyzed by Î²-ketoacyl thiolase. *Pure and Applied Chemistry* 61 (3): 303–312.

McCool, G.J. and Cannon, M.C. (1999). Polyhydroxyalkanoate inclusion body-associated proteins and coding region in Bacillus megaterium. *Journal of Bacteriology* 181 (2): 585–592.

Merrick, J.M. and Doudoroff, M. (1961). Enzymatic synthesis of poly-beta-hydroxybutyric acid in bacteria. *Nature* 189 (4768): 890–892.

Merrick, J.M. and Doudoroff, M. (1964). Depolymerization of poly-beta-hydroxybutyrate by an intracellular enzyme system. *Journal of Bacteriology* 88 (1): 60–71.

Nicholas, K.B., et al. (1997). Genedoc: a tool for editing and annoting multiple sequence alignments. Embnet.news.

Noble, M.E.M., Cleasby, A., Johnson, L.N., Egmond, M.R., and Frenken, L.G.J. (1993). The crystal structure of triacylglycerol lipase from *Pseudomonas glumae* reveals a partially redundant catalytic aspartate. *FEBS Letters* 331 (1–2): 123–128.

Oeding, V. and Schlegel, H.G. (1973). Beta-Ketothiolase from *Hydrogenomonas eutropha* H16 and its significance in the regulation of poly-beta-hydroxybutyrate metabolism. *Biochemical Journal* 134 (1): 239–248.

Okonechnikov, K., Golosova, O., Fursov, M., and the UGENE team. (2012). Unipro UGENE: a unified bioinformatics toolkit. *Bioinformatics* 28: 1166–1167.

Peoples, O.P. and Sinskey, A.J. (1989). Poly-beta-hydroxybutyrate biosynthesis in Alcaligenes eutrophus H16. Characterization of the genes encoding beta-ketothiolase and acetoacetyl-CoA reductase. *The Journal of Biological Chemistry* 264 (26): 15293–15297.

Peoples, O.P. and Sinskey, A.J.. (1989a). *The Journal of Biological Chemistry* 264: 15293–15297.

Peoples, O.P. and Sinskey, A.J.. (1989b). *The Journal of Biological Chemistry* 264: 15298–15303.

Preusting, H., Nijenhuis, A., and Witholt, B. (1990). Physical characteristics of poly(3-hydroxyalkanoates) and poly(3-hydroxyalkenoates) produced by Pseudomonas oleovorans grown on aliphatic hydrocarbons. *Macromolecules* 23 (19): 4220–4224.

Qi, Q., Rehm, B.H.A., and Steinbüchel, A. (1997). Synthesis of poly(3-hydroxyalkanoates) in Escherichia coli expressing the PHA synthase gene phaC2 from *Pseudomonas aeruginosa*: comparison of PhaC1 and PhaC2. *FEMS Microbiology Letters* 157: 155–162.

Qi, Q., Steinbüchel, A., and Rehm, B.H.A. (1998). Metabolic routing towards polyhydroxyalkanoic acid synthesis in recombinant *Escherichia coli* (fadR): inhibition of fatty acid β-oxidation by acrylic acid. *FEMS Microbiology Letters* 167 (1): 89–94.

Qi, Q., Steinbüchel, A., and Rehm, B.H.A. (2000). *In vitro* synthesis of poly(3-hydroxydecanoate): purification and enzymatic characterization of type II polyhydroxyalkanoate synthases PhaC1 and PhaC2 from *Pseudomonas aeruginosa*. *Applied Microbiology and Biotechnology* 54 (1): 37–43.

Rehm, B.H. and Steinbuchel, A. (2001). Heterologous expression of the acyl-acyl carrier protein thioesterase gene from the plant Umbellularia californica mediates polyhydroxyalkanoate biosynthesis in recombinant Escherichia coli. *Applied Microbiology and Biotechnology* 55 (2): 205–209.

Rehm, B.H.A. (2003). Polyester synthases: natural catalysts for plastics. *The Biochemical Journal* 376: 15–33.

Rehm, B.H.A., Antonio, R.V., Spiekermann, P., Amara, A.A., and Steinbuchel, A. (2002). Molecular characterization of the poly(3-hydroxybutyrate) (PHB) synthase from *Ralstonia eutropha: in vitro* evolution, site-specific mutagenesis and development of PHB synthase protein model. *Biochimica Et Biophysica Acta* 1594: 178–190.

Rehm, B.H.A., Kruger, N., and Steinbüchel, A. (1998). A new metabolic link between fatty acidde novo synthesis and polyhydroxyalkanoic acid synthesis. *Journal of Biological Chemistry* 273 (37): 24044–24051.

Rehm, B.H.A., Qi, Q., Beermann, B.B., Hinz, H., Jr, and Steinbüchel, A. (2001). Matrix-assisted *in vitro* refolding of *Pseudomonas aeruginosa* class II polyhydroxyalkanoate synthase from inclusion bodies produced in recombinant *Escherichia coli*. *Biochemical Journal* 358 (1): 263.

Rehm, B.H.A. and Steinbüchel, A. (2001). PHA synthases-the key enzymes of PHA synthesis. In: Biopolymers (ed. A. Steinbüchel and Y. Doi), 173–215. Heidelberg, Germany: Wiley-VCH.

Rehm, B.H.A. and Steinbüchel, A. (2002). Polyhydroxyalkanoate (PHA) synthases: the key enzymes of PHA synthesis. Biopolymers Online: Wiley.

Šali, A. and Blundell, T.L. (1993). Comparative protein modelling by satisdaction of spatial restraints. *Journal of Molecular Biology* 234: 779–815.

Scheuermann, R., Tam, S., Burgers, P.M., Lu, C., and Echols, H. (1983). Identification of the epsilon-subunit of Escherichia coli DNA polymerase III holoenzyme as the dnaQ gene product: a fidelity subunit for DNA replication. *Proceedings of the National Academy of Sciences* 80 (23): 7085–7089.

Schlegel, H.G., Kaltwasser, H., and Gottschalk, G. (1961). Ein Submersverfahren zur Kultur wasserstoffoxydierender Bakterien: Wachstumsphysiologische Untersuchungen. *Archiv für Mikrobiologie* 38 (3): 209–222.

Schubert, P., Krüger, N., and Steinbüchel, A. (1991). Molecular analysis of the *Alcaligenes eutrophus* poly(3-hydroxybutyrate) biosynthetic operon: identification of the N terminus of poly(3-hydroxybutyrate) synthase and identification of the promoter. *Journal of Bacteriology* 173 (1): 168–175.

Schubert, P., Steinbuchel, A., and Schlegel, H.G. (1988). Cloning of the *Alcaligenes eutrophus* genes for synthesis of poly-beta-hydroxybutyric acid (PHB) and synthesis of PHB in *Escherichia coli*. *Journal of Bacteriology* 170 (12): 5837–5847.

Senior, P.J. and Dawes, E.A. (1973). The regulation of poly-beta-hydroxybutyrate metabolism in *Azotobacter beijerinckii*. *Biochemical Journal* 134 (1): 225–238.

Shabeb, M.S., Younis, M., Amara, A.A., and Shoreit, A. (2005). Production of PHB-co-PHV from Compost Bioremediation by Purple non-sulfur bacteria. *Egyptian Journal of Biotechnology* 19: 467–478.

Slater, S., Houmiel, K.L., Tran, M., Mitsky, T.A., Taylor, N.B., Padgette, S.R., and Gruys, K.J. (1998). Multiple β-ketothiolases mediate poly(β-hydroxyalkanoate) copolymer synthesis in *Ralstonia eutropha*. *Journal of Bacteriology* 180 (8): 1979–1987.

Slater, S.C., Voige, W.H., and Dennis, D.E. (1988). Cloning and expression in Escherichia coli of the Alcaligenes eutrophus H16 poly-beta-hydroxybutyrate biosynthetic pathway. *Journal of Bacteriology* 170 (10): 4431–4436.

Solaiman, D.K.Y. and Ashby, R.D. (2005). Rapid genetic characterization of poly(hydroxyalkanoate) synthase and its applications. *Biomacromolecules* 6 (2): 532–537.

Song, J.J. and Yoon, S.C. (1996). Biosynthesis of novel aromatic copolyesters from insoluble 11-phenoxyundecanoic acid by pseudomonas putida BM01. *Applied and Environmental Microbiology* 62 (2): 536–544.

Spiekermann, P. (1999). Veranderung der eigenschaften von PHF-Synthasen durch anwendung molekulargenetischer methoden. Diplomarbeit Westfälische Wilhelms-Universitüt Münster.

Stanier, R.Y., Doudoroff, M., Kunisawa, R., and Contopoulou, R. (1959). The role of organic substrates in bacterial photosynthesis. *Proceedings of the National Academy of Sciences* 45 (8): 1246–1260.

Steinbüchel, A. (1991). Polyhydroxyalkanoic acids. In: Biomaterials, 123–213. UK: Palgrave Macmillan.

Steinbüchel, A. (2001). Perspectives for biotechnological production and utilization of biopolymers: metabolic engineering of polyhydroxyalkanoate biosynthesis pathways as a successful example. *Macromolecular Bioscience* 1 (1): 1–24.

Steinbuchel, A. and Fuchtenbusch, B. (1998). Bacterial and other biological systems for polyester production. *Trends in Biotechnology* 16 (10): 419–427.

Steinbüchel, A., Füchtenbusch, B., Gorenflo, V., Hein, S., Jossek, R., Langenbach, S., and Rehm, B.H.A. (1998). Biosynthesis of polyesters in bacteria and recombinant organisms. *Polymer Degradation and Stability* 59 (1–3): 177–182.

Steinbüchel, A. and Hein, S. (2001). Biochemical and molecular basis of microbial synthesis of polyhydroxyalkanoates in microorganisms. In: Biopolyesters, 81–123. Berlin and Heidelberg: Springer.

Stubbe, J. and Tian, J. (2003). Polyhydroxyalkanoate (PHA) homeostasis: the role of the PHA synthase. *Natural Product Reports* 20 (5): 445.

Taguchi, S., Iwata, T., Abe, H., and Doi, Y. 2012. Poly(hydroxyalkanoate)s Polymer Science: A Comprehensive Reference 9: 157–182.

Taguchi, S., Maehara, A., Takase, K., Nakahara, M., Nakamura, H., and Doi, Y. (2001). Analysis of mutational effects of a polyhydroxybutyrate (PHB) polymerase on bacterial PHB accumulation using an in vivo assay system. *FEMS Microbiology Letters* 198 (1): 65–71.

Taguchi, S., Nakamura, H., Hiraishi, T., Yamato, I., and Doi, Y. (2002). In vitro evolution of a polyhydroxybutyrate synthase by intragenic suppression-type mutagenesis. *Journal of Biochemistry* 131 (6): 801–806.

Tamura, K., Stecher, G., Peterson, D., Filipski, A., and Kumar, S. (2013). MEGA6: molecular evolutionary genetics analysis version 6.0. *Molecular Biology and Evolution* 30: 2725–2729.

Timm, A. and Steinbuchel, A. (1992). Cloning and molecular analysis of the poly(3-hydroxyalkanoic acid) gene locus of *Pseudomonas aeruginosa* PAO1. *European Journal of Biochemistry* 209 (1): 15–30.

Timm, A. and Steinbüchel, A. (1990). Formation of polyesters consisting of medium-chain-length 3-hydroxyalkanoic acids from gluconate by Pseudomonas aeruginosa and other fluorescent pseudomonads. *Applied and Environmental Microbiology* 56 (11): 3360–3367.

Tsuge, T., Watanabe, S., Shimada, D., Abe, H., Doi, Y., and Taguchi, S. (2007). Combination of N149S and D171G mutations inAeromonas caviaepolyhydroxyalkanoate synthase and impact on polyhydroxyalkanoate biosynthesis. *FEMS Microbiology Letters* 277 (2): 217–222.

Valentin, H.E. and Dennis, D. (1997). Production of poly(3-hydroxybutyrate-co-4-hydroxybutyrate) in recombinant *Escherichia coli* grown on glucose. *Journal of Biotechnology* 58 (1): 33–38.

Wallen, L.L. and Rohwedder, W.K. (1974). Poly-.beta.-hydroxyalkanoate from activated sludge. *Environmental Science & Technology* 8 (6): 576–579.

Weibull, C. (1953). Characterization of the protoplasmic constituents of *Bacillus megaterium*. *Journal of Bacteriology* 66 (6): 696–702.

Williamson, D.H. and Wilkinson, J.F. (1958). The isolation and estimation of the poly-beta -hydroxy-butyrate inclusions of Bacillus species. *Journal of General Microbiology* 19 (1): 198–209.

Yuan, W., Jia, Y., Tian, J., Snell, K.D., Müh, U., Sinskey, A.J., Lambalot, R.H., Walsh, C.T., and Stubbe, J. (2001). Class I and III polyhydroxyalkanoate synthases from ralstonia eutropha and allochromatium vinosum: characterization and substrate specificity studies. *Archives of Biochemistry and Biophysics* 394 (1): 87–98.

7

Extremozyme-Based Technology for Biofuel Generation

Bioactivity and Stability Performances

*Amal Souii, Afwa Ghorrab, Khouloud Hammami, Ahmed Slaheddine Masmoudi, Ameur Cherif, and Mohamed Neifar**

University of Manouba, ISBST, BVBGR-LR11ES31, Biotechpole Sidi Thabet, Sidi Thabet, Tunisia
**Correspondence: Mohamed Neifar, mohamed.naifar@gmail.com; mohamed.neifar@isbst.uma.tn*

7.1 Introduction

Lignocellulose is the most abundant source of carbohydrates on earth (109 tons/year) and has a great potential as a feedstock for second generation biofuel production (Aditiya et al. 2016; Gupta and Verma 2015; Spertino et al. 2018; Wang et al. 2012). The use of biofuels is an effective strategy to decrease dependence on fossil fuels and reduce greenhouse gas emissions. Compared to first-generation biofuels, which are produced from fermentation of corn, potatoes, grain (wheat, barley and rye), sugar beet, sugar cane, etc., and then compete with food crops, second- and third-generation biofuels used lignocellulosic and algal biomasses, respectively, as cost-effective, abundant, and sustainable energy resources. Among all biofuels, bioethanol is often considered the most promising alternative and/or additive for gasoline (Alalwan et al. 2019; Callegari et al. 2020; Domínguez-Bocanegra et al. 2015; Zhu et al. 2020). Table 7.1 shows the advantages and disadvantages of biofuel generations. A detailed overview on different types of biofuels and their comparative study, production and uses from an ecological perspective has been provided by Datta et al. (2019).

One of the major obstacles to the economic and industrial production of cellulosic biofuels is the high price of enzymes (Campioni et al. 2020; Klein-Marcuschamer et al. 2012; Rajeswari and Jacob 2020a). The use of cost-effective enzyme technology to degrade polysaccharides to fermentable sugars is crucial for economically viable biofuels (Bilal et al. 2017; Chettri et al. 2020, 2021; Saini et al. 2015; Wang et al. 2012, 2013a, 2013b). Several strategies have been undertaken to overcome such barriers, such as metagenomic screening and extremophile diversity exploration; microbial genome sequencing, assemby and annotation; and enzyme engineering, which plays an important role in the development of robust lignocellulolytic enzymes with improved characteristics, including higher catalytic efficiency, increased stability at elevated temperatures, and greater tolerance to end-product inhibition (Adegboye et al. 2021; Wang et al. 2015, 2019a, 2019b).

Extensive research has been conducted recently on applying oxidative and hydrolytic extremozyme producing-microbial communities/microorganisms in the pretreatment and saccharification of lignocellulosic biomass in order to reduce the production cost

Biomolecules from Natural Sources: Advances and Applications, First Edition. Edited by Vijai Kumar Gupta, Satyajit D. Sarker, Minaxi Sharma, María Elida Pirovani, Zeba Usmani, and Chelliah Jayabaskaran.

Table 7.1 Advantages and disadvantages of differents biofuels generations.

	Feedstock	Process	Biofuel products	Advantages	Disadvantages	References
First generation of biofuels	• Sugar plants: sugarcane, sugar beet, sweet sorghum • Starchy plants: wheat, corn, barley, soybean, maize, potatoes, cassava	• Fermentation	Bioethanol	• Reduces the emission of green house gases • Lower cost production than second and third generation • No lignin content	• Uses of Food as feedstock • Agricultural lands competition	Baeyens et al. 2015; Datta et al. 2019; Oropeza-De la Rosa et al. 2017; Yue et al. 2014
	• Oil crops: palm, rapeseed, soy bean, colza, tournesol	• Transesterification	Biodiesel			
Second generation of biofuels	• Agricultural crop residues: wheat straw, bagasse, corn stover • Forest residues • Paper waste • Energy corps: miscanthus, switchgrass	• Thermochemical process: gasification, methanisation, pyrolysis • Biochemical process: • enzymatic transesterifcation • Enzymatic hydrolysis and fermentation	Biohydrogen Biomethane Biodiesel Biodiesel Bioethanol	• No food competition • Low cost, available and sustainable raw materials • A high minimisation of green house gases impact • Higher biofuels yields	• High costs of biofuels production process • Necessity of efficient pretreatment for lignin removal • A complex lignocellulosic substrate	Datta et al. 2019; Jegannathan et al. 2011; Lewandowski et al. 2017; Persson et al. 2018; Tye et al. 2016; Zabed et al. 2016; Zhang et al. 2010
Third generation of biofuels	• Alagal biomass: micoalgae, macroalgae	• Thermochemical process: gasification methanisation biochemical process: • Saccharification and fermentation • Enzymatic transesterifcation	Bioethanol Biodiesel Biohydrogen	• Reduced demand for water resources • Low content of lignin and hemicellulose than lignocellulosic biomasses • No competition with land crops for agricultural soils • High algal growth potential	• High cost technology of microalgae culturing and harvesting	Datta et al. 2019; Zhang et al. 2010

and increase the biofuel yield (Ariaeenejad et al. 2019, 2020; Klippel and Antranikian 2011; Klippel et al. 2014; Lewin et al. 2017; Verma and Satyanarayana 2020; Weimann et al. 2013; Zhu et al. 2020). Numerous studies have also indicated that extremophiles with a wide variety of enzymes exhibit higher plant biomass degradation capabilities (Chettri et al. 2021; Ferrer et al. 2007; Rattu et al. 2016; Zhu et al. 2020).

7.2 Lignocellulosic Biomass as a Substrate for Second Generation Biofuels

Lignocellulosic biomass, consists of cellulose, hemicellulose, pectin and lignin (Figure 7.1). Their content in plant cell walls are most frequently analyzed using the Van Soest detergent procedure (Van Soest 1967; Van Soest et al. 1991) (Figure 7.2).

Lignocellulose is of major interest to biorefineries because of its renewable and sustainable nature as a feedstock to replace fossil fuels for the production of second generation biofuels (Binod et al. 2019; Kulkarni et al. 2015; Robak and Balcerek 2018).

Cellulose is the most abundant organic polymer produced in the biosphere. Its annual production is estimated at more than 7.5 × 1010 tons. It is a constituent polymer of the plant cell membrane composed of linear chains of D-glucose units linked by β-1,4-glycosidic bonds. The cellulose chains are organized into microfibrils. Cellulose microfibrils assemble to form strands that intertwine with each other in "a cable" or macrofibril structure. The arrangement of cellulose molecules in the microfibrils gives cellulose crystalline properties. The cellulose is embedded in the complex matrix of the plant cell

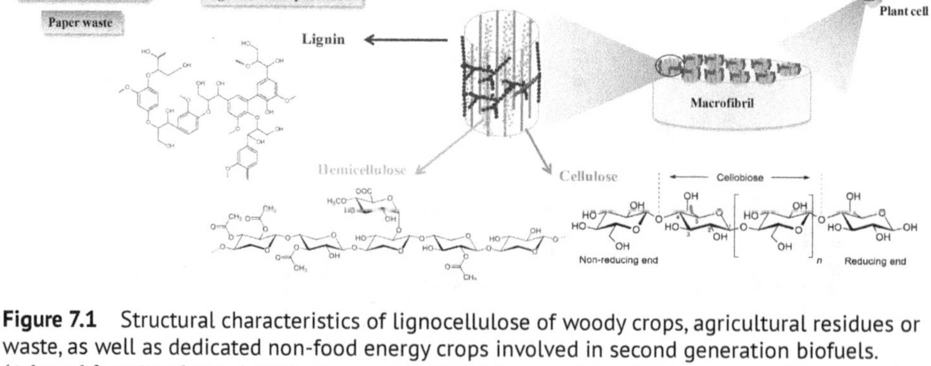

Figure 7.1 Structural characteristics of lignocellulose of woody crops, agricultural residues or waste, as well as dedicated non-food energy crops involved in second generation biofuels. (Adapted from Boudet et al. 2003; Alonso et al. 2012; Menon and Rao 2012; Nunes and Kunamneni 2018; Bertella and Luterbacher 2020.)

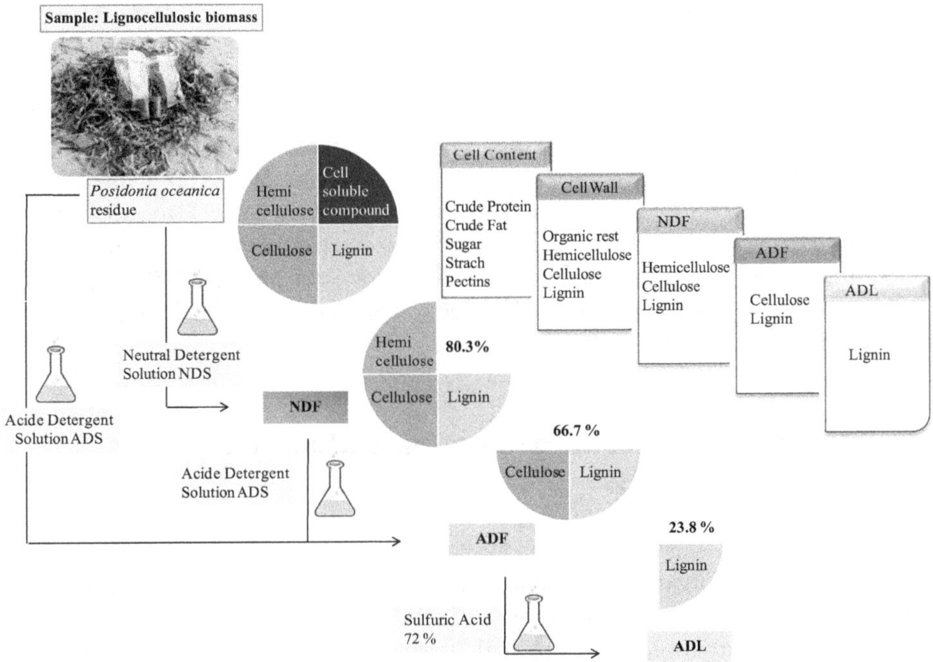

Figure 7.2 Determination of lignin, cellulose and hemicellulose contents in *Posidonia residues* using the Van Soest detergent procedure.

wall comprising hemicelluloses and lignin that limit the access of lignocellulolytic enzymes to the substrate (Benedetti et al. 2019; Callegari et al. 2020; Tingley et al. 2021).

After cellulose, hemicelluloses are second in abundance and include a variety of linear or branched polysaccharides, derived from both hexose and pentose monosaccharide units such as xylose, arabinose, mannose, glucose and galactose. The most common hemicelluloses are xylan, xyloglucan, arabinoxylan, mannan, galactomannan, arabinan, galactan and polygalacturonan. The most studied class corresponds to the xyloglucans. They consist of a glucose chain (beta 1-4) and short side chains of xylose, galactose and fucose. Because of their conformation (β 1-4 linked glucoses), xyloglucans can contract H-bonds with cellulose. Due to the presence of side chains, they are not themselves susceptible to fibril formation. They can, therefore, play a fundamental role in maintaining an organized parietal architecture by binding cellulose fibrils together. Unlike cellulose, hemicelluloses with variable and amorphous structures, are easily degraded to simple sugars by enzymatic attacks (Binod et al. 2019; Ebringerová 2005; Tingley et al. 2021).

Pectins are also one of the major components of the primary wall of higher dicotyledonous plants, accounting for about 35% of the parietal dry matter. They represent a group of heterogeneous, branched polysaccharides. Pectins consist of a backbone of α-1,4-linked D-galacturonic acid residues that can be methyl esterified or decorated with acetyl groups. Pectins are mainly grouped into homogalacturonans, xylogalacturonans and rhamnogalacturonans. Pectins interact covalently with hemicelluloses, providing structural and functional complexity to the plant cell walls (Edwards and Doran-Peterson 2012; Tingley et al. 2021; Xiao and Anderson 2013).

Cell walls are also protected by lignin, an aromatic phenylpropanoid heteropolymer that is one of the main contributors to the recalcitrance of cell walls to enzymatic attack. Lignin is derived from the oxidative polymerization of three hydroxycinnamyl alcohol derivatives named coniferyl, sinapyl, and p-coumaryl alcohols. These alcohols are incorporated into the lignin polymer as guaiacyl, syringyl and p-hydroxyphenyl units. Lignin is cross-linked with plant polysaccharides by ether or ester linkages, with ferulic acid being the predominant bridge as it is etherified to lignin, while esterified to polysaccharides (Kudanga et al., 2014; Tingley et al. 2021).

7.3 Production of Second-generation Biofuels from Lignocellulosic Biomass

Second-generation biofuels are reserved for liquid or gaseous fuels derived from the processing of non-fossil organic materials such as non-food crop feedstocks, agricultural and forestry residues, and industrial wastes (Table 7.2).

They are mainly produced using physical, thermochemical, and biochemical technologies, often after a pretreatment step of the lignocellulosic feedstock (Aditiya et al. 2016; Tingley et al. 2021). The pretreatment step is a crucial step to prepare biomass properties to facilitate conversion processes into liquid and gaseous biofuels (Binod et al. 2019; Kucharska et al. 2018; Zhu et al. 2020).

Table 7.2 Chemical composition of some abundant, inexpensive, and sustainable second-generation biofuel feedstocks.

Substrate	Cellulose (%)	Hemicellulose (%)	Lignin (%)	References
Agricultural residues				
Sugar cane bagasse	51.1	23.1	21.6	Asgher et al. 2018
Corn stover	33 – 40	13 -31	5 -26	Fang et al. 2015
Rice husk	36	15	19	Gonçalves et al. 2011
Luffa cylindrica (L.) fibers	45.80 ± 1.3	20.76 ± 0.3	13.15 ± 0.6	Zaafouri et al. 2017
Wheat straw	40.5	26.1	18.1	De La Torre et al. 2017
Rice straw	55.3 ± 2.2	22.5 ± 0.3	32.7 ± 0.6	Bhardwaj et al. 2019
Corncob biomass	34	27.38	17.6	Ganesan et al. 2020
Posidonia oceanica	42.9 ± 1.10	13.6 ± 1.86	23.8 ± 1.48	Pilavtepe et al. 2013
Lantana camara	65.83	65.83	17.56	Gujjala et al. 2016
Maize stalk	35	15	19	Gonçalves et al. 2011

(*Continued*)

Table 7.2 (Continued)

Substrate	Cellulose (%)	Hemicellulose (%)	Lignin (%)	References
Pineapple leaf	66.2	19.5	4.28	Daud et al. 2014
Agro-industrial wastes				
Oil palm empty fruit bunch	35.3 ± 2.18	29.72 ± 1.92	20.98 ± 0.91	Hafid et al. 2021
Aloe vera leaf rind biomass	42 ± 0.56	15.57 ± 1.26	16.4	Rajeswari and Jacob 2020b
Cassava peel	37.9	37	7.5	Aripin et al. 2013
Canola straw	36.6	40.9	20	Aripin et al. 2013
Pine pinaster	55.9	13.7	26.2	Aripin et al. 2013
Eucalyptus globules	53	27.7	19.9	Aripin et al. 2013
Pine needle biomass	64.12	64.12	27.79	Vaid and Bajaj 2017
Water hyacinth	19.63 ± 1.80	33.63 ± 0.81	26.13 ± 0.19	Potprommanee et al. 2017
Journal	61	16	21	Gonçalves et al. 2011
Eucalyptus grandis	38	13	37	Gonçalves et al. 2011
Coconut fibre	31.60 ± 0.51	26.33 ± 0.89	25.44 ± 0.75	Gonçalves et al. 2014
Bioenergy crops				
Napier grass	12.4	68.2	10.8	Daud et al. 2014
Miscanthus sinensis	35.2	20.5	2.6	Lee and Kuan 2015
Switchgrass	32.0	25.2	18.1	Wang et al. 2019a, 2019b

Bioethanol remains the most widely used with about 75% of the total transport biofuels. Common biochemical pathways for its production include a pretreatment step required to increase sugar yield, followed by enzymatic hydrolysis and subsequent fermentation, or a simultaneous saccharification and fermentation process (Figure 7.3).

In bioethanol production, enzyme cocktails are added to lignocellulosic biomass prior to or concurrently with microbial fermentation, or expressed from a genetically engineered microorganism for consolidated bioprocessing; whereas biogas production uses native enzyme production by anaerobic microorganisms in lignocellulosic biomass biodigesters (Abraham et al. 2020; Tingley et al. 2021). Some oleaginous microorganisms such as *Cryptococcus curvatus*, *Lipomyces starkeyi*, *Rhodosporidium toruloides*, and *Rhodococcus opacus* have been described as lipid accumulators using lignocellulosic biomass as a substrate (Tsigie et al. 2011). The accumulated lipids can be used as feedstocks for biodiesel in lignocellulosic biorefineries (Poontawee et al. 2017; Zhu et al. 2020).

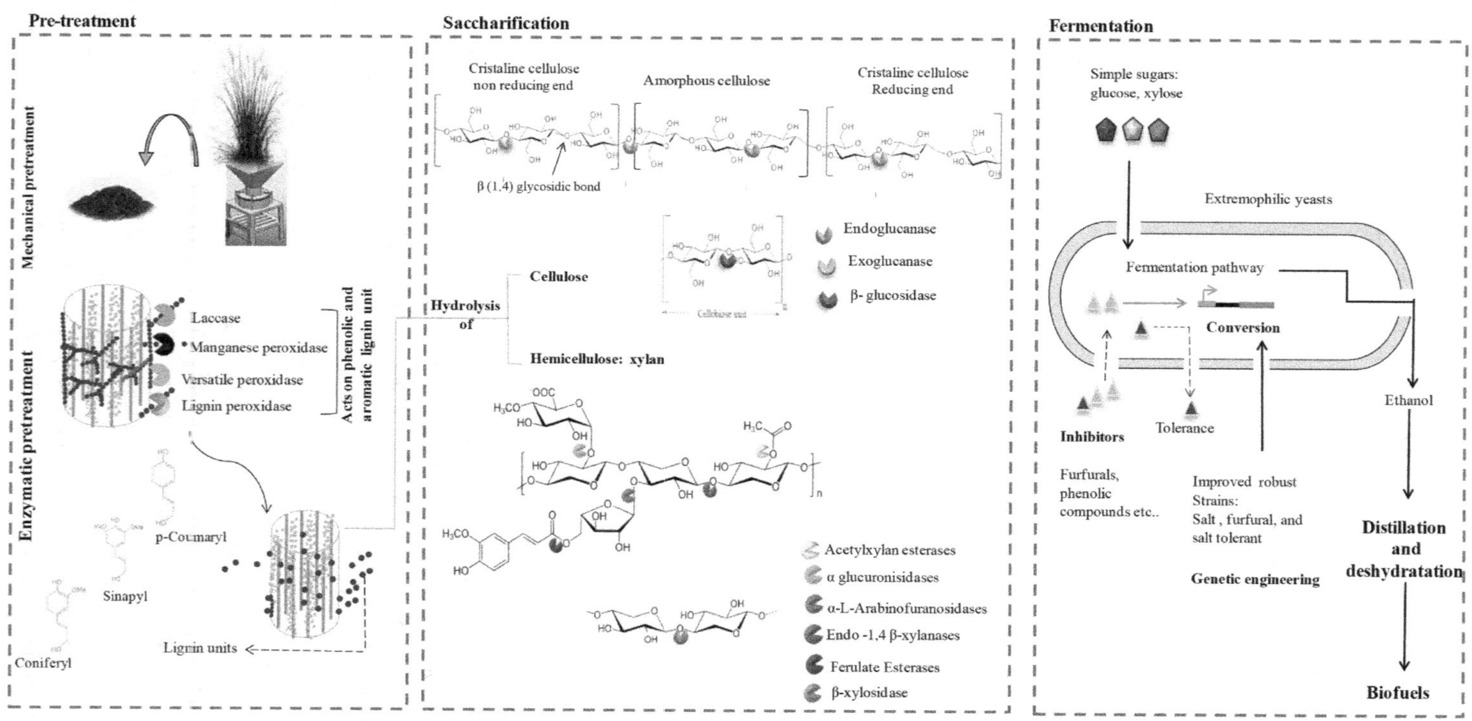

Figure 7.3 Basic steps for bioethanol production from lignocellulosic biomass and key biocatalysts used in biomass conversion.

7.4 The Carbohydrate Active enZymes (CAZymes) in Biofuel Industry

Many bacteria and fungi are capable of breaking down plant cell walls by secreting lignocellulolytic enzymes, including cellulases, xylanases, ligninases, pectinases, etc. (Arntzen et al. 2020; Ferreira Filho et al. 2017; Kim et al. 2019; Larsbrink et al. 2017; Rubio et al. 2019; Spertino et al. 2018; Tingley et al. 2021; Wu et al. 2017, 2016; Yu et al. 2021; Zerillo et al. 2013; Zhang et al. 2018). These enzymes have been classified, based on their functional amino acid sequences and structural similarities in the database of carbohydrate-active enzymes (CAZypedia, http://www.cazy.org), into glycosyl hydrolases (GH), carbohydrate esterases (CE), polysaccharide lyases (PL), and auxiliary activities (AA) (Figure 7.4). Carbohydrate active enzymes (CAZymes) play a decisive role in the pretreatment and saccharification steps of the lignocellulosic biomass for the production of second generation biofuels (Kameshwar et al. 2019; Lam et al. 2020). In 2012, Yin et al. developed the dbCAN web server integrating three state-of-the-art tools (DIAMOND, HMMER, and Hotpep) to provide a public service for automated CAZome annotation for microbial genomes and metagenomes mainly those associated with plants.

7.4.1 Cellulose-Active CAZymes

The crystalline structure of cellulose limits the access to cellulose-degrading CAZymes. However, synergistic effects can help overcome the low efficiency of these enzymes. Combined strategies involving GHs with exo- and endo-catalytic activities are applied for efficient saccharification of cellulosic biomass (Zhu et al. 2020). Endo-β-1,4-glucanases catalyze the cleavage of internal bonds of cellulose chains, whereas cellobiosidases progressively release oligo- and disaccharides from cellulose chains. The released cellooligosaccharides and cellobiose are subsequently depolymerized by cellodextrinases, endo-β-glucosidases and cellobiose phosphorylases (Nguyen et al. 2018; Wang et al.

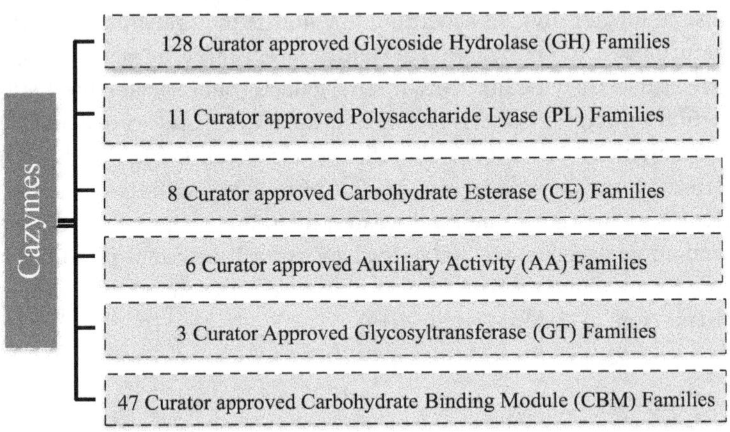

Figure 7.4 Carbohydrate-active enzymes classification into different CAZy clans and families (https://www.cazypedia.org/index.php/Sequence-based_classification).

2013a, 2013b). Endoglucanases belong to GH families 5, 6, 7, 9, 12, 44, 45, 48, 51, 74, and 124, exoglucanases to GH families 1, 3, 5, 9, and β-glucosidases to GH families 1, 3, 5, 9, 30, 116 (Lombard et al. 2014). Only endoglucanases with carbohydrate-binding modules (CBMs) can hydrolyze cellulose chains in the inner crystalline regions (Wang et al. 2012).

7.4.2 Hemicellulose- and Pectin-Active CAZymes

Many studies have focused on understanding the degradation of xylan and heteroxylan due to their abundance in plant cell walls and their potential to be enzymatically hydrolyzed into its constituent sugars for subsequent fermentation (Adhyaru et al. 2014; Bhardwaj et al. 2019; Chang et al. 2017; Da Silva et al. 2019). In addition, current efforts in emerging biofuel industries are aimed at a better understanding of xylan degradation to enhance the overall efficiency of biomass conversion to biofuels (Singh 2020; Verma and Satyanarayana 2020). Efficient depolymerization of xylan into its monosaccharide components requires an enzymatic mixture including endo-1,4-β-xylanases, β-D-xylosidases, α-L-arabinofuranosidases, α-glucuronidases, acetyl xylan esterases and ferulic/coumaric acid esterases (Dodd and Cann 2009). Enzymatic deconstruction begins with the removal of side chains by carbohydrate esterases (CEs), followed by the removal of the xylan backbone by specific glycoside hydrolases (GHs), called xylanases (Nguyen et al. 2018). Feruloyl esterase plays an important role in the disassembly of lignin-carbohydrate complexes via the cleavage of ester bonds between the alcohol and 4-O-methyl-glucuronoyl moieties of lignin and xylan, respectively. Recent studies also suggest that LPMOs are also active on xylans and xyloglucans by helping to dismantle these complex substrates (Agger et al. 2014; Li et al. 2018; Tingley et al. 2021). CAZy groups and classes involved in degradation of hemicellulose are reported by Kameshwar et al. (2019).

Pectinases comprise a heterogeneous group of enzymes that catalyze the degradation of pectic substances. Pectinases are classified into two major groups namely esterases and depolymerases (Singh et al. 2019). Esterases catalyze the removal of methyl and acetyl groups decorating the polymer backbone while depolymerases break glycosidic bonds. Depolymerases are further divided according to whether they rely on hydrolytic (hydrolases: polygalacturonases and polymethyl galacturonases) or β-elimination mechanisms (lyases: polygalacturonate lyase and poly-methyl galacturonate lyase), their substrate specificity, and their primary products (Jayani et al. 2005). Pectinases are classified into 11 PL families, namely PL1, -2, -4, -6, -7, -9, -15, -17,-19, -20 and -22, found in the CAZy database (https://www.cazypedia.org/index.php/Polysaccharide_Lyase_Families). Many agricultural and agro-industrial wastes, including sugar beet pulp, citrus pulp and peels, and apple pomace, are viable feedstocks for bioethanol production, and require pectinases for efficient saccharification due to their high pectin contents (Edwards and Doran-Peterson 2012; Singh et al. 2019).

7.4.3 Lignin-Active CAZymes

Microbial lignin-modifying enzymes (LMEs) are classified into two major families of oxidoreductases: (i) peroxidases, including lignin peroxidases (LiPs), manganese-dependent peroxidases (MnPs), polyvalent peroxidases (VPs), and dye-decolorizing peroxidases

(DyPs), and (ii) laccases (Lacs) or phenol oxidases (Neifar et al. 2016, 2017). These enzymes catalyze the oxidation of lignin generating highly reactive free radicals, which initiate the depolymerization of lignin by multiple non-enzymatic reactions. In addition, some accessory oxidative enzymes, such as aryl alcohol oxidases and quinone reductases, participate in lignin degradation by enhancing the ligninolytic activity of the main enzymes (Gonçalves et al. 2020; Janusz et al. 2017; Levasseur et al. 2013; Liu et al. 2020; Majeke et al. 2020; Park et al. 2018; Tingley et al. 2021). LMEs belong in the AA family of CAZy database (Lombard et al. 2014; Majeke et al. 2020). There are currently (April 2021) six curator AA families in CAZypedia (AA3, AA9, AA10, AA11, AA13 and AA14). Lignin degradation has been intensively studied in white rot fungi (Neifar et al. 2016, 2017). Lignin-degrading bacteria belonging to Actinomycetes, α-Proteobacteria and γ-Proteobacteria have also been reported (Bugg et al. 2011; Huang et al. 2013; Janusz et al. 2017).

7.5 Extremo-Stable Lignocellulose Active Enzymes

Industrial enzymes, such as those used in biorefineries, are considered as processing aids and technical biocatalysts. They were valued at over $1 billion in 2010, $7.81 billion in 2020 and are now expected to reach $11.92 billion by 2023 according to several research associations. This opens a new window for researchers to meet the ever-increasing global market demand, and extreme enzymes are the best candidates to consider (Desai et al. 2017; Dumorné et al. 2017; Freedonia 2018; Silveira et al. 2018; Zhu et al. 2020).

Extreme biocatalysis presents an interesting strategy for improving conventional enzyme technologies and is an attractive, sustainable, cost-effective, and environmentally friendly option for chemical catalysis. Enzymes derived from extremophiles have remarkable tolerance to inhospitable environmental conditions, making them an ideal enzyme system in biotechnology and for the degradation of recalcitrant lignocellulosic biomass in the biofuel industry (Chettri et al. 2021; Desai et al. 2017; Neifar et al. 2015a, 2015b, 2016, 2017).

Extremophiles are organisms that can grow and survive under harsh conditions, e.g., extreme conditions of temperature, pH, salinity, etc. They colonize extreme environments such as hot springs, thermal vents, glaciers, deep seas, industrial and mining wastewater effluents, salt lakes, and desert sediments (Cherif et al. 2018; Neifar et al. 2015a; Raddadi et al. 2015). Extremophilic microorganisms include hyperthermophiles (> 80°C), thermophiles (55–80 °C), psychrophiles (< 15 °C), acidophiles (growing optimally at pH < 3–4), alkalophiles (pH > 8–9), extreme halophiles (2500–5200 mM NaCl), halophiles (500–2500 mM NaCl), and slight halophiles (200–500 mM NaCl) (Bilal et al. 2017; Hess 2008; Neifar et al. 2015a, 2015b; Pikuta et al. 2007; Qurat-ul-Ain et al. 2017; Sarmiento et al. 2015). The potential application of thermophilic, psychrophilic, alkaliphilic, acidophilic, and halophilic microorganisms and their extremozymes in lignin pretreatment, saccharification and fermentation processes, has been reviewed by many authors (Chettri et al. 2021; Dumorné et al. 2017; Zhu et al. 2020).

With the development of new sequencing technologies and the variety of annotation and prediction softwares, new microorganisms and carbohydrate activity enzymes (CAZymes) for lignocellulosic biomass degradation and biofuel production have been discovered (Barnard et al. 2010; Chen et al. 2020; Chettri et al. 2020, 2021; Ferreira Filho et al. 2017; Rosnow et al. 2017; Zerillo et al. 2013; Zheng et al. 2021). To date, thousands of fungal, bacterial, and archaeal genomes are publicly available in the NCBI and GenBank databases. Genomic data for more than 120 thermophilic bacteria are available in the public database, and genomes from the genera *Pyrococcus*, *Anaerobranca*, *Thermotoga*, *Thermoplasma*, and *Thermus* have been studied in detail and provide enzymes suitable for industrial biorefinery applications (Chettri et al. 2021; Counts et al. 2017). By analyzing microbial secretomes, Tuveng et al. (2019) discovered the key enzymes involved in carbohydrate degradation processes and potentially infer the mode-of-action of biomass conversion. New enzyme activities within CAZyme families are continuously being discovered (Chettri et al. 2020; Thapa et al. 2020; Wilkens et al. 2017).

Figure 7.5 shows the cazyomes (all genomic CAZymes) of two halophilic plant growth-promoting bacteria, *H. desertis* G11 and *Oceanobacillus iheyensis* E9, isolated from hypersaline sediments of Chott El-Djerid and salt crust of Chott Douz, respectively (Neifar et al. 2019; Riahi et al. 2020). CAZyomic annotation by dbCAN2 meta server (Yin et al. 2012) revealed the presence of several genes associated with the assembly (GT) and degradation (GHs and CEs) of carbohydrate polymeres. In addition, their genomes contain a large number of genes encoding auxiliary activities (AA) and carbohydrate binding modules (CBM). Stuctural, physicochemical and biochemical properties predicted by Phyr2 (Kelley et al. 2015) and Protparam (Wilkins et al. 1999) of selected novel lignocellulose active and stable CAZymes of these two halophilic bacteria are presented in Figure 7.6. Their attractive structural and biochemical properties make them quite useful for various applications, particularly in biorefineries for lignocellulosic saline waste valorization.

The biocatalysts extracted from extremophiles are called extremozymes, and often exhibit outstanding thermal or cold adaptability, wide pH range tolerance, salt tolerance and/or pressure tolerance (Hua et al. 2018; Li and Yu 2013; Neifar et al. 2015a, 2015b; Park et al. 2017, 2018). These adaptations correspond to specific changes in the amino acid sequence, which result in variations in the structure, flexibility, charge, and/or hydrophobicity of the enzymes (Desai et al. 2017; Park et al. 2017; Viikari et al. 2007). They also allow extremozymes to extend the optimal enzymatic performance ranges, thus enabling biocatalysis under enzymatically unfavorable conditions encountered in industrial processes. In recent years, many bioinformatic models have been used as effective tools to distinguish extremozymes from non-extremozymes (Chettri et al. 2020, 2021). Some exemples of fungal and bacterial extremozymes potentially useful in pretreatment and saccharification steps of lignocellulosic biomass are illustrated in Table 7.3.

The saccharification process following pretreatment occurs at high temperatures (40–50 °C), which generally results in slow hydrolysates with low sugar yields and high risk of microbial contamination. These limitations could be addressed by using thermostable enzymes from thermophilic and thermotolerant microorganisms (Bansal et al. 2014; Bhalla et al. 2013; Faria et al. 2020; Karimi and Taherzadeh 2016; Park et al. 2018; Patel et al. 2019; Raheja et al. 2020; Sahoo et al. 2020; Sondhi et al. 2015; Ying et al. 2013). Several thermophilic fungi have been isolated and studied in recent years, such as *Talaromyces leycettanus*, *Humicola insolens*, *Acremonium thermophilum*, *Chaetomium thermophilum*, *Humicola grisea*, *Humicola insolens*, *Melanocarpus albomyces*, *Talaromyces emersonii*, *Fusarium chlamydosporum*, etc. The active and stable cellulases produced by these fungi have been purified and characterized at structural and functional levels

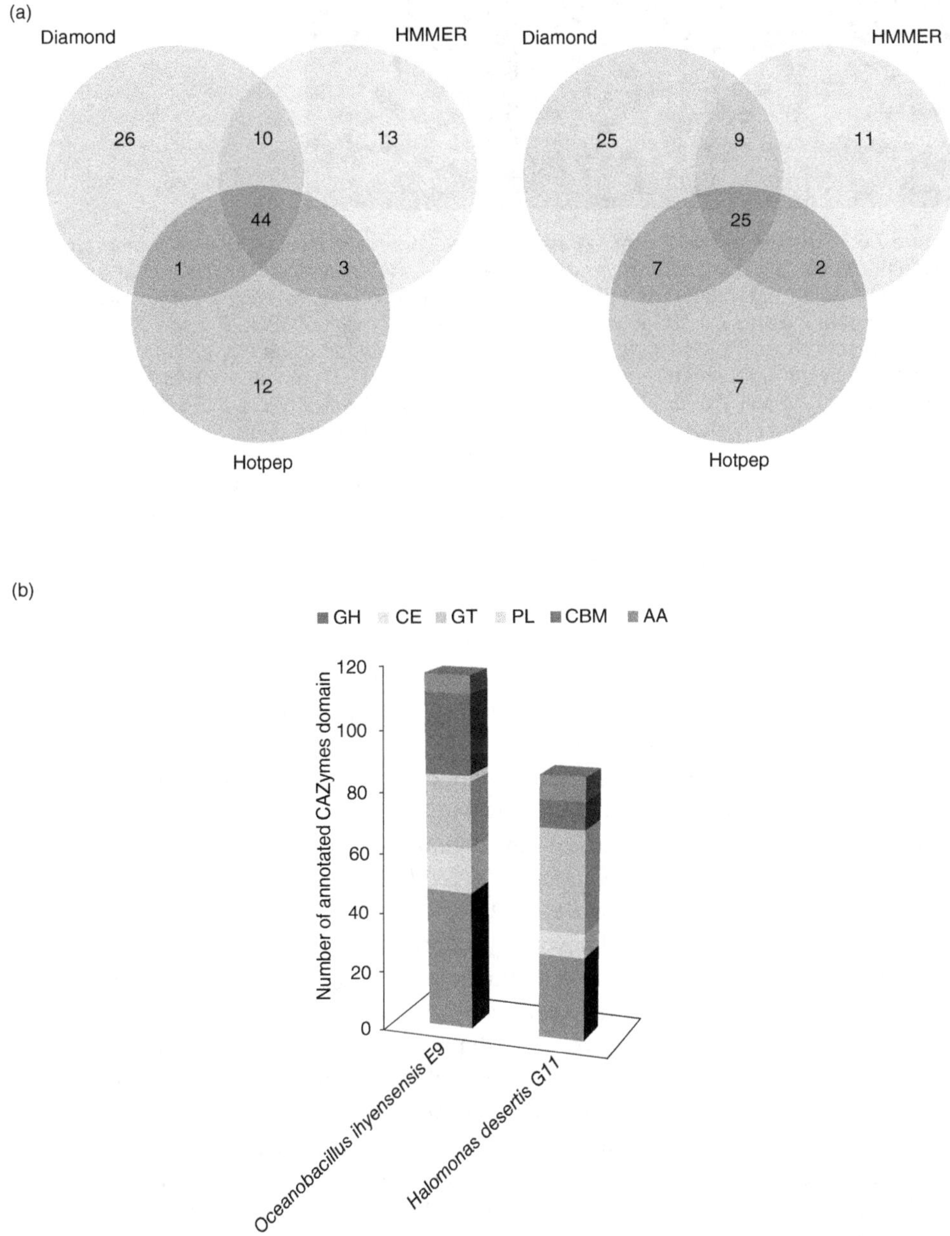

Figure 7.5 Carbohydrate-active enzymes predicted in the genome of the halophilic plant growth promoting strains *Oceanobacillus ihyensensis* E9 and *Halomonas desertis* G11 (El Hidri et al. 2013; Naili et al. 2018; Neifar et al. 2019; Riahi et al. 2020) using dbCAN2 server (Yin et al. 2012). (a) Venn diagram showed overlaps among the results of the three annotation tools; (b) repartition of CAZymes in the two bacterial genomes with AA, auxiliary activities; GH, glycoside hydrolase; GT, glycosyltransferase; CBM, carbohydrates- binding module and PL, polysaccharide lyase.

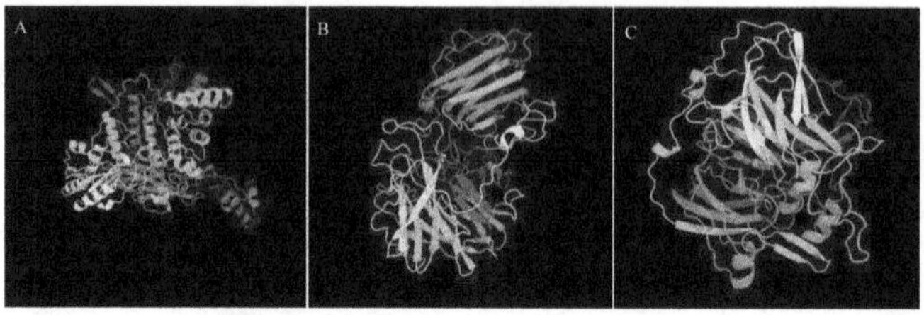

Figure 7.6 Structure properties of extremo-CAZymes produced by plant growth promoting halophilic bacteria, *Oceanobacillus iheyensis* E9 and *Halomonas desertis* G11 (Naili et al. 2018; Riahi et al. 2020), potentially useful in the delignification and saccharification of lignocellulosic *Posidonia* saline waste: (A) 3D model of GH3 β-glucosidase (WP_069686429.1: MW, 86908.38; pI, 5.41) from the strain E9. 89% (689 residues) have been modelled using Phyre2 with 100.0% confidence by the single highest scoring template c5wabD. (B) 3D model of GH43 1,4-beta-xylosidase (WP_066316589.1:MW, 60459.75; pI: 5.53) from the strain G11. 523 residues (97% of your sequence) have been modelled with 100.0% confidence by the single highest scoring templatec1yifC. (C) 3D model of AA3Laccase (WP_017796468.1: MW, 58806.77; pI, 5.32) from the strain E9. 98% (501 residues) of the sequence have been modelled with 100.0% confidence by the single highest scoring template c1ofOA. Instability index values below 40 and aliphatic index values between 77.64 and 88.96, calculated by ProtParam tool (Wilkins et al. 1999), indicated that these CAZymes are thermostable.

(Chettri et al. 2021; Gu et al. 2019; Park et al. 2017; Patel et al. 2019; Qin et al. 2020; Xia et al. 2016). In recent years, a shift towards extremozymes from extremophilic bacteria has been observed, due to their robust growth, their ability to survive in the harsh conditions of the bioconversion process, and the presence of stable multi-enzyme complexes that provide increased function and synergy (Gu et al. 2019; Sahoo et al. 2020). The most commonly used thermophilic bacteria in biorefineries belongs to the genera *Bacillus, Geobacillus, Caldibacillus, Acidothermus, Caldocellum* and *Clostridium* (Azadian et al. 2017; Qurat-ul-Ain et al. 2017; Raza et al. 2019; Tiwari et al. 2017; Zambare et al. 2011). Some thermophilic bacteria are known to produce both cellulases and xylanase, which can completely hydrolyze biomass at high temperatures (Zhu et al. 2020). For example, treatment of biomass with a thermostable cellulase produced by the thermophilic bacterium *Geobililus sp.* R7 yielded a hydrolysate that was readily fermented by Saccharomyces cerevisiae ATCC 24860T to produce 0.45–0.50 g ethanol/g glucose with 99% glucose utilization (Zambare et al. 2011). The thermophiles *Caldicellulosiruptor bescii* and *Clostridium thermocellum* have been reported for their potential to utilize cellulose, hemp, as well as pre-treated lignocellulosic biomass as a substrate to produce bioethanol (Olson et al. 2012). In addition, thermophiles *Thermoanaerobacterium thermosaccharolyticum* M18 are able to directly utilize cellulose and xylan for bioethanol production (Ábrego et al. 2017). Since non-enzymatic processes that decrease cellulose crystallinity usually involve low pH, organic solvents, chemical and oxidative reagents, and detergents, the use of multistable extremozymes that are more active on crystalline cellulose than on amorphous cellulose is highly sought after on an industrial scale. These conditions and enzymatic properties must be taken into consideration when selecting enzymes for the treatment of lignocellulosic residues (Agrawal et al. 2019; Babot et al. 2011; Hämäläinen et al. 2016; Phadtare et al. 2017; Rai et al. 2019; Rezaei et al. 2017; Schneider et al. 2020).

Table 7.3 Microbial lignocellulose degrading-enzymes and their potential in lignocellulose pretreatment and saccharification.

Enzyme type	*Microorganisme*	Substrate	Condition	Main findings	References
			Pretreatment step		
Thermotolerant laccase	*P. djamor*	Sugarcane tops	pH 7 at 40°C for 6	Delignification: 79.1%	Sherpa et al. 2018
	white rot fungi	Aloe vera leaf rind	pH 7 at 50°C for 6 h	Lignin removel: 76.67% Enhanced saccharification: 44.35% ± 1.02	Rajeswari and Jacob 2020b
	Trichoderma harzianum HZN10	Saw dust	pH 6 at 40°C for 48h		Bagewadi et al. 2017
	Myceliophthora thermophila	Ground eucalypt wood	50°C at pH 6.5 for 24h	Lignin removal with mediator: 50% Lignin without mediator: 20%	Rico et al. 2014
	Myceliophthora thermophila	Eucalypt kraft pulp	pH 6.5 at 50°C for 12 h	Reduction of Kappa number: 2.7 points Lipophilic extractives removal: 74% Tolerate wide pH range (4 -9)	Babot et al. 2011
	Geobacillus sp. WSUCF1	Corn stover bagasse	pH 6 at 50°C for 120 h	Increase in hydrolysis of corn stover by 1.31–2.28 times and bagasse by 1.32–2.02 times.	Rai et al. 2019 a
	Pantoea ananatis	Rice straw	pH 2.5 at 40°C for 12h	Rice straw delignification: 38%	Shi et al. 2015
Acidestable laccase	*Pycnoporus sanguineus H275*	Corn stover Wheat straw	pH 4 at 28°C for 24 h		Lu et al. 2010
	Trametes villosa	*Eucalyptus* wood	pH 5 at 24°C	Increase in hydrolysis: 61% Ethanol concentration: 4 g/ L	Gutiérrez et al. 2012
	Trametes versicolor	Corn stover	pH 4 at 40°C for 48 h	enhanced digestibility: 34.9%	Chen et al. 2012
	Pichia GS115	Maize straw Finger millet straw	pH 5 at 30°C for 24h	lignin loss: 1.86% in maize straw and 4.83% n finger millet straw	Kumar et al. 2018a
	Lentinula edodes	Rape straw	pH 4 at 30°C for 24h [EZ]:16.0 μg/mg substrate	lignin content was reduced by 2.0% Tolerate 15% of ethanol and methanol	Liu et al. 2020
	Myrothecium verrucaria ITCC-8447	Rice straw	pH 5 at 30 ºC for 12 h [EZ]: 20 U/g of substrate	67 % delignification of rice straw	Bhardwaj et al. 2019

(Continued)

Table 7.3 (Continued)

Enzyme type	*Microorganisme*	Substrate	Condition	Main findings	References
Thermo-alkali-stable Laccase	*Pleurotus ostreatus IBL-02*	Sugarcane bagasse	pH 4.5 at 35°C for 48 h	33.6% delignification of sugarcane bagasse	Asgher et al. 2013
	G. lucidum IBL-05	Newspapers waste	pH4.5 at 25°C for 24h	42.2% of delignification of Newspapers Waste	Bilal et al. 2017
	Myrothecium verrucaria ITCC- 8447	Wheat straw	30°C for 12 hours	Decrease in kappa number of wheat straw (58.58%)	Agrawal et al. 2019
	Caldalkalibacillus thermarum strain TA2.A1	Rice straw, corn stover,reed	pH 8 at 70°C for 24 h White laccase	Increase in phenolic monomersrelease: rice straw: 22 %, corn stover 6%, and reed 27.3%. Tolerance to higher temperature of 70-90°C and pH (8-9)	Yang et al. 2019
	Streptomyces ipomoeae CECT 3341 (SilA),	Wheat straw	pH 8 at 50°C for 24 h [EZ]: 2.5 IU/g DW substrate	35% delegnification, increase in glucose16 and xylose(6%)release ethanol yield:59.4 % ± 2.3	De La Torre et al. 2017
	Bacillus tequilensis SN4	Softwood pulp	pH 9 at 65°C for 4 h [EZ]:40U/g	Decrease in kappa number: 47 % with N-hydroxybenzotriazole as a mediator	Sondhi et al. 2015

Acido and thermostable laccase	*Pycnoporus cinnabarinus*	Wheat straw	pH 5 at 50°C for 8h	Lignin removel: 67%	Moreno et al. 2013
	G. lucidum	Sugar cane bagasse	pH 4.8 at 50°C for 24h		Sitarz et al. 2013
	Geobacillus sp.	Corn stover	pH 5 at 50°C for 8 h		Rai et al. 2019
	Trametes versicolor	Sugarcane bagasse	pH 4.5 at 50°C for 15h	Delignification: 78.29%	Asgher et al. 2018
	Trametes versicolor	Wheat straw	pH 4.8 at 50°C for 24 h	Increase in digestibility: 26%	Deng et al. 2019
	Ganoderma lucidum	Corn stover	pH 5 at 45°C for 24h	-	Fang et al. 2015
	Sordaria macrospora k-hell	Lignin	pH5 at 50°C for 24 h	-	Yang et al. 2019, 2020
	Pleurotus djamor	Lantana camara	pH 6.14 at 40.7°C for 8h	Lignin removel: 82%	Gujjala et al. 2016
	Marasmiellus palmivorus VE111	Eucalyptus globulus wood	pH 4.8 at 50°C, for 6h [Ez]:50 U/g substrate	Decrease in lignin content: 31% Increase of glucose (10%) and in xylose (15%) Increase of ethanol yield:8.5%	Schneider et al. 2020
	Trametes versicolor	Olive pomace	pH 4.5 at 40°C for 6h	Lignin removel: 77.3%	Amin et al. 2018
Acido and thermostable laccase	*Phanerochaete chrysosporium*	Corn stover	pH 4.5 at 45°C for 72 h Peroxidase	higher crystallinity portion (45.64% to 55.76%)	Wang et al. 2013a
Polyextrêmotolé-rantes enzymes	*Chromohalobacter salexigens*	Almond shell biowaste	pH 8.0 at 40°C for 72h	58% in harch condition: organic solvents, 2M NaCl	Jafari et al. 2017
	Aquisalibacillus elongatus	Sugar beet pulp	3 M NaCl pH 8 at 40°C	Extremotolerante:25–75°C and pH 5.0–10.0 78.4% of delignification after 12 h	Rezaei et al. 2017

(*Continued*)

Table 7.3 (Continued)

Enzyme type	*Microorganisme*	Substrate	Condition	Main findings	References
Thermostable enzyme	*Caldibacillus cellulovorans*	*Pinus radiata* kraft pulp	pH 6.0 at 85°C β-1,4-Mannanase	Decrease in kappa number with 95.5% of bleached pulp yield	Sunna et al. 2000
			Saccharification step		
Thermostable enzymes	*Aspergillus sp*	Aloe vera leaf rind	50°C for 72 h	Teducing sugars yield:207 ± 1.90 mg/g	Rajeswari and Jacob 2020b
	thermophilic *Bacillus subtilis K-18*	Pine needles	pH 7 at 50°C for 8h [EZ]:3.77 ± 0.11 IU/g	Saccharification yield: 54.389%	Irfan et al. 2017
	Trichoderma reesei Rut-C30	Kans grass	[cellulase and xylanase] 75 IU/g 50°C for 6 h	63.2 gL−1 of ethanol 500 mg/g:RS yield	Rajak et al. 2018
	Bacillus aerius *CMCPS1*	Laccase pretreated corncob	pH5 at 50°C for 120 h [cellulase]:40 U per gram	Reducing sugar yield: 1.37 g g−1	Ganesan et al. 2020
	Bacillus subtilis Lucky9	Corncob	xylanase (50 U/mL), pH 6.5 at 50°C for 12 h.	Yield of reducing sugars was up to 8.7 mg/mL after 8 h of incubation.	Chang et al. 2017
	B. oceanisediminis *B. halotolerans*	rice straw	50°C pH 7.4 20 min	[Sugar]:16.8 g/L for *B. oceanisediminis* For B. halotolerans: 80% of saccharification effeciency in a seawater	Indira et al. 2019
Alkalistable enzymes	*Pseudoalteromonas,sp. NJ64*	Residue clep cellulose	pH 9 at 40°C for 72h [Cellulase]: 20U/g	70 g/L glucose Ethanol yield reached 37.37% in 48 h	Wang et al. 2015

Acidotolerant enzymes	*Trichoderma harzaianum*	Sugarcane bagasse	pH 4.8 at 37°C for 24 h	69.2 and 72.9% cellulose hydrolysis of alkali and ligninolytic enzymes pretreated bagasse, respectively	Asgher et al. 2013
	Cladosporium sphaerospermum	green seaweed ulva	pH4at 40°C for 24 h Cellulase 10 U/g	Sugar yield of 112 ± 10 mg/g of substrate Ethanol yield of 0.47 g/g reducing sugar	Trivedi et al. 2015
	Galactomyces sp. CCZU11-1	Chestnut shell	pH 4.8 at 30°C [EZ]: 20 FPU/g	Ionic liquid-tolerant cellulase1,3-dimethylimidazolium dimethyl phosphate High yield of reducing sugars, 62.1%, was at 62.1% in the media containing 20% (w/v) IL	He et al. 2016
	T. reesei Rut C-30 *A. niger F38*	Office paper	pH 4.8, 10 FPU/ g, 40°C, 48 h	ethanol yield 0.38 g/g sugar	Guerfali et al. 2015
Acido thermotolerant Thermophilic acidophilic	*Clostridium saccharoperbutylace-tonicum N1-4*	Palm kernel cake	pH 4.5 at 45°C for 72 h Mannanase	71.54 ± 2.54 g/L	Shukor et al. 2016

(*Continued*)

Table 7.3 (Continued)

Enzyme type	Microorganisme	Substrate	Condition	Main findings	References
	Trichoderma reesei Rut C30	Sugarcane tops	pH 5 at 50°C for 7 h [Cellulase-xylanase] RS:508.9 mg/g	Maximum delignification: 79.1%	Sherpa et al. 2018
	Penicillium echinulatum S1M29	Eucalyptus globulus wood	pH 4.8 at 50°C for 72 h [cellulases – xylanases]: 30U/g	45.01% of glucose release and 87.09% xylose yield Ethanol yield: 115.61%	Schneider et al. 2020
	Streptomyces ipomoeae	Wheat straw	pH 5.5 at 50°C for 72 h cellulase: Celluclast 1.5 Novozyme 188	Glucose concentration 16.5 g/L. Ethanol yield59.4 ± 2.3 %	De La Torre et al. 2017
	Bacillus substilis JJBS250	Sugarcane bagasse	pH 7 at 60°C for 48h [Xylanase]: 20 U/g	Reducing sugars yield:.39 mg/g substrate.	Singh 2020
	Schizophyllum commune NAIMCC-F-03379 *Aspergillus oryzae ITCC-8571*	Rice straw	pH 5 at 50°C for 24 h [xylanase]: 100 U/g [cellulase]: 100U/g [laccase]: 20 U/g	Sugar concentration: 31.2 ± 0.7 g/L Ethanol concentration7.34 ± 0.39 g/L	Bhardwaj et al. 2019
	Schizophyllum commune *Aspergillus oryzae*	Rice straw Wheat bran	pH 4.5 at 50°C for 10h [*Xylanase*]:100 IU/g substrate	High stability in large range of temperature (25–65°C) and pH (3–10) [Sugar]: 1.162 mg/mL wheat bran 0.98 mg/mL rice straw	Kumar et al. 2018b
	Bacillus sonorensis BD92	wheat bran,	pH 4.8 50°C 24h enzyme cocktail (4.38 and 17.07 U endo-glucanase and xylanase)	glucose yield 10.70 ± 0.14 mg/g biomass xyloseyield 12.41 ± 0.34 mg/ g biomass	Raza et al. 2019
	commercial Sumizyme AP2	Luffa cylindrica (l.) Fibers	pH 4 at 60°C for 1h, [Pectinase, Cellulases, Hemicellulase]: 54000 unit/g	Reducing sugars yield 59.4 mg/g	Zaafouri et al. 2017

Acido thermo-tolerant enzymes	*Enzymatic cocktail: No. X2753 + No. C2730 + No. E-AFASE*	Pulped bamboo residues	pH 4.8 at 50°C for 48h [xylanase]: 120 IU/g and 15 IU/g [arabinofuranosidase], [cellulase]: 20 FPU/g	Increasing in saccharification efficiency from 58.98% to 83.23%.	Huang et al. 2015
	enzymatic cocktail: celluclast 1.5 L, Fungal b-glucosidase (Novozyme 188, Denmark)	Seaweed solid Wastes	pH 4.8 at 50°C for 72 h celluclast 1.5 L,:15 FPU/g substrate. b-glucosidase 52 CBU/g	Highest glucose yield: 99.8% Ethanol yield: 90.9%	Tan and Lee 2014
	commercial cellulase, Cellic CTec2	*Posidonia oceanica* residue	pH 4.8 at 50°C for 90 h [EZ]: 20 FPU/g	Reducing sugars concentration: 21.6 g/L Bioethanol yield: 62.3%	Pilavtepe et al. 2013
Acido thermotolerant enzymes	*Cellulase (Spezyme CP) Genencor (Palo Alto, CA)*	Corn stover	pH 4.8 at 50°C for 72h [EZ]: 10 FPU/g	Glucose yield: 66.61% Ethanol yield: 57.80%	Wan and Li 2010
	Aspergillus japonicus	Fruit peel waste	pH 3-4.5 at 55°C for 4h endo-xylanase	banana peel (3.36 µmol) and pineapple peel (2.85 µmol), decreasing to mango (2.69 µmol) and ponkan (2.16 µmol) peels after 15 min	Da Silva et al. 2019
	Myceliophthora thermophila	rice straw corncob	pH 5.0 60°C for 48 h, [endoxylanase]25 U/g	135.61 mg/g 48rice straw 132.61 mg/g of substrate recorded after 36 h, corncob	Dahiya et al. 2019

(*Continued*)

Table 7.3 (Continued)

Enzyme type	*Microorganisme*	Substrate	Condition	Main findings	References
alkali-thermo-halotolerant enzymes	*polyextremophilic bacterium Bacillus halodurans*	Wheat bran, corn cob	pH 8 at 60°C for 24 h [cellulase–xylanase]: 60U/g	Reducing sugars yield: 47 mg/ gwheat bran 26 mg/ g corn cobs	Rattu et al. 2016
	Bacillus mojavensis AG21 Bacillus vallismortis J77 Bacillus sonorensis 7.26 Bacillus safiensis AJ49	*Stevia rebaudiana* leaves	pH 10 at 50°C for 22 h	reducing sugar yields of *B. mojavensis*, 3620 ± 181 mg/L *B.vallismortis*, 3040 ± 164 mg/L *B. sonorensis*, 3656 ± 164 mg/L *B. safiensis*, 2654 ± 106 mg/L *Aspergillus niger*: 2646 ± 145 mg/L *Trichoderma* reesei 2253 ± 123 mg/L obtained after 2h	Souii et al. 2020
Thermo-alkali-solvent-stable	*Bacillus altitudinis*	Sorghum straw	pH 8 at 50 for 48 h xylanase	reducing sugar yield: 2.58 mg/g	Adhyaru et al. 2014
Ionic Liquid (IL) Stable enzymes	*Bacillus subtilis G2.* *enzymatic cocktail*	Pine needle biomass	(CMCase 77 IU, FPase 28 IU, and 31.5 IU xylanase) 24 h at 37°C at 180 rpm.	Saccharification efficiency of 23.57 % sugar yield: 166 mg/g	Vaid and Bajaj 2017
Thermo-acid/ alkali stable	*Aspergillus oryzae LC1*	Agro residue:rice straw, rice husk, wheat straw, wheat bran, sugarcane	100 IU 50°C 24 h xylanases	xylobiose (0.68 mg/g) from sugarcane bagasse, xylotriose (2.47 mg/g from wheat straw, without xylose duo to less β-xylosidase activity	Bhardwaj et al. 2019

Thermo-alkali-stable enzymes	*metagenome, PersiCel4*		pH = 8.5).72 hendo-β-1,4-glucanase	conditions	2019, 2020
	Geobacillus stearothermophilus	banana peel	(xylanase 39.8 IU, FPase 40 IU, pectinase 80 IU, pH 9 for 1h-4h	Sugar concentration:121.1 g/l after 29.6 min Ethanol yield 76.5% within 30 h	Prakash et al. 2018
	Myceliopthora thermophile BJA	wheat bran and corn cobs	50°C for 24 h pH 10 endoglucanase	Sugar yield: 421 mg/ g wheat bran 382 mg/ g corn cobs	Phadtare et al. 2017
	H. meridiana APCMST-KS4	rice bran wheat straw ulva	pH8 at 50°C for (30–150 min) (20–100U/ ml	saccharification rate: rice bran 68.66%, wheat straw 44.11 % Ulva 50.03%	Palavesam et al. 2015
Thermotolerant enzymes	*Acidothermus cellulolyticus* E1	corn stover.	pH 4.8 at 50°C for 72 h endo-1,4-β glucanase	Sugar concentration, up to 0.18 g/L	Ransom et al.2007

Enzyme engineering has been applied for lignocellulolytic enzymes to improve either individual cellulase properties or synergy by optimizing enzyme cocktails or by creating a more efficient multi-enzyme cellulolytic complex (cellulosome) (Gilmore et al. 2020; Rattu et al. 2016; Summers et al. 2020; Wushke et al. 2015). So far, the most notable results of protein engineering of cellulases are improvements in thermostability and, as a result, various thermostable cellulases have been constructed (Bhalla et al. 2013; Patel et al. 2019). On the other hand, the commercial devolelopment of ligninases and hemicellulases for enzymatic processing of lignocellulosic materials is not as advanced as that of cellulases, as current commercial enzyme mixtures have been developed primarily for hydrolysis of chemically pretreated biomass, which removes the majority of lignin and hemicellulose. In an effort to create thermostable enzyme complexes for biomass utilization, Gilmore et al. (2020) fused domains present in fungal cellulosomes with cellulases from *Thermotoga maritima* to create thermostable chimeric enzymes that are catalytically active at elevated temperatures (70–90 °C), and stable at moderate temperatures (50–70 °C) thus allowing for a reduction in enzyme titers required for biomass hydrolysis (Gilmore et al. 2020). CRISPR/Cas9-based systems were recently developed for the genome editing of extremophiles and is expected to be used for CAZymes production in metabolically engineered microorganisms (Chettri et al. 2020; Zhu et al. 2020)

At present, the extremophiles that can be grown in the laboratory are only a small fraction of the total diversity that exists in nature (Lee et al. 2018; Yang et al. 2016). At all phylogeny levels, uncultured microorganisms that do not grow on standard media are playing critical roles in lignocellulose degradation (Ariaeenejad et al. 2020; Zhu et al. 2020). Using a culture-independent metagenomic approach based on environmental DNA extraction, followed by direct sequencing or DNA cloning to generate a metagenomic library provides access to microbial genomes and their CAZymes (Delmont et al. 2011; Ko et al. 2013; Milshteyn et al. 2014; Montella et al. 2016; Yang et al. 2016). The study by Lewin et al. (2017) revealed that a microbial oil reservoir assemblage harbored metagenomic diversity that was exploited to screen for novel CAZymes. Their results identified novel thermostable archaeal cellulases, stable above 80 °C multi-module with high activity on different cellulose substrates. Chettri et al. (2020) used *in-silico* computational research and functional annotation coupled to enzyme assay experiments to devise cocktails of new CAZymes for use in biorefineries.

7.6 Conclusion and Future Directions

The use of extremozymes as efficient biological catalysts in the degradation of lignocellulosic by-products and wastes for the production of second generation biofuels is of great industrial and biotechnological interest. Their stability against pH, temperature and salinity makes them better suited to the harsh processing conditions required for efficient deconstruction of lignocellulose into fermentable products. Furthermore, with the advancement of microbial genome sequencing technologies, and the bioinformatics annotation tools developed to interpret CAZyomes, it has become easier to discover and deploy new, extremely stable CAZymes, adapted to the most recalcitrant lignocellulosic materials. Together, the elucidation of the structure-function of extremoCAZymes, the development of expression systems using the CRISPR-CAS system, and innovations in enzyme engineering technologies will help streamline future efforts to improve the efficiency of biofuel production, helping to unlock the energy potential of agricultural and agro-industrial waste streams. With recent advances in -omics technologies (metagenomics, metatranscriptomics, and metaproteomics) and the associated decreasing costs, the study of complex lignocellulose metabolizing communities has become increasingly accessible leading to the discovery of novel CAZymes with high catalytic efficiencies and conversion yields under extreme conditions

References

Abraham, A., Mathew, A.K., Park, H., Choi, O., Sindhu, R., Parameswaran, B., Pandey, A., Park, J.H., and Sanga, B.I. (2020). Pretreatment strategies for enhanced biogas production from lignocellulosic biomass. *Bioresource Technology* 301: 122725.

Ábrego, U., Chen, Z., and Wan, C. (2017). Consolidated bioprocessing systems for cellulosic biofuel production. *Advances in Bioenergy* 2: 143–182.

Adegboye, M.F., Ojuederie, O.B., Talia, P.M., and Babalola, O.O. (2021). Bioprospecting of microbial strains for biofuel production: metabolic engineering, applications, and challenges. *Biotechnology for Biofuels* 14 (1): 1–21.

Adhyaru, D.N., Bhatt, N.S., and Modi, H.A. (2014). Enhanced production of cellulase-free, thermo-alkali-solvent-stable xylanase from *Bacillus altitudinis* DHN8, its characterization and application in sorghum straw saccharification. *Biocatalysis and Agricultural Biotechnology* 3 (2): 182–190.

Aditiya, H.B., Mahlia, T.M.I., Chong, W.T., Nur, H., and Sebayang, A.H. (2016). Second generation bioethanol production: a critical review. *Renewable and Sustainable Energy Reviews* 66: 631–653.

Agger, J.W., Isaksen, T., Várnai, A., Vidal-Melgosa, S., Willats, W.G., Ludwig, R. et al. (2014). Discovery of LPMO activity on hemicelluloses shows the importance of oxidative processes in plant cell wall degradation. *Proceedings of the National Academy of Sciences* 111 (17): 6287–6292.

Agrawal, K., Bhardwaj, N., Kumar, B., Chaturvedi, V., and Verma, P. (2019). Process optimization, purification and characterization of alkaline stable white laccase from *Myrothecium verrucaria* ITCC-8447 and its application in delignification of agroresidues. *International Journal of Biological Macromolecules* 125: 1042–1055.

Alalwan, H.A., Alminshid, A.H., and Aljaafari, H.A.S. (2019). Promising evolution of biofuel generations. Subject review. *Renewable Energy Focus* 28: 127–139.

Amin, R., Khorshidi, A., Shojaei, A.F., Rezaei, S., and Faramarzi, M.A. (2018). Immobilization of laccase on modified Fe3O4@ SiO2@ Kit-6 magnetite nanoparticles for enhanced delignification of olive pomace bio-waste. *International Journal of Biological Macromolecules* 114: 106–113.

Ariaeenejad, S., Hosseini, E., Maleki, M., Kavousi, K., Moosavi-Movahedi, A.A., and Salekdeh, G.H. (2019). Identification and characterization of a novel thermostable xylanase from camel rumen metagenome. *International Journal of Biological Macromolecules* 126: 1295–1302.

Ariaeenejad, S., Mamaghani, A.S.A., Maleki, M., Kavousi, K., Shahraki, M.F., and Salekdeh, G.H. (2020). A novel high performance in-silico screened metagenome-derived alkali-thermostable endo-β-1, 4-glucanase for lignocellulosic biomass hydrolysis in the harsh conditions. *BMC Biotechnology* 20 (1): 1–13.

Aripin, A.M., Kassim, A.S.M., Daud, Z., and Hatta, M.Z.M. (2013). Cassava peels for alternative fibre in pulp and paper industry: chemical properties and morphology characterization. *International Journal of Integrated Engineering* 5: 1.

Arntzen, M.Ø., Bengtsson, O., Várnai, A., Delogu, F., Mathiesen, G., and Eijsink, V.G. (2020). Quantitative comparison of the biomass-degrading enzyme repertoires of five filamentous fungi. *Scientific Reports* 10 (1): 1–17.

Asgher, M., Ahmad, Z., and Iqbal, H.M.N. (2013). Alkali and enzymatic delignification of sugarcane bagasse to expose cellulose polymers for saccharification and bio-ethanol production. *Industrial Crops and Products* 44: 488–495.

Asgher, M., Wahab, A., Bilal, M., and Iqbal, H.M. (2018). Delignification of lignocellulose biomasses by alginate–chitosan immobilized laccase produced from *Trametes versicolor* IBL-04. *Waste and Biomass Valorization* 9 (11): 2071–2079.

Azadian, F., Badoei-Dalfard, A., Namaki-Shoushtari, A., Karami, Z., and Hassanshahian, M. (2017). Production and characterization of an acido-thermophilic, organic solvent stable cellulase from *Bacillus sonorensis* HSC7 by conversion of lignocellulosic wastes. *Journal of Genetic Engineering and Biotechnology* 15 (1): 187–196.

Babot, E.D., Rico, A., Rencoret, J., Kalum, L., Lund, H., Romero, J., Del Río, J.C., Martínez, A.T., and Gutiérrez, A. (2011). Towards industrially-feasible delignification and pitch removal by treating paper pulp with *Myceliophthora thermophila* laccase and a phenolic mediator. *Bioresource Technology* 102 (12): 6717–6722.

Baeyens, J., Kang, Q., Appels, L., Dewil, R., Lv, Y., and Tan, T. (2015). Challenges and opportunities in improving the production of bio-ethanol. *Progress in Energy and Combustion Science* 47: 60–88. 47: 60–88.

Bagewadi, Z.K., Mulla, S.I., and Ninnekar, H.Z. (2017). Optimization of laccase production and its application in delignification of biomass. *International Journal of Recycling of Organic Waste in Agriculture* 6 (4): 351–365.

Bansal, N., Janveja, C., Tewari, R., Soni, R., and Soni, S.K. (2014). Highly thermostable and pH-stable cellulases from *Aspergillus niger* NS-2: properties and application for cellulose hydrolysis. *Applied Biochemistry and Biotechnology* 172 (1): 141–156.

Barnard, D., Casanueva, A., Tuffin, M., and Cowan, D. (2010). Extremophiles in biofuel synthesis. *Environmental Technology* 31 (8–9): 871–888.

Benedetti, M., Locci, F., Gramegna, G., Sestili, F., and Savatin, D.V. (2019). Green production and biotechnological applications of cell wall lytic enzymes. *Applied Sciences* 9 (23): 5012.

Bhalla, A., Bansal, N., Kumar, S., Bischoff, K.M., and Sani, R.K. (2013). Improved lignocellulose conversion to biofuels with thermophilic bacteria and thermostable enzymes. *Bioresource Technology* 128: 751–759.

Bhardwaj, N., Kumar, B., Agarwal, K., Chaturvedi, V., and Verma, P. (2019). Purification and characterization of a thermo-acid/alkali stable xylanases from *Aspergillus oryzae* LC1 and its application in Xylo-oligosaccharides production from lignocellulosic agricultural wastes. *International Journal of Biological Macromolecules* 122: 1191–1202.

Bilal, M., Asgher, M., Iqbal, H.M., and Ramzan, M. (2017). Enhanced bio-ethanol production from old newspapers waste through alkali and enzymatic delignification. *Waste and Biomass Valorizatio* 8 (7): 2271–2281.

Binod, P., Gnansounou, E., Sindhu, R., and Pandey, A. (2019). Enzymes for second generation biofuels: recent developments and future perspectives. *Bioresource Technology Reports* 5: 317–325.

Bugg, T.D.H., Ahmad, M., Hardiman, E.M., and Rahmanpour, R. (2011). Pathways for degradation of lignin in bacteria and fungi. *Natural Product Reports* 28: 1883–1896.

Callegari, A., Bolognesi, S., Cecconet, D., and Capodaglio, A.G. (2020). Production technologies, current role, and future prospects of biofuels feedstocks: a state-of-the-art review. *Critical Reviews in Environmental Science and Technology* 50 (4): 384–436.

Campioni, T.S., Soccol, C.R., Libardi Junior, N., Rodrigues, C., Woiciechowski, A.L., Letti, L.A.J., and Vandenberghe, L.P.D.S. (2020). Sequential chemical and enzymatic pretreatment of palm empty fruit bunches for *Candida pelliculosa* bioethanol production. *Biotechnology and Applied Biochemistry* 67 (5): 723–731.

Chang, S., Guo, Y., Wu, B., and He, B. (2017). Extracellular expression of alkali tolerant xylanase from *Bacillus subtilis* Lucky9 in E. coli and application for xylooligosaccharides production from agro-industrial waste. *International Journal of Biological Macromolecules* 96: 249–256.

Chen, F., Ye, J., Sista Kameshwar, A.K., Wu, X., Ren, J., Qin, W., and Li, D.W. (2020). A novel cold-adaptive endo-1, 4-β-glucanase from *Burkholderia pyrrocinia* JK-SH007: gene expression and characterization of the enzyme and mode of action. *Frontiers in Microbiology* 10: 3137.

Chen, Q., Marshall, M.N., Geib, S.M., Tien, M., and Richard, T.L. (2012). Effects of laccase on lignin depolymerization and enzymatic hydrolysis of ensiled corn stover. *Bioresource Technology* 117: 186–192.

Cherif, H., Neifar, M., Chouchane, H., Soussi, A., Hamdi, C., Guesmi, A. et al. (2018). Extremophile diversity and biotechnological potential from desert environments and saline systems of southern Tunisia. In: Extremophiles: From Biology to Biotechnology (ed. V. Ravi, D. Durvasula, and V. Subba Rao), 33–64. Boca Raton, FL: CRC Publishers.

Chettri, D., Verma, A.K., Sarkar, L., and Verma, A.K. (2021). Role of extremophiles and their extremozymes in biorefinery process of lignocellulose degradation. *Extremophiles* 25 (3): 203–219.

Chettri, D., Verma, A.K., and Verma, A.K. (2020). Innovations in CAZyme gene diversity and its modification for Biorefinery applications. *Biotechnology Reports* 28: e 00525.

Counts, J.A., Zeldes, B.M., Lee, L.L., Straub, C.T., Adams, M.W., and Kelly, R.M. (2017). Physiological, metabolic and biotechnological features of extremely thermophilic microorganisms. Wiley Interdisciplinary Reviews: *Systems Biology and Medicine* 9 (3): e1377.

Da Silva, P.O., de Alencar Guimarães, N.C., Serpa, J.D.M., Masui, D.C., Marchetti, C.R., Verbisck, N.V., Zanoelo, F.F., Ruller, R., and Giannesi, G.C. (2019). Application of an endo-xylanase from *Aspergillus japonicus* in the fruit juice clarification and fruit peel waste hydrolysis. *Biocatalysis and Agricultural Biotechnology* 21: 101312.

Dahiya, S. and Singh, B. (2019). Enhanced endoxylanase production by Myceliophthora thermophila with applicability in saccharification of agricultural substrates. *3 Biotechnology* 9 (6): 1–10.

Datta, A., Hossain, A., and Roy, S. (2019). An overview on biofuels and their advantages and disadvantages. *Asian Journal of Chemistry* 31: 1851–1858.

Daud, Z., Hatta, M.Z.M., Kassim, A.S.M., Awang, H., and Aripin, A.M. (2014). Exploring of agro waste (pineapple leaf, corn stalk, and napier grass) by chemical composition and morphological study. *BioResources* 9 (1): 872–880.

De La Torre, M., Martín-Sampedro, R., Fillat, Ú., Eugenio, M.E., Blánquez, A., Hernández, M., Arias, M.E., and Ibarra, D. (2017). Comparison of the efficiency of bacterial and fungal laccases in delignification and detoxification of steam-pretreated lignocellulosic biomass for bioethanol production. *Journal of Industrial Microbiology & Biotechnology* 44 (11): 1561–1573.

Delmont, T.O., Malandain, C., Prestat, E., Larose, C., Monier, J.M., Simonet, P., and Vogel, T.M. (2011). Metagenomic mining for microbiologists. *The ISME Journal* 5 (12): 1837–1843.

Deng, Z., Xia, A., Liao, Q., Zhu, X., Huang, Y., and Fu, Q. (2019). Laccase pretreatment of wheat straw: effects of the physicochemical characteristics and the kinetics of enzymatic hydrolysis. *Biotechnology for Biofuels* 12 (1): 1–12.

Desai, C., Shaikh, M., and Markande, A.R. (2017). EXTREAMOZYMES: a review. *International Journal of Scientific Research in Science and Technology* 3: 531–535.

Dodd, D. and Cann, I.K. (2009). Enzymatic deconstruction of xylan for biofuel production. *Global Change Biology Bioenergy* 18; 1 (1): 2–17.

Domínguez-Bocanegra, A.R., Torres-Muñoz, J.A., and López, R.A. (2015). Production of bioethanol from agro-industrial wastes. *Fuel* 149: 85–89.

Dumorné, K., Córdova, D.C., Astorga-Eló, M., and Renganathan, P. (2017). Extremozymes: a potential source for industrial applications. *Journal of Microbiology and Biotechnology* 27 (4): 649–659.

Ebringerová, A. (2005). Structural diversity and application potential of hemicelluloses. *Macromolecular Symposia* 232 (1): 1–12.

Edwards, M.C. and Doran-Peterson, J. (2012). Pectin-rich biomass as feedstock for fuel ethanol production. *Applied Microbiology and Biotechnology* 95 (3): 565–575.

El Hidri, D., Guesmi, A., Najjari, A. et al. (2013). Cultivation–dependant assessment, diversity, and ecology of haloalkaliphilic bacteria in arid saline systems of southern Tunisia. *BioMed Research International* 2013: 1–15.

Fang, Z., Liu, X., Chen, L., Shen, Y., Zhang, X., Fang, W., Wang, X., Bao, X., and Xiao, Y. (2015). Identification of a laccase Glac15 from *Ganoderma lucidum* 77002 and its application in bioethanol production. *Biotechnology for Biofuels* 8 (1): 1–12.

Faria, S.P., de Melo, G.R., Cintra, L.C., Ramos, L.P., Jesuino, R.S.A., Ulhoa, C.J., and de Faria, F.P. (2020). Production of cellulases and xylanases by *Humicola grisea var. thermoidea* and application in sugarcane bagasse arabinoxylan hydrolysis. *Industrial Crops and Products* 158: 112968.

Ferreira Filho, J.A., Horta, M.A.C., Beloti, L.L., Dos Santos, C.A., and de Souza, A.P. (2017). Carbohydrate-active enzymes in Trichoderma harzianum: a bioinformatic analysis bioprospecting for key enzymes for the biofuels industry. *BMC Genomics* 18 (1): 1–12.

Ferrer, M., Golyshina, O., Beloqui, A., and Golyshin, P.N. (2007). Mining enzymes from extreme environments. *Current Opinion in Microbiology* 10 (3): 207–214.

Freedonia (2018). Global industrial enzymes – demand and sales forecasts, market share, market size, market leaders. Freedonia. www.freedoniagroup.com/World-Enzymes.html.

Ganesan, M., Vinayakamoorthy, R.M., Thankappan, S., Muniraj, I., and Uthandi, S. (2020). Thermotolerant glycosyl hydrolases-producing *Bacillus aerius* CMCPS1 and its saccharification efficiency on HCR-laccase (LccH)-pretreated corncob biomass. *Biotechnology for Biofuels* 13 (1): 1–14.

Gilmore, S.P., Lillington, S.P., Haitjema, C.H., De Groot, R., and O'Malley, M.A. (2020). Designing chimeric enzymes inspired by fungal cellulosomes. *Synthetic and Systems Biotechnology* 5 (1): 23–32.

Gonçalves, C.C., Bruce, T., Silva, C.D.O.G., Fillho, E.X.F., Noronha, E.F., Carlquist, M., and Parachin, N.S. (2020). Bioprospecting microbial diversity for lignin valorization: dry and wet screening methods. *Frontiers in Microbiology* 11: 1081.

Gonçalves, F.A., Ruiz, H.A., da Costa Nogueira, C., dos Santos, E.S., Teixeira, J.A., and de Macedo, G.R. (2014). Comparison of delignified coconuts waste and cactus for fuel-ethanol production by the simultaneous and semi-simultaneous saccharification and fermentation strategies. *Fuel* 131: 66–76.

Gonçalves, F.A., Sanjinez-Argandoña, E.J., and Fonseca, G.G. (2011). Utilization of agro-industrial residues and municipal waste of plant origin for cellulosic ethanol production. *Journal of Environmental Protection* 2: 1303–1309.

Gu, Y., Zheng, F., Wang, Y., Su, X., Bai, Y., Yao, B., Huang, H., and Luo, H. (2019). Characterization of two thermophilic cellulases from *Talaromyces leycettanus* JCM12802 and their synergistic action on cellulose hydrolysis. *PloS One* 14 (11): e0224803.

Guerfali, M., Saidi, A., Gargouri, A., and Belghith, H. (2015). Enhanced enzymatic hydrolysis of waste paper for ethanol production using separate saccharification and fermentation. *Applied Biochemistry and Biotechnology* 175 (1): 25–42.

Gujjala, L.K., Bandyopadhyay, T.K., and Banerjee, R. (2016). Kinetic modelling of laccase mediated delignification of *Lantana camara*. *Bioresource Technology* 212: 47–54.

Gupta, A. and Verma, J.P. (2015). Sustainable bio-ethanol production from agro-residues: a review. *Renewable and Sustainable Energy Reviews* 41: 550–567.

Gutiérrez, A., Rencoret, J., Cadena, E.M., Rico, A., Barth, D., José, C., and Martínez, Á.T. (2012). Demonstration of laccase-based removal of lignin from wood and non-wood plant feedstocks. *Bioresource Technology* 119: 114–122.

Hafid, H.S., Baharuddin, A.S., Mokhtar, M.N., Omar, F.N., Mohammed, M.A.P., and Wakisaka, M. (2021). Enhanced laccase production for oil palm biomass delignification using biological pretreatment and its estimation at biorefinary scale. *Biomass & Bioenergy* 144: 105904.

Hämäläinen, J., Granström, T., Mollerup, F., Wang, Y., Xiong, H., and Turunen, O. (2016). Effect of enzymatic high temperature prehydrolysis on the subsequent cellulose hydrolysis of steam-pretreated spruce in high solids concentration. *Journal of Chemical Technology and Biotechnology* 91 (6): 1844–1852.

He, Y.C., Liu, F., Gong, L., Di, J.H., Ding, Y., Ma, C.L., Zhang, D.P., Tao, Z.C., Wang, C., and Yang, B. (2016). Enzymatic in situ saccharification of chestnut shell with high ionic liquid-tolerant cellulases from *Galactomyces sp.* CCZU11-1 in a biocompatible ionic liquid-cellulase media. *Bioresource Technology* 201: 133–139.

Hess, M. (2008). Thermoacidophilic proteins for biofuel production. *Trends in Microbiology* 16 (9): 414–419.

Hua, C., Li, W., Han, W., Wang, Q., Bi, P., Han, C., and Zhu, L. (2018). Characterization of a novel thermostable GH7 endoglucanase from *Chaetomium thermophilum* capable of xylan hydrolysis. *International Journal of Biological Macromolecules* 117: 342–349.

Huang, C., He, J., Li, X., Min, D., and Yong, Q. (2015). Facilitating the enzymatic saccharification of pulped bamboo residues by degrading the remained xylan and lignin–carbohydrates complexes. *Bioresource Technology* 192: 471–477.

Huang, X.F., Santhanam, N., Badri, D.V. et al. (2013). Isolation and characterization of lignin-degrading bacteria from rainforest soils. *Biotechnology and Bioengineering* 110: 1616–1626.

Indira, D. and Jayabalan, R. (2020). Saccharification of lignocellulosic biomass using seawater and halotolerant cellulase with potential application in second-generation bioethanol production. *Biomass Conversion and Biorefinery* 10: 639–650.

Irfan, M., Mushtaq, Q., Tabssum, F., Shakir, H.A., and Qazi, J.I. (2017). Carboxymethyl cellulase production optimization from newly isolated thermophilic Bacillus subtilis K-18 for saccharification using response surface methodology. *AMB Express* 7 (1): 1–9.

Jafari, N., Rezaei, S., Rezaie, R., Dilmaghani, H., Khoshayand, M.R., and Faramarzi, M.A. (2017). Improved production and characterization of a highly stable laccase from the halophilic bacterium Chromohalobacter salexigens for the efficient delignification of almond shell bio-waste. *International Journal of Biological Macromolecules* 105: 489–498.

Janusz, G., Pawlik, A., Sulej, J., Świderska-Burek, U., Jarosz-Wilkołazka, A., and Paszczyński, A. (2017). Lignin degradation: microorganisms, enzymes involved, genomes analysis and evolution. *FEMS Microbiology Reviews* 41 (6): 941–962.

Jayani, R.S., Saxena, S., and Gupta, R. (2005). Microbial pectinolytic enzymes: a review. *Process Biochemistry* 40 (9): 2931–2944.

Jegannathan, K.R., Chan, E., and Ravindra, P. (2011). Biotechnology in biofuels-A cleaner technology. *Journal of Applied Sciences* 11 (13): 2421–2425.

Kameshwar, A.K.S., Ramos, L.P., and Qin, W. (2019). CAZymes-based ranking of fungi (CBRF): an interactive web database for identifying fungi with extrinsic plant biomass degrading abilities. *Bioresources and Bioprocessing* 6 (1): 1–10.

Karimi, K. and Taherzadeh, M.J. (2016). A critical review of analytical methods in pretreatment of lignocelluloses: composition, imaging, and crystallinity. *Bioresource Technology* 200: 1008–1018.

Kelley, L.A., Mezulis, S., Yates, C.M., Wass, M.N., and Sternberg, M.J. (2015). The Phyre2 web portal for protein modeling, prediction and analysis. *Nature Protocols* 10 (6): 845–858.

Kim, S.K., Chung, D., Himmel, M.E., Bomble, Y.J., and Westpheling, J. (2019). Heterologous co-expression of two β-glucanases and a cellobiose phosphorylase resulted in a significant increase in the cellulolytic activity of the *Caldicellulosiruptor bescii exoproteome*. *Journal of Industrial Microbiology & Biotechnology* 46 (5): 687–695.

Klein-Marcuschamer, D., Oleskowicz-Popiel, P., Simmons, B.A., and Blanch, H.W. (2012). The challenge of enzyme cost in the production of lignocellulosic biofuels. *Biotechnology and Bioengineering* 109 (4): 1083–1087.

Klippel, B. and Antranikian, G. (2011). Lignocellulose Converting Enzymes from Thermophiles. In: Extremophiles Handbook (ed. K. Horikoshi, G. Antranikian, A.T. Bull, F.T. Robb and K.O. Stetter), Vol. 1, 443–474. Springer.

Klippel, B., Sahm, K., Basner, A., Wiebusch, S., John, P., Lorenz, U., Peters, A., Abe, F., Takahashi, K., Kaiser, O., Goesmann, A., Jaenicke, S., Grote, R., Horikoshi, K., and Antranikian, G. (2014). Carbohydrate-active enzymes identified by metagenomic analysis of deep-sea sediment bacteria. *Extremophiles* 18 (5): 853–863.

Ko, K.C., Han, Y., Cheong, D.E., Choi, J.H., and Song, J.J. (2013). Strategy for screening metagenomic resources for exocellulase activity using a robotic, high-throughput screening system. *Journal of Microbiological Methods* 94 (3): 311–316.

Kucharska, K., Rybarczyk, P., Hołowacz, I., Łukajtis, R., Glinka, M., and Kamiński, M. (2018). Pretreatment of lignocellulosic materials as substrates for fermentation processes. *Molecules* 23 (11): 2937.

Kudanga, T. and Le Roes-Hill, M. (2014). Laccase applications in biofuels production: current status and future prospects. *Applied Microbiology and Biotechnology* 98 (15): 6525–6542.

Kulkarni, S.J., Shinde, N.L., and Goswami, A.K. (2015). A review on ethanol production from agricultural waste raw material. *International Journal of Scientific Research in Science, Engineering and Technology* 1: 231–233.

Kumar, B., Bhardwaj, N., Alam, A., Agrawal, K., Prasad, H., and Verma, P. (2018a). Production, purification and characterization of an acid/alkali and thermo tolerant cellulase from *Schizophyllum* commune NAIMCC-F-03379 and its application in hydrolysis of lignocellulosic wastes. *AMB Express* 8 (1): 1–16.

Kumar, V.P., Kolte, A.P., Dhali, A., Naik, C., and Sridhar, M. (2018b). Enhanced delignification of lignocellulosic substrates by *Pichia* GS115 expressed recombinant laccase. *The Journal of General and Applied Microbiology* 64 (4): 180–189.

Lam, M.Q., Oates, N.C., Thevarajoo, S., Tokiman, L., Goh, K.M., McQueen-Mason, S.J., Bruce, N.C., and Chong, C.S. (2020). Genomic analysis of a lignocellulose degrading strain from the underexplored genus *Meridianimaribacter. Genomics* 112 (1): 952–960.

Larsbrink, J., Tuveng, T.R., Pope, P.B., Bulone, V., Eijsink, V.G., Brumer, H., and McKee, L.S. (2017). Proteomic data on enzyme secretion and activity in the bacterium *Chitinophaga pinensis. Data in Brief* 11: 484–490.

Lee, J.P., Lee, H.W., Na, H.B., Lee, J.H., Hong, Y.J., Jeon, J.M., EunJuKwon, S.K.K., and Kim, H. (2018). Characterization of truncated endo-β-1,4-glucanases from a compost metagenomic library and their saccharification potentials. *International Journal of Biological Macromolecules* 115: 554–562.

Lee, W.C. and Kuan, W.C. (2015). Miscanthus as cellulosic biomass for bioethanol production. *Biotechnology Journal* 10 (6): 840–854.

Levasseur, A., Drula, E., Lombard, V., Coutinho, P.M., and Henrissat, B. (2013). Expansion of the enzymatic repertoire of the CAZy database to integrate auxiliary redox enzymes. *Biotechnology for Biofuels* 6 (1): 1–14.

Lewandowski, I. (2017). Biomass production from lignocellulosic energy crops. In: Encyclopedia of Applied Plant Sciences (ed. B. Thomas, B.G. Murray and D.J. Murphy), 2e, 159–163. Amsterdam, Netherlands: Elsevier, ISBN 978-0-12-394808-3.

Lewin, A., Zhou, J., Pham, V.T.T., Haugen, T., El Zeiny, M., Aarstad, O., Liebl, W., Wentzel, A., and Liles, M.R. (2017). Novel archaeal thermostable cellulases from an oil reservoir metagenome. *AMB Express* 7 (1): 1–17.

Li, X., Tu, T., Yao, B., Xie, X., and Luo, H. (2018). A novel bifunctional xylanase/cellulase TcXyn10A from *Thermoascus crustaceus* JCM12803. *Sheng wu gong cheng xue bao. Chinese Journal of Biotechnology* 34 (12): 1996–2006.

Li, X. and Yu, H.Y. (2013). Halostable cellulase with organic solvent tolerance from *Haloarcula sp.* LLSG7 and its application in bioethanol fermentation using agricultural wastes. *Journal of Industrial Microbiology & Biotechnology* 40 (12): 1357–1365.

Liu, C., Zhang, W., Qu, M., Pan, K., and Zhao, X. (2020). Heterologous expression of laccase from Lentinula edodes in Pichia pastoris and its application in degrading rape straw. *Frontiers in Microbiology* 11: 1086.

Lombard, V., Golaconda Ramulu, H., Drula, E., Coutinho, P.M., and Henrissat, B. (2014). The carbohydrate-active enzymes database (CAZy) in 2013. *Nucleic Acids Research* 42 (1): 490–495.

Lu, C., Wang, H., Luo, Y., and Guo, L. (2010). An efficient system for pre-delignification of gramineous biofuel feedstock in vitro: application of a laccase from *Pycnoporus sanguineus* H275. *Process Biochemistry* 45 (7): 1141–1147.

Majeke, B.M., Collard, F.X., Tyhoda, L., and Görgens, J.F. (2020). The synergistic application of quinone reductase and lignin peroxidase for the deconstruction of industrial (technical) lignins and analysis of the degraded lignin products. *Bioresource Technology* 319: 124152.

Milshteyn, A., Schneider, J.S., and Brady, S.F. (2014). Mining the metabiome: identifying novel natural products from microbial communities. *Chemistry & Biology* 21 (9): 1211–1223.

Montella, S., Amore, A., and Faraco, V. (2016). Metagenomics for the development of new biocatalysts to advance lignocellulose saccharification for bioeconomic development. *Critical Reviews in Biotechnology* 36 (6): 998–1009.

Moreno, A.D., Ibarra, D., Ballesteros, I., Fernández, J.L., and Ballesteros, M. (2013). Ethanol from laccase-detoxified lignocellulose by the thermotolerant yeast *Kluyveromyces marxianus* effects of steam pretreatment conditions, process configurations and substrate loadings. *Biochemical Engineering Journal* 79: 94–103.

Naili, F., Neifar, M., Elhidri, D., Cherif, H., Bejaoui, B., Aroua, M., Bejaoui, Z., Abassi, M., Mguiz, K., Chouchane, H., Ouzari, H.I., and Cherif, A. (2018). Optimization of the effect of PGPR–based biofertlizer on wheat growth and yield. *Biometrics and Biostatistics International Journal* 7 (3): 226–232.

Neifar, M., Chouchane, H., Belkacem, N., Jaouani, A., and Cherif, A. (2016). Ligninases and glycosyl hydrolases in developing a cellulosic bioethanol industry. *Journal of Advanced Microbiology* 2 (5): 234–248.

Neifar, M., Chouchane, H., Jaouani, A., Masmoudi, A.S., and Cherif, A. (2015a). Extremozymes as efficient green biocatalysts in bioremediation of industrial wastewaters. In: Wastewater Treatment: Processes, Management Strategies and Envirenmental/Health Impacts (ed. L.M. Barett), 191–213. New York: Nova Science Publishers Inc.

Neifar, M., Chouchane, H., Masmoudi, A.S., Jaouani, A., and Cherif, A. (2017). Biotechnological Potential of Extremophilic Laccases. Applications, Investigations and Insights, 125–142. New York: Nova Science Publishers.

Neifar, M., Chouchane, H., Najjari, A., El Hidri, D., Mahjoubi, M., Ghedira, K., Naili, F., Soufi, L., Raddadi, N., Sghaier, H., Ouzari, H.I., Masmoudi, A.S., and Cherif, A. (2019). Genome analysis provides insights into crude oil degradation and biosurfactant production by extremely halotolerant *Halomonas desertis* G11 isolated from Chott El-Djerid salt-lake in Tunisian desert. *Genomics* 111 (6): 1802–1514.

Neifar, M., Maktouf, S., Ghorbel, R.E., Jaouani, A., and Cherif, A. (2015b). Extremophiles as source of novel bioactive compounds with industrial potential. In: Biotechnology of Bioactive Compounds: Sources and Applications (ed. V.K. Gupta and M.G. Tuohy), 245–268. Hoboken: Wiley.

Nguyen, S.T.C., Freund, H.L., Kasanjian, J., and Berlemont, R. (2018). Function, distribution, and annotation of characterized cellulases, xylanases, and chitinases from CAZy. *Applied Microbiology and Biotechnology* 102 (4): 1629–1637.

Olson, D.G., Mcbride, J.E., Shaw, A.J., and Lynd, L.R. (2012). Recent progress in consolidated bioprocessing. *Current Opinion in Biotechnology* 23: 396–405.

Oropeza-De la Rosa, E., López-Ávila, L.G., Luna-Solano, G., and Cantú-Lozano, D. (2017). Bioethanol production process rheology. *Industrial Crops and Products* 106: 59–64.

Palavesam, A. (2015). Investigation on lignocellulosic saccharification and characterization of haloalkaline solvent tolerant endo-1, 4 β-d-xylanase from *Halomonas meridiana* APCMST-KS4. *Biocatalysis and Agricultural Biotechnology* 4 (4): 761–766.

Park, S., Lee, B., and Park, K. (2017). Extremophilic carbohydrate active enzymes (CAZymes). *Journal of Nutritional Health & Food Engineering* 7 (1): 00230.

Park, Y.J., Jeong, Y.U., and Kong, W.S. (2018). Genome sequencing and carbohydrate-active enzyme (CAZyme) repertoire of the white rot fungus *Flammulina elastica*. *International Journal of Molecular Sciences* 19 (8): 2379.

Patel, A.K., Singhania, R.R., Sim, S.J., and Pandey, A. (2019). Thermostable cellulases: current status and perspectives. *Bioresource Technology* 279: 385–392.

Persson, H., Han, T., Sandström, L., Xia, W., Evangelopoulos, P., and Yang, W. (2018). Fractionation of liquid products from pyrolysis of lignocellulosic biomass by stepwise thermal treatment. *Energy* 154: 346–351.

Phadtare, P., Joshi, S., and Satyanarayana, T. (2017). Recombinant thermo-alkali-stable endoglucanase of *Myceliopthora thermophila* BJA (rMt-egl): biochemical characteristics and applicability in enzymatic saccharification of agro-residues. *International Journal of Biological Macromolecules* 104: 107–116.

Pikuta, E.V., Hoover, R.B., and Tang, J. (2007). Microbial extremophiles at the limits of life. *Critical Reviews in Microbiology* 33 (3): 183–209.

Pilavtepe, M., Celiktas, M.S., Sargin, S., and Yesil-Celiktas, O. (2013). Transformation of *Posidonia oceanica* residues to bioethanol. *Industrial Crops and Products* 51: 348–354.

Poontawee, R., Yongmanitchai, W., and Limtong, S. (2017). Efficient oleaginous yeasts for lipid production from lignocellulosic sugars and effects of lignocellulose degradation compounds on growth and lipid production. *Process Biochemistry* 53: 44–60.

Potprommanee, L., Wang, X.Q., Han, Y.J., Nyobe, D., Peng, Y.P., Huang, Q., Liu, J.Y., Liao, Y.L., and Chang, K.L. (2017). Characterization of a thermophilic cellulase from *Geobacillus sp.* HTA426, an efficient cellulase-producer on alkali pretreated of lignocellulosic biomass. *PLoS One* 12 (4): e0175004.

Prakash, H., Chauhan, P.S., General, T., and Sharma, A.K. (2018). Development of eco-friendly process for the production of bioethanol from banana peel using inhouse developed cocktail of thermo-alkali-stable depolymerizing enzymes. *Bioprocess and Biosystems Engineering* 41 (7): 1003–1016.

Qin, Y., Li, Q., Luo, F., Fu, Y., and He, H. (2020). One-step purification of two novel thermotolerant β-1, 4-glucosidases from a newly isolated strain of *Fusarium chlamydosporum* HML278 and their characterization. *AMB Express* 10 (1): 1–11.

Qurat-ul-Ain, A., Yang, S.T., Manzoor, M., and Qazi, J.I. (2017). Moderate alkali-thermophilic ethanologenesis by locally isolated *Bacillus licheniformis* from Pakistan employing sugarcane bagasse: a comparative aspect of aseptic and non-aseptic fermentations. *Biotechnology for Biofuels* 10 (1): 105.

Raddadi, N., Cherif, A., Daffonchio, D., Neifar, M., and Fava, F. (2015). Biotechnological applications of extremophiles, extremozymes and extremolytes. *Applied Microbiology and Biotechnology* 99: 7907–7913.

Raheja, Y., Kaur, B., Falco, M., Tsang, A., and Chadha, B.S. (2020). Secretome analysis of *Talaromyces emersonii* reveals distinct CAZymes profile and enhanced cellulase production through response surface methodology. *Industrial Crops and Products* 152: 112554.

Rai, R., Bibra, M., Chadha, B.S., and Sani, R.K. (2019). Enhanced hydrolysis of lignocellulosic biomass with doping of a highly thermostable recombinant laccase. *International Journal of Biological Macromolecules* 137: 232–237.

Rajak, R.C. and Banerjee, R. (2018). An eco-friendly process integration for second generation bioethanol production from laccase delignified Kans grass. *Energy Conversion and Management* 157: 364–371.

Rajeswari, G. and Jacob, S. (2020a). Deciphering the aloe vera leaf rind as potent feedstock for bioethanol through enzymatic delignification and its enhanced saccharification. *Industrial Crops and Products* 143: 111876.

Rajeswari, G. and Jacob, S. (2020b). Saccharolysis of laccase delignified Aloe vera leaf rind and fermentation through free and immobilized yeast for ethanol production. *Journal of Food Process Engineering*. doi: 10.1111/jfpe.13514.

Ransom, C., Balan, V., Biswas, G., Dale, B., Crockett, E., and Sticklen, M. (2007). Heterologous Acidothermus cellulolyticus 1,4-beta-endoglucanase E1 produced within the corn biomass converts corn stover into glucose. *Applied Biochemistry and Biotechnology* 137–140 (1–12): 207–219.

Rattu, G., Joshi, S., and Satyanarayana, T. (2016). Bifunctional recombinant cellulase–xylanase (rBhcell-xyl) from the polyextremophilic bacterium *Bacillus halodurans* TSLV1 and its utility in valorization of renewable agro-residues. *Extremophiles* 20 (6): 831–842.

Raza, A., Bashir, S., and Tabassum, R. (2019). Statistical based experimental optimization for co-production of endo-glucanase and xylanase from *Bacillus sonorensis* BD92 with their application in biomass saccharification. *Folia Microbiologica* 64 (3): 295–305.

Rezaei, S., Shahverdi, A.R., and Faramarzi, M.A. (2017). Isolation, one-step affinity purification, and characterization of a polyextremotolerant laccase from the halophilic bacterium *Aquisalibacillus elongatus* and its application in the delignification of sugar beet pulp. *Bioresource Technology* 230: 67–75.

Riahi, L., Cherif, H., Miladi, S., Neifar, M., Bejaoui, B., Chouchane, H., Masmoudi, A.S., and Cherif, A. (2020). Use of plant growth promoting bacteria as an efficient biotechnological tool to enhance the biomass and secondary metabolites production of the industrial crop *Pelargonium graveolens* L'Hér. under semi-controlled conditions. *Industrial Crops and Products* 154: 112721.

Rico, A., Rencoret, J., del Río, J.C., Martínez, A.T., and Gutiérrez, A. (2014). Pretreatment with laccase and a phenolic mediator degrades lignin and enhances saccharification of Eucalyptus feedstock. *Biotechnology for Biofuels* 7: 1–14.

Robak, K. and Balcerek, M. (2018). Review of second generation bioethanol production from residual biomass. *Food Technology and Biotechnology* 56 (2): 174–187.

Rosnow, J.J., Anderson, L.N., Nair, R.N., Baker, E.S., and Wright, A.T. (2017). Profiling microbial lignocellulose degradation and utilization by emergent omics technologies. *Critical Reviews in Biotechnology* 37 (5): 626–640.

Rubio, M.V., Terrasan, C.R.F., Contesini, F.J., Zubieta, M.P., Gerhardt, J.A., Oliveira, L.C., Gonçalves, A.E. de S.S., Almeida, F., Smith, B.J., de Souza, G.H.M.F., Dias, A.H.S., Skaf, M., and Damasio, A. (2019). Redesigning N-glycosylation sites in a GH3 β-xylosidase improves the enzymatic efficiency. *Biotechnology for Biofuels* 12 (1): 1–14.

Sahoo, K., Sahoo, R.K., Gaur, M., and Subudhi, E. (2020). Cellulolytic thermophilic microorganisms in white biotechnology: a review. *Folia Microbiologica* 65 (1): 25–43.

Saini, J.K., Saini, R., and Tewari, L. (2015). Lignocellulosic agriculture wastes as biomass feedstocks for second-generation bioethanol production: concepts and recent developments. *3 Biotechnology* 5 (4): 337–353.

Sarmiento, F., Peralta, R., and Blamey, J.M. (2015). Cold and hot extremozymes: industrial relevance and current trends. *Frontiers in Bioengineering and Biotechnology* 3: 148.

Schneider, W.D.H., Fontana, R.C., Baudel, H.M., de Siqueira, F.G., Rencoret, J., Gutiérrez, A., de Eugenio, L.I., Prieto, A., Martínez, M.J., Martínez, Á.T., Dillon, A.J.P., and Camassola, M. (2020). Lignin degradation and detoxification of *eucalyptus* wastes by on-site manufacturing fungal enzymes to enhance second-generation ethanol yield. *Applied Energy* 262: 114493.

Sherpa, K.C., Ghangrekar, M.M., and Banerjee, R. (2018). A green and sustainable approach on statistical optimization of laccase mediated delignification of sugarcane tops for enhanced saccharification. *Journal of Environmental Management* 217: 700–709.

Shi, X., Liu, Q., Ma, J., Liao, H., Xiong, X., Zhang, K., Wang, T., Liu, X., Xu, T., Yuan, S., Zhang, X., and Zhu, Y. (2015). An acid-stable bacterial laccase identified from the endophyte *Pantoea ananatis* Sd-1 genome exhibiting lignin degradation and dye decolorization abilities. *Biotechnology Letters* 37 (11): 2279–2288.

Shukor, H., Abdeshahian, P., Al-Shorgani, N.K., Hamid, A.A., Rahman, N.A., and Kalil, M.S. (2016). Enhanced mannan-derived fermentable sugars of palm kernel cake by mannanase-catalyzed hydrolysis for production of biobutanol. *Bioresource Technology* 218: 257–264.

Silveira, M.H.L., Vanelli, B.A., and Chandel, A.K. (2018). Second generation ethanol production: potential biomass feedstock, biomass deconstruction, and chemical platforms for process valorization. In: Advances in Sugarcane Biorefinery (ed. A.K. Chandel and M.H.L. Silveira), 135–152. Elsevier.

Singh, R.S., Singh, T., and Pandey, A. (2019). Chapitre 1: microbial enzymes—an overview. *Advances in Enzyme Technology* 43: 1018–1091. doi: 10.1016/B978-0-444-64114-4.00001-7.

Singh, B. (2020). Enhanced production of bacterial xylanase and its utility in saccharification of sugarcane bagasse. *Bioprocess and Biosystems Engineering* 1–11.

Sitarz, A.K., Mikkelsen, J.D., Højrup, P., and Meyer, A.S. (2013). Identification of a laccase from *Ganoderma lucidum* CBS 229.93 having potential for enhancing cellulase catalyzed lignocellulose degradation. *Enzyme and Microbial Technology* 53 (6–7): 378–385.

Sondhi, S., Sharma, P., George, N., Chauhan, P.S., Puri, N., and Gupta, N. (2015). An extracellular thermo-alkali-stable laccase from *Bacillus tequilensis* SN4, with a potential to biobleach softwood pulp. *3 Biotechnology* 5 (2): 175–185.

Souii, A., Guesmi, A., Ouertani, R., Cherif, H., Chouchane, H., Cherif, A., and Neifar, M. (2020). Carboxymethyl cellulase production by extremotolerant bacteria in low-cost media and application in enzymatic saccharification of Stevia biomass. *Waste and Biomass Valorization* 11: 2111–2122.

Spertino, S., Boatti, L., Icardi, S., Manfredi, M., Cattaneo, C., Marengo, E., and Cavaletto, M. (2018). *Cellulomonas fimi secretomes*: in vivo and in silico approaches for the lignocellulose bioconversion. *Journal of Biotechnology* 270: 21–29.

Summers, S., Kraft, C., Alamdari, S., Pfaendtner, J., and Kaar, J.L. (2020). Enhanced activity and stability of *Acidothermus cellulolyticus* endoglucanase 1 in ionic liquids via engineering active site residues and non-native disulfide bridges. *ACS Sustainable Chemistry & Engineering* 8 (30): 11299–11307.

Sunna, A., Gibbs, M.D., Chin, C.W., Nelson, P.J., and Bergquist, P.L. (2000). A gene encoding a novel multidomain β-1, 4-mannanase from *Caldibacillus cellulovorans* and action of the recombinant enzyme on kraft pulp. *Applied and Environmental Microbiology* 66 (2): 664–670.

Tan, I.S. and Lee, K.T. (2014). Enzymatic hydrolysis and fermentation of seaweed solid wastes for bioethanol production: an optimization study. *Energy* 78: 53–62.

Thapa, S., Mishra, J., Arora, N., Mishra, P., Li, H., OHair, J., Bhatti, S., and Zhou, S. (2020). Microbial cellulolytic enzymes: diversity and biotechnology with reference to lignocellulosic biomass degradation. *Reviews in Environmental Science and Biotechnology*. doi: 10.1007/s11157-020-09536-y.

Tingley, J.P., Low, K.E., Xing, X., and Abbott, D.W. (2021). Combined whole cell wall analysis and streamlined in silico carbohydrate-active enzyme discovery to improve biocatalytic conversion of agricultural crop residues. *Biotechnology for Biofuels* 14 (1): 1–19.

Tiwari, R., Singh, P.K., Singh, S., Nain, P.K., Nain, L., and Shukla, P. (2017). Bioprospecting of novel thermostable β-glucosidase from *Bacillus subtilis* RA10 and its application in biomass hydrolysis. *Biotechnology for Biofuels* 10 (1): 1–18.

Trivedi, N., Reddy, C.R.K., Radulovich, R., and Jha, B. (2015). Solid state fermentation (SSF)-derived cellulase for saccharification of the green seaweed Ulva for bioethanol production. *Algal Research* 9: 48–54.

Tsigie, Y.A., Wang, C.Y., Truong, C.T., and Ju, Y.H. (2011). Lipid production from *Yarrowia lipolytica* Po1g grown in sugarcane bagasse hydrolysate. *Bioresource Technology* 102: 9216–9222.

Tuveng, T.R., Eijsink, V.G., and Arntzen, M.Ø. (2019). Proteomic detection of carbohydrate-active enzymes (CAZymes) in microbial secretomes. *Functional Proteomics* 1871: 159–177.

Tye, Y.Y., Lee, K.T., Abdullah, W.N.W., and Leh, C.P. (2016). The world availability of non-wood lignocellulosic biomass for the production of cellulosic ethanol and potential pretreatments for the enhancement of enzymatic saccharification. *Renewable and Sustainable Energy Reviews* 60: 155–172.

Vaid, S. and Bajaj, B.K. (2017). Production of ionic liquid tolerant cellulase from *Bacillus subtilis* G 2 using agroindustrial residues with application potential for saccharification of biomass under one pot consolidated bioprocess. *Waste and Biomass Valorization* 8 (3): 949–964.

Van Soest, P.J. (1967). Development of a comprehensive system of feed analyses and its application to forages. *Journal of Animal Science* 26 (1): 119–128.

Van Soest, P.V., Robertson, J.B., and Lewis, B. (1991). Methods for dietary fiber, neutral detergent fiber, and nonstarch polysaccharides in relation to animal nutrition. *Journal of Dairy Science* 74 (10): 3583–3597.

Verma, D. and Satyanarayana, T. (2020). Xylanolytic extremozymes retrieved from environmental metagenomes: characteristics, genetic engineering, and applications. *Frontiers in Microbiology* 11: 551109.

Viikari, L., Alapuranen, M., Puranen, T., Vehmaanperä, J., and Siika-Aho, M. (2007). Thermostable enzymes in lignocellulose hydrolysis. *Biofuels* 108: 121–145.

Wan, C. and Li, Y. (2010). Microbial pretreatment of corn stover with *Ceriporiopsis subvermispora* for enzymatic hydrolysis and ethanol production. *Bioresource Technology* 101 (16): 6398–6403.

Wang, M., Li, Z., Fang, X., Wang, L., and Qu, Y. (2012). Cellulolytic enzyme production and enzymatic hydrolysis for second-generation bioethanol production. *Biotechnology in China III: Biofuels and Bioenergy* 128: 1–24.

Wang, F.Q., Xie, H., Chen, W., Wang, E.T., Du, F.G., and Song, A.D. (2013a). Biological pretreatment of corn stover with ligninolytic enzyme for high efficient enzymatic hydrolysis. *Bioresource Technology* 144: 572–578.

Wang, H.C., Chen, Y.C., Huang, C.T., and Hseu, R.S. (2013b). Cloning and characterization of a thermostable and pH-stable cellobiohydrolase from Neocallimastix patriciarum J11. *Protein Expression and Purification* 90 (2): 153–159.

Wang, Y.B., Gao, C., Zheng, Z., Liu, F.M., Zang, J.Y., and Miao, J.L. (2015). Immobilization of cold-active cellulase from antarctic bacterium and its use for kelp cellulose ethanol fermentation. *BioResources* 10 (1): 1757–1772.

Wang, H., Hart, D.J., and An, Y. (2019a). Functional metagenomic technologies for the discovery of novel enzymes for biomass degradation and biofuel production. *BioEnergy Research* 12 (3): 457–470.

Wang, Z., Zheng, F., and Xue, S. (2019b). The economic feasibility of the valorization of water hyacinth for bioethanol production. *Sustainability* 11 (3): 905.

Weimann, A., Trukhina, Y., Pope, P.B., Konietzny, S.G., and McHardy, A.C. (2013). De novo prediction of the genomic components and capabilities for microbial plant biomass degradation from (meta-) genomes. *Biotechnology for Biofuels* 6 (1): 1–13.

Wilkens, C., Andersen, S., Dumon, C., Berrin, J.-G., and Svensson, B. (2017). GH62 arabinofuranosidases: structure, function and applications. *Biotechnology Advances* 35: 792–804.

Wilkins, M.R., Gasteiger, E., Bairoch, A., Sanchez, J.C., Williams, K.L, Appel, R.D., Hochstrasser, D.F. (1999). Protein identification and analysis tools on the ExPASy server. *Methods in Molecular Biology* 112: 531–552.

Wu, W., Davis, R.W., Tran-Gyamfi, M.B., Kuo, A., LaButti, K., Mihaltcheva, S., Hundley, H., Chovatia, M., Lindquist, E., Barry, K., Grigoriev, I.V., Henrissat, B., and Gladden, J.M. (2017). Characterization of four endophytic fungi as potential consolidated bioprocessing hosts for conversion of lignocellulose into advanced biofuels. *Applied Microbiology and Biotechnology* 101 (6): 2603–2618.

Wu, W.H., Hung, W.C., Lo, K.Y., Chen, Y.H., Wan, H.P., and Cheng, K.C. (2016). Bioethanol production from taro waste using thermo-tolerant yeast *Kluyveromyces marxianus* K21. *Bioresource Technology* 201: 27–32.

Wushke, S., Levin, D.B., Cicek, N., and Sparling, R. (2015). Facultative anaerobe *Caldibacillus debilis* GB1: characterization and use in a designed aerotolerant, cellulose-degrading coculture with *Clostridium thermocellum*. *Applied and Environmental Microbiology* 81 (16): 5567–5573.

Xia, W., Bai, Y., Cui, Y., Xu, X., Qian, L., Shi, P., Zhang, W., Luo, H., Zhan, X., and Yao, B. (2016). Functional diversity of family 3 β-glucosidases from thermophilic cellulolytic fungus *Humicola insolens* Y1. *Scientific Reports* 6 (1): 1–11.

Xiao, C. and Anderson, C.T. (2013). Roles of pectin in biomass yield and processing for biofuels. *Frontiers in Plant Science* 4: 67.

Yang, C., Xia, Y., Qu, H., Li, A.D., Liu, R., Wang, Y., and Zhang, T. (2016). Discovery of new cellulases from the metagenome by a metagenomics-guided strategy. *Biotechnolology for Biofuels* 9 (1): 138.

Yang, X., Gu, C., and Lin, Y. (2020). A novel fungal laccase from *Sordaria macrospora k-hell*: expression, characterization, and application for lignin degradation. *Bioprocess and Biosystems Engineering* 43: 1–7.

Yang, Y., Song, W.Y., Hur, H.G., Kim, T.Y., and Ghatge, S. (2019). Thermoalkaliphilic laccase treatment for enhanced production of high-value benzaldehyde chemicals from lignin. *International Journal of Biological Macromolecules* 124: 200–208.

Yin, Y., Mao, X., Yang, J., Chen, X., Mao, F., and Xu, Y. (2012). dbCAN: a web resource for automated carbohydrate-active enzyme annotation. *Nucleic Acids Research* 40: 445–451.

Ying, Y., Meng, D., Chen, X., and Li, F. (2013). An extremely thermophilic anaerobic bacterium *Caldicellulosiruptor sp.* F32 exhibits distinctive properties in growth and xylanases during xylan hydrolysis. *Enzyme and Microbial Technology* 53 (3): 194–199.

Yu, H.W., Im, J.H., Kong, W.S., and Park, Y.J. (2021). Comparative analysis of carbohydrate active enzymes in the *Flammulina velutipes var. lupinicola* genome. *Microorganism* 9 (1): 20.

Yue, D., You, F., and Snyder, S.W. (2014). Biomass-to-bioenergy and biofuel supply chain optimization: overview, key issues and challenges. *Computers & Chemical Engineering* 66: 36–56.

Zaafouri, K., Ziadi, M., ben Hassen-Trabelsi, A., Mekni, S., Aïssi, B., Alaya, M., and Hamdi, M. (2017). Enzymatic saccharification and liquid state fermentation of hydrothermal pretreated Tunisian *Luffa cylindrica* (L.) fibers for cellulosic bioethanol production. *Renewable Energy* 114: 1209–1213.

Zabed, H., Sahu, J.N., Boyce, A.N., and Faruq, G. (2016). Fuel ethanol production from lignocellulosic biomass: an overview on feedstocks and technological approaches. *Renewable and Sustainable Energy Reviews* 66: 751–774.

Zambare, V.P., Bhalla, A., Muthukumarappan, K., Sani, R.K., and Christopher, L.P. (2011). Bioprocessing of agricultural residues to ethanol utilizing a cellulolytic extremophile. *Extremophiles* 15: 611.

Zerillo, M.M., Adhikari, B.N., Hamilton, J.P., Buell, C.R., Lévesque, C.A., and Tisserat, N. (2013). Carbohydrate-active enzymes in Pythium and their role in plant cell wall and storage polysaccharide degradation. *PLoS One* 8 (9): e72572.

Zhang, H., Yohe, T., Huang, L., Entwistle, S., Wu, P., Yang, Z., Busk, P.K., Xu, Y., and Yin, Y. (2018). dbCAN2: a meta server for automated carbohydrate-active enzyme annotation. *Nucleic Acids Research* 46 (1): 95–101.

Zhang, Y.P., Sun, J., and Zhong, J.J. (2010). Biofuel production by in vitro synthetic enzymatic pathway biotransformation. *Current Opinion in Biotechnology* 21 (5): 663–669.

Zheng, Y., Maruoka, M., Nanatani, K., Hidaka, M., Abe, N., Kaneko, J., Sakai, Y., Abe, K., Yokota, A., and Yabe, S. (2021). High cellulolytic potential of the Ktedonobacteria lineage revealed by genome-wide analysis of CAZymes. *Journal of Bioscience and Bioengineering* 3: 622–630, S1389-1723(21)00025-6.

Zhu, D., Adebisi, W.A., Ahmad, F., Sethupathy, S., Danso, B., and Sun, J. (2020). Recent development of extremophilic bacteria and their application in biorefinery. *Frontiers in Bioengineering and Biotechnology* 8: 483.

8

The Role of Divalent Cations in Antibiotic Sensitivity

A Molecular Aspect

Amro Abd Al Fattah Amara[1,2]

[1]*Protein Research Department, Genetic Engineering and Biotechnology Research Institute, City of Scientific Research and Technological Applications, Alexandria, Egypt*
[2]*Head of Scientific Publishing office, City of Scientific Research and Technological Applications, Alexandria, Egypt*
Corresponding author: amroamara@web.de

8.1 Introduction

It is important to understand the nature of the atoms, ions, metals, and the other described chemical compounds included in this chapter in more detail; it is recommended that a textbook is referred to to gain more information about their nature, biological effect, chemical reactions, catalytic effect, etc. However, information concerning the topic will be included. One of the most important parts of a living organism is metal. Apparently, early humans knew the effect of such metals and used them in different forms, like rings, in contact with their skin. They knew that such alloys transferred to their body some health benefits. The ancient Egyptians used lead to treat eye infections. Another well known example is the use of silver in China. Nowadays, such beliefs have proved to be correct, the body needs only a trace of such elements and the best way to transfer this (at least in ancient times) was through direct contact with the skin. Many treatments of this type are still secrets in some ethnic groups and are not yet completely understood. Some differentiated between silver and gold and prevented people from wearing gold or other kinds of rings. Some understood the many differences between different types of agate, different agates contain different types of elements and they believed that their contact with the skin had different effects on us. For example, they believed that some types of agate increased anxiety, and that others had a calming effect. They believed that wearing rings made of silver would protect from some diseases. It was a folk belief that Noble people are healthier due to their blue blood. Nowadays science has clarified that they were healthy because they were wearing a lot of silver objects. Some have linked the many incorrect decisions made by Roman Emperors to their wine drinking, but science has proved that the wine was stored in containers made of lead which interferes with the memory and thinking. Japanese, Indian and Arabic swords were stronger due to the type of iron used, the processing steps, include subjecting the sword to a source of carbon during their heating/cooling processes. Egyptians used

Biomolecules from Natural Sources: Advances and Applications, First Edition. Edited by Vijai Kumar Gupta, Satyajit D. Sarker, Minaxi Sharma, María Elida Pirovani, Zeba Usmani, and Chelliah Jayabaskaran.

carbon just because they knew it was a good solution. Nanoparticles of some metals were formed in ancient times due to a specific process when making glass or using clay with a different source of natural metals, this caused changes in their color due to the change in the light, a practice used nowadays in the new science investigating nanoparticle. Modern science has benefitted from early human knowledge and many of those unexplained experiences have become scientifically clear.

A metal is an element that has a characteristic luster, conducts heat and electricity well, and deforms without breaking when pounded. Cations are positively charged ions; an atom or a group of atoms that has lost one or more electrons. Divalent cations have two valences. This can be explained by a lower total coordination of the divalent atoms. The nomenclature of ionic compounds is based on some rules. Most cations are monoatomic ions formed by metal atoms. The name of the ion is then the same as the name of the metal (Na^+ is the sodium ion). Some other metals form more than one type of cation. Then the name of the ion is shown by its charge using a Roman numeral (Cu^+ is named copper (I) ion and Cu^{2+} is named copper (II) ion). Another style of nomenclature is by adding -ous to the name of the ion of the pair that has the lower positive charge (Cu^+ is named cuprous ion) and the ending -ic is used to name the ion that has the higher positive charge (Cu^{2+} is named cupric ion). A polyatomic ion is one that is formed from several atoms held together by covalent bonds ($NH_4{}^{2+}$ the ammonium ion). It is very important to highlight here that some cations can be changed from monovalent cation to divalent cations and *vice versa*. This is an important phenomenon particularly in biology and molecular biology (Cu^+ and Cu^{2+}). Ions, when changing from one to two positive charges enable electron loss and gain easily which is an important reaction in catalysis. Gaining or losing electrons will effect the targeted exchanger and might destroy it or convert it from one compound to another, however, the ion itself will not be affected.

Modern biological research to some extent neglects the role of ions in the cells and tissues during the experimental design and wrongly correlates the *in vitro* results to the *in vivo* ones. The differences between the *in vivo* and the *in vitro* application must be well clarified. Some experiments use *in vitro* distilled or bi-distilled-water or buffers did not represent the proper *in vivo* bio-physiological solution. For example, experiments that concluded the death of cancer cells because of the existence of a microbe (or a chelating agent) probably are wrong. That is because microbes will utilize the nutrient, ions and other essential elements essential for the survival of the cancer cells. This practice is hard to achieve in the human body while the blood will provide the cells with new cations.

This chapter covers five main elements which are:

1) The ions including the divalent cations.
2) The antibiotics particularly those affected by the presence of the divalent cations and the ionophor antibiotics.
3) The enzymes involved in the antibiotic resistance.
4) The macromolecules and their interaction with the ions.
5) The role of the divalent in the catalysis and in the antibiotic resistances' different mechanisms.

The nature of each element will be briefly described to help in understanding the topic. A general discussion of the different factors correlated to the divalent cations and the antibiotic resistance will be included. The main aim of this chapter is to clarify the role of ions in the biological system particularly their interaction with macromolecules. More description will be for the divalent cations. The relationship between the divalent cation and the antibiotic resistance through different related mechanisms will be described. Some case studies concerning the role of the divalent cations in the resistance and the sensitivity of the bacteria during the treatment with the antibiotics will be addressed.

8.2 The Elements of the Macromolecules

All of the biological system molecules and macromolecules are built from a number of elements. However, oxygen, hydrogen, carbon and nitrogen are more representative than the other atoms. Their atoms bond with one another by chemical bonds through the interactions of the electrons in their outer shells. Compounds are structures that are bigger in size than molecules. Macromolecules are even larger, usually in polymeric form, such as protein, DNA, or polyhydroxyalkanoate. The name of a molecule is given for two or more atoms chemically bonded together. The new molecules react as one unit, such as O_2. So, macromolecules could be composed of thousands of atoms. The rich structure of different molecules is connected together to form the macromolecules (polymers). Macromolecules in the shape of polymers are composed of repeated homogenous or heterogeneous units, which are named homo or hetero polymers, respectively.

8.2.1 Bonds and Forces

Because they are adapted to the biological system, special forms of chemical bonds are more common, such as covalent, ionic, and hydrogen bonds. Other natural forces exist such as van der Waal forces, which occur between two or more atoms existing in close proximity to one another (3–4 Å) (such as substrate and enzyme binding). Another force is the hydrophobic interaction, generated by the existence of the molecule or the macromolecules in water. Water-repelling molecules tend to cluster in an aqueous environment, leading to association of nonpolar parts. Such a phenomenon could be found in phospholipid membrane, protein (folding), and substrate/enzyme binding. Non-bond forces contribute an essential role in the biological system while they show the importance of water and its content of ions, charges, and molecules in the protein three-dimensional (3D) structure.

8.2.2 The Dynamicity of the Macromolecules

One representative example is the protein. Concerning its function, proteins are highly functional macromolecules responsible for catalyzing the biological biochemical reactions (Amara 2015a, 2019). They carry out all functions of an organism (Amara 2015a, 2019), the single-stranded nature of protein molecules, together with their composition of 20 (or more) different amino acid building blocks with different side chains and different physicochemical properties, allows them to fold into a massive number of different 3D folding forms. The correct structure and folding should provide binding pockets through which they can specifically interact. Because of the difference in their amino acid types and order within the backbone, each protein is designed to bind and to react with a specific substrate or a group of substrates. Protein's 3D structure enables it to act as an enzyme, active protein and/or functional protein. Protein cofactors and coenzymes are smaller molecules that can donate the protein with specific activities away from those associated with the polypeptide chain alone. One protein could exist in many 3D structures, particularly as a response to the environment. From the different types of macromolecules proteins are unique structures that have inconceivably different types of ability to carry out functions. In fact, they are responsible for nearly all the functions in our bodies. One should not neglect the external forces, which have an effect on the protein structure/function /specificity. In fact, every change that is able to affect the protein folding will have an effect on its function. More simply, for the protein structure function one could revise, for example, the effect of the different environmental, chemical factors and trace elements on a protein such as the alkaline proteases used with powder detergent. Moreover, to understand the function of a protein one should understand the function and the structure of their amino acid constituents as well as the function of the other macromolecules, ions, water chemistry, the ionic forces, the thermodynamic, etc. In fact, the protein is at the core of life science and connected with chemistry and physics. It is very dynamic and variable. Such a dynamic macromolecule could catalyze a reaction significantly faster than any chemical catalyst. The explanation of being better than chemical catalyst is based on its ability to carry out several steps of chemical reactions in order. For that it can carry out polymerization for the homo-polymer of thousands of subunits and produce a highly homogenized polymer.

8.2.3 Protein as Functional Unique Macromolecules

The biological system is rich with an endless number of molecules and macromolecules. Each has its own criteria. However, they can be classed into groups. Each group has similarities and differences. Proteins differ from one another mainly in their sequence of amino acids. Proteins are folded into different specific 3D structures that determine their activity. A protein contains at least one long polypeptide. Short polypeptides, containing less than 20–30 residues, are rarely considered to be proteins and are commonly called peptides, or sometimes oligopeptides. Even short peptide can be involved in catalysis (El-Baky et al. 2018a, 2021). Some short peptides prove to have a highly antimicrobial activities against virulence strains of Gram-positive and Gram-negative bacteria as well as antiviral and anticancer activities as reported by El-Baky (2021) and El-Baky et al. (2021). The individual amino acid residues are bonded together by peptide bonds and adjacent amino acid residues. The sequence of amino acid residues in a protein is defined by the sequence of its gene. In general, the genetic code specifies 20 standard amino acids; but in certain organisms the genetic code can include selenocysteine and in some archaea, pyrrolysine. The residues in a protein are often chemically modified by post-translational modification. Some proteins have non-peptide groups attached, which can be called prosthetic groups or cofactors. Proteins can also work together to reach a particular function, and they often associate to form stable protein complexes. Proteins can carry out the catalysis without external aid. Or they could coordinate with some metals or other residue of another subunit, similar/different protein, or even another macromolecule. The metal ions are not out of the equation. They are involved widely in the catalysis. Many of the reactions of enzymes need metal ions, however, others do not. One feature of the small molecules in each of the biological and designed chemical catalysts such as N_2, CH_4, NO, CO, H_2O_2, O_2 is that they are activated by metal ions. Such small molecules can hardly be activated except by electronic perturbation through metal ions. Metal ions are the most aggressive reagent, i.e. attacking groups, in the biochemical reactions. A unique feature of the metalloenzyme (as an example) is that the metal can be replaced often without loss of activity (Frausto de Silva and Williams 1993). While some proteins, mainly the enzymes, need at least one ion to carry out catalysis and some need more than one. Others are working as chelating agents. One example of proteins that are attracting great interest is lactoferrin particularly the camel lactoferrin which is an iron chelating protein (El-Baky 2018).

8.3 Examples of the Sensitivity of the Macromolecules

The enzyme as a catalyst is more active than the non-protein-based catalyst. Being fast does not mean being suitable for every case, in particular due to the response of the immune system. To avoid the use of a protein which might enter the blood during a pharmaceutical preparation it is wise to use the chemical catalyst represented in a formula that contains ions which might work alone or might activate the internal enzymes and proteins. In many cases, apparently, the problem is not in the protein as macromolecule but in the ions that activate this protein or support its 3D structure. The research concerning ions in biology will open opportunities for treatment where the ions will switch on/off the catalytic cavity of the protein or complement its function. Recently some chemical compounds have shown a universal approach for the removal of biological cells (the viruses were also removed) (Amara 2015b; Amara et al. 2013a, 2013b; Amro et al. 2014; El-Baky and Amara 2014).

As above the structure is responsible for the function. The huge structure compositions of the biological system enable a global dynamic and variation. Ions flourish and fortify the structure dynamic of the biological systems. However, the biological system and the ions interact with each other without degradation of the ions and the charges derived from the different energy supplied either biologically or from the surrounding environment or through the various chemical reactions re-enable a finished reaction to restart again (a pathway repeated by the same enzymes). Remarkably in many cases the energy is derived from the main steps of the process and even yields some energy that could support other reactions. A representative example is the pathway or the anaerobic metabolism of the carbohydrates. The metabolic flux, enzymes, metals, energy, structure/functions and others are all involved in processes that could be simple, such as in the case of "from a gene to a protein process", which in fact is not that simple because of the formation of the components of each of the responsible gene or the responsible protein. On the other hand, it could be a really complicated process such as building an antibiotic by the polyketides.

In general, the biological system is sensitive to the chemical structure, particularly the metals. Other forms could also be sensitive. For example, in the red blood cells (RBCs), the porphyrin ring which affects the different coordination of the globin protein could differentiate between O_2 and CO_2. This uncountable specificity proves the unique dynamicity of the biological structure as well as the proteins that carry them.

Different conditions affect on the RBCs function (where they gain O_2 and where they lose it). Minor changes in RBCs can be responsible for serious disease; sickle cell anemia is an example.

Another example of the role of changing molecular structure on function is that the analogues have a structure similar to certain compounds, with minor differences. Some analogues are more stable or even nondegradable. Substrates are degraded by enzymes. IPTG is a good example. Analogues can show us that our biological system even though highly specific, is governed by chemical and the physical laws. Similar compounds to a certain substrate could interfere with its related enzymes and/or wrongly activate certain biopathway(s). Such a phenomenon could be helpful in the research but might be harmful in real life. The biological system works mechanically, perfectly but blindly.

The monomeric backbone of biopolymers and the macromolecules is important. One should note that the monomeric structure builds the overall 3D structure, which is responsible for the function. Additionally, such biopolymers and macromolecules upon their degradation will be divided into their monomeric constituents. Protein monomeric structures in their simple or single form (e.g., amino acids) are also important. For example, food rich in cysteine is accepted differently by the cell membrane if compared with that rich in cystine. The cell membrane transports cysteine ten times faster than cystine. Our understanding of such properties will enable us to select the type of foods that will benefit us (Amara 2013).

Understanding the effect of the structure on the function will let us understand that improbable compounds could wrongly act as analogues and activate some pathways mistakenly. Another remarkable observation from the biological system is the proteases which did not digest themselves. They are well designed to digest other proteins but not to digest themselves. It might be their folding structure that is responsible for that. I personally believe that the protein as a catalytic macromolecule will continue to amaze us with new information which deepens our understanding of the role of their structure. Proinsuilin is not active until after being digested. Some known proteins have properties not discovered yet.

8.4 Unique Examples Concerning the Role of Divalent Cations

8.4.1 The Ion Channels

The ion channels could be defined as membrane proteins that function as pores, with the additional ability to distinguish between different types of ions (Hinrichsen 1993; Sarkadi 1983). The most famous type of channel is the Ca^{2+}-dependent-K^+ channels which belong to a class of potassium channels that increase the potassium permeability of a cell following a rise in intracellular calcium. In the case of calmodulin a ubiquitous calcium-binding protein, Ca^{2+} is preferable to Cd^{2+}, Sr^{2+}, Mn^{2+}(Hinrichsen 1993). The calcium-binding protein involved in many cellular functions, has a general selectivity sequence of Ca^{2+} > Cd^{2+}> Hg^{2+}> Sr^{2+} > Mn^{2+}> Zn^{2+} > Pb^{2+}> Co^{2+}> Mg^{2+}> Ni^{2+}> Ba^{2+}(Hinrichsen 1993). For some cases the order changes such as in the case of troponin C, a calcium-binding protein found in muscles, the sequence is Ca^{2+}, Cd^{2+}, Sr^{2+}, Pb^{2+}, Mn^{2+}>> Mg^{2+}, Ni^{2+}, Zn^{2+}, Co^{2+}, Ba^{2+}(Hinrichsen 1993). Chelating of other divalent cations will interfere with the different channels functions. The channels play a crucial role in transporting ions to the cells: ions activate both of the macro and the micro molecules of the cells.

8.4.2 Coenzymes

Many coenzymes are vitamins or compounds derived from vitamins. Some enzymes require the presence of metal ions (such as Co^{2+}, Fe^{2+}, Zn^{2+}, Cu^{2+}, and Mg^{2+}) to function. The action of enzymes can be inhibited, or retarded, in several ways. An inhibitor is a substance that combines with an enzyme at the active site and prevents the enzyme from functioning. Divalent cations are also involved in the control of signal transduction by protein kinases. Mg^{2+}, Zn^{2+}, and Mn^{2+} assist in substrate binding and phosphoryl transfer by protein kinase A (Jaffe and Cohn 1978). NMR studies with Cd(II) and Zn(II) chelates of adenosine 5'-O-(3-thiotriphosphate) (ATPgammaS) and the Cd(II) chelate of adenosine 5'-O-(2-thiotriphosphate) (ATPbetaS) indicate that these metal ions chelate to the sulfur atom of the thiophosphate group. Since Mg(II) chelation to oxygen of the thiophosphate group of diastereoisomer is equivalent to the configuration of the Cd(II) chelation of the opposite diastereoisomer, an inversion of the stereospecificity is observed when Cd(II) substitutes Mg(II) in the phosphoryl transfer reactions (yeast hexokinase and rabbit muscle pyruvate kinase catalyzed the reaction). When Co(II) is the activating ion for yeast hexokinase with ATPbetaS as substrate, no stereospecificity is observed. The absolute stereochemistry of the Mg(II) complex of the B isomer of ATPbetaS is found out by its stereospecificity in the hexokinase reaction (Jaffe and Cohn 1978).

8.4.3 Antibiotics as Ionophores

A variety of antibiotics are able to form lipid-soluble complexes that provide a vehicle for inorganic cations to traverse lipid barriers and cell membranes. This characteristic ion-bearing property led to their being named ionophores (Pressman et al. 1967). Four groups of macrocyclicionophore antibiotics were recognized: cyclodepsipeptides, polypeptides, macrotetrolides and polyethers. They are widely used in studies of biological transport systems. These antibiotics differ from the surfactant-type antibiotics, such as the polymyxins or polyenes, which cause general loss of cell contents by disruption of membranes. Ionophores are sometimes referred to as "complexones" (Ovchinnikov et al. 1974). It was shown by Pressman (1965) and other workers that the valinomycin structure has several characteristic properties. Ionophore functioning as a molecule of valinomycin must meet the following requirements: the ability to form enough stable K^+ complexes; the ability to partition between the surface and the internal zone of the membrane; enough lipophilicity of the K^+ complex for it to be able to pass through the internal zone of the membrane; and enough rapid exchange of the potassium ion between two ionophore molecules (Pressman 1965). Interestingly, Monensin forms stable complexes with many monovalent cations, making these ions soluble in ordinary organic solvents. The ion specificity is Ag^+ > Na^+ > K^+ > Rb^+ > Ca^+ > Li^+ > NH^{4+}. This property allows Monensis to be used as a specific ion exchange electrode. It can also be used for the concentration of solutes against their concentration gradients. A23187 greatly raises the transport of divalent cations across biological membranes although it binds many divalent cations. A23187 strongly binds a Mn^{2+}, Ca^{2+} and Mg^{2+}. For more details refer to Betine (1983).

8.5 Like the Protein Some Ions are Unique

Magnifying the role of the antioxidant disturbs the body balance,degrading the indigenous antioxidant system and leaving the body under attack. The ions, free radicals, oxidants should be maintained at a certain level to protect the body from pathogens and other forms of miscellaneous attacks (Amara 2010, 2013; Hussein and Amara 2006). Nature is highly balanced (Amara 2010, 2013; Hussein and Amara 2006), for the microbiologist Fe^{2+}, Cu^{2+}, Zn^{2+}, Mn^{2+}, Co^{2+} and Ni^{2+} are transition metals while Ni^{2+}, Mg^{2+}, Ca^{2+} are alkaline earth metals, they are responsible for many vital activities in prokaryotic and in eukaryotic cells. Their deficiency is responsible for many disorders and malfunctions. For example, Fe^{2+} is found in the porphyrin ring of hemoglobin, myoglobin, in the cytochrome proteins, myeloperoxidase, NO synthase, coenzyme Q10, respiratory complex, etc. Its deficiency causes many diseases such as anemia (Dev and Babitt 2017). Another example Cu^{2+} is a catalytic cofactor in redox chemistry of proteins, important in growth and development its deficiency causes multisystem effects due to altered energy production, metabolism, and oxidative damage (Ackerman and Chang 2018). The role of the divalent cations in the RNA polymerase is well proved (Helfman et al. 1981; Koren and Mildvan 1977). Ions participate in 70% of binding sites and active sites of intracellular enzymes (Petukh 2014). Cu^{2+} is the major stabilizer of the smallest soluble oligomers that induce neuron death, whereas Zn^{2+} would induce the formation of larger and less toxic amyloid Aß aggregates. Six histidines are engineered in many proteins to facilitate their purification using Ni^{2+}-column (Hoffmann et al. 2002).

8.5.1 Some are Preferable to Others

Acinitobacter spp., differentiate between two carbon sources. Are the cells, the atoms, the compounds and the macromolecules in competition? My first experimental design study concerned degrading phenol using *Acinitobacter spp,*. The experiment aimed to optimize the phenol degradation. I added glucose because I thought that it might help the microbe to form a biomass. The experiment proved that the microbe did not utilize phenol in the presence of the glucose. How could the microbe do that? For any circumstances, it must have a control mechanism that enables the entry of glucose to the cytoplasm but not the phenol. Or, there is a repressor for the phenol degrading pathway that works in the presence of glucose. Or, glucose as a chelating agent that prevents the supply of essential ions or metals to the phenol degrading pathway. In fact, it might be a combination of all of this working together.

The microbe utilizes certain pathways to produce the protein which will either be the product in some biotechnological applications such as the lipases and the proteases or will be the producer in other such as the PhaC synthases which produce the polyhydroxyalkanoates. So, the raw crude element will either be the amino acids to produce the protein (e.g., lipase) or hydrocarbon compounds to produce the polyhydroxyalkanoates. So, why are the trace elements influencing the process? How could trace elements optimize the production of certain biotechnological macromolecules? Trace elements are reported as being essential in most microbial media. In fact, many of the roles of ions are not well proved. The synthetic medium should be prepared with an adequate amount of trace elements to study the effect of each ion on the microbial growth. Many buffers either contain elements or contain chelating agents that suppress the action of the divalent cations. There are many experimental design experiments that prove the role of trace elements in optimizing complicated experiments.

Example of a medium containing ions and trace elements: mineral medium (MM) (Schlegel 1961) Na_2HPO_4. 12 H_2O g, KH_2PO_4 1.5 g, NH_4Cl 1 g, $MgSO_4$. 7 H_2O 0.2 g, $CaCl_2$. 2 H_2O 0.02 g, Fe(III)NH_4- citrate 1.2 mg, trace elements solution 6 0.1 ml, H_2O_{bidest} ad 1000 ml, pH 6.9; trace elements solution 6: $ZnSO_4$. 7 H_2O 10 mg, $MnCl_2$. 4 H_2O 3 mg, H_3BO_3 30 mg, $CoCl_2$. 6 H_2O 20 mg, $CuCl_2$. 2 H_2O 1 mg, $NiCl_2$. 6 H_2O 2 mg, Na_2MoO_4. 2 H_2O 3 mg, H_2O_{bidest} ad 1000 ml. The trace element solution was sterilized with a 0.22 μm filter

8.5.2 The Protein Charges and What Could Charges Do

Protein is resisting the run on the SDS polyacrylamide gel because of its 3D structure (folding form) and the charges in its backbone. The SDS-denaturing buffer, which contains the SDS and the β-mercaptoethanol, enables charging of the protein with negative charge and breaking any cystin-cystin bonds. The buffer also contains EDTA which will chelate the divalent cations. In fact, it is an ideal buffer to readjust the folded native protein to be more linear with negative charge on its surface. That enables it to move from the negative current to the positive one. The temperature gives the required energy to peform such a reaction. It must be that Laemmli (1970) is able to recognize the role of the ions and charges in the protein folding.

8.6 Bacterial Cell Wall

The term peptidoglycan was first introduced by Weidel and Pelzer (1964) to describe a "rigid bag of the volume and shape of the cell". The formation of bacterial peptidoglycan is done through a complex pathway. Peptidoglycan is composed of a polysaccharide backbone made up of 1,4-linked polysaccharide of alternating N-acetylglucosamine (GlcNAc) and N-acetylmuramic acid (MurNAc) units. A pentapeptide side chain, of general structure l-Ala-γ-d-Glu-X-d-Ala-d-Ala is attached to the l actyl moiety of the MurNAc units. Peptidoglycan is cross-linked via transpeptidation between the terminal amino group of the charged amino acid at position 3 and the d-Ala residue at position 4 of a second strand, after removal of the fifth residue (van Heijenoort and Gutmann 2000). Bacteria contain an osmotic pressure that is between 5 and 20 atmospheres and is thus greater than that of the surrounding medium. The peptidoglycan layer is the structure that facilitates maintenance of this pressure difference with sufficient plasticity to allow the various types of transformation of nutrients essential for cell growth, division, secretion, extra-enzymatic activities and so on. The degree of cross-linking and its position also differs between species of bacteria with Gram-positive organisms having a higher degree of cross-linking than Gram-negative organisms, which have the added protection of the outer membrane (Labischinski et al. 1985; Naumann et al. 1982). EDTA, Tris, triethylamine, polymyxin B, aminoglycosides, and nitrilotriacetate remove Mg^+ from the bacterial membrane. That result in dissociation of the lipopolysaccharide cross bridge and destabilization of the outer membrane (Leive 1974). Recently, Amara et al. (2013a, 2013b, 2015) proved that different strains of *Escherichia coli* have different cell wall sensitivity to SDS, NaOH, H_2O_2 during the preparation of bacterial ghosts using a sponge like protocol for bacterial evacuation. Later different bacteria from different species show different cell wall sensitivity to the used compound beside $NaHCO_3$ and the lysozyme and Proteinas K. The sponge like protocol is proposed to be used also for strain differentiation.

8.7 The Different Mechanisms of Antibiotic Resistance

It is not the aim of this study to describe the different mechanisms of the antibiotic resistance other than those linked to the divalent cations. However, there are many proposed mechanisms for bacterial resistance to the antibiotics. To my knowledge one could propose a classification based on the cause of the resistance. They could be summarized as follows:

1) acquiring foreigner resistance factor
2) developing a new resistance factor
3) mutating an existing resistance factor
4) adding new resistance genes in an existing resistance factor
5) change the structure of the antibiotic
6) degrade part of the structure of the antibiotic
7) producing chelating agents that could inhibit the antibiotic activity (Mulcahy et al. 2008)
8) biofilm formation or secrete polysaccharide (or any other structure) that protect the cell (Amara 2011)
9) up regulation of pumps (efflux) that expel the drug from the cell (Morita et al. 2012)
10) changing the outer membrane permeability (Ghai and Ghai 2018)
11) changing the antibiotic binding protein which transfers it to the cell
12) using metal cations to degrade the antibiotics
13) spore formation (Krueger and Kolodziej 1978; Lenz and Vogel 2014)
14) develop a static condition with a slow rate of the cell's metabolism
15) acquiring a resistance factor following by rebuilding the damaged part of the cell wall (Amara 2011)
16) hidden under the dead cells biomass and biofilms as proposed by Amara (2011).

8.7.1 Ions are Involved in the Resistance

According to Hussein and Amara (2006) and Amara and Hussein (2006) divalent cations are involved in nearly all types of antibiotic resistance. In detailed studies concerning the analysis of many isolates of multi-antibiotics resistance *P. aeruginosa* from Tanta

University Hospital based on the concept of using different concentrations of the different antibiotics (related to the different mode of actions). Most of the resistant strains turned out to be sensitive to the antibiotic after applying LC_{50} and LC_{10} dosage of the EDTA and different concentrations of the used antibiotics. The study recommended using the EDTA combination with antibiotics of different classes for superficial treatment and with different disinfectants or with Na- citrate to control the *P. aeruginosa* (see details below).

8.7.2 Antibiotics, Divalent Cations, and the Bacterial Outer Membrane

Colistin-resistant bacteria develop improved fitness by altering their outer membrane composition (Powers and Trent 2018). LPS/LOS at the cell's surface support the membrane against antibiotics and detergents (Nikaido 2003). Polymyxins, the class of antibiotics that includes colistin, directly damage the outer membrane (Band and Weiss 2015).

Metallo-β-lactamases (MBLs) are usually found in Gram-negative organisms (Sommer et al. 2009). MBLs are using zn^{2+} in the catalysis processes. MBLs do not hydrolyze aztreonam, a monobactam, at significant rates. These enzymes also require divalent cation(s), mainly zinc, for activity and are thus inhibited by metal chelators such as EDTA and EGTA (Bush et al. 1995). They are classified as molecular class B enzymes (Ambler 1980). Most MBLs, Zn1 is tightly coordinated and Zn^{2+} is loosely coordinated (Fabiane et al. 1998; Hernandez Valladares et al. 1997; Li et al. 1989). In addition, two zinc binding sites are present in the metallo-enzymes. It is generally proposed that one zinc ion (Zn1) is coordinated by residues His116, His118 and His196, and the other zinc (Zn2) is coordinated by residues Asp120, Cys221 and His263 (Fabiane et al. 1998). Amara and Hussein (2006) and Hussein and Amara (2006) suggested focusing on using compounds that chelate the divalent cations to control antibiotic resistant microbes particularly *P. aeruginosa*. Perhaps one of the first experiments that combined the antibiotic with a chelating agent (EDTA), or the disinfectant that really succeeded to control virulence multidrug resistance strains were proposed by Amara and Hussein (2006) and Hussein and Amara (2006), however they did not specify if the resistance to the ampicillin was due to Zn^{2+} dependent beta lactamase or something else (Amara and Hussein 2006; Hussein and Amara 2006). At that time the fear that EDTA might cause problems for *in vivo* treatment is a reason for limiting the application to other safe places. It is also important to highlight here that other types of resistance to other antibiotics were found, which suggest the presence of other metallo enzymes from groups other than the cell wall synthesis inhibiting antibiotics. Some studies aimed to produce a drug that could chelate divalent cations particularly to chelate the Zn^{2+}. For example Aminocarboxylic acids related to aspergillomarasmine A (AMA) and ethylenediamine-N,N'-disuccinic acid (EDDS) are strong zinc-binders and inhibitors of the metallo-beta-lactamase NDM-1 (Tehrani et al. 2020). Zinc chelators as carbapenem adjuvants for metallo-beta-Lactamase-producing bacteria were investigated *in vitro* and *in vivo* (Principe et al. 2020).

Some others described inhibition for New Delhi Metallo-beta-lactamase 1 without removing of the Zn^{2+}. Compounds such as Benzimidazole, Benzoxazole, Cephalosporin based on their chelating activity were proposed (Jackson et al. 2020a, 2020b).

In general, macromolecules perform their different function through the active sites in their backbone by inducing chemical reactions. They are in need of an electron, a proton or both which are either gained from activating the protein specific amino acid's side chain or from the surrounding ions. Both the molecules with positive and negative charges are in need of finalizing the biological systems different reactions, particularly catalysis. The cells are able to control the ion amounts by different mechanisms. For the antibiotic resistant proteins, they use ions to deactivate the antibiotic in the reflux system, by changing the cell wall permeability, in the ionophor antibiotic, in the biofilm formation, and in many other biological activities related to antibiotic resistance. Using a chelating agent such as EDTA could cut the processes and keep the antibiotic structure safe and the microbes from becoming sensitive (even if harbouring an antibiotic resistant factor). Ion management will support controling the microbial resistance.

During the discussions of the various examples in this chapter, some general roles should be recognized; they are:

1) The elements of the biological system are governed by the laws of physics and chemistry. They perform their reactions mechanically but more specifically and in order of coordination of their components such as the amino acids involved in one multistep catalytic process.
2) The biological system is dynamic and variable.
3) We do not have all the required information but we can build on some proved results and observations.
4) Non-clear results which have not yet been explained are not necessarily wrong. For example, responsibility for the resistance to the antibiotics itself was first attributed to the R factor only. Nowadays the picure is clearer. The early non-proven suggestions have today become facts.
5) Some microbes are really tricky, such as *P. aeruginosa* (Adekunle et al. 2020; Afshari et al. 2020; Algammal et al. 2020). They could harbor many virulence factors other than their ability to produce alginate or the existence of multi-antibiotic and drug resistance factors. From hundreds of isolates from the same geographical area one might not find two atypically similar strains. They are either highly mutated or highly diverse.
6) Nutrients contribute to an epigenetic-like factor in the similarity and differences of the microbes as well as other factors.
7) The microbes are different genetically and environmentally. Macromolecules are sensitive to the environmental factors so the overall cell structure is changed by the change in the nutrient and the environment as proposed (Amara and Shibl 2015; Amara et al. 2012). A clear example of opportunistic pathogens might be *Candida albicans* which change their behavior by changing the environment. Such morphological changes will change their sensitivity to different factors including the antibiotic.
8) Proteins are responsible for functions, so similar proteins with minor changes do different functions. A clear example is the β and α globin mutants and their role in sickle cell anemia. A single change of the protein structure will cause a change (even minor) in their function.
9) β lactamases is a clear example of using electrons or protons in the side chains residue of the catalytic amino acids (serine β lactamase) or importing such help from the ions around its environment to complement the reaction (e.g., such as the z^{2+}metallo β lactamase).

10) The biological system requires a case-by-case study as proposed by Hussein and Amara (2006) where each microbial strain is diverse from other one.
11) Using simple inhibition zone experiments is not enough. *P. aeruginosa* which are sensitive in low antibiotic concentration change resistance at higher dosage as described by Hussein and Amara (2006). That could be explained because high concentration might elevate a new or non-working yet particular resistant mechanism. For that, I proposed using a serial dilution experiment instead of an inhibition zoom experiment for more precise results.

8.8 *P. aeruginosa* Outer Membrane

P. aeruginosa is always reported as a multidrug resistant microbe. Each year there are many publications describing the different multidrug mechanisms for which *P. aeruginosa* are used (Amsalu et al. 2020; de Lacerda Coriolano et al. 2020; Hedayati Ch et al. 2020; Li et al. 2020; Olsson et al. 2020). *P. aeruginosa* shows inherent multidrug resistance mechanisms by its outer membrane low-permeability and some specific multidrug efflux (Mix), like Mix XY-OprM (Aires et al. 1999). Mutants of the nalB gene of *P. aeruginosa* lead to overexpression of the outer membrane protein OprM (49 kDa), with increase in the resistance to quinolones, cephems, penams, meropenem, tetracycline, chloramphenicol and erythromycin (Sanchez et al. 2002). Plasmid and phage DNA materials mediated resistance to various antibiotics and other antimicrobial drugs (Shahid et al. 2003). Induction of the outer membrane protein HI in *P. aeruginosa* results in decreased susceptibility to aminoglycosides, polymyxin B, and EDTA. Protein HI can become the major cellular protein in cells grown in low (0–02 mM) Mg^{2+}. Protein HI induction was prevented by supplementing low Mg^{2+} medium. Cells grew in Ca^{2+}, but not in Mg^{2+}. Cells grown in Mg^{2+} Ca^{2+}, Mn^{2+}, or Zn^{2+} showed enhanced levels of these cations respectively as their major cell envelope-associated cation. HI replace divalent cations at a specific outer membrane site attacked by cationic antibiotics and chelators of divalent cations. Alterations in growth conditions can cause substantial changes in the protein composition of the outer membrane of *P. aeruginosa* (Hancock et al. 1982; Nicas and Hancock 1983). Some of these changes affect susceptibility to antimicrobial agents. Growth of susceptible *P. aeruginosa* in Mg^{2+}-limited medium has long been known to result in resistance to EDTA and polymyxin (Brown and Melling 1969a, 1969b, 1969c; Hansotia et al. 1969). The growth in low Mg^{2+} leads to a large increase in the outer membrane protein HI (Nicas and Hancock 1980). Protein HI acts to replace divalent cations at a site on the lipopolysaccharide, and protects this site from attack by chelators of divalent cations and cationic antibiotics, such as polymyxin B and gentamicin, which may compete for this site (Hancock 1981a, 1981b; Hancock et al. 1981; Nicas and Hancock 1980). A significant alteration in the hydrophilic uptake pathway for antibiotics was presented (Hancock et al. 1981). *P. aeruginosa* flourish the industry of the antibiotic with many ideas aiming to control such virulence opportunistic pathogens (Baek et al. 2020; Idowu et al. 2020; Ren et al. 2020; Rundell et al. 2020; Schaible et al. 2020).

8.9 The Effect of the Removal of Divalent Cations

Divalent cations have an important role in prokaryote cells especially in cell protection or in infection adaptation (Sarkisova et al. 2005). It is important here to highlight that the microbes defend themselves against the antibiotics and the other antimicrobial agents to survive. For that purpose microbes have developed some mechanisms to repel such attacks.

Removal of divalent cations will first block the most important sensors for the bacteria. EDTA was used traditionally to prove the positive and negative effects of the presence or the absence of divalent cations (Hafer et al. 2020). The experiments proved that the presence of divalent cations used in the different signalling mechanisms help *P. aeruginosa* produce alginate biofilm as a fast protection against H_2O_2 or antibiotics (Price et al. 2020). Sulfane Sulfur is an intrinsic signal activating MexR-regulated antibiotic resistance in *P. aeruginosa* (Xuan et al. 2020). An interesting study described that the oxygen levels at the site of *P. aeruginosa* infection can strongly influence virulence and antibiotic resistance in this pathogen, although the oxygen-sensing and signalling mechanisms underpinning this response have remained unknown (Schaible et al. 2020). Controlling such sensors might lead to a better control to the *P. aeruginosa* pathogenisity (Schaible et al. 2020). A protein signal acting, as a switch for controlling flagella rotation where a phosphotransfer occurs from CheW to CheY and in which autodephosphorylation of CheW phosphate can be fulfilled by Co^{2+}, Mg^{2+}, Mn^{2+}, or Zn^{2+}, or a lesser Cd^{2+} (Adamo et al. 2004). *P. aeruginosa* use divalent cations in infection adaptation (Chan et al. 2005). The role of Ca^{2+} is clear in cystic fibrosis patients infected with *P. aeruginosa* where, Ca^{2+} ions cause mucoid *P. aeruginosa* to have a more compact gelatinous appearance (Govan 1988). EDTA, which chelates Ca^{2+} and other divalent cations, could improve the antibiotic effectiveness *in vivo* (Wood et al. 1980). Magnesium reported, as an important divalent cation for *P. aeruginosa* like in protease virulence factor mechanisms (Guina 2003; Guina et al. 2003). In antibiotic treatment Ca^{2+} and Mg^{2+} can influence bacterial sensitivity to antibiotics, therefore media like Muller–Hinton broth, which lacks Ca^{2+} and Mg^{2+}, can lead to therapeutic failures with strains reported as sensitive. Root et al. (1988) suggest that using EDTA alone might be effective in wiping out catheter-associated biofilms (Root et al. 1988). Percival et al. (2005) describe the tetrasodium EDTA, as a novel catheter lock solution against biofilms where *P. aeruginosa*, *Klebsiella pneumoniae*, *E. coli*, *Staphylococcus epidermidis*, and *Candida albicans* were used to evaluate an experimental protocol designed to be used as a model for treatment of a catheter colonized by biofilm microorganisms. Variable effects of EDTA on pathogenic bacteria have been reported by many authors (Adachi et al. 2002; Nicas and Hancock 1980).

8.10 Case Studies

8.10.1 Case Study I Antibiotic-EDTA Combination

Aiming to close the gap between the World Health Organization (WHO) and the National Committee for Clinical Laboratory Standards Performance standards for antimicrobial susceptibility testing (NCCLS) methods and that divalent cations will give wrong result during sensitivity tests particularly when using a Mueller–Hinton medium, Hussein and Amara (2006) made a study in sequence to discover the best way to look at the role of divalent cations in the presence of EDTA and various antibiotics. Studies contained antibiotics from different groups against six virulence multidrug resistant strains which showed treatment problems in hospital. To evaluating the role of divalent cations in antibiotic resistance the study included a medium containing Ca^{2+} and Mg^{2+}. Various antibiotics were tested with or without the LC_{10} and LC_{50} equivalent EDTA against each strain. The LC_{10} and LC_{50} values were determined by Probit analysis and identified, as mM EDTA, which can inhibit the growth of 10% and 50% of *P. aeruginosa*. The conventional agar disc diffusion method was performed, as described in NCCLS documents (NCCLS 1999), to determine the antibiotic sensitivity. The treatment was performed in the absence and presence of LC_{10} and LC_{50} equivalent EDTA. A–Z medium was used rather than Muller–Hinton agar and the inhibition zone diameter were determined after 24 hr incubation at 30 °C. The used antibiotics were protein synthesis inhibitor [Tetracycline 30 μg ml^{-1}, Chloramphenicol 30 μg ml^{-1}, Garamycin 10 μg ml^{-1}, Erythromycin 15 μg ml^{-1}], Cell wall synthesis inhibitor [Ampicillin 10 μg ml^{-1}, Durcef 30 μg ml^{-1}, Unasyn 15 μg ml^{-1}]; DNA synthesis inhibitor [Nalidixic acid 30 μg ml^{-1}, Ciprofloxacin 5 μg ml^{-1}]; Folic acid synthesis analogue [Sulphamethazole 25 μg ml^{-1}] and RNA synthesis inhibition [Rifampicin 30 μg ml^{-1}]. The study found that after adding EDTA, 70% of the strains changed from resistant to sensitive. EDTA will improve the antibiotic performance during superficial treatment or any other treatment where EDTA or EDTA-like drugs could be used. The divalent cations represented in Ca^{2+} and Mg^{2+}contribute to a major role in antibiotic resistant mechanisms. Chelating Ca^{2+} and Mg^{2+} significantly increases the sensitivity of the antibiotic. The effect of EDTA and other chelating agents could be evaluated using their LC_{10} and LC_{50} equivalent amount with various antibiotics. The successful antibiotics after combination with EDTA and other EDTA-like drugs could be used in the treatment of other pathogens. *P. aerginosa* strains from the same origin have the same resistant mechanism even they are different in their phenotypic characteristic, which makes the physicians choice easier and could be used in sensitive bacterial typing. It is important to highlight here that research that is aimed at controlling certain ions in the body might be faced with the blood recovering any losses. And if all ions are removed the treated organism will not be able to survive because other essential processes are in need of those ions. To solve that problem some scientists have suggested using such successful chelating agents in superficial treatment or in a perfectly targeted area. From the many macromolecules that exist in nature and using ions in their catalysis this chapter will focus on reporting the divalent cations and the antibiotic sensitivity. Other studies which investigate other types of divalent cations are also reported (Bilinskaya et al. 2020).

8.10.2 Case Study II Disinfectants-EDTA Combination

Amara and Hussein (2006) selected three *P. aeruginosa* strains resistant to antibiotics as above and tested them against Na-citrate and disinfectants (out of the six strains in the previous case study). The strains were successfully grown in the presence of Na-citrate, disinfectants and completely resistant to protein synthesis inhibitor antibiotics; Tetracycline, Chloramphenicol and Erythromycine, cell wall synthesis inhibitor antibiotics; Duracef and

Unasyn (partially to the ampicillin), and to RNA synthesis inhibitor antibiotic; Rifampicin. The statistical analysis of the data proved the significant effect of adding EDTA to disinfectants or Na-citrate (Hussein and Amara 2006).The offline strategy is a sanitization method for controlling *P. aeruginosa* resistant strains before they infect susceptible patients where the antibiotics treatment in this case will not be efficient. Using EDTA *in vivo* could not be easily used to investigate many biological aspects, such as its toxicity, so controlling pathogens, which have powerful protection mechanisms such as alginate production in *P. aeruginosa*, This will be improved *in vitro* by using EDTA (Amara and Hussein 2006; Percival et al. 2005).

8.10.3 Case Study III *P. aeruginosa* Alginate

H_2O_2 induce alginate production of mutants in *P. aeruginosa*, the alginate will protect the microbe as a biofilm and will capture the divalent cations. Additionally, other ions, oxidants and free electrons produced against *P. aeruginosa* are prevented from harming the cells, e.g., the production of alginate in the lung. Moreover, the microbes, upon becoming part of the biofilm slowed their growth (as described above). Very slow growth in it simply means that the cells consume less. So, the effective dosage of any antimicrobial agents will not reach the cells. For that, higher concentrations of antibiotics are required to kill the biofilm's microbes. The biofilm, particularly the alginate, requires the Ca^{2+} to form a well-found out biofilm. A variable number of divalent cations cause variable biofilm rigidity. This phenomenon is utilized in forming different density of alginates. Some resistant enzymes use divalent cations in the resistance mechanism to the antibiotics. But apparently the Ca^{2+} used by the biofilm is replaced from the environment, e.g., the lung which is supplied with the nutrient from the blood. However, in Petri dishes these amounts might not be enough to activate the biofilm as one type of the microbial resistance mechanism. Some *P. aeruginosa* apparently lost their ability to mutate by the lode of the H_2O_2 and were suggested to be double mutated (produced alginate and then losing their ability to produce it). The variation between various *P. aeruginosa* isolates are an interesting subject of investigation. H_2O_2 is one of those elements that could make a return to the *P. aeruginosa* history of mutation and alginate production (Amara et al. 2011). The alginate which is a divalent chelating agent has a significant role in the antibiotic resistance not only in the alginate producing microbes but also in the coexisted ones (Amara et al. 2013; El-Shanshoury et al. 2018). Additionally, it allows mechanical protection to attached microbes that are still viable. Viable microbes which are partially attacked are a target for free mobile elements or to DNA free fragments emerging from completely cell wall ruptured microbes as proposed by Amara (2011). The process could be named "static antibiotic natural transformation" which happened in the case when using static antibiotic, particularly ampicillin.

8.11 Other Ways to Break the Microbial Cell Wall

The sponge like protocol represents perfectly the ability of both the chemical compounds and the enzymes to do similar targeted processes. I believe that the used chemical compounds (NaCl, NaOH, SDS, $CaCO_3$, $NaHCO_3$ and H_2O_2) might also collaborate with some of the microbe macromolecules to cause self-degradation to cell walls (Amara 2015c, 2016; Amara et al. 2013a, 2013b, 2014; El-Baky et al. 2018b; Menisy et al. 2017; Sheweita et al. 2019). Or, they are doing that by activating certain other ions (Amara 2011). For that, I did not involve the EDTA during my design of the protocol. I designed the sponge like protocol to evacuate microbial cells to mimic what happened naturally (Amara 2015c; Amara et al. 2013a, 2013b; Menisy et al. 2017). That could also be understood by our understanding of the power of the enzymes, active proteins and the active macromolecules which use ions (and their charges) in catalysis (Amara 2015a). The macromolecules even if well-structured finally they peform their function by terminal charges in the catalytic active residue sites such as the catalytic amino acids (Amara 2015a, 2019). However, the charges and ions are localized in highly specific orientation around their specific target. They are like a multistep machine each one does a step in order to finalize the product perfectly. Free ions and charges are like the parts of this machine but are separate from each other so they seldom, or weakly, perform the same job. Organization, coordination, step-fitting, and the function/structure/specificity are critical issues in the interaction of the macromolecules and their substrates. Eliminating a single final step from the process (e.g., the divalent cations) will corrupt all the processes. That also might explain the behavior of some nanoparticles which they are larger in size than the atomic divalent cations but show good results (Amara 2005a, 2005b, 2011).

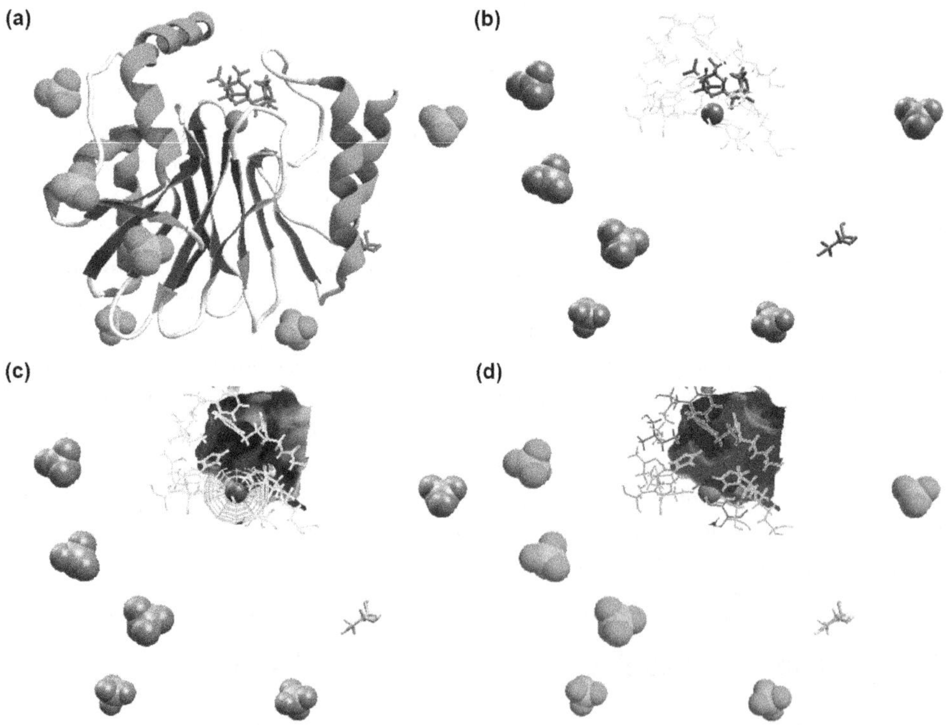

Figure 8.1 1x8i pdb file of the 3D protein of the crystal structures of the monozinc carbapenemase cpha and its complex with biapenem (Garau et al. 2005). The 3D structure of the enzyme and the biapenem, (b) shows the organic view (c) the electrostatic interaction, (d) the hydrophobicity.

It might be interesting to report that the trace element is effective in the production of the alginate (El-Shanshoury et al. 2018). The role of the nutrient in the elevation of the mutants has been proposed by Amara et al. (Amara et al. 2012; Amara and Shibl 2015). It is based on screening the similarity and differences between the four different classes based on their protein sequences. The investigation shows that a single strain, (*E. coli* ATCC 8739 found in the protein database) has different β-lactamases. By investigating the BLAST protein database for the existence of one of the different *E. coli* β-lactamases, the results show that this protein can be found in hundreds of different microbes with 100% identity (Amara et al. 2012). Some basic evidence about the stability of the biological system and the elements involved as well as a discussion about why similar or different mutants exist in the same gene and the factors involved in their distribution (Amara 2013, 2014; Amara and Shibl 2015) is required. Additionally, the role of the nutrient in the stability of the gene structure in a particular strain and in its change in another one was discussed and proposed (Amara 2013, 2014; Amara and Shibl 2015). Or, it could exist in the same species as well as in different species in different identities. The question is, are the differences and the similarities between the β-lactamase because of mutations, host adaptation, its mobility, or something else. The used strain *E. coli* ATCC 8739 which hs been published separately elsewhere shows different β-lactamase protein sequences from the same classes. This strongly supports the hypothesis that environmental effects such the presence of the alginate and nutritional factors could mutate the β-lactamase genes.

8.12 Conclusion

The biological system is highly balanced. Ions are a crucial part of the structure/function of macromolecules. Changing the macromolecule's environment or their backbone structure will affect their specificity or their functions. The structure/function/specificity is highly sensitive to physical, chemical and biological changes. This specificity could be imagined by understanding the role of the globins and the porphyrin ring in O_2/CO_2 transport. The living cells requiring unique ions as well as other cell components. Divalent cations are an essential part of the biological system. They are responsible for different mechanisms in the cells such as transport, the signal, the enzyme's activity, the folding properties, etc. Eliminating the divalent cations from the surrounding content of the biological cells proved to be a significant different side effect. To highlight the importance of the divalent cations in a well-studied biological mechanism the role divalent cation in the antibiotic resistance is one example of the many vital roles of the divalent cations was selected. *P. aeruginosa* is a well-studied microbe due to its multivirulence and multiresistance factors. The role of the divalent cation sensitivity to the antibiotic concerning *P. aeruginosa* is well proved. Different *P. aeruginosa* resistant strains to different antibiotics were made sensitive to the antibiotics used when the divalent cation was eliminated from their surrounding medium. Metallo-β-lactamase, as an example of the role of a divalent cation (Zn^{2+}) in deactivating the lactam ring in the lactam antibiotics is addressed. Our understand of the role of ions including the divalent cations will enable better smart drugs to be designed as well as enabling the use of simple methods for treating many illnesses. Experiments that concluded in the death of cells because of the existence of other cells (such as killing cancer cells) with microbes (or a chelating agent) probably are wrong that because microbes will utilize the nutrient essential for surviving of the cancer cells. The resistance to antibiotics could be summarized in acquiring, developing and mutating a resistant factor or adding new resistance genes to an existing resistance factor; changing or breaking the structure of the antibiotic; producing chelating agents; biofilm formation, the protection under dead microbes, upregulation of pumps (efflux), changing the antibiotic binding protein, and using metal cations to deactivate and degrade the antibiotics. Morphology change such as spore formation, developing a static condition and slowing down the cell's metabolism and rebuilding the damaged part of the cell wall after acquiring a resistance factor are other ways for antibiotic resistance. Ions, oxidants (H_2O_2) and free radicals produced by the lung cells against *P. aeruginosa* are prevented from harming the cells by producing the alginate. Besides being involved in the microbe virulence factors and, in their resistance, different mechanisms to the antibiotic divalent cations and other ions are essential elements in many other important reactions concerning the cell's viability, the enzymes and proteins reactions. Chelating the divalent cations will impair the biological system particularly the cell membrane, the membrane channels, the enzymes and proteins that are needed by them. A better understanding of the role of the divalent cations in antibiotic resistance will enable better pathogen control.

Acknowledgement

The author acknowledges Professor Dr. Mohamed Zakaria Hussein, Medicinal Collage, Tanta University, for his support with many *P. aeruginosa* medical isolates which were the starting point for conducting some of the research concerning the role of divalent cations in the bacterial antibiotic resistance which started in 2003.

Conflict of interest

The author declares that there is no conflict with any concerning this chapter

References

Ackerman, C.M. and Chang, C.J. (2018). Copper signaling in the brain and beyond. *The Journal of Biological Chemistry* 293: 4628–4635.

Adachi, Y., Akamatsu, H., and Horio, T. (2002). The effect of antibiotics on the production of superantigen from Staphylococcus aureus isolated from atopic dermatitis. *Journal of Dermatological Science* 28 (1): 76–83.

Adamo, R., Sokol, S., Soong, G., Gomez, M.I., and Prince, A. (2004). *Pseudomonas aeruginosa* flagella activate airway epithelial cells through asialo GM1 and toll-like receptor 2 as well as toll-like receptor 5. *American Journal of Respiratory Cell and Molecular Biology* 30 (5): 627–634.

Adekunle, C., Mustapha, A., Odewale, G., and Richard, O. (2020). Detection of antibiotic resistance genes among multiple drug resistant *Pseudomonas aeruginosa* isolated from clinical sources in selected health institutions in kwara state. *Infectious Disorders Drug Targets* 21 (6): e170721187999.

Afshari, H., Maleki, M., Hakimian, M., Tanha, R.A., and Salouti, M. (2020). Immunogenicity evaluating of the SLNs-alginate conjugate against Pseudomonas aeruginosa. *Journal of Immunological Methods* 488: 112938.

Aires, J.R., Köhler, T., Nikaido, H., and Plésiat, P. (1999). Involvement of an active efflux system in the natural resistance of *Pseudomonas aeruginosa* to aminoglycosides. *Antimicrobial Agents and Chemotherapy* 43 (11): 2624–2628.

Algammal, A.M., Mabrok, M., Sivaramasamy, E., Youssef, F.M., Atwa, M.H., El-Kholy, A.W., Hetta, H.F., and Hozzein, W.N. (2020). Emerging MDR-*Pseudomonas aeruginosa* in fish commonly harbor oprL and toxA virulence genes and bla_{TEM}, $bla_{CTX\text{-}M}$, and tet_A antibiotic-resistance genes. *Scientific Reports* 10 (1): 15961.

Amara, A.A. (2005a). New strategy for controlling resistant strain of Pseudomonas aeruginosa. In Proceeding of Tagungsband zum 2. Gemeinsamen kongress der DGHM und VAAM Göttingen, 25-28.9.2005. in Göttingen, Germany. *BioSpectrum*: Abstract No. KHP002, 74.

Amara, A.A. (2005b). Simple slot lysis and plasmid curing for determination the transfer of resistant factors within the pathogenic microbial strains. In Proceeding of Tagungsband zum 2. Gemeinsamen kongress der DGHM und VAAM Göttingen, 25-28.9.2005. in Göttingen, Germany. *BioSpectrum*: Abstract No. KHP004, 74.

Amara, A.A. (2010). The philosophy behind exo/endo/existing antioxidants and our built-in oxidant and antioxidant system. *Pharmazie* 65 (10): 711–719.

Amara, A.A. (2011). Opportunistic pathogens and their biofilm "food for thought". In: *Science against Microbial Pathogens: Communicating Current Research and Technology Advances* (ed. A. Mendez-Vilas), Vol. 2, 813–824. FORMATEX Microbiology Series No 3 2.

Amara, A.A. (2013). The inevitability of balanced lives: genes and, foods in action and, interactions. *IIOAB Journal* 4 (2): 1–27.

Amara, A.A. (2014). Proteases and their adaptation. *IIOAB Journal* 5 (4): 12–29.

Amara, A.A. (2015a). An overview of the molecular and cellular interactions of some bioactive compounds. In: *Biotechnology of Bioactive Compounds* (ed. V.K. Gupta and M.G. Tuohy), Chapter 21, 525–554. Chichester, UK: John Wiley & Sons, Ltd. doi: 10.1002/9781118733103.ch21.

Amara, A.A. (2015b). *Kostenlose Viral Ghosts, Bacterial Ghosts, Microbial Ghosts and More.* Germany: Schüling Verlage Book. ISBN 978365232691.

Amara, A.A. (2015c). *Saccharomyces cerevisiae* ghosts using the sponge-like re-reduced protocol. *SOJ Biochemistry* 2 (1): 1–4.

Amara, A.A. (2016). Lysozymes, proteinase K, bacteriophage E lysis proteins, and some chemical compounds for microbial ghosts preparation: a review and food for thought. *SOJ Biochemistry* 2 (1): 1–16.

Amara, A.A. (2019). *An Overview about Our Understanding of Protein Activity, Specificity, Structure and Function* (ed. S.M. Abdel-Aziz, N. Garg, A. Aeron, C.K. Jha, S.C. Nayak and V.K. Baijapi). Nova publishers. Microbial Catalysts. Chapter 14, Vol. 1.

Amara, A.A., Hassan, M.Z., Abulhamd, A.T., and Haroun, B.M. (2011). algs genes: H_2O_2 map their viability and validity in *Pseudomonas aeruginosa*. *International Journal of Biotechnology & Biochemistry* 7 (3): 321–330.

Amara, A.A. and Hussein, Z.M. (2006). Offline controlling of *Pseudomonas aeruginosa* resistant to protein inhibitor antibiotics using combination of EDTA and Na-citrate or disinfectant(s). *Microbiology* (China) 33: 95–100.

Amara, A.A., Salem-Bekhit, M.M., and Alanazi, F.K. (2013a). Sponge-like: a new protocol for preparing bacterial ghosts. *The Scientific World Journal* 2013: 1–7. ID 545741.

Amara, A.A., Salem-Bekh, M.M., and Alanazi, F.K. (2013b). Preparation of bacterial ghosts for *E. coli* JM109 using sponge-like reduced protocol. *Asian Journal of Biological Sciences* 6 (8): 363–369.

Amara, A.A., Salem-Bekhit, M.M., and Alanazi, F.K. (2013c). Plackett Burman randomization method for Bacterial Ghosts preparation form *E. coli* JM109. *Saudi Pharmaceutical Journal* 22 (3): 273–279.

Amara, A.A. and Shibl, A.M. (2015). A hypothesis about the variation within B-lactamases: an epigenetic-like mechanism. *IIOAB Journal* 6 (1): 1–11.

Amara, A.A., Shibl, A.M., and Tawfik, A.F. (2012). Deeper in the antibiotics paradigm: microbes react individually and collectively for spreading the resistance. *African Journal of Microbiology Research* 6 (9): 2010-2019.

Ambler, R.P. (1980). The structure of beta-lactamases. *Philosophical Transactions of the Royal Society of London. Series B, Biological Sciences* 289 (1036): 321–331.

Amro, A.A., Neama, A.J., Hussein, A., Hashish, E.A., and Sheweita, S.A. (2014). Evaluation the surface antigen of the *Salmonella typhimurium* ATCC 14028 ghosts prepared by "SLRP". *The Scientific World Journal* 2014: 1–6. ID 840863.

Amsalu, A., Sapula, S.A., De Barros Lopes, M., Hart, B.J., Nguyen, A.H., Drigo, B., Turnidge, J., Leong, L.E., and Venter, H. (2020). Efflux pump-driven antibiotic and biocide cross-resistance in Pseudomonas aeruginosa isolated from different ecological niches: a case study in the development of multidrug resistance in environmental hotspots. *Microorganisms* 8 (11).

Baek, M.S., Chung, E.S., Jung, D.S., and Ko, K.S. (2020). Effect of colistin-based antibiotic combinations on the eradication of persister cells in *Pseudomonas aeruginosa*. *The Journal of Antimicrobial Chemotherapy* 75 (4): 917–924.

Band, V.I. and Weiss, D.S. (2015). Mechanisms of antimicrobial peptide resistance in Gram-negative bacteria. *Antibiotics* (Basel) 4: 18–41.

Betine, V. (1983). Chapter 11 Mechaism of action of antibiotics. In: *The Chemistry and Biology of Antibiotics* (ed. V. Betine), Vol. 5, 349–429. Elsevier Scientific Publishing Company.

Bilinskaya, A., Buckheit, D.J., Gnoinski, M., Asempa, T.E., and Nicolau, D.P. (2020). Variability in zinc concentration among Mueller-Hinton broth brands: impact on antimicrobial susceptibility testing of metallo-beta-lactamase-producing enterobacteriaceae. *Journal of Clinical Microbiology* 58: 12 e02019-20.

Brown, M.R. and Melling, J. (1969a). Loss of sensitivity to EDTA by Pseudomonas aeruginosa grown under conditions of Mg-limitation. *Journal of General Microbiology* 55 (3): 439-444.

Brown, M.R. and Melling, J. (1969b). Role of divalent cations in the action of polymyxin B and EDTA on *Pseudomonas aeruginosa*. *Journal of General Microbiology* 59 (2): 263–274.

Brown, M.R. and Melling, J. (1969c). Sensitivity of Pseudomonas aeruginosa to polymyxin B sulphate and EDTA: role of the divalent cation of the growth medium. *Journal of General Microbiology* 58 (3): xii.

Bush, K., Jacoby, G.A., and Medeiros, A.A. (1995). A functional classification scheme for beta-lactamases and its correlation with molecular structure. *Antimicrobial Agents and Chemotherapy* 39 (6): 1211–1233.

Chan, P.C., Huang, L.M., Wu, P.S., Chan, P.Y., Yang, T.T., Lu, C.Y., Lee, P.I., Chen, J.M., Lee, C.Y., and Chang, L.Y. (2005). Clinical management and outcome of childhood lung abscess: a 16-year experience. *Journal of Microbiology, Immunology, and Infection* 38: 183–188.

de Lacerda Coriolano, D., de Souza, J.B., Bueno, E.V., Medeiros, S., Cavalcanti, I.D.L., and Cavalcanti, I.M.F. (2020). Antibacterial and antibiofilm potential of silver nanoparticles against antibiotic-sensitive and multidrug-resistant *Pseudomonas aeruginosa* strains. *Brazilian Journal of Microbiology* 52 (1): 267–278.

Dev, S. and Babitt, J.L. (2017). Overview of iron metabolism in health and disease. *Hemodialysis International* 21: S6–S20.

El-Baky, N.A. (2018). Differential antimicrobial effectiveness of camel lactoferrin-oleic acid and bovine lactoferrin-oleic acidcomplexs against several pathogens. *SOJ Biochemistry* 4 (1): 1–9.

El-Baky, N.A. and Amara, A.A. (2014). Newcastle disease virus (LaSota strain) as a model for virus Ghosts preparation using H_2O_2 bio-critical concentration. *International Science and Investigation Journal* 3 (5): 38–50.

El-Baky, N.A., Elkhawaga, M.A., Abdelkhalek, E.S., Sharaf, M.M., Redwan, E.M., and Kholef, H.R. (2021). *de novo* expression and antibacterial potential of four lactoferricin peptides in cell-free protein synthesis system. *Biotechnology Reports* 29: e00583.

El-Baky, N.A., Uversky, V.N., and Redwan, E.M. (2018b). Virucidal activity of cell-penetrating peptides of viral origin. *Journal of Biomolecular Structure & Dynamics* 36 (7): 1739–1746.

El-Shanshoury, A.E.-R.R., Kenawy, E.-R., Amara, A.A., Mohamed, S.F., Salama, A.F., and Kishk, S. (2018). Optimization of the culture medium using the Burman and Box-Behnken experimental designs for an enhanced production of alginate by Azotobacter vinelandii. *Biotechnologia* 99 (3): 185–196.

Fabiane, S.M., Sohi, M.K., Wan, T., Payne, D.J., Bateson, J.H., Mitchell, T., and Sutton, B.J. (1998). Crystal structure of the zinc-dependent beta-lactamase from Bacillus cereus at 1.9 A resolution: binuclear active site with features of a mononuclear enzyme. *Biochemistry* 37 (36): 12404–12411.

Frausto de Silva, J.J.R. and Williams, R.J.P. (1993). *The Biological Chemistry of the Elements: The Inorganic Chemistry of Life*. Oxford: Clarendon Press.

Garau, G., Bebrone, C., Anne, C., Galleni, M., Frere, J.M., and Dideberg, O. (2005). A metallo-beta-lactamase enzyme in action: crystal structures of the monozinc carbapenemase cpha and its complex with biapenem. *Journal of Molecular Biology* 345: 785-795.

Ghai, I. and Ghai, S. (2018). Understanding antibiotic resistance via outer membrane permeability. *Infection and Drug Resistance* 20 (8): 523–530.

Govan, J.R.W. (1988). Alginate biosynthesis and other unusual characteristics associated with the pathogenesis of *Pseudomonas aeruginosa* in cystic fibrosis. In: *Bacterial Infections of Respiratory and Gastrointestinal Mucosae* (ed. E. Griffiths, W.E. Donachie and J. Stephen), 67–96. Oxford: IRL Press.

Guina, T., Wu, M., Miller, S.I., Purvine, S.O., Yi, E.C., Eng, J., Goodlett, D.R., Aebersold, R., Ernst, R.K., and Lee, K.A. (2003). Proteomic analysis of *Pseudomonas aeruginosa* grown under magnesium limitation. *Journal of the American Society for Mass Spectrometry* 14 (7): 742–751.

Hafer, E., Holzgrabe, U., Kraus, K., Adams, K., Hook, J.M., and Diehl, B. (2020). Qualitative and quantitative (1) H NMR spectroscopy for determination of divalent metal cation concentration in model salt solutions, food supplements, and pharmaceutical products by using EDTA as chelating agent. *Magnetic Resonance in Chemistry* 58 (7): 653–665.

Hancock, R.E. (1981a). Aminoglycoside uptake and mode of action–with special reference to streptomycin and gentamicin. I. Antagonists and mutants. *The Journal of Antimicrobial Chemotherapy* 8 (4): 249–276.

Hancock, R.E. (1981b). Aminoglycoside uptake and mode of action-with special reference to streptomycin and gentamicin. II. Effects of aminoglycosides on cells. *The Journal of Antimicrobial Chemotherapy* 8 (6): 429–445.

Hancock, R.E., Poole, K., and Benz, R. (1982). Outer membrane protein P of Pseudomonas aeruginosa: regulation by phosphate deficiency and formation of small anion-specific channels in lipid bilayer membranes. *Journal of Bacteriology* 150 (2): 730–738.

Hancock, R.E., Raffle, V.J., and Nicas, T.I. (1981). Involvement of the outer membrane in gentamicin and streptomycin uptake and killing in *Pseudomonas aeruginosa*. *Antimicrobial Agents and Chemotherapy* 19 (5): 777–785.

Hansotia, P., Peters, H., Bennett, M., and Brown, R. (1969). Chelation therapy in Wegener's granulomatosis. Treatment with EDTA. *The Annals of Otology, Rhinology, and Laryngology* 78 (2): 388–402.

Hedayati Ch, M., Abolhassani Targhi, A., Shamsi, F., Heidari, F., Salehi Moghadam, Z., Mirzaie, A., Behdad, R., Moghtaderi, M., and Akbarzadeh, I. (2020). Niosome-encapsulated tobramycin reduced antibiotic resistance and enhanced antibacterial activity against multidrug-resistant clinical strains of Pseudomonas aeruginosa. *Journal of Biomedical Materials Research Part A* 109 (6): 966-980.

Helfman, W.B., Hendler, S.S., and Smith, D.W. (1981). Divalent cation-activated RNA synthesis in toluene-treated *Escherichia coli*. *Canadian Journal of Biochemistry* 59 (7): 511–518.

Hernandez Valladares, M., Felici, A., Weber, G., Adolph, H.W., Zeppezauer, M., Rossolini, G.M., Amicosante, G., Frere, J.M., and Galleni, M. (1997). Zn(II) dependence of the Aeromonas hydrophila AE036 metallo-beta-lactamase activity and stability. *Biochemistry* 36 (38): 11534–11541.

Hinrichsen, R.D. (ed.) (1993). *Molecular Biology Intelligence Unit: Calcium-dependent Potassium Channels*. Austin, TX: R.G. Landes Company. 1–147. ISBN-10: 1570590117.

Hoffmann, N., Amara, A.A., Beermann, B.B., Qi, Q., Hinz, H., and Rehm, B. (2002). Biochemical characterization of the *Pseudomonas putida* 3-hydroxyacyl ACP:CoA transacylase, which diverts intermediates of fatty acid *de novo* biosynthesis. *Journal of Biological Chemistry* 277 (45): 42926–42936.

Hussein, M.Z. and Amara, A.A. (2006). Case-by-case study using antibiotic-EDTA combination to control *Pseudomonas aeruginosa*. *Pakistan Journal of Pharmaceutical Sciences* 19 (3): 236–243.

Idowu, T., Ammeter, D., Brizuela, M., Jackson, G., Alam, S., and Schweizer, F. (2020). Overcoming beta-Lactam resistance in *Pseudomonas aeruginosa* using non-canonical tobramycin-based antibiotic adjuvants. *Bioorganic & Medicinal Chemistry Letters* 30 (21): 127575.

Jackson, A.C., Pinter, T.B.J., Talley, D.C., Baker-Agha, A., Patel, D., Smith, P.J., and Franz, K.J. (2020a). Benzimidazole and benzoxazole zinc chelators as inhibitors of metallo-beta-lactamase NDM-1. *ChemMedChem* 16: 654–661.

Jackson, A.C., Zaengle-Barone, J.M., Puccio, E.A., and Franz, K.J. (2020b). A cephalosporin prochelator inhibits New Delhi metallo-beta-lactamase 1 without removing zinc. *ACS Infectious Diseases* 6 (5): 1264–1272.

Jaffe, E.K. and Cohn, M. (1978). Divalent cation-dependent stereospecificity of adenosine 5′-O-(2-thiotriphosphate) in the hexokinase and pyruvate kinase reactions. The absolute stereochemistry of the diastereoisomers of adenosine 5′-O-(2-thiotriphosphate). *The Journal of Biological Chemistry* 253 (14): 4823–4825.

Koren, R. and Mildvan, S. (1977). Magnetic resonance and kinetic studies of the role of the divalent cation activator of RNA polymerase from *Escherichia coli*. *Biochemistry* 16 (2): 241–249.

Krueger, W.B. and Kolodziej, B.J. (1978). Divalent cation mobility throughout exponential growth and sporulation of Bacillus megaterium. *Microbios* 18 (73–74): 159–167.

Labischinski, H., Barnickel, G., Naumann, D., and Keller, P. (1985). Conformational and topological aspects of the three-dimensional architecture of bacterial peptidoglycan. *Annales de l'institut Pasteur. Microbiologie* 136A(1): 45–50.

Laemmli, U.K. (1970). Cleavage of the structural proteins during the assembly of the head of bacteriophage T4. *Nature* 227: 680–685.

Leive, L. (1974). The barrier function of Gram-negative envelope. *Annals of the New York Academy of Sciences* 235: 109–127.

Lenz, C.A. and Vogel, R.F. (2014). Effect of sporulation medium and its divalent cation content on the heat and high pressure resistance of *Clostridium botulinum* type E spores. *Food Microbiology* 44: 156–167.

Li, F., Chen, D., Li, L., Liang, D., Wang, F., and Zhang, B. (2020). Analysis of metallo-beta-lactamases, oprD mutation, and multidrug resistance of beta-lactam antibiotic-resistant strains of *Pseudomonas aeruginosa* isolated from Southern China. *Current Microbiology* 77 (11): 3264–3269.

Li, M., Wang, D.C., and Liang, D.C. (1989). Crystal structure determination of 4-zinc bovine insulin at 1.9 Å resolution. *Science in China. Series B, Chemistry, Life Sciences & Earth Sciences* 32 (11): 1308–1317.

Menisy, M., Hussein, A., Abd El Kader, G.A., Sheweita, S., and Amara, A.A. (2017). *Klebsiella pneumoniae* ghosts as vaccine using sponge like reduced protocol. *Cellular & Molecular Medicine: Open Access* 03 (02): 1-8: 11. doi: 10.21767/2573-5365.100034.

Morita, Y., Tomida, J., and Kawamura, Y. (2012). MexXY multidrug efflux system of *Pseudomonas aeruginosa*. *Fronteirs in Microbiology* 3: 1–13.

Mulcahy, H., Charron-Mazenod, L., and Lewenza, S. 2008. Extracellular DNA chelates cations and induces antibiotic resistance in *Pseudomonas aeruginosa* biofilms. 4 (11): e1000213.

Natioal Committee for Clinical Laboratory Standards. (1999). Performance standards for antimicrobial susceptobility testing. Ninth informational supplement. Wayne, Pennsylvania: NCCLS; document M100-S9; Vol. 19. No. 1, Table 21.

Naumann, D., Barnickel, G., Bradaczek, H., Labischinski, H., and Giesbrecht, P. (1982). Infrared spectroscopy, a tool for probing bacterial peptidoglycan. Potentialities of infrared spectroscopy for cell wall analytical studies and rejection of models based on crystalline chitin. *European Journal of Biochemistry* 125 (3): 505–515.

Nicas, T.I. and Hancock, R.E. (1980). Outer membrane protein H1 of Pseudomonas aeruginosa: involvement in adaptive and mutational resistance to ethylenediaminetetraacetate, polymyxin B, and gentamicin. *Journal of Bacteriology* 143 (2): 872–878.

Nicas, T.I. and Hancock, R.E. (1983). Alteration of susceptibility to EDTA, polymyxin B and gentamicin in *Pseudomonas aeruginosa* by divalent cation regulation of outer membrane protein H1. *Journal of General Microbiology* 129 (2): 509–517.

Nikaido, H. (2003). Molecular basis of bacterial outer membrane permeability revisited. *Microbiology and Molecular Biology Reviews* 67: 593–656.

Olsson, A., Wistrand-Yuen, P., Nielsen, E.I., Friberg, L.E., Sandegren, L., Lagerback, P., and Tangden, T. (2020). Efficacy of antibiotic combinations against multidrug-resistant Pseudomonas aeruginosa in automated time-lapse microscopy and static time-kill experiments. *Antimicrobial Agents and Chemotherapy* 64 (6): e02111.

Ovchinnikov, Y.A., Ivanov, V.T., Evstratov, A.V., Mikhaleva, I.I., Bystrov, V.F., Portnova, S.L., Balashova, T.A., Meshcheryakova, E.N., and Tulchinsky, V.M. (1974). The enniatin ionophores. Conformation and ion binding properties. *International Journal of Peptide and Protein Research* 6 (6): 465–498.

Percival, S.L., Kite, P., Eastwood, K., Murga, R., Carr, J., Arduino, M.J., and Donlan, R.M. (2005). Tetrasodium EDTA as a novel central venous catheter lock solution against biofilm. *Infection Control & Hospital Epidemiology* 26 (6): 515–519.

Petukh, M. and Alexov, E. (2014). Ion binding to biological macromolecules. *Asian Journal of Physics* 23: 735–744.

Powers, M.J. and Trent, M.S. (2018). Phospholipid retention in the absence of asymmetry strengthens the outer membrane permeability barrier to last-resort antibiotics. *Proceedings of the National Academy of Sciences of the United States of America* 115: E8518–E8527.

Pressman, B.C. (1965). Induced active transport of ions in mitochondria. *Proceedings of the National Academy of Sciences of the United States of America* 53 (5): 1076–1083.

Pressman, B.C., Harris, E.J., Jagger, W.S., and Johnson, J.H. (1967). Antibiotic-mediated transport of alkali ions across lipid barriers. *Proceedings of the National Academy of Sciences of the United States of America* 58 (5): 1949–1956.

Price, C.E., Brown, D.G., Limoli, D.H., Phelan, V.V., and O'Toole, G.A. (2020). Exogenous alginate protects staphylococcus aureus from killing by Pseudomonas aeruginosa. *Journal of Bacteriology* 202 (8): e00559.

Principe, L., Vecchio, G., Sheehan, G., Kavanagh, K., Morroni, G., Viaggi, V., di Masi, A., Giacobbe, D.R., Luzzaro, F., and Luzzati, R. and others. (2020). Zinc chelators as carbapenem adjuvants for metallo-beta-lactamase-producing bacteria: in vitro and in vivo evaluation. *Microbial Drug Resistance* 26 (10): 1133–1143.

Ren, Z., Zheng, X., Yang, H., Zhang, Q., Liu, X., Zhang, X., Yang, S., Xu, F., and Yang, J. (2020). Human umbilical-cord mesenchymal stem cells inhibit bacterial growth and alleviate antibiotic resistance in neonatal imipenem-resistant Pseudomonas aeruginosa infection. *Innate Immunity* 26 (3): 215–221.

Root, J.L., McIntyre, O.R., Jacobs, N.J., and Daghlian, C.P. (1988). Inhibitory effect of disodium EDTA upon the growth of Staphylococcus epidermidis in vitro: relation to infection prophylaxis of Hickman catheters. *Antimicrobial Agents and Chemotherapy* 32 (11): 1627–1631.

Rundell, E.A., Commodore, N., Goodman, A.L., and Kazmierczak, B.I. (2020). A screen for antibiotic resistance determinants reveals a fitness cost of the flagellum in Pseudomonas aeruginosa. *Journal of Bacteriology* 202 (6): e000682.

Sanchez, P., Linares, J.F., Ruiz-Diez, B., Campanario, E., Navas, A., Baquero, F., and Martinez, J.L. (2002). Fitness of in vitro selected Pseudomonas aeruginosa nalB and nfxB multidrug resistant mutants. *The Journal of Antimicrobial Chemotherapy* 50: 657–664.

Sarkadi, B. and Gandos, G. (1983). Potassiumtransport incell membrane. In: *The Enzyme of Biological Membranes*, 2e (ed. A. Martonsi), 30: 193–243.

Sarkisova, S., Patrauchan, M.A., Berglund, D., Nivens, D.E., and Franklin, M.J. (2005). Calcium-induced virulence factors associated with the extracellular matrix of mucoid *Pseudomonas aeruginosa* biofilms. *Journal of Bacteriology* 187 (13): 4327–4337.

Schaible, B., Crifo, B., Schaffer, K., and Taylor, C.T. (2020). The putative bacterial oxygen sensor Pseudomonas prolyl hydroxylase (PPHD) suppresses antibiotic resistance and pathogenicity in Pseudomonas aeruginosa. *The Journal of Biological Chemistry* 295 (5): 1195–1201.

Schlegel, H.G. (1961). Die Speicherstoffe von *Chromatium okenii*. *Archiv fur Mikrobiologie* 42: 110–116.

Shahid, M., Malik, A., and Sheeba. (2003). Multidrug-resistant *Pseudomonas aeruginosa* strains harbouring R-plasmids and AmpC beta-lactamases isolated from hospitalised burn patients in a tertiary care hospital of North India. *FEMS Microbiology Letters* 228: 181–186.

Sheweita, S.A., Batah, A.M., Ghazy, A.A., Hussein, A., and Amara, A.A. (2019). A new strain of *Acinetobacter baumannii* and characterization of its ghost as a candidate vaccine. *Journal of Infection and Public Health* 12 (6): 831–842.

Sommer, M.O., Dantas, G., and Church, G.M. (2009). Functional characterization of the antibiotic resistance reservoir in the human microflora. *Science* 325: 1128–1131.

Tehrani, K., Fu, H., Bruchle, N.C., Mashayekhi, V., Prats Lujan, A., van Haren, M.J., Poelarends, G.J., and Martin, N.I. (2020). Aminocarboxylic acids related to aspergillomarasmine A (AMA) and ethylenediamine-N,N'-disuccinic acid (EDDS) are strong zinc-binders and inhibitors of the metallo-beta-lactamase NDM-1. *Chemical Communications* (Camb) 56 (20): 3047–3049.

van Heijenoort, J. and Gutmann, L. (2000). Correlation between the structure of the bacterial peptidoglycan monomer unit, the specificity of transpeptidation, and susceptibility to beta-lactams. *Proceedings of the National Academy of Sciences of the United States of America* 97 (10): 5028–5030.

Weidel, W. and Pelzer, H. (1964). Bagshaped macromolecules—a new outlook on bacterial cell walls. *Advances in Enzymology and Related Areas of Molecular Biology* 26: 193–232.

Wood, R.E., Kilinger, J.D., Thomassen, M.J., and Cash, H.A. (1980). The effect of EDTA and antibiotics on *Pseudomonas aeruginosa* isolated from cystic fibrosis patients: a new chemotherapeutic approach. In: *Perspectives in Cystic Fibrosis* (ed. J.M. Sturgess), 365–369. Tornto: Canadian Cystic Fibrosis Foundation.

Xuan, G., Lu, C., Xu, H., Chen, Z., Li, K., Liu, H., Xia, Y., and Xun, L. (2020). Sulfane Sulfur is an intrinsic signal activating MexR-regulated antibiotic resistance in Pseudomonas aeruginosa. *Molecular Microbiology* 114 (6): 1038–1048.

9

Biomolecules from Vegetable Wastes

Begoña de Ancos and Concepción Sánchez-Moreno

Department of Characterization, Quality and Safety, Institute of Food Science, Technology and Nutrition, Spanish National Research Council, (ICTAN-CSIC), Madrid (28040), Spain ancos@ictan.csic.es
**Corresponding author: ancos@ictan.csic.es*

9.1 Introduction

Over the past century, enormous progress has been achieved worldwide in improving human welfare. Societies have changed radically thanks to quantum leaps in technology, rapid urbanization, and innovations in production systems. In fact, never in the history of mankind have we had access to such a quantity of high-quality food, although the eradication of hunger in much of the world remains a goal to be achieved. However, this development has had very high environmental costs, seen in numerous impacts, such as the decrease of biodiversity in large areas, climate changes, the proliferation of the use of food packaging systems, the lengthening of distribution chains as well as the changes in dietary patterns that have a high environmental footprint, and increased food production that generates high amounts of waste.

The world's population will have more than 9 billion by 2050, 10.8 billion by 2080 and 11.2 by 2100 (FAO 2018), and two out of every nine will live in urban areas. This growth in urban population requires solutions that improve the quality of life, especially by reduction of air pollution and CO_2 emissions, replacing non-renewable energy sources with renewable ones and prevention and proper management of food waste.

The circular economy offers an avenue to sustainable growth, good health, and decent jobs, while saving the environment and its natural resource, being one of the most important milestones to reduce the food waste. The circular economy is a general term that encompasses all activities that reduce, reuse, and recycle materials in the processes of production, distribution, and consumption. In the circular economy it is expected that the food industry will function as a natural eco-system, and waste products from one industry will become a raw material to another industry (Rajkovic et al. 2020). Further, the change from a linear economy (take, make, dispose) to a circular economy (renew, remake, share) is expected to support significantly the attainment of the Sustainable Development Goals (SDGs), particularly SDG 12 on responsible consumption and production (UN 2015).

Fruit and vegetable wastes and by-products not only represent the wasting of food commodities and critical resources such as land, water, fertilizers, chemicals, energy, and labor, but also indirectly include a broad range of environmental impacts, such as soil erosion, deforestation, water and air pollution, as well as greenhouse gas emissions

Biomolecules from Natural Sources: Advances and Applications, First Edition. Edited by Vijai Kumar Gupta, Satyajit D. Sarker, Minaxi Sharma, María Elida Pirovani, Zeba Usmani, and Chelliah Jayabaskaran.

that occur during the processes of food production, storage, transportation, and waste management (Martharu et al. 2016).

Thus, food waste and food losses are some of the main concerns of our time. Food waste and food losses can be considered distinct, but they are sometimes difficult to distinguish. While food losses are commonly the result of technical limitations of infrastructure and handling, such as production, post-harvest, processing stages, packaging, storage, and marketing, food waste is commonly the result of negligence or a conscious decision to throw away food that has been more related to losses at the end of the supply chain, during retail and final consumption (Sagar et al. 2018). Also, the boundary between waste and by-product is not fixed. By-products have been related with those products created during the production process of the main product and which are not the primary goal of the production (Rajkovic et al. 2020).

One third of global food production estimated in 2017 by FAO to be 8.7 billion tons, is lost or wasted every year (FAO 2017, 2018) and the losses and wastes of horticultural commodities are the highest of all types of foods, reaching up to 60% depending the commodity and the country (Gustavsson et al. 2011).

Fruit and vegetable losses, wastes, and by-products occur during all phases of the supply and handling chain, including harvesting, post-harvest, transport to packing houses or markets, classification and grading, storage, transporting, marketing, processing, and at home before or after preparation (Figure 9.1) (Sagar et al. 2018).

From a global perspective, over 40% of the initial weight of vegetables and fruits is lost or discarded through the food supply chain and is more than 50% in less industrialized countries (e.g., Sub-Saharan Africa, North Africa, and Central and East Asia) (Gustavsson et al. 2011) (Figure 9.2). It is remarkable that 15–20% of losses occur during the agricultural production and postharvest phases in industrialized regions (e.g., Europe, North America), mostly due to fruit and vegetable grading due to quality standards set by retailers. In fact, more than 50 million tons of fruit and vegetables are discarded every year in Europe due to a combination of the high quality standards of supermarkets, strict government regulations and the high expectations that consumers have when buying these foods in terms of size, shape, and color (Porter et al. 2018). Also, in the industrialized regions, between 15–30% of waste are caused by the consumer due to incorrect food handling at home (Gustavsson et al. 2011) (Figure 9.2).

Fruit and vegetables are processed for economical and logistical reasons in order to improve their commercial shelf-life and digestibility, in accordance with the consumer

Figure 9.1 Fruit and vegetable losses and waste. (Victoria M and artemidovna, Adobe Stock.)

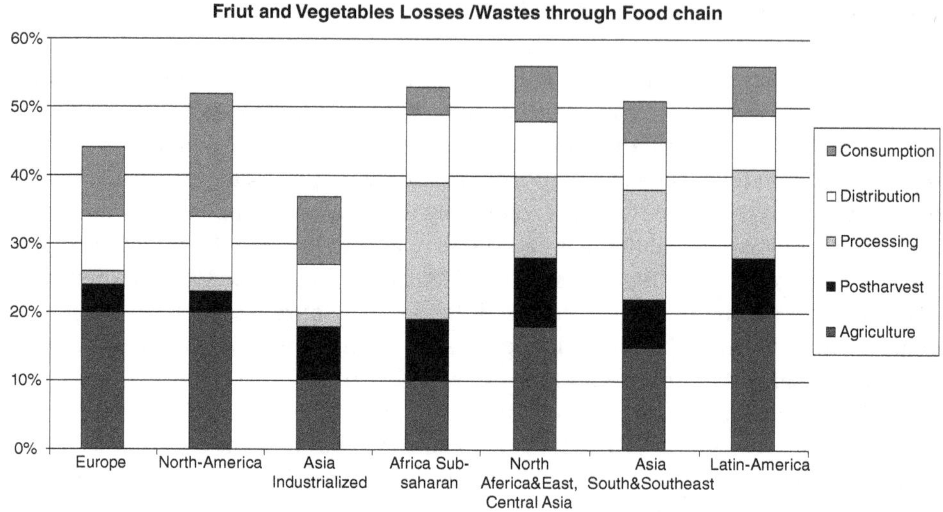

Figure 9.2 Fruit and vegetable losses and wastes through the food chain. *Source*: Gustavsson et al. (2011).

habits of each country or to facilitate the consumption by special groups (children, pregnant, elderly, patients with certain pathologies, etc.). This has generated millions of tons of waste and by-products such us peel, seeds, stones, residual pulp, and discarded pieces that generate important environmental management problems for the industry. Many fruit and vegetables, such as oranges, pineapples, apples, onions, tomatoes, potatoes, carrots, green peas, artichokes, and asparagus, that are used to obtain juices, frozen pulps and frozen products, dehydrated and canned products, etc., generate significant amounts of waste and by-products (Sagar et al. 2018).

Some of the major contributors to fruit and vegetable waste during processing, distribution, and consumption were China, the USA, Philippines, and India, which generate approximately 32, 15, 6.5 and 1.8 million tons of waste, respectively (Wadhwa and Bakshi 2013) (Table 9.1). While in China and the USA the main percentage of fruit and vegetable waste is produced in the consumption phase (15 and 28%, respectively), in the Philippines and India it is in the processing phase (25%) (Table 9.1).

In the European Union (EU) around 638 million tons of food commodities were available for human consumption in 2011, generating approximately 129 million tons of food waste along the whole food supply chain. Hence, food waste accounts for 20% of food produced. The total fruit and vegetable waste along the supply chain in the EU was calculated to be 28.1 million tons of fruit and 31.3 million tons of vegetables, that correspond with 41% and 46% of the total amount of fruit and vegetables available for consumption, respectively (Table 9.2). The largest amount of fruit and vegetable waste is generated during the primary production stage (16% and 19.5%, respectively) followed by consumption at house (12.6 and 17.8%, respectively) (Table 9.2) (Caldeira et al. 2019).

Table 9.1 Fruit (F) and Vegetable (V) waste generated after processing, packaging, distribution, and consumption.

Country	Production (Mt)			F and V Processed (%)	F and V Processed (million tons)	Waste (%)			Waste generated (Mt)
	F	V	Total			Proce-ssing	Distri-bution	Con-sumption	
China	122.19	473.06	595.25	23.00	136.91	2	8	15	31.98
USA	25.38	35.29	60.67	65.0	39.43	2	12	28	14.95
Philip-pines	16.18	6.30	22.48	78.0	17.53	25	10	7	6.53
India[1]	74.88	146.55	221.43	2.2	4.87	25	10	7	1.81

Source: Wadhwa and Bakshi (2013); Mt, million tons.

Table 9.2 Plant food waste generated in the European Union in 2011 along the food supply chain.

Product	Plant Food available in the EU (Mt)	Waste (Mt)					Total Waste generated (Mt)
		Primary Production	Processing and Manufac-toring	Retail and Distribution	Consump-tion Households	Consump-tion Food Services	
Cereals	78.2	1.2	2.5	1.7	8	2.2	15.6
Oil Crops	35.4	32.2	10	0.1	1.4	0.3	12.7
Potatoes	42.8	1.2	2.1	0.3	4.9	0.8	9.4
Sugar Beet	118.7	3.1	0.0	0.4	1.3	0.3	5.1
Fruit	67.9	11.1 (16%)	6.1 (9%)	0.8 (1.17%)	8.6 (12.6%)	1.5 (2.3%)	28.1 (41%)
Vegetables	68.5	13.4 (19.5%)	2.6 (3.8%)	0.9 (1.2%)	12.2 (17.8%)	2.2 (3.2 %)	31.3 (45.7%)

Source: Caldeira et al. (2019); Mt, million tons; Food available includes food consumed, water content and by-products (e.g., for sugar beet water that is evaporated in the sugar processing process and molasses used for animal feed is included).

9.2 Vegetable Waste and By-products as a Source of Bioactive Compounds

Generally, the health benefits of the consume of fruit and vegetables are usually studied together (Benetou et al. 2008; Crowe et al. 2011; Hu et al. 2014; Kalmpourtzidou et al. 2020) although it would be better to study these food groups separately. Fruit and vegetables share beneficial health effects due to both bioactive compounds (carotenoid and phenolic compounds, vitamins, minerals, fibers, etc.) present in their composition although they differ widely in the nature and ratio of them (Liu 2013a; Sharma et al. 2014).

Moreover, fruits usually have a higher concentration of sugars than vegetables, while vegetables are more likely to have a higher concentration of fibers and proteins (Appleton et al. 2016). In fact, studies that have separated fruit and vegetable groups have shown differences in health effects (Appleton et al. 2016; Villegas et al. 2008; Wang et al. 2015). In general, low vegetable intake is associated with higher incidence of non-communicable diseases (mainly cardiovascular diseases (CVDs), cancer, chronic respiratory disease, and diabetes) (Zhan et al. 2017).

Nowadays, in line with the transition of a linear to a circular economy for a sustainable 21st century, vegetable waste and by-products are considered to be very important materials that can be used to recover highly valuable biomolecules (pigments, carotenoids, phenolic compounds, terpenes, phytosterol, dietary fibers, pectins, polisaccharides, organic acids, enzymes, proteins, essential oils, etc.) that have been associated with health-promoting properties including antimicrobial, antibiotic, antioxidant, analgesic, antiplatelet, antitrombotic, antiinflammatory, antidiabetic, anticarcinogenic, hypolipidemic, antihypertensive, hepatoprotective, and immunoprotective effects, among others (Liu 2013b).

The most produced vegetables in the world in 2019 are shown in Table 9.3 with potatoes, tomatoes, and onions in the top three positions (FAO 2019). According to different studies in different parts of the world, the vegetables that produced the highest amounts of wasted mass were lettuce, tomatoes, onions, potatoes, carrots, artichokes, cucumbers, peppers, and spinach (Alzate et al. 2017; Mattson et al. 2018).

9.2.1 Tomato (*Solanum lycopersicum L.*)

The tomato is one of the most popular vegetables in the world (Table 9.3), the annual tomato production in 2019 was 180.77, 62.87, 16.52 and 11.35 million tons in the whole world, China, EU, and North America, respectively. In addition to that to be consumed

Table 9.3 The most produced vegetables in the world in 2019.

Rank	Vegetable	Production (Mt)
1	Potato	370.44
2	Tomato	180.76
3	Onion (dry)	99.96
4	Cucumber and gherkin	87.80
5	Cabbage and other brassicas	70.15
6	Eggplant	55.19
7	Carrot and Turnip	44.76
8	Chilies and Peppers	30.03
9	Spinach	30.11
10	Garlic	30.71
11	Lettuce and Chicory	29.13

FAO (2019); Mt, million tones.

as a fresh vegetable, approximately 25% of world tomato production is destined for industrial production of juice, paste, sauce, purée, and ketchup as the main derived tomato products (FAO 2019). The yield of tomato processing could be between 95–97% depending on the variety and maturity of tomatoes as well as the type of the processing employed. Thus, tomato processing produces approximately 3–5% (w/w) of waste called tomato pomace which mainly consists of peel, seeds, and a small amount of pulp (Lu et al. 2019). This tomato pomace is a rich source of bioactive compounds such as carotenoids (mainly lycopene), phenolic compounds, organic acids, vitamins, dietary fibre, protein, and unsaturated fatty acids (mainly linoleic acid) (Szabo et al. 2018). Regarding phenolic compounds, tomato by-products contain flavanones (naringenin glycosylated derivatives) and flavonols (mainly quercetin, rutin, and kaempferol glycoside derivatives). Tomato waste is considered to be one of the best sources of lycopene that has been associated with reduced risk of degenerative conditions such as cancer and cardiovascular related diseases (Burton-Freeman et al. 2012; Fuentes et al. 2013; Giovannucci 2002).

Figure 9.3 shows the main bioactive compounds that can be found in tomato pomace.

When the by-products of tomato processing were analysed, together with the unprocessed tomatoes (Kalogeropoulos et al. 2012), it was found that tomato waste (peel + seeds) contained significantly lower amounts of lycopene and increased amounts of ß-carotene, tocopherols, sterols, and terpenes, while their fatty acid profile was similar to that of unprocessed tomatoes (Table 9.4). Regarding polyphenols, hydroxycinnamic acids predominated in whole tomatoes, while flavonoids predominated in tomato waste with naringenin comprising 87% of flavonoids. Also, tomato waste contained similar amounts of total polyphenols and exhibited similar DPPH radical scavenging activity and ferric reducing power to unprocessed tomatoes (Table 9.4).

However, when the concentration of lycopene in the different parts of the tomato fruit was calculated in fresh weight, the tomato peel had a concentration of lycopene (486 μg g^{-1} fresh weight – fw) significantly higher than the whole tomato (29 μg g^{-1} fw). Moreover, the use of enzymes capable of hydrolyzing the cell walls such as cellulases and

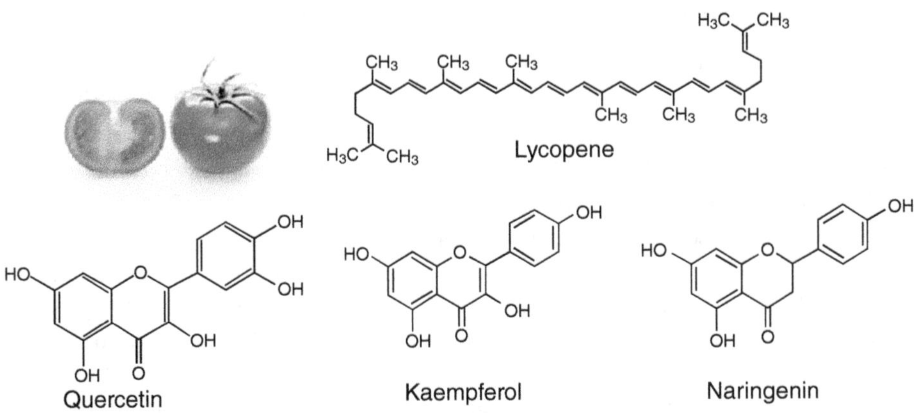

Figure 9.3 Main bioactive compounds in the by-products and waste of tomatoes. (Serhiy Shyullye, Adobe Stock.)

Table 9.4 Phytochemical compounds of whole raw tomato and its corresponding by-product (peel + seeds) formed during tomato paste production.

Compound	Unprocessed Whole Tomato	By-product	Significance
Carotenoids (g kg^{-1} dw)			
Lycopene	1013.2±89	413.7±80	**
β-Carotene	86.1±4.4	149.8±86	**
α-Tocopherol (g kg^{-1} dw)	85.8±5.9	155.7±10	**
Sterols (g kg^{-1} dw)			
β-Sitosterol	91.5±2.2	378.8±53	**
Stigmasterol	67.3±2.5	151.7±19	**
Campesterol	10.8±0.8	65.6±5.8	**
Radical scavenging activity (mMol TE[a] kg^{-1} dw)	3014.1±217	2990.4±164	n.s.
Ferric reduced power (mMol AAE[b] kg^{-1} dw)	1000.8±921	953.7±77.9	n.s.
Total polyphenols (mg GAE[c] kg^{-1} dw)	9779.3±72	9452.8±477	n.s.
Phenolic acids (mg kg^{-1} dw)			
Chlorogenic acid	25.3±1.8	51.7±13.8	n.s.
Caffeic acid	26.6±2.2	33.5±13.1	n.s.
Total phenolic acids	120.3±2.3	128.1±7.6	n.s.
Total hydroxycinnamic acids	105.5±2.4	120.8±2.3	n.s.
Flavonoids (mg kg^{-1} kg dw)			
Naringenin	14.8±2.2	328.6±50.5	**
Quercetin	3.70±0.1	6.80±1.10	n.s.
Kaempherol	2.4±0.41	5.5±1.2	*
Total flavonoids	51.8±2.3	378.7±62.3	**

dw, dry weight; n.s., no significative; [a] TE, Trolox Equivalents; [b] Ascorbic Acid Equivalents; [c] GAE, Gallic Acid Equivalents
Source: Kalogeropoulos et al. (2012).

pectinases, can increase by 107% and 206%, respectively, the extraction of lycopene from tomato peels (Choudhari and Ananthanarayan 2007).

Besides traditional solvent extraction, several innovative extraction methods have been explored for the recovery of carotenoids from tomato processing wastes and by-products, such as ultrasound assisted, microwave assisted, enzyme-assisted, high pressure, and pulsed electric field extraction (Andreou et al. 2020). Also, extraction systems using supercritical extraction with carbon dioxide (CO_2 flow rate = 4 kg h^{-1} at 55 °C 300 bar) and the use of a cosolvent (5% of ethanol) was able to extract up to 50% the lycopene and ß-carotene from tomato peel (initial content 309.6±15 and 29.6±3 μg g^{-1} dry weight, respectively) (Baysal et al. 2000; Zuknic et al. 2012). Also, dry tomato pomace can be used as ingredient to functionalize different foods such as meat products (hamburger, sausages, ham, etc.), bread, cookies, pasta, and tomato products (purée, pasta, and ketchup) (Lu et al. 2019).

9.2.2 Onion (Allium cepa L.)

Onion (*Allium cepa L.*) is the third most important horticultural crop worldwide (Table 9.3), with 99.98 million tons in 2019 and showing a steadily increased production within the last years (FAO 2019). Approximately the 66.4% of the worldwide production is located in Asia, the highest producing country being China with 24.96 million tons. The Americas, EU, and Spain produced 9.75, 6.46, 1.45 million tons, respectively (FAO 2019). Every year, onion processing generates more than 0.5 million tons of onion solid waste in the EU including skins (the outermost layers), roots and stalks, and bulbs unfit for consumption or commercial use (Choi et al. 2015). This has become an environmental concern because onion residues are not suitable for fodder because of their characteristic smell, nor can they be used as an organic fertilizer as is traditionally done, due to the development of phytopathogenic agents. Onion by-products, particularly the non-edible brown skin and external layers, are rich in flavonoids mainly quercetin and quercetin glycosides such as quercetin-4′-O-glycoside, quercetin 3,4′-O-diglycoside, and quercetin-3-O-glycoside, which represent about 80% of the total flavonoid content (González-Peña et al. 2013). Also, kaempferol and isorhamnetin glycosides can be found in certain onion varieties. Anthocyanins, mainly cyanidin-3-O-(6′-malonilglucoside), are also present in the red onions plus significant amounts of flavonols (Nile et al. 2021).

The distribution and type of flavonoids and other bioactive compounds in the onions strongly depend on the cultivar, the agronomic practices, environmental characteristics, and the layer of the onion. Thus, the outermost layers are rich in quercetin aglycone, minerals, and dietary fibre, whereas the inner parts are rich in quercetin glycosides, fructans, and sulphoxides of S-alk(en)yl-L-cysteine as shown on Table 9.5 for onion cv Recas (Benítez et al. 2011; Rodríguez Galdón et al. 2008).

Onion is recognized as the major dietary source of quercetin and its O-glycosylated derivatives (Guo and Bruno 2015; Lee and Mitchell 2012). Numerous published studies have investigated the potential of such by-products as raw materials for the production of bioactive ingredients (Benítez et al. 2011; Colina-Coca et al. 2014; González-Peña et al. 2013; Roldán-Marín et al. 2008, 2009a, 2009b). There is supporting evidence from *in vitro*, *in silico* and *in vivo* studies regarding the potential use of onion bioactive compounds obtained from onion by-products or onion in different forms (peel extract, onion slices, onion powder, etc.) for obtaining effective food ingredients with specific health-beneficial effects such as antioxidant, antimicrobial, antidiabetic, analgesic, antiplatelet, antithrombotic, anti-inflammatory, antidiabetic, anticarcinogenic, hypolipidemic, antihypertensive, hepatoprotective, and immunoprotective effects (Colina-Coca et al. 2017; Galavi et al. 2021; González-Peña et al. 2014, 2017; Khajah et al. 2021; Marrelli et al. 2018; Michalak-Majewska et al. 2020; Nile et al. 2021; Salamatullah et al. 2021; Teshika et al. 2019).

Figure 9.4 shows the main bioactive compounds that can be found in onion waste.

Onion by-products may be used as sources of bioactive ingredients to obtain functional foods such as pasta, breads, biscuits, etc. (Celano et al. 2021; Jiang et al. 2021; Masood

Table 9.5 Phytochemical from whole onion cv. Recas and its by-products.

	Total Phenols (mg GAE g^{-1} dw)[1]	Total Flavonoids (mg g^{-1} dw)[2]	Total ACSOs (μmol g^{-1} dw)	Total Dietary Fibre (mg g^{-1} dw)	AA (FRAP) (μmol Fe^{2+} g^{-1} dw)[3]
Whole onion	17.3±1.3	10.3±0.3	23.8	291	83.5±1.8
Brown outer peel	52.7±0.9	43.1±41.8	4.6	750	227.8±3.2
Outer flesh layers	19.7±1.6	19.5±0.7	29.9	312	105.1±0.6
Inner flesh layers	9.4±0.6	7.0±0.1	54.2	222	28.7±1.7
Upper and lower cut	30.5±2.0	25.9±0.7	22.2	667	156.1±1.6

[1]GAE, Gallic acid equivalents
[2]Quercetin equivalents; dw, dry weight; ACSOs, S-alk(en)yl-L-cysteine sulphoxides.
[3]AA; Antioxidant activity
Source: Benítez et al. (2011).

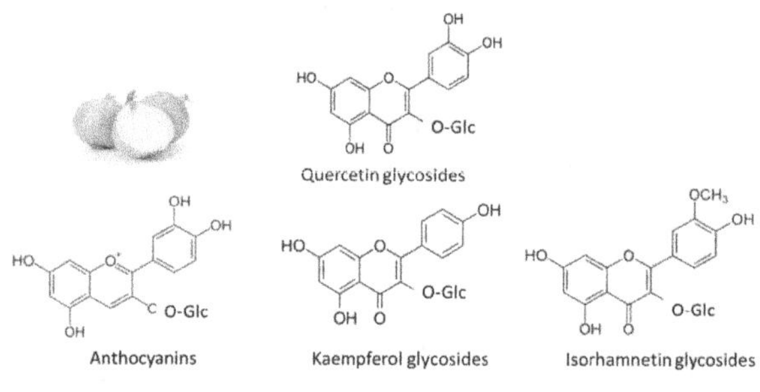

Figure 9.4 Main bioactive compounds in onion by-products and waste. (berganibt, Adobe Stock.)

et al. 2021; Michalak-Majewska et al. 2020). Also, quercetin-derived products obtained from onion by-products could be employed as commercial, health dietary supplements (nutraceuticals) (Tomé-Carneiro and Visioli 2016).

9.2.3 Lettuce (*Lettuca sativa L.*)

Lettuce (*Lactuca sativa L.*) is most often grown as a leaf vegetable, but also sometimes for its stem and seeds. World production of lettuce in 2019 was 29.13 million tones, with China being the highest producer with 16.31million tons. Other important lettuce producers in 2019 were North America and the EU with 3.77 and 3.66 million tons, respectively (FAO 2019). Lettuce is most often used for salads, although it is also found in other types of food, such as soups, sandwiches, and wraps; it can also be grilled. At present, in the context of the increasing demand from consumers of fresh-cut vegetables, lettuce salads represent 50% of the entire fresh-cut market in Europe and USA.

The production of fresh-cut lettuce from whole heads requires the preliminary removal of the external leaves and core, leading to waste amounts up to 50% of the initial lettuce head weight (Plazzota and Manzzoco 2018). Discarded external leaves of lettuce have been reported to show higher phenol content than the edible portions, due to the intense secondary metabolism activated by external influences. Also, fresh-cut processing further promotes polyphenol production in lettuce tissue, as a response to cell injury such as leaf cutting or shredding (Plazzota and Manzzoco 2018).

Lettuce by-products are rich in phenolic compounds and their composition varies with the lettuce variety as well as with the climatic conditions and agricultural practices employed (frequency of irrigation, type of fertilizer, etc.). At present, there is a growing consumer demand for salads of a single variety of lettuce (Iceberg, Batavia, Trocadero, Lollo Roso, Oak Leaf, or Romana), or mixtures of leaves of different varieties. There is also a growing consumer demand for baby leaf salads mainly lettuce or other leafy vegetables such as spinach, chard, watercress, and rocket salad. This increase in the consumption of lettuce salads has caused a significant rise of the amounts of lettuce by-products.

Table 9.6 shows the phenolic composition of by-products obtained from lettuce processing shown according to the main phenolic families, variety of lettuce, and solvent used in the extraction procedure (Llorach et al. 2004).

Generally, the major phenolic fraction of lettuce is made up of caffeic acid derivatives (90%), mainly represented by caffeoylquinic and caffeoyltartaric acid derivatives, among them, the main derivative identified was dicaffeoyltartaric acid (chicoric acid), followed by caffeoyl tartaric acid. In addition, chlorogenic acid (3-O-caffeoylquinic acid) and

Table 9.6 Phenolic compounds in the by-products of different varieties of lettuce and escarole.

Variety (extractant)	Total Phenols ($\mu g\ g^{-1}$ fw)	Total Flavonols ($\mu g\ g^{-1}$ fw)	Total Flavones ($\mu g\ g^{-1}$ fw)
Romana			
(water)	496.00	84.22	46.33
(methanol)	221.00	85.15	30.42
Iceberg			
(water)	211.05	21.84	7.14
(methanol)	108.10	24.38	9.20
Baby			
(water)	1088.00	157.70	5.80
(methanol)	1215.20	320.23	21.70
Escarole			
(water)	420.50	346.32	nd
(methanol)	415.43	407.00	nd

fw, fresh weight; nd, not detected.
Source: Llorach et al. (2004).

isochlorogenic acid were also identified (Llorach et al. 2008; Plazzota and Manzzoco 2018). Flavonoid compounds in lettuce leaves represent a minor fraction (5%) of the total phenolic compounds and are mainly composed of flavonols and flavones. The most common flavonols found in lettuce are derivatives of quercetin-3-O-glucuronide and kaempferol-3-O-glucuronide. Flavone compounds, principally luteolin-7-O-glucuronide, have also been identified (Plazzota and Manzzoco 2018).

Additionally, anthocyanin compounds have been identified in the pigmented leaves of red lettuce, mainly derivatives of cyanidin such as cyanidin-3-O-(6-malonilglucoside) and cyanidin-3-O-glucoside (Llorach et al. 2008).

Figure 9.5 shows the main bioactive compounds that can be found in lettuce waste and by-products.

Although the concentration of phenolic compounds in lettuce is relatively low, the high consumption of this vegetable worldwide makes it one of the vegetables that most contributes to food waste. However, the extraction of phenolic compounds from lettuce waste could be improved using different technologies such as ultrasound (400 W, 24 kHz for 120 s) that are able to increase up to 47% the phenolic compounds extracted in comparison with a traditional solid–liquid extraction (50 °C for 15 min) using green solvents such as aqueous ethanol solutions (Plazzota and Manzzoco 2018).

9.2.4 Potato *(Solanum tuberosum L.)*

Potato is a starchy, tuberous crop than belongs to the *Solanaceae* family. The word "potato" can refer to the plant itself, in addition to the edible tuber. Potatoes are one of the most important basic crops for human consumption, together with wheat, rice, and corn. The world production of potatoes in 2019 was about 370.44 million tons with China and India being the largest producers of potatoes in the world (88.4 and 42.3 Mt). USA and Germany rank 4th and 5th in the world with 19.18 and 10.62 million tons respectively (FAO 2019).

Potatoes can be prepared in many ways: wholes or peeled, cut up, with seasoning or without. Generally, the potato has to be peeled and cut before being processed for consumption with the aim of swelling the starch granules (boiled, steamed, baked, grilled, fried, dehydrated, etc.). The most common potato dishes consist of boiled potatoes (served

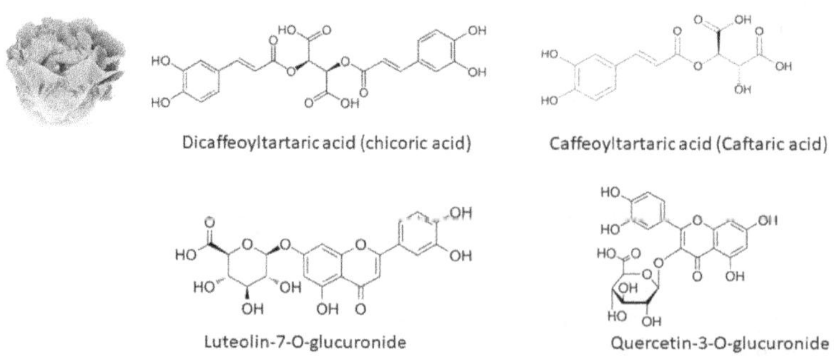

Figure 9.5 The main bioactive compounds in lettuce by-products and waste. (khumthong, Adobe Stock.)

hot or cold in salads), mashed potatoes, or fried potatoes as chips. By-products derived from potato processing can be divided into two major categories: discarded potatoes (whole or cut potatoes not destined for human consumption), and potato processing wastes (derived from the manufacture of potato ingredients or potato-based food products). Both discarded tubers and potato by-products represent a disposal problem to the potato industry, since the wet wastes constitute a source of plant spoilage and pathogenic infections (Joly et al. 2021).

Potato peel is the major component of processing waste of the potato industry, and contains different bioactive compounds such us phenolic compounds (chlorogenic acids, flavonoids, etc.), glycoalkaloids, vitamin C, and dietary fibre. Potato peel contains a high content of polyphenols, which was reported to be 10 times higher than the levels in the flesh, accounting for approximately 50% of all polyphenols in the potato tuber where concentrations vary depending on the potato variety (Table 9.7). In potato peel, the largest proportion of phenolic acids consists of chlorogenic acids (caffeoyl-quinic acids), which reached 90% (Mattila and Hellstrom, 2007). Thus, potato flesh contained a much lower concentration of chlorogenic acid (30–900 $\mu g\ g^{-1}$ fw) and other phenolic acids such as flavonoids (0–30 $\mu g\ g^{-1}$ fw) than potato skins which showed much higher levels (1000–4000 $\mu g\ g^{-1}$ fw) of chlorogenic acid (Lewis et al. 1998).

Phenolic compounds in potatoes are mostly in soluble form (free phenols, soluble esters, and glycosides) and to a lesser degree in the insoluble form due to the phenols attached to cell walls. 90% of the potato phenolic compounds in soluble form in the pulp and skin of the potato are hydroxycinnamic acid derivatives, fundamentally chlorogenic acid derivatives. Other phenolics found in potatoes are 4-O-caffeoylquinic acid (cryptochlorogenic acid), 5-O-caffeoylquinic (neo-chlorogenic acid), 3,4-dicaffeoylquinic and 3,5-dicaffeoylquinic acids. The purple-colored varieties have a higher content of anthocyanins and flavonoids than white-fleshed varieties. It is noteworthy that the peel and the adjacent flesh have phenolic compound concentrations and antioxidant activity up to 50% higher than the rest of the pulp (Albishi et al. 2013). The main soluble phenolic compound in potato peel is chlorogenic acid and its derivatives (Mattila and Hellström 2007) (Figure 9.6). The phenolic compounds concentration in boiled potato peel varies from 230–450 $\mu g\ g^{-1}$ fw. This concentration was significantly higher than that found in the boiled whole potato (100–170 $\mu g\ g^{-1}$ fw) (Table 9.7).

Traditional extraction technologies have employed conventional solid–liquid extraction techniques, such as soxhlet, heat reflux, and maceration to extract phenolic compounds from potato peel. Recently, novel extraction techniques such as microwave-assisted

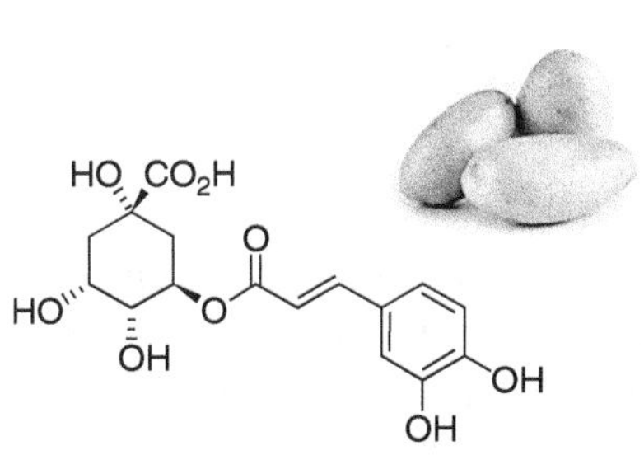

Chlorogenic acid

Figure 9.6 The main bioactive compounds in potato by-products and waste.

Table 9.7 Phenolic compounds from whole potatoes and their corresponding by-products.

Potato variety Product	Chlorogenic acid ($\mu g\ g^{-1}$ fw)*	Total phenols ($\mu g\ g^{-1}$ fw)
Van Gogh		
Whole/boiled/peeled	41±2	100
Fresh peel	260±25	340
Boiled peel	230±0.3	440
Rosamunda		
Whole/boiled/peeled	8.6±1,5	19
Fresh peel	150±11	250
Boiled peel	130±2.3	230
Nicola		
Whole/boiled/peeled	91±2.8	170
Fresh peel	230±15	350
Boiled peel	270±11	450

*Caffeic acid equivalents; fw, fresh weight. Source: Mattila and Hellstrom (2007).

extraction (MAE), ultrasound-assisted extraction (UAE), and supercritical fluid (SF) have been widely used in the extraction of phenolic compounds from potato peel (Joly et al. 2021; Samarian et al. 2012; Singh et al. 2011) and have provided significantly higher yield than conventional methods for the extraction or chlorogenic acid. Thus, by-products of the processing of potatoes, constituted mainly by peel, are excellent raw materials for obtaining functional ingredients (Mattila and Hellström 2007) (Table 9.7).

9.2.5 Carrot (*Daucus carota L.*)

Carrot (*Daucus carota L*) is a root vegetable that it is mostly consumed in the Mediterranean diet (fresh, converted to juice, frozen, canned, dehydrated, etc.). This vegetable is known for its characteristic orange colour, although there are also some purple, red, yellow, and white varieties. World production of carrots and turnips in 2019 was 44.76 million tons and more than half were grown in China. Production in the EU and USA was 5.56 and 2.2 million tons in 2019, respectively (FAO 2019). Carrots are used in a variety of ways, in salads, and soups, or form part of a famous vegetable soup known us "Julienne" together with onion and celery. In recent years, the consumption of carrot juice has increased due to its high content of antioxidants, particularly β-carotene, vitamins, and minerals. Also, carrots are extensively commercialized as a fresh-cut product or minimally processed vegetable such as mini carrots or strips and sticks that have been peeled, washed, sliced, or diced. Also, it is one of the principal components of purées destined for baby food and healthy beverages combined with other vegetables.

Carrots are considered to be a rich source of β-carotene and vitamins (thiamine, riboflavin, folic acid, and vitamin B-complex), and minerals (calcium, copper, magnesium,

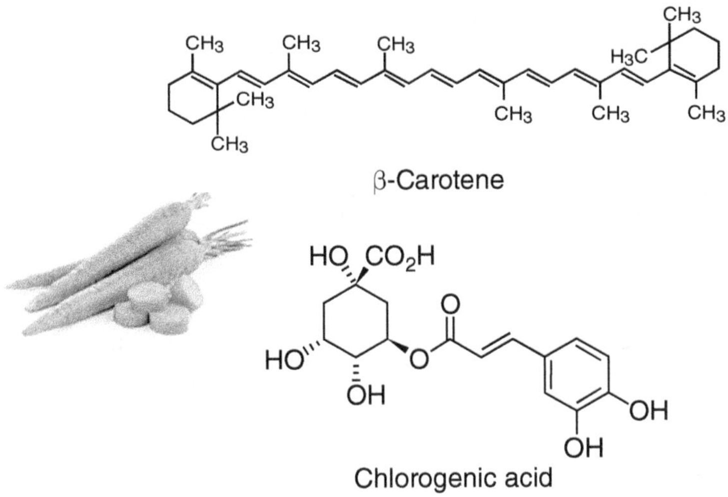

Figure 9.7 The main bioactive compounds in carrot by-products and waste. (atoss, Adobe Stock.)

potassium, phosphorus and iron). Also, carrots have phenolic compounds and the concentration of these bioactive compounds varies depending on the variety (Table 9.8) but also on the growing (irrigation, fertilizer, etc.) and processing conditions. All carrot varieties have a very high concentration of chlorogenic and caffeic acids derivatives, including 3-O- and 5-O-caffeoylquinic acid, 3-O-p-cumorylquinic and 5-O-feruoylquinic acid, and also 3,5-dicaffeoylquinic acid. Derived compounds from p-hydroxybenzoic and ferulic acids have also been identified. It is noteworthy that purple carrot varieties have a higher total phenolic concentration than the other varieties (Table 9.8) with 5-O-caffeoylquinic acid (540 $\mu g\ g^{-1}$ dry weight) concentration ten times higher than other varieties. Moreover, purple carrot varieties showed anthocyanins in their phenolic composition and also higher antioxidant capacity (DPPH and ABTS) than other varieties (Sun et al. 2009). Furthermore, orange carrot varieties have higher concentrations of total carotenes (lutein plus α-carotene plus lycopene plus β-carotene) than other varieties. Generally, β- and α-carotene are the major carotenoids in orange carrots ranging between 13–40% and 44–79% of the total, respectively, while lycopene is the principal carotenoid (419 $\mu g\ g^{-1}$ dw) in red varieties (Table 9.8) (Sun et al. 2009).

The main by-product of carrot is carrot pomace, generated during juice extraction. The peel removed prior to processing (epidermis and lateral roots mainly) and discarded carrots due to inferior quality can be utilized for the extraction of important bioactive compounds. In fact, up to 80% of the initial β-carotene content in the carrot can be left in the pomace (Sharma and Kumar 2018). The carrot pomace also has high dietary fibre content. Thus, the 63.6% of the dry weight of the solid residue is composed of total fibre, being soluble fibre the 50% of the total (Chau et al. 2004).

Carrot pomace is a good source of β-carotene, dietary fiber, pectins, and chlorogenic acid that need to be stabilized by eliminating the water as a step prior to the extraction of the bioactive compounds (Figure 9.7). Table 9.9 shows the effect of different drying methods on bioactive compounds. While drying methods reduce the concentration of bioactive compounds, dried carrot pomace maintains high levels of them (Hernández-Ortega

Table 9.8 Bioactive compounds extracted from different carrot varieties.

Color variety	Total phenols (mg GAE g^{-1} fw)*	Chlorogenic acid (µg g^{-1} dw)	Total carotenoids (µg g^{-1} dw)	β-carotene (µg g^{-1} dw)	α-carotene (µg g^{-1} dw)	Lycopene (µg g^{-1} dw)
Purple–orange	38.7±5.4a	18790±38a	771.0±22e	128±17d	18.9±2c	3.68±2.03b
Purple–yellow	15.0±1.1b	7661±4.9b	334.2±12d	239±8c	83.7±4c	2.02±0.28b
Red	2.27±0.1c	1347±2.3c	610.1±40c	187±18c	1.74±0.7c	419.4±49a
Dark orange	1.66±0.1c	631±0.56c	1334.7±71a	940±54a	382±18a	7.76±0.89b
Orange	2.34±0.05c	1150±0.30c	816.3±17b	579±79b	228±141b	5.09±1.52b
Yellow	1.97±0.6c	306±1.59c	52.0±17f	30±8e	1.86±1.2c	0.32±0.10b
White	2.35±0.32c	978±2.78c	17.6±11f	2.8±4e	0.5±0.33c	0.35±0.0

*GAE, Gallic acid equivalents; fw, fresh weight; dw, dry weight. Different letters in the same column indicate significant differences ($p < 0.05$).
Source: Sun et al. 2009.

et al. 2013). It also significant that carrot pomace has the advantage of not transferring undesirable flavours to the food to which it is added (O'Shea et al., 2012). Also, dried carrot pomace containing different bioactive compounds can be incorporated into extrudates, cakes, bread, biscuits etc., to extrudates, biscuits, cookies, wheat rolls, wheat breads, fish sausages, and several other products, to enhance their nutritional and functional values (Hernández-Ortega et al. 2013; Sharma and Kumar 2018).

9.2.6 Artichoke (*Cynara scolymus L.*)

Artichoke (*Cynara scolymus L.*) is a flowering plant cultivated as a food that belongs to the *Asteraceae* family. Artichokes are traditionally consumed in the Mediterranean diet in different preparations, fresh, canned, roasted, or baked, among others (Aguilera et al. 2016). The world production of artichokes in 2019 was 1.59 million tons, and the Mediterranean countries were the major producers: Italy (378 820 tons), Egypt (269 899 tons), Spain (199 940 tons), and Algeria (119 636 tons) (FAO 2019). The edible portion of this plant includes the receptacle of immature flowers and the inner bracts, named "capitula" or heads. During artichoke processing, the residues, principally external leaves or stems, represent approximately 60–80% of the total harvested plant material, this gives rise to more than 460 000 tons of waste generated annually. Nevertheless, these by-products possess a high content of bioactive phenolic compounds, inulin, fibers, and minerals (El Mesallamy et al. 2020; López-Salas et al. 2021; Pandino et al. 2013; Sánchez-Rabaneda et al. 2003).

Furthermore, some processing technologies require prior blanching to inactivate spoilage enzymes (polyphenoloxidase, peroxidase, etc.). This pre-treatment results in a high amount of solid and liquid (water blanching) wastes with a high concentration of valuable phenolic compounds and dietary fibre (Table 9.9) (Femenia et al. 1998; Larrosa et al. 2002).

Table 9.9 Phytochemicals extracted from artichoke by-products.

Vegetable	By-product	Phytochemical	Concentration*
Artichoke	Blanching leaves	Phenols	4320 μg g^{-1} fw
	Blanching liqueur	Phenols	6380 μg mL^{-1}
	Blanching leaves	Phenols	4400 μg g^{-1} fw
	Blanching liqueur	Phenols	6600 μg mL^{-1}

*Chlorogenic acid equivalents; fw, fresh weight. *Source*: Larrosa et al. (2002).

Consumption of artichokes or derived products has been associated with important health benefits in humans. Particularly hepatoprotective, anticancer, and hypocholesterolemic, antibacterial, and diuretic effects have described (Kukic et al. 2008; Llorach et al. 2002; Shallan et al. 2020). Artichokes also have significant antioxidant properties due to their high content of caffeic acid derivatives such as chlorogenic acid (5-O-caffeoylquinic). This vegetable also contains a high content of flavonoids, especially glucosides and rutinosides derivatives of apigenin and luteolin, and derivatives of cyanidin-caffeoylglucoside (Shen et al. 2010). Therefore, solid and liquid by-products from artichoke processing are an important source of phenolic compounds with antioxidant properties and a protective effect on health. However, due to the heat treatment applied to the artichoke, some phenolic compounds are converted into their isomers. Thus, cynarin (1,3-O-dicaffeoylquinic acid) and neochlorogenic acid (3-O-caffeoylquinic acid) found in artichoke derived products come from the isomerization of 1,5-O-dicaffeoylquinic acid and chlorogenic acid, respectively.

The presence of bioactive compounds in artichoke by-products has aroused greater interest in its use as a functional ingredient (Llorach et al. 2002; López-Salas et al. 2021).

9.3 Extraction Systems to Recover Bioactive Compounds from Vegetable Wastes

The increase in the world population and consumer awareness of the health benefits of consuming fruit and vegetables has caused the consumption of raw and processed fruits and vegetables to increase considerably in recent years. The growing demand for raw and processed fruit and vegetables had led to a higher production of plant waste (approximately 50% of their weight) in the form of peel, cores, pomace, unripe, and/or damaged fruit and vegetables. The disposal of fruit and vegetable waste and by-products represent a growing social, environment, and economic problem. Fruit and vegetable waste and by-products are usually reutilized for animal feed or processed into biogas or compost to obtain biofertilizer. Nevertheless, these materials can be considered as a largely available, low-cost source of value-added molecules as bioactive compounds (carotenoids, phenolic compounds, terpenes, phytosterol, dietary fibers, pectins, policsaccharides, organic acids, enzymes, proteins, essential oils, etc.) with multiple biological properties (Liu 2013a). Nowadays, there is an increasing interest in recovering these bioactive compounds from fruit and vegetable wastes, losses and by-products by using green technologies that represent more efficient and sustainable processes (Panzella et al. 2020).

However, fruit and vegetable waste and by-products are highly fermentable and perishable materials usually prone to microbial spoilage because of high moisture (80–90%), free soluble sugars (6–64%), and crude protein (10–24%) contents. Also, the waste obtained from fruit and vegetable consumption and processing industry is extremely diverse due to the use of the wide variety of vegetables, the broad range of processes, and the multiplicity of products (Kodogoda et al. 2017). Thus, an important step in the extraction of bioactive compound from this material is the stabilization and preparation of them for the extraction (O'Shea et al. 2012).

9.4 Stabilization

The stabilization of the by-products before subjecting them to extraction is a critical phase that requires the study of the most suitable conditions to prevent the degradation of bioactive compounds. Various pre-treatments could be used such as wet grinding (for reducing the particle size of wet waste); dryng (oven, lyophilization), and grinding the dried extract to reduce the particle size (O'Shea et al. 2012).

Due to the high water content of plant waste, losses and by-products (71% in tomato, 90% in artichoke and 82% in beet), drying process (85–100°C) prior to the extraction of bioactive compounds from fruit and vegetables waste and by-products may be useful to inactivate the enzymes that cause various degradative processes such as enzymatic browning and to reduce the microbial load (Peschel et al. 2006). However, thermal treatments can be harmful to the bioactive compounds. As an alternative to the traditional thermal treatments, different non-thermal technologies have been studied as pre-treatments to the extraction process such as ohmic heating, pulsed electric fields (PEF), and high-pressure processing (HPP). These treatments not only favor the stabilization of the fruit and vegetable wastes and by-products but also prevent the degradation of the bioactive compounds and improve their subsequent extraction. Ohmic heating could be used to produce a homogeneous and faster electrical heating which reduces the loss of heat-sensitive bioactive compounds, such as vitamin C (Icier 2010). Also, combined technologies such as pasteurization followed by freezing and freeze/drying has been studied to preserve the bioactive compounds of onion waste (Roldán et al. 2008).

9.5 Extraction Techniques for Recovery of Bioactive Compounds

Traditional extraction techniques like solid–liquid extraction by using organic solvents and novel extraction methods like ultrasound-assisted extraction (UAE), microwave-assisted extraction (MAE), pressurized liquid extraction (PLE), supercritical fluid extraction (SFE), and enzyme-assisted extraction (EAE) techniques could be used to obtain the target bioactive compound from each fruit and vegetable waste and/or by-product (Bromberger Soquetta et al. 2018; Panzella et al. 2020; Ran et al. 2019).

9.5.1 Conventional Techniques

The traditional method of solid–liquid extraction based on the use of organic solvents is a simple process that does not require expensive equipment, but has the disadvantages that it requires a prolonged time of extraction, agitation, mild temperature, and large amounts of potentially toxic organic solvents. The first improvement needed for this technology is increasing the extraction temperature for better dissolution and lower solvent viscosity, as done, e.g., in heated reflux extraction or in a Soxhlet apparatus (which has the further advantage of separating the soluble from the insoluble fractions). Decreasing the particle size is also a factor that has long been taken into consideration as it facilitates mass transfer and thus increases extraction speed and yield. These procedures require a further concentration process (evaporation) for the recovery of the bioactive compounds (Lizárraga-Velázquez et al. 2020; Zhang et al. 2018).

Hydrodistillation has been employed to extract volatile organic and non-organic compounds with distillated water. The procedure consists of three different phases: hydrodiffusion, hydrolysis, and decomposition by heat. This technique has been usually employed to obtain essential oils (e.g., terpenes). The heat causes the breakdown of the cell structure and the releasing of the aromatic compounds but also could favor their degradation. To reduce this undesirable effect and to increase the yield of volatile compounds extracted, hydrodistillation has been used in combination with non-thermal technologies. Thus, the extraction of essential oils from citrus peels is increasing significantly using electrofluidic as pre-treatment before hydrodistillation (Wu et al. 2017) and also combined treatment of microwave-assisted hydrodistillation (Bustamante et al. 2016).

9.5.2 Non-Conventional Extraction Technologies (Green Technologies)

The objective of green extraction processes is to reduce the use of harmful organic solvents, to achieve a faster extraction rate, more effective energy use, increase mass and heat transfer, improve the extraction yield, reduce the equipment size and the number of processing steps. The application of these technologies also helps to preserve the natural environment and its resources (Panzella et al. 2020; da Silva et al. 2016). Some of these techniques are described below:

9.5.2.1 Enzyme-Assisted Extraction

The enzyme assisted extraction (EAE) of bioactive compounds from fruit and vegetable wastes have been employed to reduce the use of conventional solvents (Nadar et al. 2018). Enzymatic treatments with enzymes such as pectinase, cellulose, alcalase, among others, are able to degrade the cell wall constituents and to facilitate the extraction of bioactive compounds such as the carotenoid lycopene from tomato peel (Leyva-López et al. 2017), flavonoids from grape skins (Tomaz et al. 2016), and bioactive peptides from olive seeds (Sánchez and Vázquez 2017).

The key to an efficient enzyme-assisted extraction is to select the optimal conditions such as type and concentration of enzyme, temperature, time, and pH. Also, the action of the EAE could be improved by the combination with other technologies such as ultrasound, microwave, or supercritical fluid extraction (Marathe et al. 2019).

9.5.2.2 Ultrasound-assisted Extraction

Ultrasound (> 20 KHz) improves extraction through acoustic cavitation: above certain energy levels, the acoustic waves interact with the solvent and dissolved gas by creating free bubbles that can expand to a maximum size and violently collapse, locally generating extreme heat and pressures. Due to this cavitation phenomenon, the cell walls can be ruptured, providing channels for solvent access, and mass transfer is improved. The efficiency of an extraction method is affected by many process parameters, e.g., solvent composition, the ratio of solvent to solid, ultrasonication time and temperature. Low frequency (20–40 kHz) generates large cavitation bubbles, increasing the number of microwaves and ensuring higher solvent penetration and extraction rate. Several ultrasound-assisted extraction (UAE) methods have been considered as promising technologies for the extraction of bioactive compounds from fruit and vegetable wastes and by-products (Zou et al. 2014). Temperatures below 50 ºC are recommended to avoid bioactive compound degradation. The type of solvent is a critical factor in UAE and generally a mixture of aqueous and organic solvent is used in different ratios. The most frequently employed organic solvents are ethanol, methanol, acetone, and isopropanol. Also, the vegetal material–solvent ratio has to be considered for the optimal extraction of bioactive compounds from fruit and vegetable waste and by-products (Lizárraga-Velázquez et al. 2020). Thus, UAE has been employed for the extraction of phenolic compounds from citrus peel waste (Gómez-Mejía et al. 2019), grape pomace (Drosou et al. 2015), and artichoke residues (Rabelo et al. 2016), among others, in different UAE conditions. In general, UAE conditions must be optimized for each vegetal source and chemical structure of the bioactive compound to be extracted (Lizárraga-Velázquez et al. 2020).

9.5.2.3 Microwave Assisted Extraction

lMicrowave assistedextraction (MAE) can be classified as a green extraction technique due to the fact that it reduces the use of solvent and shortens the extraction time. Microwaves are electromagnetic fields in the range 300 MHz to 300 GHz. MAE is based is dielectric heating, that is the process in which a microwave electromagnetic radiation heats a dielectric material by molecular dipole rotation of the polar components present in the matrix (Ran et al. 2019). The solvent penetrates inside the solid matrix by diffusion and the solute is dissolved to reach a concentration that is limited by the solid's characteristics. Several parameters should be considered to optimize the MAE process: solvent, solid–solvent ratio, microwave power, temperature, and time. Ethanol, alone or in combination with water, is one of the most common solvents used in MAE because it has a good capacity to absorb microwave energy and exhibits good solubilizing properties toward phenolic compounds (Bromberger-Soqueta et al. 2019).

MAE has been studied for the extraction of different bioactive compounds (carotenoids, polyphenol, etc.) from a wide variety of fruit and vegetable by-products. As an example, MAE has been employed for extracting lycopene from tomato peel (Coelho et al. 2019; Ho et al. 2015), polyphenols from eggplant peel (Doulabi et al. 2020), apple pomace (Grigoras et al. 2012), mango peel (Dorta et al. 2013), kiwi pomace (Carbone et al. 2020), raw solid waste of orange juice (Kontantinos et al 2021), pomegrate fruit peel (Prodomos et al. 2020), and other compounds such as bioactive inositols from lettuce (Zuluoga et al. 2020), and carbohydrates from potato peel (Gaudino et al. 2020).

9.5.2.4 Pressurized Liquid Extraction

Pressurized liquid extraction (PLE) is considered to be a green extraction technology due to its not requring the use of organic solvents allowing water to be employed and reducing the amount of solvent needed for the extraction of bioactive compounds from the fruit and vegetable waste. PLE extraction is a solid–liquid extraction that employs pressurized solvents at high temperature (>100 °C). The application of high pressure and temperature to the solvent allows it to remain in a liquid state beyond its normal boiling point. The high pressure and temperature increases the solubility of the bioactive compounds and the mass transfer rate. PLE produces more efficient extractions than the conventional extraction methods and requires smaller solvent amounts (Lizárraga-Velázquez et al. 2020). PLE has been efficiently used to extract bioactive compounds, mainly phenolic compounds, from different fruit and vegetable waste. For example, PLE conditions were optimized for the extraction of phenolic compounds from pomegranate fruit peel using water at 3 MPa, 126.1 °C, solvent–solid 54.8 mL g^{-1} for an extration time of 18.5 min (Yan et al. 2017). Also, different PLE conditions were studied for the extraction of saponins from ginseng waste water. The optimal conditions were 43.45 bar, 207 °C, solid–solvent ratio of 0.04 g mL^{-1}, with an extraction time of 15 min and an agitation speed of 199 rpm (Saravana et al. 2016). PLE has been studied for the extraction of phenolic compounds and flavonoids from onion skin waste using waster as solvent. The optimum PLE extraction conditions were 30 bar, 170–230 °C, pH 6, particle size 200–500 nm and time of treatment 30 min (Munir et al. 2018). These PLE examples show that it is essential to select the process conditions for each fruit and vegetable waste to avoid degradation of the bioactive compounds.

9.5.2.5 Supercritical Fluid Extraction

Supercritical fulid extraction (SFE) is considered to be a green technology due to its avoidance of the use of toxic organic solvents. The use of supercritical fluids for recovery of bioactive compounds from fruit and vegetable waste has the advantage that a supercritical fluid has lower viscosity and a higher diffusion coefficient than a liquid solvent. These characteristics allow greater penetration of the fluid into the matrix and, therefore, a more efficient extraction of bioactive compounds even at room temperature. Also, the extracted compounds are easily recovered by reducing the pressure and thus allowing the supercritical fluid to return to the gas phase (da Silva et al. 2016; Gallego et al. 2019). Carbon dioxide (CO_2) is the fluid commonly used in SFE since it is present at a low critical temperature (31.1 °C) and pressure (73.7 bar) so it reduces the degradation of the thermal sensitive compounds. Supercritical CO_2 extraction (SC-CO_2) is limited to non-polar or mid-polar compounds such as terpenes (essential oils and carotenoids) although co-solvents such as ethanol have been used to extract more polar solvents (da Silva et al. 2016). An efficient extraction of bioactive compounds from fruits and vegetable waste depends not only the physicochemical parameters of the SC-CO_2 (temperature, pression time, solvent flow rate, and amount of co-solvent) but also the nature of the waste matrix and the processing rquired before extraction (drying, lyophilization, particle size, etc.). The optimization of these parameters for each type of waste and bioactive compound to be extracted is essential to obtain efficient and selective extraction and good yield of the final product (Piechowiak et al. 2020). For example, SC-CO_2 was used to extract carotenoids from carrot peel using ethanol as co-solvent and the optimum conditions for the maximum carotenoid recovery (86.1%), were: temperature 59 °C, pressure 349 bar, and co-solvent percentage 15.5% (De Andrade Lima et al. 2018). Also, SC-CO_2 were used for recovering luteolin and chlorophylls from spinach waste using the optimal conditions of: temperature 56 °C, pressure 39 MPa, and 10% of ethanol as co-solvent (Derrien et al. 2018). A supercritical fluid extraction (SC-CO2) was used to extract high-quality oleoresin rich in carotenoids (lycopene and β-carotene) and phenolic compounds from tomato skin by-products using the most efficient conditions of: temperature 60 °C, pressure 550 bar, fluid flow rate 2 mL min^{-1} and time of 80 min (Pellicanco et al. 2020). Nowadays, supercritical fluid extraction of carotenoids from vegetable waste matrices has been extensively studied (De Andrade Lima et al. 2019).

9.5.2.6 Deep Eutectic Solvent Extraction

Deep eutectic solvents (DES) are a new type of green solvents that hae been developed to replace common organic solvents that present high toxicity and volatility, leading to evaporation of volatile organic compounds to the atmosphere (Dai et al. 2013). DES are easily prepared by mixing a hydrogen bond acceptor (HBA) and a hydrogen bond donor (HBD). DES compared to organic solvents are biodegradable, with very low toxicity and low price. Generally, DES are prepared with the salt choline chloride as HBA and using as HBD, ethylene glycol, glycerol, but alsoalcohols, amino acids (alanine, proline, serine, etc.), carboxylic acids (lactic acid, citric acid, ascorbic acid, etc.) and sugars (glucose, fructose, xylose, etc.) (Figure 9.8). When these last natural metabolites are employed, the solvent is called natural deep eutectic solvent (NADES).

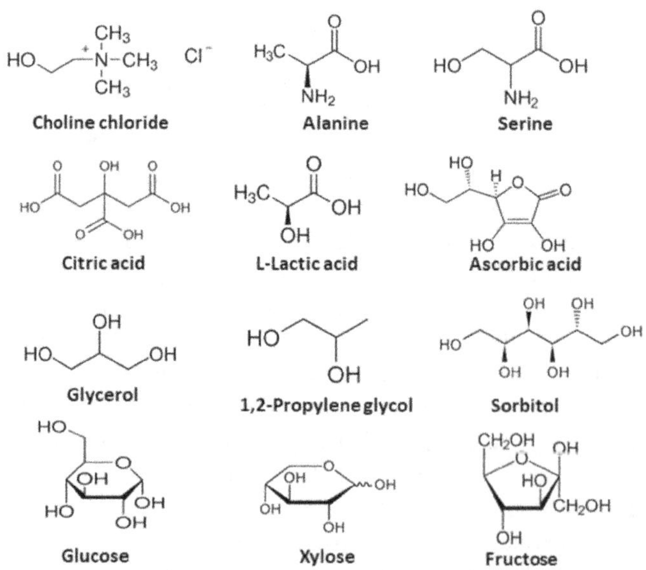

Figure 9.8 Different compounds with the ability to form natural deep eutectic solvents (NADES).

Different DES and NADES such as lactic acid–glucose, glucose–choline chloride or fructose–glucose–sucrose, have demonstrated a high ability to extract phenolic compounds from different plant materials such as safflower, green coffee beans, etc. (Paiva et al. 2014). NADES has good physicochemical characteristics which can be used as alternative extraction solvents. They are liquid at room temperature, their viscosity can be adjusted easily and present low toxicity, and are considered to be sustainable solvents. NADES can dissolve both polar and nonpolar compounds and can be used as solvents for the extraction of different types of bioactive compounds from fruit and vegetable by-products and waste, although the type of NADES employed and the extraction conditions will depend on the nature of the bioactive compound and the type of plant matrix (Kalhor and Ghandi 2019). NADES are being investigated for the extraction of different bioactive compounds, mainly phenolic compounds, from the agro-food industrial by-products such as red grape and tomato pomace, onion waste, olive cake, among other (Dordevic et al. 2018; Fernández et al. 2018; Makris 2018).

9.6 Conclusion

Vegetable waste and by-products has been demonstrated to be a good source of valuable biomolecules (pigments, carotenoids, phenolic compounds, terpenes, phytosterol, dietary fibers, pectins, polisaccharides, organic acids, enzymes, proteins, essential oils, etc.) that has been associated with health-promoting properties in humans. The application of green technologies for the recuperation of vegetable waste and by-product has been very efficient in the extraction of these bioactive compounds and is also an interesting way to reduce industrial waste, cost, and environmental impact generated by the habitual destruction of vegetable by-products, with the added value of obtaining phytochemicals with health beneficial properties.

However, additional studies are needed regarding toxicology, before using functional ingredients obtained from vegetable processing by-products, to ensure that the ingredient is free of pesticides and other undesired or toxic substances. Bioactivity studies are also required to enable us to determine the bioaccessibility and bioavailability of the bioactive compounds extracted from these by-products. Therefore, the correct characterization of vegetable by-products and their corresponding extracts is critical for potential commercialization. The industrial production of bioactive compounds from the by-products obtained in the processing of vegetables and their use as functional ingredients in foods, requires the coordination of interdisciplinary studies from food technologists, food chemists, nutritionists and toxicologists.

References

Aguilera, Y., Martin-Cabrejas, M.A., and Gonzalez de Mejia, E. (2016). Phenolic compounds in fruits and beverages consumed as part of the Mediterranean diet: their role in prevention of chronic diseases. *Phytochemistry Reviews* 15: 405–423.

Albishi, T., John, J.A., Abdulrahman, S.A.K., and Shahidi, F. (2013). Phenolic content and antioxidant activities of selected potato varieties and their processing by-products. *Journal of Functional Foods* 5 (2): 590–600.

Alzate, L.M., González, D., Hincapié, S., Cardona, B.L., Londozo-Londoño, J., and Jiménez-Cartagena, C. (2017). The profile of bioactive substances in ten vegetable and fruit by-products from a food supply chain in Colombia. *Sustainable Production and Consumption* 9: 37–43.

Andreou, V., Dimopoulos, G., Dermesonlouoglou, E., and Taoukis, P. (2020). Application of pulsed electric fields to improve product yield and waste valorization in industrial tomato processing. *Journal of Food Engineering* 270: 109778.

Appleton, K.M., Hemingway, A., Saulais, L., Dinnella, C., Monteleone, E., Depezay, L., Morizet, D., Armando Perez-Cueto, F.J., Bevan, A., and Hartwell, H. (2016). Increasing vegetable intakes: rationale and systematic review of published interventions. *European Journal of Nutrition* 55: 869–896.

Baysal, T., Ersus, S., and Starmans, D.A.J. (2000). Supercritical CO2 extraction of β-carotene and lycopene from tomato paste waste. *Journal of Agricultural and Food Chemistry* 48 (11): 5507–5511.

Benetou, V., Orfanos, P., Lagiou, P., Trichopoulos, D., Boetta, P., and Trichopoulou, A. (2008). Vegetables and fruits in relation to cancer risk: evidence from the Greek EPIC cohort study. *Cancer Epidemiology Biomarkers and Prevention* 17: 387–392.

Benítez, V., Mollá, E., Martín-Cabrejas, M.A., Aguilera, Y., López-Andréu, F.J., Cools, K., Terry, L.A., and Esteban, R.M. (2011). Characterization of industrial onion wastes (*Allium cepa* L.): dietary fibre and bioactive compounds. *Plant Foods for Human Nutrition* 66 (1): 48–57.

Bromberger Soquetta, M., Terra, L.M., and Bastos, C.P. (2018). Green technologies for the extraction of bioactive compounds in fruits and vegetables. *CYTA: Journal of Food* 16 (1): 400–412.

Burton-Freeman, B., Talbot, J., Park, E., Krishnankutty, S., and Edirisinghe, I. (2012). Protective activity of processed tomato products on postprandial oxidation and inflammation: a clinical trial in healthy weight men and women. *Molecular Nutrition and Food Research* 56 (4): 622–631.

Bustamante, J., van Stempvoort, S., García-Gallarreta, M., Houghton, J.A., Briers, H.K., Budarin, V.L., Matharu, A.S., and Clark, J.H. (2016). Microwave assisted hydro-distillation of essential oils from wet citrus peel waste. *Journal of Cleaner Production* 137: 598–605.

Caldeira, C., De Laurentiis, V., Corrado, S., van Holsteijn, F., and Sala, S. (2019). Quantification of food waste per product group along the food supply chain in the European Union: a mass flows analysis. *Resources, Conservation and Recycling* 149: 479–488.

Carbone, K., Tiziana, A., and Iadecola, A. (2020). Exploitation of kiwi juice pomace for the recovery of natural antioxidants through microwave-assisted extraction. *Agriculture* 10: 435.

Celano, R., Docimo, T., Piccinelli, A.L., Gazzerro, P., Tucci, M., Di Sanzo, R., Carabetta, S., Campon, L., Russo, M.T., and Rastrelli, L. (2021). Onion peel: turning a food waste into a resource. *Antioxidants* 10: 304.

Chau, C.V., Chen, C.H., and Lee, M.H. (2004). Comparison of the characteristics, functional properties, and in vitro hypoglycemic of various carrot insoluble fiber rich fractions. *LWT – Food Science and Technology* 37 (2): 155–160.

Choi, I.S., Cho, E.J., Moon, J.H., and Bae, H.J. (2015). Onion skin waste as a valorization resource for the by-products quercetin and biosugar. *Food Chemistry* 188: 537–542.

Choudhari, S. and Ananthanarayan, L. (2007). Enzyme aided extraction of lycopene from tomato tissues. *Food Chemistry* 102 (1): 77–81.

Coelho, M., Pereira, R., Rodigues, A.S., Teixeira, J.A., and Pintado, M.E. 2019. Extraction of tomato by-products' bioactive compounds using ohmic technology. *Food and Bioproducts Processing* 11: 329–339.

Colina-Coca, C., González-Peña, D., de Ancos, B., and Sánchez-Moreno, C. (2017). Dietary onion ameliorates antioxidant defence, inflammatory response, and cardiovascular risk biomarkers in hypercholesterolemic Wistar rats. *Journal of Functional Foods* 36: 300–309.

Colina-Coca, C., Rodríguez-Alcalá, L., Fontecha, J., González-Peña, D., De Ancos, B., and Sánchez-Moreno, C. (2014). Effects of hypercholesterolemic diet enriched with onion 518 as functional ingredient on fatty acid metabolism in Wistar rats. *Food Research International (Ottawa, Ont.)* 64: 546–552.

Crowe, F.L., Roddam, A.W., Key, T.J., Appleby, P.N., Overvad, K., Jakobsen, M.U. et al. (2011). Fruit and vegetable intake and mortality from ischaemic heart disease: results from the European prospective investigation into cancer and nutrition (EPIC)-heart study. *European Heart Journal* 32: 1235–1243.

da Silva, R.P.F.F., Rocha-Santos, T.A.P., and Duarte, A.C. (2016). Supercritical fluid extraction of bioactive compounds. *TrAC – Trends in Analytical Chemistry* 76: 40–51.

Dai, Y., van Spronsen, J., Witkamp, G.J., Verpoorte, R., and Choi, Y.H. (2013). Natural deep eutectic solvents as new potential media for green technology. *Analytica Chimica Acta* 766: 61–68.

De Andrade Lima, M., Charalampopoulos, D., and Chatzifragkou, A. (2018). Optimisation and modelling of supercritical CO2 extraction process of carotenoids from carrot peels. *The Journal of Supercritical Fluids* 133: 94–102.

De Andrade Lima, M., Kestegoglou, I., and Charalampopoulos, D. (2019). Supercritical fluid extraction of carotenoids from vegetable waste matrices. *Molecules* 24 (3): 466.

Derrien, M., Aghabararnejad, M., Gosselin, A., Desjardins, Y., Angers, P., and Boumghar, Y. (2018). Optimization of supercritical carbon dioxide extraction of lutein and chlorophyll from spinach by-products using response surface methodology. *LWT – Food Science and Technology* 93: 79–87.

Dordević, B.S., Todorović, Z.B., Troter, D.Z., Stanojević, L.P., and Veljković, V.B. (2018). The extraction of quercetin from waste onion (*Allium cepa* L.): tunic by the aqueous solutions of different deep eutectic solvents. *Advanced Technologies* 7 (2): 05–10.

Dorta, E., Lobo, M.G., and González, M. (2013). Improving the efficiency of antioxidant extraction from mango peel by using microwave-assisted extraction. *Plant Foods for Human Nutrition* 68: 190–199.

Doulabi, M., Golmakani, M.T., and Ansari, S. (2020). Evaluation and optimization of microwave-assisted extraction of bioactive compounds from eggplant peel by-product. *Journal of Food Processing and Preservation* 44: e14853.

Drosou, C., Kyrialopouloy, K., Bimpilas, A., Simogiannis, D., and Krokida, M.A. (2015). Comparative study on different extraction technique to recover red grape pomace polyphenols from vinification byproducts. *Industrial Crops and Products* 75: 141–149.

El-Mesallamy, A.M.D., Abdel-Hamid, N., Srour, L., and Hussein, S.A.M. (2020). Identification of polyphenolic compounds and hepatoprotective activity of artichoke (Cynara scolymus L.) edible part extracts in rats. *Egyptian Journal of Chemistry* 63: 2273–2285.

FAO. (2017). Strategic work of FAO for sustainable food and agriculture of the United States. http://www.fao.org/3/a-i6488e.pdf (accessed 10 May 2021).

FAO. (2018). The future of food and agriculture. Alternative pathways to 2050. Food and Agriculture of the United States. http://www.fao.org/3/I8429EN/i8429en.pdf (accessed 11 May 2021).

FAO. (2019) Food and agriculture organization of the United Nations. Onion Annual Production. http://www.fao.org/faostat/es/#data/QC (accessed 9 May 2021).

Femenia, A., Robertson, J.A., Waldron, K.W., and Sevendran, R.R. (1998). Cauliflower (*Brassica oleracea* L), globe artichoke (*Cyanara scolymus*) and chicory witloof (Cichorium intybus) processing byproducts as sources of dietary fibre. *Journal of the Science of Food and Agriculture* 77 (4): 511–518.

Fernández, M.L.A., Espino, M., Gomez, F.J.V., and Silva, M.F. (2018). Novel approaches mediated by tailor-made green solvents for the extraction of phenolic compounds from agro-food industrial by-products. *Food Chemistry* 239: 671–678.

Fuentes, E.R., Carle, L., Astudillo, L., Guzmán, M., Gutiérrez, G., Carrasco, G., and Palomo, I. (2013). Antioxidant and antiplatelet activities in extracts from green and fully ripe tomato fruits (*Solanum licopersicum*) and pomace from industrial tomato processing. *Evidence-based Complementary and Alternative Medicine* 2013: 867578.

Galavi, A., Hosseinzadeh, H., and Razavi, B.M. (2021). The effects of *Allium cepa* L. (onion) and its active constituents on metabolic syndrome: a review. *Iranian Journal of Basic Medical Sciences* 24: 3–16.

Gallego, R., Bueno, M., and Herrero, M. (2019). Sub-and supercritical fluid extraction of bioactive compounds from plants, food-by-products, seaweeds and microalgae–an update. *TrAC – Trends in Analytical Chemistry* 116: 198–213.

Gaudino, E.C., Colletti, A., Grillo, G., Tabasso, S., and Cravotto, G. (2020). Emerging processing technologies for the recovery of valuable bioactive compounds from potato peels. *Foods* 9 (11): 1598.

Giovannucci, E. (2002). A review of epidemiologic studies of tomatoes, lycopene, and prostate cancer. *Experimental Biology and Medicine* 227 (10): 852–859.

Gómez-Mejía, E., Rosales-Conrado, N., León-González, M.E., and Madrid, Y. (2019). Citrus peels waste as a source of value-added compounds: extraction and quantification of bioactive polyphenols. *Food Chemistry* 295: 289–299.

González-Peña, D., Angulo, J., Vallejo, S., Colina-Coca, C., De Ancos, B., Sánchez-Ferrer, C.F., Peiró, C., and Sánchez-Moreno, C. (2014). High-cholesterol diet enriched with onion affects endothelium-dependent relaxation and NADPH oxidase activity in mesenteric microvessels from Wistar rats. *Nutrition and Metabolism (Lond)* 11: 57.

González-Peña, D., Checa, A., de Ancos, B., Wheelock, C.E., and Sánchez-Moreno, C. (2017). New insights into the effects of onion consumption on lipid mediators using a diet-induced model of hypercholesterolemia. *Redox Biology* 11: 205–212.

González-Peña, D., Colina-Coca, C., Char, C.D., Cano, M.P., de Ancos, B., and Sánchez-Moreno, C. (2013). Hyaluronidase inhibiting activity and radical scavenging potential of flavonols in processed onion. *Journal of Agricultural and Food Chemistry* 61: 4862–4872.

Grigoras, E.D., Lazar, G., and Elfakir, C. (2012). Bioactive compounds extraction from pomace of four apple varieties. *Journal of Engineering Research and Studies* 18: 96–103.

Guo, Y. and Bruno, R.S. (2015). Endogenous and exogenous mediators of quercetin bioavailability. *The Journal of Nutritional Biochemistry* 26 (3): 201–210.

Gustavsson, J., Cederberg, S., Sonesson, U., Van Otterdijk, R., and Meybeck, A. (2011) Global food losses and food waste: extent, causes and prevention. FAO. Rome. Food and Agriculture Organization of the United Nations. http://www.fao.org/3/a-i2697e.pdf (accessed 20 May 2021).

Hernández-Ortega, M., Kissangou, G., Necoechea-Mondragón, H., Sánchez-Pardo, M.E., and Ortiz-Moreno, A. (2013). Microwave dried carrot pomace as a source of fiber and carotenoids. *Food and Nutrition Sciences* 4: 1037–1046.

Ho, K.K.H.Y., Ferruzzi, M.G., Liceaga, A.M., and San Martín-González, M.F. (2015). Microwave-assisted extraction of lycopene in tomato peels: effect of extraction conditions on all-trans and cis-isomer yields. *LWT – Food Science and Technology* 62: 160–168.

Hu, D., Huang, J., Wang, Y., Zhang, D., and Qu, Y. (2014). Fruits and vegetables consumption and risk of stroke. *Meta-analysis of Prospective Cohort Studies* 45: 1613–1619.

Icier, F. (2010). Ohmic blanching effects on drying of vegetable byproduct. *Journal of Food Process Engineering* 33 (4): 661–683.

Jiang, G., Wu, Z., Ameer, K., and Song, C. (2021). Physicochemical, antioxidant, microstructural, and sensory characteristics of biscuits as affected by addition of onion residue. *Journal of Food Measurement and Characterization* 15: 817–825.

Joly, N., Souidi, K., Depraetere, D., Wils, D., and Martin, P. (2021). Potato by-products as a source of natural chlorogenic acids and phenolic compounds: extraction, characterization, and antioxidant capacity. *Molecules* 26: 177.

Kalhor, P. and Ghandi, K. (2019). Deep eutectic solvents for pretreatment, extraction, and catalysis of biomass and food waste. *Molecules* 24: 4012.

Kalmpourtzidou, A., Eilander, A., and Talsma, E.F. (2020). Global vegetable intake and supply compared to recommendations: a systematic review. *Nutrients* 12: 1558.

Kalogeropoulos, N., Chiou, A., Pyriochou, V., Peristeraki, A., and Karathanos, V.T. (2012). Bioactive phytochemical in industrial tomatoes and their processing byproducts. *LWT – Food Science and Technology* 49 (2): 213–216.

Khajah, M.A., EL-Hashim, A.Z., Orabi, K.Y., Hawai, S., and Sary, H.G. (2021). Onion bulb extract can both reverse and prevent colitis in mice via inhibition of proinflammatory signaling molecules and neutrophil activity. *PLoS ONE* 15 (10): e0233938. doi: https://doi.org/10.1371/journal.pone.0233938.

Kodagoda, K.H.G.K. and Marapana, R.A.U.J. (2017). Utilization of fruit processing by-products for industrial applications: a review. *International Journal of Food Sciences and Nutrition* 2: 24–30.

Konstantinos, P., Ioannis, G., Konstantinos, G. et al. (2021). Optimization of vacuum-microwave-assisted extraction of natural polyphenols and flavonoids from raw solid waste of the orange juice producing industry at industrial scale. *Molecules* 26: 246.

Kukić, J., Popović, V., Petrović, S., Mucaji, P., Ćirić, A., Stojkovic′, D., and Soković, M. (2008). Antioxidant and antimicrobial activity of Cynara cardunculus extracts. *Food Chemistry* 107: 861–868.

Larrosa, M., Llorach, R., Espín, J.C., and Tomás-Barberán, F.A. (2002). Increase of antioxidant activity of tomato juice upon functionalisation with vegetable byproduct extracts. *LWT – Food Science and Technology* 35 (6): 532–542.

Lee, J. and Mitchell, A.E. (2012). Pharmacokinetics of quercetin absorption from apples and onions in healthy humans. *Journal of Agricultural and Food Chemistry* 60 (15): 3874–3881.

Lewis, C.E., Walker, J.R.L., Lancaster, J.E., and Sutton, K.H. (1998). Determination of anthocyanins, flavonoids and phenolic acids in potatoes I: coloured cultivars of solanum tuberosum L. *Journal of the Science of Food and Agriculture* 77: 45–57.

Leyva-López, N., Valdez-Torres, B., Delgado-Vargas, F., García-Estrada, R.S., and Heredia, J.B. (2017). Optimized enzyme-aided extraction enhances recovery of carotenoids from tomato peel and improves the biological activity. *International Journal of Pharma and Bio Sciences* 8: 721–727.

Liu, R.H. (2013a). Health-promoting components of fruits and vegetables in the diet. *Advances in Nutrition* 4 (3): 384S–392S.

Liu, R.H. (2013b). Dietary bioactive compounds and their health implications. *Journal of Food Science* 78: A18–A25.

Lizárraga-Velázquez, C.E., Leyva-López, N., Hernández, C., Gutiérrez-Grijalva, E.P., Salazar-Leyva, J.A., Osuna-Ruíz, I., Martínez-Montaño, E. et al. (2020). Antioxidant molecules from plant waste: extraction, techniques and biological properties. *Processes* 8: 1566.

Llorach, R., Espín, J.C., Tomás-Barberán, F.A., and Ferreres, F. (2002). Artichoke (*Cynara scolymus* L.) byproducts as a potential source of health-promoting antioxidants phenolics. *Journal of Agricultural and Food Chemistry* 50 (12): 3458–3464.

Llorach, R., Martínez-Sánchez, A., Tomás-Barberán, F.A., Gil, M.I., and Ferreres, F. (2008). Characterisation of polyphenols and antioxidant properties of five lettuce varieties and escarole. *Food Chemistry* 108 (3): 1028–1038.

Llorach, R., Tomás-Barberán, F.A., and Ferreres, F. (2004). Lettuce and chicory byproducts as a source of antioxidant phenolic extracts. *Journal of Agricultural and Food Chemistry* 52: 5109–5116.

López-Salas, L., Borrá-Linars, I., Quintin, D., García-Gómez, P., Giménez-Martínez, R., Segura-Carretero, A., and Lozano-Sánchez, J. (2021). Artichoke by-products as natural source of phenolic food ingredient. *Applied Sciences* 11: 3788.

Lu, Z., Wang, J., Gao, R., Yea, F., and Zhao, G. (2019). Sustainable valorisation of tomato pomace: a comprehensive review. *Trends in Food Science and Technology* 86: 172–178.

Makris, D.P. (2018). Green extraction processes for the efficient recovery of bioactive polyphenols from wine industry solid wastes: recent progress. *Current Opinion in Green and Sustainable* 13: 50–55.

Marathe, S.J., Jadhav, S.B., Bankar, S.B., Dybey, K.K., and Singhai, R.S. (2019). Imrovements in the extraction of bioactive compounds by enzymes. *Current Opinion in Food Science* 25: 62–72.

Marrelli, M., Amodeo, V., Statti, G., and Conforti, F. (2018). Biological properties and bioactive components of Allium cepa L.: focus on potential benefits in the treatment of obesity and related comorbidities. *Molecules* 30: 119.

Martharu, A.S., de Melo, E.M., and Houghton, J. (2016). Opportunity for high value-added chemicals from food supply chain wastes. *Bioresource Technology* 215: 123–130.

Masood, S., Rehman, A., Bashir, S., Shazly, M.E., Imran, M., Khalil, P., Ifthikar, F., Jaffar, H.M., and Khursheed, T. (2021). Investigation of the anti-hyperglycemic and antioxidant effects of wheat bread supplemented with onion peel extract and onion powder in diabetic rats. *Journal of Diabetes and Metabolic Disorders*. doi: 10.1007/s40200-021-00770-x.

Mattila, P. and Hellström, J. (2007). Phenolic acids in potatoes, vegetables, and some of their products. *Journal of Food Composition and Analysis* 20 (3–4): 152–160.

Mattson, L., Williams, H., and Begher, J. (2018). Waste of fresh fruit and vegetables at retailers in Sweden – measuring and calculation of mass, economic cost and climate impact. *Resources, Conservation, and Recycling* 130: 118–126.

Michalak-Majewska, M., Złotek, U., Szymanowska, U., Szwajgier, D., Stanikowski, P., Matysek, M., and Sobota, A. (2020). Antioxidant and potentially anti-inflammatory properties in pasta fortified with onion skin. *Applied Sciences* 10: 8164.

Munir, M.T., Kheirkhah, H., Baroutian, S., Quek, S.Y., and Young, B.R. (2018). Subcritical water extraction of bioactive compounds from waste onion skin. *Journal of Cleaner Production* 183: 487–494.

Nadar, S.S., Rao, P., and Rathod, V.K. (2018). Enzyme assisted extraction of biomolecules as an approach to novel extraction technologies. *Food Research International (Ottawa, Ont.)* 108: 309–330.

Nile, A., Gansukh, E., Park, G.S., Kim, D.H., and Nile, S.H. (2021). Novel insights on the multi-functional properties of flavonol glucosides from red onion (*Allium cepa* L) solid waste–In vitro and in silico approach. *Food Chemistry* 335: 127650.

O'Shea, N., Arendt, E.K., and Gallagher, E. (2012). Dietary fibre and phytochemical characteristics of fruit and vegetable by-products and their recent applications as novel ingredients in food products. *Innovative Food Science and Emerging Technologies* 16: 1–10.

Paiva, A., Craveiro, R., Aroso, I., Martins, M., Reis, R.L., and Duarte, A.R. (2014). Natural deep eutectic solvents – solvents for the 21st century. *ACS Sustainable Chemistry and Engineering* 2: 1063–1071.

Pandino, G., Lombardo, S., and Mauromicale, G. (2013). Globe artichoke leaves and floral stems as a source of bioactive compounds. *Industrial Crops and Products* 44: 44–49.

Panzella, L., Moccia, F., Nasti, R., Marzorati, S., Verotta, L., and Napolitano, A. (2020). Bioactive phenolic compounds from agri-food wastes: an update on green and sustainable extraction methodologies. *Frontiers in Nutrition* 7: 60.

Pellicanco, T.M., Sicari, V., Loizzo, M.R., Lepporini, M., Falco, T., and Poiana, M. (2020). Optimizing the supercritical fluid extraction process of bioactive compounds from processed tomato skin by-products. *Food Science and Technology* 40 (3): 692–697.

Peschel, W., Sánchez-Rabaneda, F., Diekmann, W., Plescher, A., Gartzía, I., Jiménez, D., Lamuela-Raventós, R., Buxaderas, S., and Codin, C. (2006). An industrial approach in the search of natural antioxidants from vegetable and fruit wastes. *Food Chemistry* 97 (1): 137–150.

Piechowiak, T., Grzelak-Błaszczyk, K., Bonikowski, R., and Balawejder, M. (2020). Optimization of extraction process of antioxidant compounds from yellow onion skin and their use in functional bread production. *LWT – Food Science and Technology* 117: 108614.

Plazzota, S. and Manzzoco, L. (2018). Effect of ultrasounds and high-pressure homogenization on the extraction of antioxidant polyphenols from lettuce waste. *Innovative Food Science and Emerging Technologies* 50: 11–19.

Porter, S.D., Reay, D.S., Bomberg, E., and Higgins, P. (2018). Avoidable food losses and associated production-phase greenhouse gas emissions arising from application of cosmetic standards to fresh fruit and vegetables in Europe and the UK. *Journal of Cleaner Production* 201: 869–878.

Prodromos, S., Stefanos, L., Konstantinos, P., and Giovasis, I. (2020). Optimization of vacuum microwave-assisted extraction of pomegranate fruits peels by the evaluation of extracts' phenolic content and antioxidant activity. *Foods* 9 (11): 1655.

Rabelo, R.S., Machado, M.T., Martínez, J., and Hubinger, M.D. (2016). Ultrasound assisted extraction and nanofiltration of phenolic compounds from artichoke solid wastes. *Journal of Food Engineering* 178: 170–180.

Rajkovic, M.B., Minic, D.P., Milincic, D., and Zdravkovic, M. (2020). Circular economy in food industry. *Zastita Materijala* 6: 229–250.

Ran, X.L., Zhang, M., Wang, Y., and Adhikari, B. (2019). Novel technologies applied for recovery and value addition of high value compounds from plant byproducts: a review. *Critical Reviews in Food Science and Nutrition* 59: 450–461.

Rodríguez Galdón, B., Rodríguez Rodríguez, E.M., and Díaz Romero, C. (2008). Flavonoids in onion cultivars (Allium cepa L.). *Journal of Food Science* 73: C599–C605.

Roldán, E., Sánchez-Moreno, C., De Ancos, B., and Cano, M.P. (2008). Characterisation of onion (*Allium cepa* L.) by-products as food ingredients with antioxidant and antibrowning properties. *Food Chemistry* 108 (3): 907–916.

Roldán-Marín, E., Krath, B.N., Poulsen, M., Binderup, M.L., Nielsen, T.H., Hansen, M., Barri, T., Langkilde, S., Cano, M.P., Sánchez-Moreno, C., and Dragsted, L.O. (2009a). Effects of an onion by-product on bioactivity and safety markers in healthy rats. *The British Journal of Nutrition* 102 (11): 1574–1582.

Roldán-Marín, E., Sánchez-Moreno, C., Lloría, R., De Ancos, B., and Cano, M.P. (2009b). Onion high-pressure processing: flavonol content and antioxidant activity. *LWT – Food Science and Technology* 42 (4): 835–841.

Sagar, N.A., Pareek, S., Sharma, S., Yahia, E.M., and Lobo, M.G. (2018). Fruit and vegetable waste: bioactive compounds, their extraction, and possible utilization. *Comprehensive Reviews in Food Science and Food Safety* 17 (3): 512–531.

Salamatullah, A.M., Uslu, N., Özcan, M.N., Alkaltham, M.S., and Hayat, K. (2021). The effect of oven drying on bioactive compounds, antioxidant activity, and phenolic compounds of white and red-skinned onion slices. *Journal of Food Processing and Preservation* 45 (2): e15173.

Samarin, A., Poorazarang, H., Hematyar, N., and Elhamirad, A. (2012). Phenolics in potato peels: extraction and utilization as natural antioxidants. *World Applied Sciences* 18: 191–195.

Sánchez, A. and Vázquez, A. (2017). Bioactive peptides: a review. *Food Quality and Safety* 1: 29–46.

Sanchez-Rabaneda, F., Jauregui, O., and Lamuela-Raventos, R.M. (2003). Identification of phenolic compounds in artichoke waste by high performance liquid chromatography—Tandem mass spectrometry. *Journal of Chromatography. A* 1008: 57–72.

Saravana, P.S., Getachew, A.T., Ahmed, R., Cho, Y.J., Lee, Y.B., and Chun, B.S. (2016). Optimization of phytochemicals production from the ginseng by-products using pressurized hot water: experimental and dynamic modelling. *Biochemical Engineering Journal* 113: 141–151.

Shallan, M.A., Ali, M.A., Meshrf, W.A., and Marrez, D.A. (2020). In vitro antimicrobial, antioxidant and anticancer activities of globe artichoke (Cynara cardunculus var. scolymus L.) bracts and receptacles ethanolic extract. *Biocatalysis and Agricultural Biotechnology* 29: 101774.

Sharma, H.K. and Kumar, P. (2018). Utilization of carrot pomace. In: Food Processing By-Products and Their Utilization. Ch. 10 (ed. A.K. Anal), 207–229. Hoboken (NJ): John Wiley & Sons Ltd.

Sharma, S., Sheehy, T., and Kolonel, L. (2014). Sources of vegetables, fruits, and vitamins A, C and E among five ethnic groups: results from the Multiethnic Cohort Study. *European Journal of Clinical Nutrition* 68: 384–391.

Shen, Q., Dai, Z., and Lu, Y. (2010). Rapid determination of caffeoylquinic acid derivatives in Cynara scolymus L. by ultra-fast liquid chromatography/tandem mass spectrometry based on a fused core C18 column. *Journal of Separation Science* 33: 3152–3158.

Singh, A., Sabally, K., Kubow, S., Donnelly, D.J., Gariepy, Y., Orsat, V., and Raghavan, G.S.V. (2011). Microwave-assisted extraction of phenolic antioxidants from potato peels. *Molecules* 16: 2218–2232.

Sun, T., Simon, P.W., and Tanumihardjo, S.A. (2009). Antioxidant phytochemicals and antioxidant capacity of biofortified carrots (*Daucus carota* L.) of various colors. *Journal of Agricultural and Food Chemistry* 57: 4142–4147.

Szabo, K., Catoi, A.F., and Vodnar, D.C. (2018). Bioactive compounds extracted from tomato processing by-products. *Plant Foods for Human Nutrition* 73 (4): 268–277.

Teshika, J.D., Zakariyyah, A.M., Zaynab, T., Zengin, G., Rengasamy, K.R.R., Pandian, S.K., and Fawzi, M.M. (2019). Traditional and modern uses of onion bulb (*Allium cepa* L.): a systematic review. *Critical Reviews in Food Science and Nutrition* 59: S39–S70.

Tomaz, L., Maslov, L., Sypic, D., Preiner, D., Asperger, D., and Lontic, J.K. (2016). Recovery of flavonoids from grape skins by enzyme-assisted extraction. *Separation Science and Technology* 51: 255–268.

Tomé-Carneiro, J. and Visioli, F. (2016). Polyphenol-based nutraceuticals for the prevention and treatment of cardiovascular disease: review of human evidence. *Phytomedine* 23 (11): 1145–1174.

UN (United Nations). (2015). Sustainable Development Goals (SDGs). https://www.un.org/sustainabledevelopment/sustainable-development-goals (accessed 20 May 20121).

Villegas, R., Shu, X.O., Gao, Y.T., Yang, G., Elasy, T., Li, H., and Zheng, W. W. (2008). Vegetable but not fruit consumption reduces the risk of type 2 diabetes in Chinese women. *The Journal of Nutrition* 138: 574–580.

Wadhwa, M. and Bakshi, S.P.M. (2013). Utilization of Fruit and Vegetable Wastes as Livestock Feed and as Substrates for Generation of Other Value-Added Products. RAP publication 2013-04 (FAO).

Wang, M., Qin, S., Zhang, T., Song, X., and Zhang, S. (2015). The effect of fruit and vegetable intake on the development of lung cancer: a meta-analysis of 32 publications and 20 414 cases. *European Journal of Clinical Nutrition* 69: 1184–1192.

Wu, F., Jin, Y., Xu, X., and Yang, N. (2017). Electrofluidic pretreatment for enhancing essential oil extraction from citrus peel waste. *Journal of Cleaner Production* 159: 85–94.

Yan, L., Cao, Y., and Zheng, G. (2017). Optimization of subcritical water extraction of phenolic antioxidants from pomegranate (*Punica granatum* L.) peel by response surface methodology. *Analytical Methods* 9: 4647–4656.

Zhan, J., Liu, Y.J., Cai, L.B., Xu, F.R., Xie, T., and He, Q.Q. (2017). Fruit and vegetable consumption and risk of cardiovascular disease: a meta-analysis of prospective cohort studies. *Critical Reviews in Food Science and Nutrition* 57: 1650–1663.

Zhang, Q.W., Lin, L.G., and Ye, W.C. (2018). Techniques for extraction and isolation of natural products: a comprehensive review. *Chinese Medicine* 13: 20.

Zou, T.B., Xia, E.Q., He, T.P., Huang, M.Y., Jia, Q., and Li, H.W. (2014). Ultrasound assisted extraction of mangiferin from mango (*Mangifera indica* L.) leaves using response surface methodology. *Molecules* 19: 1411–1421.

Zuknic, M.H., Norulaini, N.A.N., and Omar, A.K.M. (2012). Supercritical carbon dioxide extraction of lycopene: a review. *Journal of Food Engineering* 112 (4): 253–262.

Zuluaga, A.M., Mena-Garcia, A., Chito-Trujillo, D., Rada-Mendoza, M., Sanz, M.L., and Ruiz-Matute, A.I. (2020). Development of a microwave-assisted extraction method for the recovery of bioactive inositols from lettuce (*Lactuca sativa*) byproducts. *Electrophoresis* 41 (20): 1804–1811.

10

Retention of Natural Bioactive Compounds of Berry Fruits during Surface Decontamination Using an Eco-friendly Sanitizer

María P. Méndez-Galarraga[1,2], *Franco Van de Velde*[1,2], *Andrea M. Piagentini*[1], *and María Elida Pirovani*[1]

[1]*Instituto de Tecnología de Alimentos, Facultad de Ingeniería Química, Universidad Nacional del Litoral, Santiago del Estero 2829, 3000, Santa Fe, Argentina.*
mpirovan@fiq.unl.edu.ar
[2]*Consejo de Investigaciones Científicas y Técnicas (CONICET), Argentina. Godoy Cruz 2290 (C1425FQB) CABA, Argentina.*

10.1 Introduction

Berries are small fleshy fruits, which are cultivated and commonly consumed in fresh and processed forms. This group includes, among others, strawberries (*Fragaria* spp.) and blackberries (*Rubus* spp.). Both fruits are popularly consumed due to the attractiveness of their color and taste and are recognized as a very rich source of vitamin C and phenolic compounds with human health benefits (Jung et al. 2015). In Argentina, the cultivated area, the production, and the market position of these fruits have increased in the last ten years. Production is labor-and capital-intensive, which generates high profitability in small areas, and is a mobilizer of local and regional economies (Gómez Riera et al. 2013).

On the other hand, strawberries and blackberries have an extremely short post-harvest life, since they are susceptible to mechanical damage, physiological and microbiological deterioration, and water loss (Ramos et al. 2013). The spoilage microorganisms may cause up to 15% post-harvest fruit losses. Yeasts and molds are often the majority microflora of raw fruits due to them having a lower pH than vegetables. The main yeast and mold genera present in blackberries and strawberries are *Alternaria, Aspergillus, Botrytis, Fusarium, Penicillium, Rhizopus, and Saccharomyces* (Ramos et al. 2013; Vardar et al. 2012).

In addition to spoilage microorganisms that cause loss of quality, human-pathogenic microorganisms may also be present, affecting the safety of these products. Contamination of fresh fruits and vegetables is of special concern because such products are likely to be consumed raw, thus posing a potential safety problem. The microorganisms most

Biomolecules from Natural Sources: Advances and Applications, First Edition. Edited by Vijai Kumar Gupta, Satyajit D. Sarker, Minaxi Sharma, María Elida Pirovani, Zeba Usmani, and Chelliah Jayabaskaran.

frequently associated with outbreaks related to the consumption of fruits and vegetables include bacteria (*Salmonella* spp., *Listeria monocytogenes*, *Escherichia coli*, and *Shigella* spp.), viruses, and parasites (Ramos et al. 2013). Safe production methods and proper disinfection/decontamination procedures are therefore critical steps in ensuring the safety of fresh and ready-to-eat fruits and vegetables (Artés et al. 2009).

On the other hand, berries provide significant health benefits because of their high levels of polyphenols, vitamins, minerals, and fibers (Giampieri et al. 2012; Nile and Park 2013). Anthocyanins, ellagitannins, proanthocyanidins, and other flavonoids are the main phenolic compounds in strawberries and blackberries. These compounds are not only responsible for some sensorial characteristics, but they also impart relevant biological properties to the consumers such as antioxidant, anti-inflammatory, anti-atherosclerotic, and anti-carcinogenic (Crecente-Campo et al. 2012; Giampieri et al. 2015; Nowicka et al. 2019). There is evidence of the role of polyphenols as signaling molecules involved in modulation of signal pathways, and thereby affecting cellular function and gene expression (Aaby et al. 2012; Van de Velde et al. 2019).

Anthocyanins are responsible for the red–purple color of these berries. Pelargonidin and cyanidin, glycoside derivatives from the anthocyanins, are the main flavonoids found in strawberries with reported concentrations of up to 65 mg 100 g^{-1} of fresh weight (FW) (Giampieri et al. 2012). Meanwhile, cyanidin-based anthocyanins have been reported in blackberries, being cyanidin-3-O-glucoside the main phenolic compound, representing around 71% of the total polyphenolic content of the fruit (Van de Velde et al. 2020; Zia-Ul-Haq et al. 2014).

Strawberries are important sources of vitamin C, with concentrations in the range 40–60 mg 100 g^{-1} FW (Giampieri et al. 2015). Although blackberries are not among the principal suppliers of vitamin C, they can contribute an appreciable amount of ascorbic acid to a balanced diet (~ 15 mg 100 g^{-1} FW) (Davey et al. 2000).

Researchers, nutritionists, and even at a governmental level promoted the consumption of fruit and vegetable products, and particularly of berries. However, despite their well-known benefits, safety is an issue which needs to be taken into account, and consequently, decontamination methods using disinfectant agents are usually applied to these products (Artés and Allende 2015; Lafarga et al. 2019). In this regard, during the processing and the subsequent storage, the bioactive compounds such as polyphenols and vitamins could change, and the healthy potential of fruits and vegetables could be different from the non-processed products.

Therefore, this chapter aims to review the impact of different washing and/or disinfection methods on the bioactive compounds of strawberries and blackberries.

10.2 Fruit and Vegetable Washing and/or Disinfection Techniques

Fruit and vegetables contain a great diversity of microorganisms. As these products can be consumed raw or minimally-processed, minimizing the contamination risk is an important issue. It is preferably to prevent contamination in the first place. The best way to do this is to avoid contamination throughout harvest and the post-harvest manipulation steps. However, this is not always possible and the use of techniques that reduce/eliminate microorganisms is of extreme importance. The washing and/or disinfection is one of the most critical processing steps, especially in fresh-cut produce, due to it affecting the quality, safety, and shelf-life of the product (Nicolau-Lapeña et al. 2019).

Several chemical compounds are commercially available for washing and/or disinfection of fruits and vegetables. Chlorine and its derivatives are the most common and economical alternatives. Chlorinated compounds are widely used in the whole and minimally processed fruit and vegetable industry to control the number of microorganisms in the wash water and reduce surface contamination, in that way improving the microbiological quality of the products (Pirovani et al. 2006). The spectrum of microorganisms killed or inhibited by chlorine-based compounds is probably broader than that by any other approved disinfectant. The effectiveness of chlorine depends on the concentration of the sanitizer as well as pH, temperature, and organic matter of the wash-water, but is also affected by the surface properties of the fruit and vegetables (Chaidez et al. 2012; Pirovani et al. 2006).

Furthermore, chlorine dosage below the recommended setpoint does not provide any effect on the microbial population, while its excessive use could interfere with physiological, sensory, nutritional, and phytochemical properties of fresh and minimally processed products, or generate toxic by-products, like trihalomethanes and haloacetic acids, in the processing effluents (Chaidez et al. 2012; Silveira et al. 2008).

Based on these concerns, researchers have studied different sanitizers that do not have these adverse effects (Alexandre et al. 2012; López-Gálvez et al. 2010; Silveira et al. 2010; Sun et al. 2014; Vardar et al. 2012).

In this sense, peracetic acid (PAA) has gained interest as a sanitizing agent for whole or minimally processed fresh fruit and vegetables (Nicolau-Lapeña et al. 2019; Vandekinderen et al. 2008; Van de Velde et al. 2010). Commercially, PAA is available as a quaternary equilibrium of acetic acid,hydrogen peroxide, PAA, and water. The peracetic acid does not react with proteins to produce toxic or carcinogenic compounds. Its decomposition products are only oxygen and acetic acid, making its use sustainable and eco-friendly (Silveira et al. 2008; Vandekinderen et al. 2009). Among PAA advantages are its activity independence with organic matter when compared with chlorine and a broader temperature and pH range of action. Moreover, the use of PAA does not cause the formation of toxic or carcinogenic by-products (Lee and Huang 2019; Nicolau-Lapeña et al. 2019). PAA is a powerful oxidant agent, with an oxidation potential of 1.81 eV, higher than those of chlorine dioxide (1.57 eV) and sodium hypochlorite (1.36 eV) (Pechacek et al. 2015).

The US Food and Drug Administration (FDA) approved the use of PAA for the sanitization of certain food products, including fruit and vegetables, at concentrations not exceeding 80 ppm in the wash water (CFR 2018). The disinfection efficiency of PAA toward microorganisms can be ranked as follows on a general basis: bacteria > viruses > bacterial spores > protozoan cysts (Kitis 2004).

In summary, PAA presents significant advantages over other disinfectants: it is an effective water soluble germicide that requires low concentrations of use with a moderate cost (higher cost compared with traditional sanitizers such as sodium hypochlorite), it does not affect the environment, it is active in the presence of organic matter and hard water, and its concentrated solution is stable in storage for a long time (Kunigk and Almeida 2001). Furthermore, PAA can be used over a broad spectrum of pH (3.0–7.5), and disinfection processes can be carried out at room temperature, being active even at temperatures as low as 0 °C (Kunigk and Almeida 2001).

The selection of the appropriate technology for washing/disinfection of fruit and vegetables depends on many factors, such as the initial microbial load, the product surface, the type, concentration, time and exposure temperature of the disinfectant (Beuchat et al. 2004). Additionally, it should be considered that when a washing/disinfection operation is not designed and/or carried out properly, it could create lesions on the plant tissue surface, producing cross-contamination and/or internalization of microbiological and chemical contaminants, as well as potentially generating losses by oxidation or lixiviation of pigments, vitamins and important compounds for human health.

There are different techniques for brining the fruit and the sanitizing agent into contact: immersion, spraying, and fogging. In each one of these techniques, the right combination of operational variables should be determined that allows the desired disinfection to be achieved without loss of quality or healthy potential.

10.2.1 Washing/disinfection by Immersion

Washing/disinfection by immersion or dipping is the most frequently used technique for whole and minimally processed fruits and vegetables. This step aims to reduce the produce surface microorganisms and foreign material (soil, insects, plant debris, etc.). In the case of fresh-cut products, it is also used for reducing cellular fluids that could have been generated after cutting (Pirovani et al. 2006). The design of the washing equipment (agitation, water volume/product weight ratio, etc.) solves issues related to removal of foreign materials or cellular fluids. However, it is necessary to use sanitizing agents to reduce the microbial load and to keep the microorganism concentration levels low in the washing water that is into contact with the products. As mentioned before, PAA is an appropriate alternative disinfectant agent, considering its effectiveness in reducing the microbial load of fruit and vegetables, and its minimal or no production of toxic or carcinogenic compounds during operation (Lee and Huang 2019; Silveira et al. 2008; Vandekinderen et al. 2008).

Van de Velde et al. (2013) modeled the changes in the total anthocyanin content, ascorbic acid and vitamin C contents, and color of quartered fresh-cut strawberries from two cultivars ("Selva" and "Camarosa") as a consequence of the washing/disinfection by immersion in solutions of PAA at different concentrations (0–100 mg L^{-1}), contact times (10–120 s) and temperatures (4–40 °C). The reduction of anthocyanin and ascorbic acid contents were principally affected by PAA concentration and processing time, affecting both cultivars in the same way. Previously, authors reported the same behavior for the reductions of total phenolic content and the antioxidant capacity of strawberries after the same washing/disinfection conditions (Van de Velde et al. 2012). As indicated above, commercial PAA solution is a quaternary equilibrium of PAA, acetic acid, hydrogen peroxide, and water, and an oxidizing effect due to PAA and hydrogen peroxide could explain the losses in both bioactive compounds. Applying the obtained models at 80 mg L^{-1} PAA, 120 s, and 22 °C, the reductions of the contents of total anthocyanins, ascorbic acid, total phenolic, and antioxidant capacity were 30, 37, 11.6, and 13.2%, respectively. These predicted values demonstrated the possible losses of bioactive compounds after the washing/disinfection operation for both strawberry cultivars. Concerning vitamin C (ascorbic acid plus dehydroascorbic acid), the effect of washing/disinfection variables was cultivar dependent. The "Camarosa" cultivar lost approximately 10% vitamin C at any condition in the experimental domain assayed. However, for the "Selva" cultivar, the vitamin C reduction was affected by the operation variables. Working at 80 mg L^{-1} PAA, 120 s and 25 °C, resulted in a predicted vitamin C reduction of approximately 30%.

In contrast, Nicolau-Lapeña et al. (2019) reported significant increases of anthocyanin values for whole strawberries washed by immersion with a solution of PAA 20 mg L^{-1} for 2 min and with 80 mg L^{-1} for 1 min. However, this tendency was not always consistent with the previous results, that is, the authors found a reduction of around 13% in the anthocyanin content of strawberries washed with PAA 80 mg L^{-1} for 2 min. In addition, the authors reported no changes in the total phenolic content and in the antioxidant capacity measured through DPPH and FRAP methods of strawberries sanitized with PAA (20–80 mg L^{-1}) for 1 and 2 min compared with initial values. The washing/disinfection of whole strawberries may avoid bioactive compound loss/degradation, as observed when the washing/disinfection operation is done on fresh-cut fruit, leading to lixiviation of water-soluble phytochemicals, such as anthocyanins, and/or their further oxidation by a strong oxidant as the PAA mixture.

Continuing with their previous study, Van de Velde et al. (2014) optimized the washing/disinfection operation of quartered fresh-cut strawberries based on the retention of bioactive compounds (ascorbic acid and total anthocyanins) and the microbial load reduction. It was concluded that, when the objective was to maximize the total microbial reduction with losses in ascorbic acid and total anthocyanins lower than 10%, the values of the operative variables should be 100 mg L^{-1} PAA, 24 °C, and 50 s. Under these conditions, the washed fresh-cut strawberries showed 1.8 log CFU g^{-1} microbial count reduction, but reduction of anthocyanins and ascorbic acid were 19.2 and 22.5%, respectively. On the other hand, when the objective was to maximize the retention of ascorbic acid and total anthocyanins with moderate microbial reduction, the operative variable values should be set at 20 mg L^{-1} PAA, 18 °C, and 52 s. Washed fresh-cut strawberries under these conditions showed 1.0 log CFU g^{-1} microbial count reduction, 12.8% reduction of total anthocyanins and 6.7% reduction of ascorbic acid. The latter results confirm that it is possible to prioritize the bioactive compound retention during the washing/disinfection of fruit with an acceptable microbiological reduction.

10.2.2 Spray Washing/disinfection

The spray washing technique is an alternative to dipping and could be considered as an appropriate method for sanitizing fresh fruit and vegetables because its effectiveness is not only due to the action of the disinfectant agent but also to the physical removal produced by the spray pressure (Chang and Schneider 2012). In addition, spray washing disinfection is an eco-friendly technology alternative because it generates less wash water than dipping or immersion.

Méndez-Galarraga et al. (2018) modeled the spray washing/disinfection with PAA at different concentrations (1–240 mg L^{-1}) and spraying times (11–138 s), and optimized the operation based on microbiological, nutritional, and changes in sensorial quality of fresh-cut strawberries. Response surface methodology was used to study this operation, and the responses were evaluated on the day of processing (day zero) and after seven days of storage at 2 °C (day seven). The maximum total mesophilic, mold and yeast reduction on day zero were 1.3 (at 183 mg L^{-1} and 109 s), 2.3 (at 240 mg L^{-1} for any treatment time within the experimental range assayed), and 3.3 (at 240 mg L^{-1} and 90 s) log, respectively. After seven days, the impact of the spray washing operation on the microbiological load of the fruit was maintained. As described above, Van de Velde et al. (2014) maximized ascorbic acid and total anthocyanin retention with a considerable microorganism reduction 1 log in washing/disinfection of fresh-cut strawberries by immersion with 20 mg L^{-1} of PAA and 52 s. At these operation conditions, Méndez-Galarraga et al. (2018) reported a retention of approximately 79% of total anthocyanins, 74% of ascorbic acid, and around 84% of vitamin C retention. This result shows the partial oxidation of ascorbic acid to L-dehydroascorbic acid due to PAA, keeping the biological function of vitamin C (Hernández et al. 2006). Simultaneously, Méndez-Galarraga et al. (2018) reported retention of about 98% of the antioxidant capacity, probably due to the high retention of the phenolic compounds (~ 96%), which remain constant through refrigerated storage (Table 10.1).

Méndez-Galarraga et al. (2018) employed the multiple response optimization procedure based on Derringer's desirability function to find the optimal conditions for spray washing/disinfection of fresh-cut strawberries. For this, the authors maximized the total mesophilic, molds, and yeasts reductions on the processing day (day zero), and the optimal conditions obtained were 240 mg L^{-1} and 97 s.

Table 10.1 Bioactive attributes retention (%) of fresh-cut strawberries, with respect to unwashed cut fruit, after spray washing/disinfection with PAA at day zero and after seven days of storage at 2 °C.

Parameters (%)	Time (d)	
	Zero	Seven
TPRi	95.3[a]	97.1[a]
TARi	78.8[a]	79.4[a]
ACRi	98.2[a]	97.3[a]
AARi	74.1[a]	76.5[a]
VitCRi	83.0[a]	85.5[a]

Different letters in the same row indicate significant differences ($P \leq 0.05$) by t-test

TPRi: total phenolic retention at day i; TARi: total anthocyanin retention at day i; ACRi: antioxidant capacity retention at day i; AARi: ascorbic acid retention at day i; VitCRi: vitamin C retention at day i; i: analysis day, zero or seven.

Based on the results above, spray washing with PAA could be considered a good alternative or complementary disinfection method for fresh-cut strawberries due to appropriate reductions of microbial loads with minimal changes in quality and nutritional attributes.

10.2.3 Disinfection by Fogging

Fogging or aerosolization is a dispersion in the air of a disinfectant agent in the form of a fine mist in the air and its forward application over products. The application of sanitizer agents by fogging could be a promising technology for berry disinfection due to # handling and generation of moisture on the surface of the fruits being minimized (Oh et al. 2005). In addition, sanitization by fogging would allow the limitation that sometimes washing/disinfection operations may have for reaching and eliminating pathogens located in inaccessible places of the vegetables (Lee et al. 2004) to be overcome. However, gaseous sanitizers have several disadvantages, for example, the need for special apparatus for gas generation and the number of applicable gaseous sanitizers is limited (Oh et al. 2005). Fogging with peracetic acid (PAA) emerges as a promising option for the control of the microbial population and the extension of the shelf life of berries. The application of sanitizers in food by fogging is little studied. Oh et al. (2005) studied the aerosolization of PAA over lettuce leaves inoculated with *Escherichia coli* O157:H7, *Salmonella typhimurium*, and *Listeria monocytogenes* for 10 to 60 min. The authors reported high reductions (near to 4 log) in the inoculated populations at maximum time. However, results of the impact of treatments on the nutrient and phytochemical content of the plant material are not available.

Van de Velde et al. (2016a) modeled and optimized the disinfection operation of fresh strawberries by fogging with PAA, trying to achieve good microbiological reductions on the surface of the fruits, without affecting their general quality, bioactive compound content, and antioxidant capacity. Response surface methodology was used to study the fogging operation, and the concentration of PAA (3.4–116.6 μL PAA per L of air chamber) and the time of treatment (5.7–69.3 min) were the variables studied. The higher the PAA fogging concentration and treatment time, the higher the reduction in total mesophilic microbial yeast and mold counts on the surface of strawberries after fogging. The authors found satisfactory results, not only on the processing day, but also after seven days of storage. Moreover, the authors reported that the total mesophilic reduction was, in general, lower after seven days of storage, keeping good levels of reduction (~2 log). On the other hand, the reduction of yeasts and molds was higher after seven days of storage, which would indicate a residual action of the ecological fogging sanitizer; similarly to that reported by Méndez-Galarraga et al. (2018) for the fresh-cut strawberries washed by spray.

Furthermore, Vaccari (2017), using the same experimental design used by Van de Velde et al. (2016a), reported that the strawberries disinfection by fogging with intermediate PAA concentrations and shorts periods of times (e.g. 60 $\mu L\ L^{-1}$, 30 min) achieved 4 log reduction of *E. coli* ATCC 25922 inoculated on the strawberries surface, and keeping this reduction during the refrigerated storage. Méndez-Galarraga et al. (2018) found similar results for washing/disinfection of fresh-cut strawberries by spraying with

240 mg L^{-1} PAA and 138 s. These results correlated with those obtained for *E. coli* O157:H7 (Oh et al. 2005) due to the similarities that these microorganisms present related to their growth characteristics, cold tolerance, fixation to fresh produce surfaces, and inactivation by heat and antimicrobials (Kim and Harrison 2009).

With respect to the impact of the fogging with PAA on bioactive compounds, Van de Velde et al. (2016a) reported that the retentions of vitamin C, total anthocyanins, total phenolic compounds, and antioxidant capacity of fogged strawberries were affected by both the concentration of PAA and the treatment time. Increasing the magnitude of both experimental variables produced lower retention of bioactive compounds. According to the predictions obtained by working at the maximum PAA concentration (116.6 μL L^{-1}) and treatment time (69.3 min), retention higher than 65% of vitamin C, total phenolic compounds, and antioxidant capacity can be achieved after seven days of storage. However, these conditions could cause losses of around 70% of total anthocyanin content in the same fruit (Van de Velde et al. 2016a). Anthocyanins are unstable compounds and sensitive to degradation during the processing and storage of vegetable products (Crecente-Campo et al. 2012). Therefore, taking into account the high sensitivity of anthocyanins, their relevance to the color of strawberries, and their contribution to the antioxidant capacity, their retention should be prioritized in the fogging disinfection operation, in addition to ensuring a good microbiological reduction.

In a later study, Van de Velde et al. (2016b) reported that fogging with PAA caused oxidation to individual phenolic compounds of strawberries to different degrees, according to their chemical nature. As previously observed, individual anthocyanins were the most vulnerable phenolic compounds to oxidation with PAA fogging, followed by low-polymerized proanthocyanidins, hydroxycinnamic acid derivatives, and ellagitannin sanguiin H-6. The higher the PAA concentration and treatment time, the lower the retention of individual anthocyanins. For example, retention of cyanidin-3-*O*-glucoside, pelargonidin-3-*O*-glucoside, and pelargonidin-3-*O*-rutinoside were 6.6, 15.0 and 43.6%, respectively, for strawberries fogged with 116.6 μL L^{-1} PAA and 37.5 min (Van de Velde et al. 2016b).

The oxidant effects of PAA and hydrogen peroxide on anthocyanins and other bioactive compounds of strawberries were described by Özkan et al. (2005). Hydrogen peroxide present in the sanitizing mixture with PAA, and its derivative products, are strong nucleophiles, which can be added to anthocyanin molecules causing its breakdown and/or formation of colorless chalcones (Özkan et al. 2005).

On the other hand, Vaccari (2017) studied fogging disinfection with PAA on fresh blackberries. In this case, for all PAA concentrations (3.4–116.6 μL PAA L^{-1}), the treatment showed a reduction of the total mesophilic microorganism of 2 log for both short (5.7 min) and long times (69.3 min). Moreover, increasing PAA concentration significantly reduced the levels of molds and yeasts, with the maximum reductions corresponding to the higher PAA concentrations and treatment times used (116.6 μLPAA L^{-1} and 69.3 min). After storage at 2 °C, the reduction values of molds and yeasts were similar or even higher, indicating a residual action of PAA in fogged blackberries. Van de Velde et al. (2016b) studied the effect of the fogging operation with PAA concentration (3.4–116.6 μL PAA per L of air chamber) and time of treatment (5.7–69.3 min) on bioactive compounds of fresh blackberries. They reported that fogging operation, at determined PAA concentrations and times, caused high losses in the bioactive potential of blackberries. Flavonoids like anthocyanins, which represent around 80% of the total phenolic compounds in blackberries, were affected for some treatment conditions. The predicted anthocyanin retentions in the most extreme fogging conditions used (116.6 μL PAA L^{-1} and 69.3 min) were 55.1, 71.8, 58.2, and 52.9% for cyanidin-3-O-glucoside, cyanidin-3-O-xyloside, cyanidin-3-O-(6-O-malonyl glucoside), and cyanidin-3-O-dioxaliglucoside, respectively. On the other hand, ascorbic acid and vitamin C were also affected by process variables. The predicted retention in the most extreme fogging conditions (116.6 μL PAA L^{-1} and 69.3 min) were 71.3 and 62.2% for ascorbic acid and vitamin C, respectively. In these conditions, the predicted retention antioxidant capacity was 54.7%, in agreement with the predicted losses in anthocyanin and vitamin C contents.

According to all the above results, although PAA fogging treatment proved to be effective in controlling the native and pathogenic microbiological load of berries, the oxidizing properties of the disinfectant, in some conditions, produced losses of the bioactive potential of the fruits.

Therefore, the use of the optimization methodology of multiple responses allowed the optimal conditions to be determined to maximize the microbiological reduction and to minimize the sensory changes and losses in the bioactive potential of the fruits. For example, for strawberries, the disinfection treatment by fogging at optimal conditions(10.1 μL L^{-1} and 29.6 min) achieves reductions of about 1 log in total mesophilic microorganisms, yeasts, and molds, with retention of the bioactive compounds of 92%, and without changes in the color of the fruit. Moreover, the fogged strawberries (under those conditions) stored for seven days at 2 °C remained with appropriate microbiological reductions (~1.3 log) and good retention of bioactive compounds (greater than 73.8%).

In short, the fogging of berry fruits (strawberries and blackberries) using the ecological PAA sanitizer could be a promising option to extend the post-harvest life of these fruits without compromising their general appearance.

10.3 Conclusions

This chapter expands on the information available on how different surface decontamination techniques (immersion, spraying, or fogging) can affect the healthy potential of strawberries and blackberries.

All the techniques required process optimization to find the conditions that allow achieving surface decontamination, without losing the healthy potential of the fruits.

In the case of strawberries, it is concluded that the spraying and fogging techniques were the ones that allowed higher retention of healthy potential expressed through the retained antioxidant capacity to be achieved. Regarding the immersion technique, this alternative allowed the greatest retention of vitamin C.

For blackberries, it is concluded that the fogging technique tested, even under optimized conditions, causes a great loss of antioxidant capacity (63%), although retaining a high level of vitamin C and phenolic compounds of fruit.

References

Aaby, K., Mazur, S., Nes, A., and Skrede, G. (2012). Phenolic compounds in strawberry (Fragaria x ananassa Duch.) fruits: composition in 27 cultivars and changes during ripening. *Food Chemistry* 132 (1): 86–97. doi: 10.1016/J.FOODCHEM.2011.10.037.

Alexandre, E.M.C., Brandão, T.R.S., and Silva, C.L.M. (2012). Efficacy of non-thermal technologies and sanitizer solutions on microbial load reduction and quality retention of strawberries. *Journal of Food Engineering* 108 (3): 417–426. doi: 10.1016/j.jfoodeng.2011.09.002.

Artés, F. and Allende, A. (2015). Processing lines and alternative preservation techniques to prolong the shelf-life of minimally lines and alternative preservation processing long the shelf-life of minimally fresh processed to pro techniques leafy vegetables. *European Journal of Horticultural Science* 70 (5): 231–245.

Artés, F., Gómez, P., Aguayo, E., Escalona, V., and Artés-Hernández, F. (2009). Sustainable sanitation techniques for keeping quality and safety of fresh-cut plant commodities. *Postharvest Biology and Technology* 51: 287–296. doi: 10.1016/j.postharvbio.2008.10.003.

Beuchat, L.R., Adler, B.B., and Lang, M.M. (2004). Efficacy of chlorine and a peroxyacetic acid sanitizer in killing listeria monocytogenes on iceberg and romaine lettuce using simulated commercial processing conditions. *Journal of Food Protection* 67 (6): 1238–1242. doi: 10.4315/0362-028x-67.6.1238.

[CFR] C of FR. (2018). Online reference included in secondary direct food additives permitted in food for human consumption: chemicals used in washing or to assist in the peeling of fruits and vegetables. Chlorine dioxide. https://www.accessdata.fda.gov/scripts/cdrh/cfdocs/cfcfr/cfrsearch.cfm?fr=173.315 (accessed May 2019).

Chaidez, C., Castro-del Campo, N., Heredia, J.B., Contreras-Angulo, L., González–Aguilar, G., and Ayala–Zavala, F. (2012). Chlorine. In: *Decontamination of Fresh and Minimally Processed Produce (First Edit)* (ed. V. Gómez-López). New Jersey, EEUU: John Wiley and Sons, Ltd., Publication.

Chang, A.S. and Schneider, K.R. (2012). Evaluation of overhead spray-applied sanitizers for the reduction of salmonellaon tomato surfaces. *Journal of Food Science* 77 (1). doi: 10.1111/j.1750-3841.2011.02486.x.

Crecente-Campo, J., Nunes-Damaceno, M., Romero-Rodríguez, M.A., and Vázquez-Odériz, M.L. (2012). Color, anthocyanin pigment, ascorbic acid and total phenolic compound determination in organic versus conventional strawberries (Fragaria×ananassa Duch, cv Selva). *Journal of Food Composition and Analysis* 28 (1): 23–30. doi: 10.1016/j.jfca.2012.07.004.

Davey, M.W., Montagu, M.V., Inzé, D., Sanmartin, M., Kanellis, A., Smirnoff, N., ... Fletcher, J. (2000). Review Plant L -ascorbic acid: chemistry, function, metabolism, bioavailability and effects of processing. *Journal of the Science of Food and Agriculture* 860 (December 1999): 825–860. doi: 10.1002/(sici)1097-0010(20000515)80:7<825::aid-jsfa598>3.0.co;2-6.

Giampieri, F., Forbes-Hernandez, T.Y., Gasparrini, M., Alvarez-Suarez, J.M., Afrin, S., Bompadre, S., ... Battino, M. (2015). Strawberry as a health promoter: an evidence based review. *Food and Function* 6 (5): 1386–1398. doi: 10.1039/C5FO00147A.

Giampieri, F., Tulipani, S., Alvarez-Suarez, J.M., Quiles, J.L., Mezzetti, B., and Battino, M. (2012). The strawberry: composition, nutritional quality, and impact on human health. *Nutrition* 28 (1): 9–19. doi: 10.1016/j.nut.2011.08.009.

Gómez Riera, P., Bruzone, I., and Kirschbaum, D. (2013). Visión Prospectiva de la Cadena de Frutas Finas Al 2030. Ministerio Ciudad Autónoma de, Buenos Aires.

Hernández, Y., Lobo, M.G., and González, M. (2006). Determination of vitamin C in tropical fruits: a comparative evaluation of methods. *Food Chemistry* 96 (4): 654–664. doi: 10.1016/j.foodchem.2005.04.012.

Jung, H., Lee, H.J., Cho, H., Lee, K., Kwak, H.-K., and Hwang, K.T. (2015). Anthocyanins in Rubus fruits and antioxidant and anti-inflammatory activities in RAW 264.7 cells. *Food Science and Biotechnology* 24 (5): 1879–1886. doi: 10.1007/s10068-015-0246-1.

Kim, J. and Harrison, M. (2009). Surrogate selection for Escherichia coli O157: H7 Based on cryotolerance and attachment to Romaine lettuce. *Journal of Food Protection* 72 (7): 1385–1391. doi: 10.4315/0362-028X-72.7.1385.

Kitis, M. (2004). Disinfection of wastewater with peracetic acid: a review. *Environment International* 30 (1): 47–55. doi: 10.1016/S0160-4120(03)00147-8.

Kunigk, L. and Almeida, M.C.B. (2001). Action of peracetic acid on Escherichia coli and Staphylococcus aureus in suspension or settled on stainless steel surfaces. *Brazilian Journal of Microbiology* 32 (1): 38–41. doi: 10.1590/S1517-83822001000100009.

Lafarga, T., Colás-medà, P., Abadías, M., Aguiló-aguayo, I., and Bobo, G. (2019). Strategies to reduce microbial risk and improve quality of fresh and processed strawberries: a review. *Innovative Food Science and Emerging Technologies* 52: 197–212. doi: 10.1016/j.ifset.2018.12.012.

Lee, J., Koo, N., and Min, D.B. (2004). Reactive oxygen species, aging, and antioxidative nutraceuticals. *Comprehensive Reviews in Food Science and Food Safety* 3: 21–33. doi: 10.1111/j.1541-4337.2004.tb00058.x.

Lee, W.-N. and Huang, C.-H. (2019). Formation of disinfection byproducts in wash water and lettuce by washing with sodium hypochlorite and peracetic acid sanitizers. *Food Chemistry: X* 1: 100003. doi: 10.1016/j.fochx.2018.100003.

López-Gálvez, F., Gil, M.I., Truchado, P., Selma, M.V., and Allende, A. (2010). Cross-contamination of fresh-cut lettuce after a short-term exposure during pre-washing cannot be controlled after subsequent washing with chlorine dioxide or sodium hypochlorite. *Food Microbiology* 27 (2): 199–204. doi: 10.1016/j.fm.2009.09.009.

Méndez-Galarraga, M.P., Salsi, M.S., Piagentini, A.M., and Pirovani, M.É. (2018). Spray washing disinfection with peracetic acid in the processing of fresh-cut strawberries: an alternative for dipping techniques. *International Journal of Fruit Science* 19 (3): 258–275. doi: 10.1080/15538362.2018.1502722.

Nicolau-Lapeña, I., Abadias, M., Bobo, G., Aguiló-aguayo, I., Lafarga, T., and Viñas, I. (2019). Strawberry sanitization by peracetic acid washing and its effect on fruit quality. *Journal Of Food Microbiology* 83 (April): 159–166. doi: 10.1016/j.fm.2019.05.004.

Nile, S. and Park, S. (2013). Edible berries: review on bioactive components and their effect on human health. *Nutrition* 1–11. doi: 10.1016/j.nut.2013.04.007.

Nowicka, A., Kucharska, A.Z., Sokó, A., and Fecka, I. (2019). Comparison of polyphenol content and antioxidant capacity of strawberry fruit from 90 cultivars of Fragaria × ananassa Duch. *Food Chemistry* 270 (June 2018): 32–46. doi: 10.1016/j.foodchem.2018.07.015.

Oh, S.-W., Gray, P.M., Dougherty, R.H., and Kang, D.-H. (2005). Aerosolization as novel sanitizer delivery system to reduce food-borne pathogens. *Letters in Applied Microbiology* 41 (1): 56–60. doi: 10.1111/j.1472-765X.2005.01711.x.

Özkan, M., Yemenicioğlu, A., and Cemeroğlu, B. (2005). Degradation of various fruit juice anthocyanins by hydrogen peroxide. *Food Research International* 38: 1015–1021. Elsevier. doi: 10.1016/j.foodres.2005.03.013.

Pechacek, N., Osorio, M., Caudill, J., and Peterson, B. (2015). Evaluation of the toxicity data for peracetic acid in deriving occupational exposure limits: a minireview. *Toxicology Letters* 233 (1): 45–57. doi: 10.1016/j.toxlet.2014.12.014.

Pirovani, M., Güemes, D., and Piagentini, A. (2006). Procesamiento de Vegetales Frescos Cortados. Santa Fe: Universidad Nacional del Litoral.

Ramos, B., Miller, F.A., Brandão, T.R.S., Teixeira, P., and Silva, C.L.M. (2013). Fresh fruits and vegetables—an overview on applied methodologies to improve its quality and safety. *Innovative Food Science and Emerging Technologies* 20: 1–15. doi: 10.1016/J.IFSET.2013.07.002.

Silveira, A.C., Aguayo, E., and Artés, F. (2010). Emerging sanitizers and Clean Room packaging for improving the microbial quality of fresh-cut "Galia" melon. *Food Control* 21 (6): 863–871. doi: 10.1016/j.foodcont.2009.11.017.

Silveira, A.C., Conesa, A., Aguayo, E., and Artes, F. (2008). Alternative sanitizers to chlorine for use on fresh-cut "galia" (cucumis melo var. catalupensis) melon. *Journal of Food Science* 73 (9): 405–411. doi: 10.1111/j.1750-3841.2008.00939.x.

Sun, X., Bai, J., Ference, C., Wang, Z., Zhang, Y., Narciso, J., and Zhou, K. (2014). Antimicrobial activity of controlled-release chlorine dioxide gas on fresh blueberries†. *Journal of Food Protection* 77 (7): 1127–1132. doi: 10.4315/0362-028X.JFP-13-554.

Vaccari, M.C. (2017). Desinfección de frutillas y zarzamoras frescas por nebulización con ácido peracético. Universidad Nacional del Litoral.

Van de Velde, F. (2012). Changes in the healthy potential and nutritional and sensory quality of minimally processed strawberries as a consequence of processing conditions, temperature, and storage time. Doctoral thesis. Universidad Nacional del Litoral.

Van de Velde, F., Esposito, D., Overall, J., Méndez-Galarraga, M.P., Grace, M., Pirovani, M.É., and Lila, M.A. (2019). Changes in the bioactive properties of strawberries caused by the storage in oxygen - and carbon dioxide - enriched atmospheres. *Food Science and Nutrition* 1–10. doi: 10.1002/fsn3.1099.

Van de Velde, F., Güemes, D.R., and Pirovani, M.É. (2014). Optimization of the peracetic acid washing disinfection of fresh-cut strawberries based on microbial load reduction and bioactive compounds retention. *International Journal of Food Science and Technology* 49 (2): 634–640. doi: 10.1111/ijfs.12346.

Van de Velde, F., Méndez-Galarraga, M.P., and Pirovani, M.É. (2020). Effect of enriched O2 and CO2 atmospheres on the overall quality and the bioactive potential of fresh blackberries. *Postharvest Biology and Technology* 164: 111166. doi: 10.1016/j.postharvbio.2020.111166.

Van de Velde, F., Piagentini, A.M., Güemes, D.R., and Pirovani, M.E. (2013). Modelling changes in anthocyanins, total vitamin C and colour as a consequence of peracetic acid washing disinfection of two cultivars of strawberries for fresh-cut processing.

International Journal of Food Science and Technology 48 (5): 954–961. doi: 10.1111/ijfs.12047.

Van de Velde, F., Tavella, A., Piagentini, A., Güemes, D., and Pirovani, M. (2010). Retención De Compuestos Bioactivos En El Lavado-Desinfección De Frutillas Mínimamente Procesadas Con Ácido Peracético. *Revista Iberoamericana de Tecnología Postcosecha* 11: 162–170.

Van de Velde, F., Vaccari, M.C., Piagentini, A.M., and Pirovani, M.E. (2016a). Optimization of strawberry disinfection by fogging of a mixture of peracetic acid and hydrogen peroxide based on microbial reduction, color and phytochemicals retention. *Food Science and Technology International*. doi: 10.1177/1082013215625696.

Van de Velde, F., Grace, M., Pirovani, M., and Lila, M. (2016b). Impact of a new postharvest disinfection method based on peracetic acid fogging on the phenolic profile of strawberries. *Postharvest Biology and Technology* 117: 197–205. doi: 10.1016/j.postharvbio.2016.03.005.

Vandekinderen, I., Devlieghere, F., De Meulenaer, B., Ragaert, P., and Van Camp, J. (2009). Optimization and evaluation of a decontamination step with peroxyacetic acid for fresh-cut produce. *Food Microbiology* 26 (8): 882–888. doi: 10.1016/j.fm.2009.06.004.

Vandekinderen, I., Van Camp, J., Devlieghere, F., Veramme, K., Denon, Q., Ragaert, P., and De Meulenaer, B. (2008). Effect of decontamination agents on the microbial population, sensorial quality, and nutrient content of grated carrots (Daucus carota L.). *Journal of Agricultural and Food Chemistry* 56: 5723–5731. doi: 10.1021/jf800681a.

Vardar, C., Ilhan, K., and Karabulut, O.A. (2012). The application of various disinfectants by fogging for decreasing postharvest diseases of strawberry. *Postharvest Biology and Technology* 66: 30–34. doi: 10.1016/j.postharvbio.2011.11.008.

Zia-Ul-Haq, M., Riaz, M., De Feo, V., Jaafar, H., and Moga, M. (2014). Rubus fruticosus L.: constituents, biological activities and health related uses. *Molecules* 19 (8): 10998–11029. doi: 10.3390/molecules190810998.

11

Biomolecules from Basil – Pharmacological Significance

Ivayla Dincheva and Ilian Badjakov

Department of Agrobiotechnology, Agrobioinstitute, Agricultural Academy, Sofia, Bulgaria
**Corresponding author: ibadjakov@gmail.com*

11.1 Introduction

The genus *Ocimum L.* (Lamiaceae family) includes approximately 150 species, based on variation in morphological characteristics such as growth habit, leaf and flower color, size and shape, and aromatic composition. In the Mediterranean region, one of the most important and frequently consumed species is *Ocimum basilicum L.*, commonly known as sweet basil. From ancient times to the present day, this plant is referred to as the "king of the herbs". The name basil is probably derived from the Greek word "βασιλικού" meaning "royal". A Latin word "basiliscus" relates to a mythical fire-breathing dragon that can kill at a glance. According to a Roman legend, basil was the antidote to the basilisk's poison (Makri and Kintzios 2008).

Due to the presence of secondary metabolites such as terpenes, flavonoids, tocopherols, steroids, *Ocimum* species have been reported to exhibit antibacterial, antifungal, antiproliferative, anti-inflammatory, antioxidant, hypoglycaemic, hypolipidemic, antithrombotic, insecticidal, and larvicidal activities. It has also shown inhibitory effects on platelet aggregation.

As a fresh herb, basil is widely used in cooking. It is famous for its use in Italian dishes, such as pesto, and adds a delicious taste to olive oil for salad dressings. It is also used to flavour vegetables, poultry, fish, tea, honey, and liquor (Marotti et al. 1996). The leaves and flowers of basil are used in folk medicine as a tonic, anthelmintic antidiarrheal and hepatoprotective agent (Ademiluyi et al. 2016; Bilal et al. 2012; Mahajan et al. 2013). The oil of the plant is beneficial for the relief of spasm, mental weariness, cold, and rhinitis. It has been used as a remedy for boredom and convulsion. Basil cures headache, improves digestion and is also good for toothache and earache. The plant is used to treat gonorrhea, chronic diarrhea, fever, otitis, coughs, acne, and constipation (Mathews et al. 1993; Miraj and Kiani 2016; Rahman et al. 2011). At present antimicrobial resistance is a global health problem according to the WHO and requires urgent multisectoral action in order to achieve sustainable development goals. This has resulted in an increased interest in

Biomolecules from Natural Sources: Advances and Applications, First Edition. Edited by Vijai Kumar Gupta, Satyajit D. Sarker, Minaxi Sharma, María Elida Pirovani, Zeba Usmani, and Chelliah Jayabaskaran.

medicinal plants since 30–50% of current pharmaceuticals and nutraceuticals are plant-derived (Anand et al. 2019). The disadvantages of plant derived bioactive molecules is the low content of these important secondary metabolites and its derivatives and high demand in pharmaceutical industries so it is necessary to increase quality and productivity. Biotechnologies, metabolome engineering as well as genome editing strategies offer a wide range of procedures modulating bioactive compound synthesis.

11.2 Cultivar and Chemotaxonomic Classification

Synonyms and detailed morphological features of the well-studied species are listed in Table 11.1.

Most commercial basil cultivars available in the market belong to the species *O. basilicum* and are classified in seven types (Darrah 1974): I – the tall slender type (includes the sweet basil group); II – large-leafed ("lettuce leaf basil"); III – dwarf types ("bush" basil); IV – compact types (*O. basilicum* var. *thyrsiflora*); V – purple-colored basil types with traditional sweet basil flavour; VI – purple types ("Dark Opal", "Purple Ruffles", "Red Rubin") with a sweet basil and clove-like aroma; VII – citriodorum types (include lemon-flavoured basils).

The marker constituents within the *Ocimum* species, which play a major role in the taxonomic identification, were produced by two different pathways: phenylpropanoids by the shikimic acid pathway and regular terpenes by the mevalonic acid pathway (Gang

Table 11.1 The most common species of basil with description.

Latin name	Common name	Description
Ocimum basilicum	Sweet basil	white flowers; bright green,1 cm long leaves; clove-like scent
Ocimum basilicum "Genovese"	Genovese basil	dark green leaves up to 5 cm long
Ocimum basilicum crispum	Lettuce-leaf basil	large, wide leaves
Ocimum basilicum minimum	Greek basil	dwarf with less than 1 cm long leaves; white flowers.
Ocimum thrysiflora	Thrysiflora basil	ornamental seed head; purple flowers
Ocimum basilicum purpurascens	Purple basil	lavender like flowers with the same shape and size leaf as sweet basil
Ocimum basilicum odoratum	Scented basil	cinnamon, lemon, licorice, and anise scent
Ocimum sanctum	Holy basil	small leaves with a clove-like fragrance; violet or white flowers
Ocimum kilimandscharicum	Camphor basil	strong, medicinal scent with gray-green leaves
Ocimum micranthemum	Peruvian basil	sweet flavour

et al. 2001). The complexity in *Ocimum* essential oil resulted from differences in the dominance of these species as well as their mixed pathways, morpho- and chemotypes. Also, the differences in chemical composition might be due to a number of factors including climatic conditions, geographical locations, stages of development, and post-harvest processing before oil extraction (Javanmardi et al. 2002).

11.3 Bioactive Constituents in Basil

Ocimum species have been known for their healthful properties for a long time due to the presence of plant secondary metabolites. According to Zahran et al. (2020) percentages of different classes of compounds distributed in the genus Ocimum are as follows: terpenoids (39%), flavonoids (22%), simple phenolics (15%), miscellaneous (7.3%), fatty acids (4.9%), lignans and neolignans (3.4%), anthocyanins (3%), sterols (1.9%), and coumarins (1.5%). The chemical composition of *Ocimum sp.* is presented in Table 11.2. Basil possesses a high content of vitamins and minerals, most notably vitamin A, vitamin C, β-carotene, calcium, and phosphorus (Filip 2017). The basil plant has macro- and micro-nutritional values including carbohydrates (2.65 g 100 g^{-1}), protein (3.15 g 100 g^{-1}), fat (0.64 g 100 g^{-1}), dietary fiber (1.60 g 100 g^{-1}), energy (23 Kcal) (Marwat et al. 2011).

11.4 Pharmacological Activities

Due to its actual and perceived health-promoting effects, there is an increasing tendency to include basil derivatives with a robust folk medicine tradition in therapeutic and preventive strategies. The ethnopharmacologically documented effects of *O. basilicum* making it useful for protection against or treating several diseases.

11.4.1 Antimicrobial Activities

Antibiotic resistance is a natural process caused by mutations in bacterial genes. However, the excessive and improper use of antibiotics accelerates the emergence and spread of resistant bacteria. One of the best methods for resolving this serious problem is to look for new therapeutic agents from plants with antimicrobial activities against the pathogenic microorganisms.

Due to their constituents, especially terpenes, essential oils derived from several *Ocimum* species have been reported to be active against several Gram-positive and Gram-negative bacteria as well as against yeasts and fungi. Janssen et al. (1989) investigated the antimicrobial potential of *O. canum*, *O. gratissimum*, *O. trichodon*, and *O. urticifolium* grown in Rwanda against *E. coli*, *B. subtilis*, *S. aureus*,and *Trichophyton mentagrophytes* var. *interdigitale*. *O. canum* was the most effective amongst the studied oils. *O. sanctum* essential oil has been found to inhibit *in vitro* growth of *B. anthracis* and *P. aeruginosa* which showed its antibacterial activity.

Table 11.2 Phytochemical constituents in *Ocimum sp.*

Chemical constituents	Plant organ	Instrumental method	References
Flavonoids			
quercetin; isoquercetrin; kaempferol; quercetin-3,5-diglycoside; quercetin-3-O-galactoside; quercetin-3-O-glucoside; quercetin-3-O-glucoside-2”-gallate; quercetin-3-O-rhamnoside; quercetin-3-O-(2”-o-galloyl)-rutinoside; kaempferol-3-O-rutinoside; kaempferol-3-O-glucoside; rutin; apigenin; apigenin glucoronide	leaves, whole plant	HPLC-APcI-MS; LC-DAD-ESI-TOF-MS; HPLC–(DAD-ESI)-MS^n	(Jayasinghe et al. 2003)
Phenolic acids			
caffeic acid; caftaric acid; p-coumaric acid; rosmarinic acid; chicoric acid; gallic acid, 3,4-dihydroxybenzoic acid, ferulic acid, vanillic acid, 4-hydroxybenzoic acid, ellagic acid; chlorogenic acid; 2,5-dihydroxybenzoic acid	leaves	^{1}H NMR; ^{13}C NMR; HPLC-PAD; HPLC-UV; HPLC–DAD	(Jayasinghe et al. 2003; Rezzoug et al. 2019; Touiss et al. 2019)
Steroids			
basilimoside; daucosterol; β-sitosterol; stigmasterol; basilol; ocimol	aerial parts, whole plant	1H- and 13C-NMR; HR EIMS	(Siddiqui et al. 2007)
Coumarins			
coumarin; aesculetin;	leaves, whole plant	HPLC-DAD	(Matos et al. 2015)
Monosaccharides	leaves	HPLC-RID	(Fernandes et al. 2019)
fructose; glucose; trehalose; sucrose			
Organic acids			
ascorbic acid, citric acid, malic acid, oxalic acid, shikimic acid, succinic acid, fumaric acid and quinic acid	leaves	UPLC-DAD	(Fernandes et al. 2019)
Tocopherols			
α-, β-, γ- and δ-tocopherols	leaves	HPLC-FD; HPLC-RF	(Fernandes et al. 2019; Jayasinghe et al. 2003)

(Continued)

Table 11.2 (Continued)

Chemical constituents	Plant organ	Instrumental method	References
Aromatic Compounds			
4-allylphenol; anethole; anisaldehyde; benzyl alcohol; cuminaldehyde; estragole; ethyl cinnamate; methyl benzoate; methyl cinnamate; methyl eugenol; methyl salicylate; p-methoxycinnamaldehyde;phenethyl alcohol; phenyl acetaldehyde; safrole; benzaldehyde; cis-hex-3-enyl acetate	leaves	GC-MS	(Lee et al. 2005; Vani et al. 2009)
Aliphatic ketones			
β-ionone; cis-jasmone; 3-hydroxy-2-butanone; 6-methyl-5-heptenone; 6-methyl-(E,E)-3,5-heptadien-2-one; trans-β-ionone-5,6-epoxide	leaves	GC-MS	(Lee et al. 2005; Nascimento et al. 2020)
Aliphatic aldehyde			
(E)-3-hexenal; n-hexanal;	leaves	GC-MS	(Lee et al. 2005)
Aliphatic alcohols			
cyclohexanol; hexanol; octanol; 3-octanol; 1-octen-3-ol; 1-penten-3-ol; (Z)-2-pentenol; (Z)-3-hexanol	leaves	GC-MS	(Lee et al. 2005)
Triterpenoids			
alphitolic acid; betulin; betulinic acid; 3-epimaslinic acid; euscaphic acids; oleonolic acid; pomolic acid; ursolic acid; basilol; ocimol	flowers, leaves, roots, seeds		(Pandey et al. 2015; Siddiqui et al. 2007)
Monoterpene hydrocarbons			
α-phellandrene; β-phellandrene; α-pinene; β-pinene; α-terpinene; terpinolene; myrcene; camphene; (z)-β-ocimene; 3-carene; (E)-β-ocimene; limonene; p-cymene; 2-thujene; γ-terpinene	aerial parts, flowering parts, leaves	GC-MS	(Chenni et al. 2016; Gebrehiwot et al. 2015; Rezzoug et al. 2019)

Oxygenated monoterpenes			
α-citral; α-fenchyl acetate; α-terpineol; borneol; bornyl acetate; camphor; carvacrol; carvone; 1-8-cineole; (Z)-linalool oxide; citronellol; endo-fenchol; eugenol; exo-2-hydroxycineole-acetate; fenchone; geranial; geraniol; geranyl acetate; hotrienol; lavandulol; linalool; linalool cis-furanoid; linalool trans-furenoid; trans-linalool oxide; linalyl acetate; menthol; menthone; hin; myrtenal; myrtenol; iso-neomenthol; neral; nerol; pinocarvone; p-menth-1,8-dien-4-ol; p-menth-2-en-1-ol; piperitone; pulegone; (E)-sabinene hydrate; iso-pinocamphone; (E)-pinocamphone; (Z)-rose oxide; (Z)-sabinene hydrate; terpinen-4-ol; terpinyl formate; thymol; verbenone	aerial parts, flowering parts, leaves	GC-MS	(Chenni et al. 2016; Gebrehiwot et al. 2015; Kathirvel and Ravi 2012; Rezzoug et al. 2019)
Sesquiterpene hydrocarbons			
α-acoradiene; β-acoradiene; α-amorphene; aromadendrene; α-(Z)-bergamotene; (E)-β-bergamotene; (E)-β-bisabolene; α-bulnesene; β-bourbonene; bicycloelemene; bicyclogermacrene; β-caryophyllene; iso-caryophyllene; β-cedrene; β-copaene; β-cubebene; cadina-3,5-diene; α-cadinene; γ-cadinene; δ-cadinene; (Z)-calamenene; α-cedrene; α-copaene; α-cubebene; dehydroaromadendrene; β-elemene; (E)-β-farnesene; β-guaiene; α-guaiene; α-gurjunene; γ-gurjunene; germacrene-A; germacrene-B; Germacrene-D; guaia-1(10),11-dieneα-humulene; iso-ledene; longifolene; γ-muurolene; (Z)-muurola-4(14),5-diene; α-7-epi-selinene; β-selinene; α-ylangene; valencene; α-zingiberene;	aerial parts, flowering parts, leaves	GC-MS	(Chenni et al. 2016; Gebrehiwot et al. 2015; Kathirvel and Ravi 2012; Rezzoug et al. 2019)
Oxygenated sesquiterpenes			
β-basibolol; α-cadinol; tau-cadinol; cubenol; caryophyllene oxide; 1,10-di-epi-cubenol; dihydroactinidiolide; β-eudesmol; α-humulene oxide; isospathulenol; muurolol; spathulenol; viridiflorol	aerial parts, flowering parts, leaves, roots	GC-MS	(Chenni et al. 2016; Gebrehiwot et al. 2015; Rezzoug et al. 2019)

The potency of basil essential oil as a antibacterial agent suggests its potential activity against *S. mutans*, which is the main cause of dental caries. Attached to the tooth surface these bacteria could excrete a slime layer. The accumulation of microbial community leads to the formation of a biofilm known as plaque. A recent report showed that basil essential oil exhibited antibacterial activity against *S. mutans* with IC_{90} of 0.23%, biofilm formation inhibition with IC_{50} of 0.68%, and biofilm degradation by 32.15% at 1.00%. Also the antibacterial activity of basil essential oil within its micro-emulsion mouthwash formula was maintained at 100% after three months storage at room temperature and sustained at 99.62% at accelerated storage condition (Astuti et al. 2016). In another study basil displayed a greater potential of antibacterial activity against *B. cereus, B. subtilis, B. megaterium, S. aureus, Listeria monocytogenes, E. coli, Shigella boydii, S. dysenteriae, Vibrio, parahaemolyticus, V. mimicus*, and *Salmonella typhi* (Hussain et al. 2008). The methanol extract of *O. basilicum*, suspended with 5 mL deionized water showed antimicrobial activity against the visible inhibition zones against the strains of *Pseudomonas aeruginosa, Shigella* sp., *Listeria monocytogenes, Staphylococcus aureus* and two different strains of *Escherichia coli*. It was observed that the treated cell walls of sensitive bacteria were damaged by using the scanning electron microscope (Kaya et al. 2008). The work of Araújo Silva et al. (2016) demonstrated that basil oil in association with existing antibiotics – imipenem and ciprofloxacin may increase their antibacterial activity through molecular synergism against bacterial strains *Staphylococcus aureus* and *Pseudomonas aeruginosa*.

Salmonellosis is characterized by gastrointestinal disorders manifested predominantly by diarrhea and abdominal cramps. The essential oil from basil leaves had strong antimicrobial activity against *S. enteritidis* SE3, MIC 20.0 $\mu g\ mL^{-1}$ *in vitro*.

A number of studies report the strong antifungal action of basil essential oil. Using the agar plate method, Dube et al. (1989) showed that, in a dose of 1.5 ml L^{-1}, basil oil completely inhibited the growth of 22 species of molds, including the mycotoxin-producing strains of *Aspergillus flavus* and *Aspergillus parasiticus*. Molds and their toxicogenic metabolites represent a diffuse and dangerous cause of food contamination. In a study, the complete inhibition of *A. flavus* growth was noted at 1000 ppm basil oil concentration; whereas marked inhibition of aflatoxin B1 production was observed at all concentrations tested – 500, 750 and 1000 ppm, respectively (Abou El-Soud et al. 2015).

Basil has also been documented as an effective natural antiparasitic against drug-resistant strains of *Plasmodium falciparum* (Ntonga et al. 2014).

Many researchers reported that there is some relationship between the chemical structures of the most abundant compounds in the essential oils and the antimicrobial or antifungal activities (Lis-Balchin et al. 1998). However, the synergistic or antagonistic effect of compounds in minor percentage in the mixture has to be considered.

Essential oils from *O. basilicum* L. (cvs. German and Mesten) and *O. sanctum* L. (cv. Local), grown in Mississippi showed *in vitro* activity against *Leishmania donovani* with IC_{50} values from 37.3 to 49.6 $\mu g\ mL^{-1}$, which was comparable to the activity of commercial oil (IC_{50} = 40–50 $\mu g\ mL^{-1}$). Minor basil oil constituents such as δ-cadinene, 3-carene, α-humulene, citral, and caryophyllene had antileishmanial activity (Zheljazkov et al. 2008).

11.4.2 Antioxidant Activities

The disturbance in the oxidant–antioxidant balance induces an oxidative stress that leads to increase and consequent attack of reactive oxygen species (ROS), such as superoxide anion ($O_2^•$), alkoxyl ($RO^•$), hydroxyl ($HO^•$), and peroxyl ($RO_2^•$) radicals against the structural cell's components and nitrogen species (RNS) – nitric oxide ($NO^•$) and peroxynitrite ($ONOO^•$) (Weidinger and Kozlov 2015). The oxidative stress is related to the development of several human disorders, such as cancer and Alzheimer's disease (Poprac et al. 2017). Some varieties of *Ocimum* spp. were compared on the basis of their chemical composition. The major compound for all varieties, with contents from 51.1% to 5% was linalool. In Italian large leaf, purple and green purple ruffles, methyl chavicol was the predominant constituent (46.4–59.5%). Holy and cinnamon basil were characterized by a high percentage of methyl eugenol (74.7%) and *trans*-methyl cinnamate (45.9%), respectively. The antioxidant activities were conducted by ABTS and FRAP assays. The highest radical scavenging activity (996.7 μmol Trolox ml^{-1} ABTS; 1262.9 μmol AA mL^{-1} FRAP) showed purple EO, which was associated with the highest levels of eugenol (14.7%) – an important component due to its high antibacterial and antioxidant activities.

Different authors have employed the DPPH assay to evaluate the antioxidant activities of basil oil. Stanojevic et al. (2017) described the chemical composition and antioxidant activity of basil EO which was characterized by the highest content of linalool (31.6%) and methyl chavicol (23.8%). The investigated oil has shown the best antioxidant properties after 90 minutes of incubation with EC_{50} value of 2.38 mg mL^{-1}. The DPPH scavenging activity assay indicated that basil oil may be an alternative antioxidant, with applications in food and pharmaceutical industries. DPPH, ABTS, Phosphomolybdenum, and FRAP assay were used to elucidate the antioxidant activities of oil and ethanolic extract from *O. basilicum*. The results demonstrated excellent radical-scavenging and higher antiradical capacity (Rezzoug et al. 2019).

A study investigated the biological activities of oil from sweet basil leaves via its effect on α-amylase, α-glucosidase and angiotensin-I-converting enzyme (ACE) activities, inhibition of Fe^{2+} and sodium nitroprusside (SNP)-induced lipid peroxidation in the pancreas and heart homogenates of rats. The results obtained described a dose-dependent inhibition of α-amylase ($IC_{50} = 3.21$ mg mL^{-1}), α-glucosidase ($IC_{50} = 3.06$ mg mL^{-1}) and ACE ($IC_{50} = 0.89$ mg mL^{-1}) activities *in vitro*. The essential oil also inhibited both Fe^{2+}- and SNP-induced lipid peroxidation in the pancreas and hearts of rats. The antioxidant and enzyme inhibitory effects of the essential oil could be attributed to the presence of phytochemicals, which could be responsible for the antidiabetic and antihypertensive properties of the oil (Ademiluyi et al. 2016).

Oils from the aerial parts of basil exhibited good antioxidant activity as measured by DPPH free radical-scavenging ability, inhibition of linoleic acid oxidation and bleaching β-carotene in linoleic acid system (Hussain et al. 2008). Various solvent extracts from the leaves and flowers of purple basil were tested *in vitro* for their antioxidant ability. The results showed a significant antioxidant activity which may be due to its lipid peroxidation inhibition, metal chelating activities and radical scavenging properties (Yeşiloğlu and Sit 2012). The antioxidant properties of different extracts (Et_2O, $CHCl_3$, EtOAc, n-BuOH, and H_2O) of basil oil were studied. Among all of them, the aqueous, ethyl acetate and n-butanol extracts showed strong free radical scavenging activity. The levels of phenolics and flavonoids in the extracts could explain the observed differences in antioxidant activity (Kaurinovic et al. 2011).

11.4.3 Anti-inflammatory Activities

The progression of inflammation leads to detrimental effects such as tissue damage, vascular alterations, loss of function, and pain (Maroon et al. 2010). Basil extracts, rich in rosmarinic acid, possess considerable potential in suppressing rheumatoid arthritis, and atherosclerosis- and asthma-associated inflammation. Thus, activity depends on multiple mechanisms such as inhibition of cyclooxygenase-2 (COX-2), reduction of pro-inflammatory -cytokines and -mediators levels including interleukin 6, PGE2 and tumor necrosis factor (TNF)-α, increase in anti-inflammatory cytokines levels like interleukin 10 (Mueller et al. 2010; Umar et al. 2014).

In a paper the anti-inflammatory properties of basil oil and an extract were evaluated. The mechanism is based on a composed interaction between the inhibition of pro-inflammatory mediators and the stimulation of anti-inflammatory cytokines.

Cytokines are subdivided in pro-inflammatory, which initiate defence against pathogens (IL–1, IL-2, IL-6, IL-8, and TNF-α) and anti-inflammatory (IL–1 antagonist receptor, IL-4, IL–10, and IL–13) which regulate the inflammatory process helping to balance the inflammatory response (Goldstein et al. 2006). The obtained results noted that the percentage of production of IL–10, the anti-inflammatory cytokine, shows an increase of more than 60% at the highest dose of the extract when compared with the positive control (ibuprofen) (Güez et al. 2017).

Raina et al. (2016) compared the anti-inflammatory activity of aqueous and methanolic extracts of aerial parts of basil in macrophage (RAW264.7), human chondrosarcoma (SW1353) cell lines, and human primary chondrocytesin order to correlate their efficacy in terms of management of osteoarthritis. In RAW264.7, aqueous extract decreased nitric oxide (35%) and prostaglandin E2 (70.8%) production, in SW1353 and chondrocytes – PGE2 (76.11%) and leukotriene B4 (59.6%), in chondrocytes – production of matrix metalloproteinase -2 (58.49%), -9 (43.13%) and –13 (54.54%) significantly more than methanolic extract. All these data suggest that aqueous extract of *O. basilicum* could be explored for its potential applications in the management of inflammatory conditions associated with osteoarthritis.

11.4.4 Antiplatelet Activities

Antiplatelet activity of aqueous extract from basil aerial parts was studied by using thrombin (0.5 U mL^{-1}) and ADP (5 μM) as agonists. The obtained results showed an inhibition of ADP-induced platelet aggregation by 13.0%, 28.2%, 30.5%, 44.7% and 53.0% at a dose of 1, 2, 3, 4 and 5 g L^{-1}, respectively. Thrombin-induced platelet activation was also reduced by 15.0%, 23.0%, 40.0%, 38.4%, and 42.0% at the same doses of extract (Amrani et al. 2009). The *in vitro* antiplatelet activity of *O. basilicum* methanolic extract was determined by Naidu et al. (2015). The inhibition of platelet aggregation was found to be 63.79%, IC_{50} value = 8.43 mg mL^{-1}, which is probably due to the high levels of polyphenols such as quercetin, rutin, kaempferol, and caffeic acid.

11.4.5 Antithrombotic Activities

The effects of aqueous extract of sweet basil on platelet aggregation and experimental thrombus were studied by Tohti et al. (2006). Data proved a strong inhibitory effect on a platelet aggregation induced by ADP and thrombin (dose-dependent) and an antithrombotic effect *in vivo* develops progressively over 7 days and disappears over 3–7 days.

The antithrombotic activity of ethanolic extract from *Ocimum basilicum L.* was investigated in mice. Fibrin degradation products were assayed. Results showed that the extract activates fibrinolytic which induces the generation of Farnesyl Diphosphate Synthase (FDPs) and inhibits thrombosis process (Kuerban et al. 2017).

11.4.6 Antihypertensive Activities

Hypertension is one of the most important risk factors of arterial thrombosis. Its clinical effects are heart failure, acute coronary syndrome, and ischemic stroke (Kamińska et al. 2005).

Umar et al. (2010) tested the effects of basil extract on a rat model of hypertension, on determinants of hypertension (plasma angiotensin II, endothelin –1) and its consequences (renal function, myocardial hypertrophy). It was found that the extract reduced systolic and diastolic blood pressure by about 20 and 15 mm Hg, respectively, compared with 35 and 22 mm Hg for captopril, from the lowest dose tested with no dose dependency. Cardiac hypertrophy, angiotensin II and endothelin –1 were reduced also.

11.4.7 Antihyperlipidemic and Antiulcerative Activities

The hypolipidemic activity of aqueous extract from sweet basil was studied in mice models. Animals were injected intraperitoneally with Triton WR–1339 at a dose of 200 mg kg^{-1} body weight. 24 h after treatment, the oral administration of the extract exerts a significant reduction of total cholesterol, triglycerides and LDL-cholesterol concentrations by 56%, 63% and 68%, respectively, but HDL-cholesterol was not increased markedly. The basil extract also prevented plasma lipid oxidation by 16%, 20%, 32% and 44% at doses of 10, 25, 50 and 100 μg mL^{-1}, respectively. These results indicate that the phenolic extract might be beneficial in lowering hyperlipidemia and preventing atherosclerosis (Touiss et al. 2017).

Aqueous and ethanolic extracts of *O. basilicum* whole plant were evaluated in order to establish the effect on gastric and duodenal ulceration in rat models by using cysteamine hydrochloride. Both the extracts produced significant activity in cysteamine induced duodenal ulceration as the aqueous extract showed more potent activity than the ethanolic one. The extracts also increased healing of gastric ulceration *in vivo* (Prabhat et al. 2010). In another experimental study, *O. basilicum*'s fixed oil presented strong antiulcer activity against aspirin, indomethacin, alcohol, histamine, reserpine, serotonin, and stress-induced ulceration. It was found that the lipoxygenase inhibiting, histamine antagonistic and anti-secretory effects of the oil have a significant impact on antiulcer activity (Singh 1999).

Antiulcerogenic activities of methanolic and aqueous extracts from basil aerial parts were studied against aspirin-induced gastric ulcers in rats. The effects on output of gastric acid, pepsin and hexosamine concentrations in gastric fluid were recorded in ulcerated and non-ulcerated rats. The acid output was decreased by the methanolic extract while hexosamine secretion was enhanced which suggests that the antiulcerogenic effect is due to decreases of acid and pepsin outputs which enhance gastric mucosal strength (Akhtar and Munir 1989).

11.4.8 Hypoglycemic and Hepatoprotective Activities

In vitro hypoglycemic activity of *O. basilicum* aqueous extract in mice was investigated. The extract showed significant dose-dependent inhibition against rat intestinal maltose, sucrose, and porcine pancreatic α-amylase (IC_{50} 21.31, 36.72 and 42.50 mg mL^{-1}, respectively) inhibitory activities. The results showed that basil extract offered positive benefits to control diabetes via antioxidant properties, α-glucosidase and α-amylase inhibiting activities (El-Beshbishy and Bahashwan 2012).

The effect of methanolic leaf extract of basil, after benzene-induced hematotoxicity in Swiss albino mice was evaluated by hematological parameters (i.e., hemoglobin (Hb), red blood cell (RBC) and white blood cell (WBC) counts), cell cycle regulatory proteins expression and DNA fragmentation analysis of bone marrow cells. It was determined by an up-regulation of p53 and p21 and down regulation of levels of CDK2, CDK4, CDK6 and cyclins D1 and E. DNA was less fragmented. The secondary metabolites of basil leaf extract, comprising essential oil monoterpene geraniol and its oxidized form citral as major constituents, have modulatory effect in haematological abnormalities and cell cycle deregulation induced by benzene in mice (Saha et al. 2012).

The ethanolic extract of leaves from *O. basilicum* was studied for its hepatoprotective effects against H_2O_2 and CCl_4 induced liver damage. The extract showed significant anti-lipid peroxidation activity *in vitro*, as well as exhibiting a strong effect in superoxide radical and nitric oxide radical scavenging, indicating their potent antioxidant effects (Meera et al. 2009).

11.4.9 Anticonvulsant Activities

The anticonvulsant effects and possible CNS depressant activity of *O. basilicum* (access "Maria Bonita") leaf essential oil in different experimental models were analyzed. The results revealed that all doses of basil oil showed the depressant CNS activity with the decrease of spontaneous activity, ptosis, ataxia, and sedation. The o il also increased the latency for development of convulsions in pentylenetetrazol (PTZ) and picrotoxin tests ($p < 0.05$). For PTZ, the effect of essential oil was reversed by flumazenil. EO did not interfere with the convulsions induced by strychnine ($p > 0.05$) (Oliveira et al. 2009).

Ismail (2006) proved that the basil essential oil increased in a dose-dependent manner the latency of convulsion and percent of animals exhibiting clonic seizures after intraperitoneal administration. The ED_{50} values of the investigated oil were as follows: 0.61 mL kg^{-1}, 0.43 mL kg^{-1}, and 1.27 mL kg^{-1}, against convulsions induced by pentylenetetrazole, picrotoxin, and strychnine, respectively. The observed anticonvulsant and hypnotic activities could be related to the presence of a variety of terpenes such as linalool, 1,8-cineol, and eugenol. It has been reported that these constituents exert anticonvulsant activity, potentiate phenobarbitone sleeping time, and have an inhibitory effect on locomotor activity (Santos and Rao 2000; Szabadics and Erdélyi 2000; Wie et al. 1997).

11.4.10 Immunomodulatory Activities

The aqueous extract from aerial parts of *O. basilicum* cv. cinnamon showed significant inhibitory effects against HIV–1 induced cytopathogenicity in MT-4 cells with an effective dose ED = 16 μg mL^{-1}. The results indicate its immunomodulatory action taking place in the cellular level, including platelet anti-aggregant property and inhibitory activity to counter HIV–1 reverse transcriptase (Yamasaki et al. 1998).

In another study, the authors evaluated the immunomodulatory activities of aqueous extract of *O. basilicum* on human peripheral blood mononuclear cells (PBMC) by lymphoproliferation test, and defined the responding cells by flow cytometry, secretion of various cytokines by ELISA, and expression of mRNA by quantitative at concentrations tested (0.135, 0.270, and 0.540 μg mL^{-1}) of extract were capable of dose-dependently stimulating DNA synthesis of human PBMC. Basil extract suppressed also cytokines produced by TH1 (IL-2, IFN-γ, and TNF-β), TH2 (IL-5, IL–10) as well as regulatory T (TGF-β) cells, and expression of ERK2 mRNA in PBMC. These data convincingly demonstrate that basil extract possesses direct immunomodulatory effects on the basic functional properties of human immune cells, possibly mediated by\n the ERK2 MAP-kinase signal pathway (Tsai et al. 2011).

11.4.11 Cytotoxicity Effect

The chemical composition and an *in vitro* anticancer activity of the essential oil from *O. basilicum*, cultivated in the Western Ghats of South India were examined. The major constituents in oil were found to be methyl cinnamate (70.1%), linalool (17.5%), β-elemene (2.6%) and camphor (1.5%) which revealed that the investigated plant belongs to the methyl cinnamate and linalool chemotype. A methyl thiazol tetrazolium assay was used for *in vitro* cytotoxicity screening against the human cervical cancer cell line (HeLa), human laryngeal epithelial carcinoma cell line (HEp-2) and NIH 3T3 mouse embryonic fibroblasts. On the basis of the obtained IC_{50} values (90.5 and 96.3 μg mL^{-1}, respectively), the authors concluded that the basil oil has a potent cytotoxicity (Kathirvel and Ravi 2012).

In a study, the methanolic extract of basil showed strong cytotoxic activity against colon (HCT116) and liver (HEPG2) carcinoma cell line, where IC_{50} of the two human cell line were 27 μg mL^{-1} and 34.5 μg mL^{-1}, respectively. The extract was found to include 12 active phenolic compounds such as ferulic acid, gallic acid, chlorogenic acid, caffeic acid, cinnamic acid, ellagic acid, kaempferol, catechin, quercetin (Abd El-Azim et al. 2015).

The cytotoxic activity of basil essential oil was tested against breast cancer cell lines (MCF-7) using MTT assay. The results showed a significant cytotoxic activity of oil in concentration ranges between 4 and 100 μg mL^{-1} with IC_{50} value of 60 μg mL^{-1} (Tamil Selvi et al. 2015).

Bhagat et al. (2020) studied, the cytotxicity of basil essential oil against six cancer cell lines: prostate (DU–145), liver (HEP-2), neuroblastoma (IMR-32), and colon (HT-29, 502713, SW-620) and dose-dependent cytotxic response was observed against all the cancer cell lines. The highest ability to reduce MTT to the formazan derivative was observed against colon (HT-29) cancer cell line with IC_{50} value of 0.034 μl mL^{-1} which indicates for the prominent cytotoxic/antiproliferative effect, probably through induction of irreparable DNA damage leading to subsequent cell death in apoptotic manner.

11.4.12 Anticancer Activities

The anticancer activity of methanolic extract from basil and its fractions were evaluated, using human cancer cell lines and the mechanism of their antiproliferative action was explored. The extract was fractionated into petroleum ether soluble (PE-S) and insoluble (PE-I) fractions, which were tested on HT–144, MCF-7, NCI-H460 and SF-268 cell lines by Sulforhodamine B assay and the effects on the cytoskeleton and nuclei of MCF-7 cells were studied by immunofluorescence microscopy. The further fractionation of PE-I (GI_{50} 5 μg mL^{-1}; LC_{50} 71 μg mL^{-1} against MCF-7) showed the presence of four compounds, mainly ursolic acid (LC_{50} 18.6 μg ml^{-1}). It was found that ursolic acid (100 μM) induced a significant decrease in the percentage of cells in anaphase/telophase stages along with F-actin aggregation and mitotic spindle distortion. The obtained results prove antiproliferative activity of basil extract against MCF-7 cells which may partly be due to effects of ursolic acid on F-actin and microtubules (Qamar et al. 2010).

The antiproliferative and cytotoxic activities of extracts from basil cultivars (*O. b.* "purple ruffle", *O. b.* "dark opale", *O. b.* "genovese", *O. b.* "anise", *O. b.* "bush green" and *O. b.* L. – OBL) against line HeLa, MCF-7, Jurkat, HT-29, T24, MIAPaCa-2 cancer cells and one normal human cell line HEK-293 were examined. OBL exhibited the highest antioxidant and antiproliferative activities. All tasted extracts had certain anticancer activity against diverse cancer cells due to the presence of compounds such as rosmarinic acid, chicoric acid and caftaric acid (Elansary and Mahmoud 2015).

The antitumor activity of the essential oil obtained from leaves of *O. basilicum* "Cinnamon" was examined against three different cancer cell lines including MDA-MB-231, MCF7 and U-87 MG. The IC_{50} values were 432.3 μg mL^{-1}, 320.4 μg ml^{-1} and 431.2 μg mL^{-1}, respectively. The chemical composition revealed the presence of linalool, eugenol, eucalyptol, hinesol, trans-α-bergamotene and γ-cadinene as major constituents (Aburjai et al. 2020).

11.4.13 Insecticidal and Larvicidal Activities

The bean weevil, *Acanthoscelides obtectus*, causes significant post-harvest losses in the common bean, *Phaseolus vulgaris* L. The control of this insect involves the use of conventional insecticides, but there is an increasing demand in the search for new active products for pest control for the reduction of adverse effects on human health and the environment. Basil essential oil reduced the bean weight losses and the number of beans damaged by *A. obtectus*, Oil also exhibited an insecticidal activity on the insect, both directly in adult insects or indirectly on bean seeds. The results proved the potential of oil as active substances against *A. obtectus* in environmentally low risk pest control strategies (Rodríguez-González et al. 2019).

The insecticidal activity of basil oil and its three major components (linalool, estragole and trans-anethole) were tested on three tephritid fruit fly species *Ceratitis capitata* (Wiedemann), *Bactrocera dorsalis* (Hendel), and *Bactrocera cucurbitae* (Coquillett). The lethal times for 90% mortality of the three fly species to 10% of the test chemicals were between 8 and 38 min as the toxic action of basil oil in *C. capitata* occurred significantly faster than in *B. cucurbitae* and *B. dorsalis*. Linalool and Estragole acted faster in *B. dorsalis* than in *C. capitata* and *B.cucurbitae*, whereas trans-anethole action was similar to all three species. In a mixture with cuelure (attractant to *B. cucurbitae* male), the potency of linalool decreased as the concentration of cuelure increased, due to linalool hydrolysis catalyzed by acetic acid from cuelure degradation. If estragole, linalool, methyl eugenol, trans-anethole, and basil oil are mixed with attractants such as paraphermones in an appropriate formula, they may be used as a natural insecticide (Chang et al. 2009).

Benelli et al. (2019) evaluated the insecticidal activity of white wild basil (*Ocimum gratissimum*) essential oil, ethanolic and aqueous extracts against insect pests and vectors, i.e., the tobacco cutworm *Spodoptera littoralis*, the housefly *Musca domestica*, and the filariasis vector *Culex quinquefasciatus*. The toxicity of the oil and extracts against the non-target earthworm *Eisenia fetida* was assessed and it was found that they are not toxic over the positive control α-cypermethrin. The essential oil was significantly more active on tested insects than extracts with values as follows: LC_{50}/LD_{50} of 39.6 mg L^{-1} on *C. quinquefasciatus*, 72.2 μg per adult on *M. domestica* and 30.2 μg per larva on *S. littoralis*. Additionally, the oil and ethanolic extract at sublethal doses (10 and 70 μg cm^{-2}, respectively) affected the survival of *S. littoralis* larvae from the third day on. In another study the essential oils obtained from *Ocimum americanum*, *O. basilicum*, *O. basilicum* fa. *citratum*, *O. gratissimum* and *O. tenuiflorum*, were tested for mosquito repellent and larvicidal activities. The obtained results proved that all the investigated oils exhibited both activities as *O. basilicum* showed the strongest larvicidal activity ($EC_{50} = 81$, $EC_{90} = 113$ ppm) and *O. gratissimum* exhibited the longest duration of action for mosquito repellent activity (more than two hours). The chemical composition of the oils indicated the presence of camphor, caryophyllene oxide, 1,8-cineole, methyleugenol, limonene, myrcene, and thymol, all known insect repellents (Chokechaijaroenporn et al. 1994).

The mosquito repellent activities of basil oil against *Anopheles gambiae* and *Culex quinquefasciatus* were investigated. At 50% concentration, oil exhibited significant repellent potential on *Anopheles gambiae* with a protection time of 183 min and at 100% concentration – the highest protection time against the two species of mosquito tested (Baba et al. 2012).

A significant toxic effect was observed by Govindarajan et al. (2013) when the activity of basil essential oil was tested against late third-stage larvae of *Culex tritaeniorhynchus*, *Aedes albopictus* and *Anopheles subpictus* with LC_{90} values of 23.44, 21.17 and 18.56 ppm and LC_{50} values of 14.01, 11.97 and 9.75 ppm respectively.

11.5 Alteration of Bioactive Content in Basil

Basil is rich in bioactive components such as linalool, eugenol, 1,8-cineole, caffeic, rosmarinic, chicorinc, and caftaric acids, which provide carminative, galactogogue, digestive, antispasmodic, antibacterial, anticonvulsant, and anticarcinogenic activities. The level of bioactive components is influenced by genetic and environmental factors.

Among environmental factors, light provides essential energy for growth and development of plants and acts as an external signal to trigger specific responses in the plants. The quality and quantity of light also affects the accumulation of flavonoids, carotenoids, tocopherols and biosynthesis of essential oil. Hosseini et al. (2018) investigated the effects of light spectrum on essential oil composition, antiradical activity and phenolic content of green and purple varieties of *Ocimum basilicum* using different spectral light compositions – red (R), blue (B), white (W), 50% red + 50% blue (50:50% R:B) and 70% red + 30% blue (70:30% R:B) provided by LED modules with the same light intensity for 30 days. Data showed that growing both basil varieties under 70:30% R:B light resulted in the highest antiradical activity and phenolic content. However, in green basil, the chemical composition of the essential oil was improved by growing under 70:30% R:B light, while in the purple variety R light induced production of limonene, α-pinene and β-myrcene. In conclusion, as the effects of light spectra on bioactive constituents depended on variety and type of compounds, through exposure to specific light spectrum it is possible to alter the bioactive compound composition in basil. In another study, Simon et al. (1992) investigated the possible alteration of leaf oil composition in a sweet basil by mild and moderate plant water stress. After 21 days of plant water deficit, the authors found that the oil content increased from 3.1 to 6.2 $\mu L\ g^{-1}$ leaf dry weight and the xylem water potential decreased from −0.30 to −1.12 MPa. Also, significant decreases in leaf and stern dry weights were observed when the plant water deficit increased. The leaf area from plants subjected to a mild water deficit (−0.68 MPa) was not significantly reduced compared to the control (non-stressed plants). Additionally, water stress altered the oil composition. When the water stress increased, the quantities of linalool and estragol increased too, while the relative proportion of sesquiterpenes decreased.

Efforts are underway to develop sweet basil as a high-value essential oil crop due to the increasing demands for natural products with potential health benefits and industrial purposes. Selecting an appropriate fertilizer may lead to improved crop yields and quality. Fertilization with organic soil amendments such as biosolids, arbuscular mycorrhizal fungi (AMF) is central in organic farming. The effects of biological and conventional fertilizers on plant growth, EO composition and yield of two Romanian sweet basil cultivars – the green-leafed "Aromat de Buzau" (AB) and the purple-leafed "Violet de Buzau" (VB) were evaluated. The results demonstrated that conventional fertilizers increased fresh yield (116 v% and 68% for the AB and VB, respectively), whilc biological fertilizers positively alter the EO composition, leading to increased crop quality (Burducea et al. 2018).

Biotechnologies offer a wide range of opportunities related to secondary metabolite production. In the work of (Kintzios et al. 2004) nodal explants with lateral buds and leaf-derived suspension cultures of sweet basil, *Ocimum basilicum L.*, were set up for cultivation in an airlift bioreactor system in an attempt to scale up rosmarinic acid accumulation. In bioreactors, enhanced growth was highly positive correlated with rosmarinic acid accumulation (r2 = 0.99), which was 7 times (suspension cultures) or 44 times (regenerants) higher than in suspension cultures in 250 mL flasks (4 μg g^{-1} DW). In addition, culture fresh biomass growth was positively correlated with ROS production (r2 = 0.99), but negatively correlated with soluble protein and carbohydrate concentration (r2 = − 0.79 and − 0.57, respectively). Since ROS are produced by mitochondria, chloroplasts and peroxisomes as byproducts of normal cellular metabolism, their increased concentration might have been associated with the rapid turn-over of primary metabolites (Kintzios et al. 2004).

Mehring et al. (2020) established for the first time *Ocimum basilicum* cambial meristematic cells as for triterpenoids (oleanolic acid and ursolic acid) production. In this study the authors tested productivity of cambial meristematic cells and dedifferentiated cell cultures cultivated in shake flasks and wave-mixed disposable bioreactors including methyl jasmonate as elicitor. The highest accumulation of oleanolic acid (2.32-fold) and ursolic acid (1.92-fold) were determined in cambial meristematic cells cultures (Mehring et al. 2020).

Development and applications of genetic linkage maps based on DNA markers linked to important traits as well as quantitative trait locus (QTL), are becoming essential tools of modern plant breeding programs. The first genetic linkage map including downy mildew resistance QTL for sweet basil (*Ocimum basilicum*) were constructed by Pyne et al. (2017). In order to improve available genetic resources, expressed sequence tag simple sequence repeat (EST-SSR) and single nucleotide polymorphism (SNP) markers were developed and used to genotype the MRI x SB22 F2 mapping population, which segregates for response to *downy mildew*. A densely constructed linkage map is formed by 42 EST-SSR and 1847 SNP markers spanning 3030.9 cM. The authors clearly demonstrate and evidence provided for an additive effect between the two minor QTL and the major QTL dm11.1 increasing downy mildew susceptibility (Pyne et al. 2017).

Recently, developments of efficiency systems for manipulating the biosynthesis of bioactive compounds reveal clustered regularly interspaced short palindromic repeats (CRISPR)/CRISPR-associated9 (Cas9) endonuclease system as a powerful RNA-guided genome editing tool for transcriptional modulation.

11.6 Conclusion

Plants have been used since ancient times for the treatment of an enormous number of diseases. *O. basilicum L.* has a vast spectrum of pharmacological activities. Extracts and essential oils of the various parts have been used for their antibacterial, antioxidant, antidiabetic, anticancer, anticonvulsant, antihyperlipidemic, anti-inflammatory, hepatoprotective, and immunomodulatory activities. Thanks to its diverse biological potential, basil has great scope for further new areas of investigation. Future research should be emphasized on *O. basilicum L.* for evaluation of its pharmacological properties for control of various diseases especially in cancer, cardiac, and neuropsychological disorders.

Developments in the field of biotechnologies in the context of metabolome engineering provide effective procedures such as modulate biosynthesis of bioactive molecules through biomass production by bioreactor systems. Genetic linkage maps supported a powerful platform into modern plant breeding programmes. Clustered regularly interspaced short palindromic repeat (CRISPR)/Cas9 gene editing technology has revolutionized engineering of natural product biosynthesis, crop breeding and functional genomics.

References

Abd El-Azim, M.H., Abdelgawad, A.A., El-Gerby, M., Ali, S., and El-Mesallamy, A.M. (2015). Phenolic compounds and cytotoxic activities of methanol extract of basil (Ocimum basilicum L.). *Journal of Microbial and Biochemical Technology* 7: 182–185.

Abou El-Soud, N.H., Deabes, M., El-Kassem, L.A., and Khalil, M. (2015). Chemical composition and antifungal activity of *Ocimum basilicum* L. Essential Oil. *Open Access Macedonian Journal of Medical Sciences* 3: 374–379.

Aburjai, T., Mansi, K., Azzam, H., Alqudah, D.A., Alshaer, W., and Abuirjei, M. (2020). Chemical compositions and anticancer potential ofessential oil from greenhouse-cultivated Ocimum basilicum leaves. *Indian Journal of Pharmaceutical Sciences* 82: 179–184.

Ademiluyi, A.O., Oyeleye, S.I., and Oboh, G. (2016). Biological activities, antioxidant properties and phytoconstituents of essential oil from sweet basil (*Ocimum basilicum* L.) leaves. *Comparative Clinical Pathology* 25: 169–176.

Akhtar, M.S. and Munir, M. (1989). Evaluation of the gastric antiulcerogenic effects of Solanum nigrum, *Brassica oleracea* and *Ocimum basilicum* in rats. *Journal of Ethnopharmacology* 27: 163–176.

Amrani, S., Harnafi, H., Gadi, D. et al. (2009). Vasorelaxant and anti-platelet aggregation effects of aqueous *Ocimum basilicum* extract. *Journal of Ethnopharmacology* 125: 157–162.

Anand, U., Jacobo-Herrera, N., Altemimi, A., and Lakhssassi, N. (2019). A comprehensive review on medicinal plants as antimicrobial therapeutics: potential avenues of biocompatible drug discovery. *Metabolites* 9: 258. doi: 10.3390/metabo9110258.

Araújo Silva, V., Pereira da Sousa, J., de Luna Freire Pessôa, H. et al. (2016). *Ocimum basilicum*: antibacterial activity and association study with antibiotics against bacteria of clinical importance. *Pharmaceutical Biology* 54: 863–867.

Astuti, P., Saifullah, T., Wulanjati, M., Yosepine, A., and Ardianti, D. (2016). Basil essential oil (*Ocimum basilicum* L.) activities on streptococcus mutans growth, biofilm formation and degradation and its stability in micro-emulsion mouthwash formula. *International Journal of Pharmaceutical and Clinical Research* 8: 26–32.

Baba, G., Lawal, A.O., and Sharif, H.B. (2012). Mosquito repellent activity and phytochemical characterization of essential oils from Striga hermonthica, Hyptis spicigera and Ocimum basilicum leaf extracts. *British Journal of Pharmacology and Toxicology* 3 (2): 43–48.

Benelli, G., Pavela, R., Maggi, F. et al. (2019). Insecticidal activity of the essential oil and polar extracts from *Ocimum gratissimum* grown in Ivory Coast: efficacy on insect pests and vectors and impact on non-target species. *Industrial Crops and Products* 132: 377–385.

Bhagat, M., Sangral, M., Kumar, A., Rather, R.A., and Arya, K. (2020). Chemical, biological and in silico assessment of *Ocimum viride* essential oil. *Heliyon* 6: e04209.

Bilal, A., Jahan, N., Ahmed, A., Bilal, S.N., Habib, S., and Hajra, S. (2012). Phytochemical and pharmacological studies on *Ocimum basilicum* Linn – a review. *International Journal of Current Research and Review* 4: 73–83.

Burducea, M., Zheljazkov, V.D., Dincheva, I. et al. (2018). Fertilization modifies the essential oil and physiology of basil varieties. *Industrial Crops and Products* 121: 282–293.

Chang, C.L., Cho, I.K., and Li, Q.X. (2009). Insecticidal activity of basil oil, trans-anethole, estragole, and linalool to adult fruit flies of Ceratitis capitata, Bactrocera dorsalis, and Bactrocera cucurbitae. *Journal of Economic Entomology* 102: 203–209.

Chenni, M., El Abed, D., Rakotomanomana, N., Fernandez, X., and Chemat, F. (2016). Comparative study of essential oils extracted from Egyptian basil leaves (*Ocimum basilicum* L.) using hydro-distillation and solvent-free microwave extraction. *Molecules (Basel, Switzerland)* 21: E113.

Chokechaijaroenporn, O., Bunyapraphatsara, N., and Kongchuensin, S. (1994). Mosquito repellent activities of ocimum volatile oils. *Phytomedicine* 1: 135–139.

Darrah, H.H. (1974). Investigation of the cultivars of the basils (Ocimum). *Economic Botany* 28: 63–67.

Dube, S., Upadhyay, P.D., and Tripathi, S.C. (1989). Antifungal, physicochemical, and insect-repelling activity of the essential oil of *Ocimum basilicum*. *Canadian Journal of Botany* 67: 2085–2087.

Elansary, H.O. and Mahmoud, E.A. (2015). *In vitro* antioxidant and antiproliferative activities of six international basil cultivars. *Natural Product Research* 29: 2149–2154.

El-Beshbishy, H. and Bahashwan, S. (2012). Hypoglycemic effect of basil (*Ocimum basilicum*) aqueous extract is mediated through inhibition of α-glucosidase and α-amylase activities: an *in vitro* study. *Toxicology and Industrial Health* 28: 42–50.

Fernandes, F., Pereira, E., Cri, A. et al. (2019). *Ocimum basilicum* var. purpurascens leaves (red rubin basil): a source of bioactive compounds and natural pigments for the food industry. *Food and Function* 10: 3161–3171.

Filip, S. (2017). Basil (Ocimum basilicum L.) a source of valuable phytonutrients. *International Journal of Clinical Nutrition & Dietetics* 3 (118). doi: 10.15344/2456-8171/2017/118.

Gang, D.R., Wang, J., Dudareva, N. et al. (2001). An investigation of the storage and biosynthesis of phenylpropenes in sweet basil. *Plant Physiology* 125: 539–555.

Gebrehiwot, H., Bachetti, R., and Dekebo, A. (2015). Chemical composition and antimicrobial activities of leaves of sweet basil (*Ocimum basilicum* L.) herb. *International Journal of Basic and Clinical Pharmacology* 4: 869–875.

Goldstein, S.L., Leung, J.C., and Silverstein, D.M. (2006). Pro- and anti-inflammatory cytokines in chronic pediatric dialysis patients: effect of aspirin. *Clinical Journal of the American Society of Nephrology* 1: 979–986.

Govindarajan, M., Sivakumar, R., Rajeswary, M., and Yogalakshmi, K. (2013). Chemical composition and larvicidal activity of essential oil from *Ocimum basilicum* (L.) against Culex tritaeniorhynchus, Aedes albopictus and Anopheles subpictus (Diptera: Culicidae). *Experimental Parasitology* 134: 7–11.

Güez, C.M., Souza, R.O.d., Fischer, P. et al. (2017). Evaluation of basil extract (*Ocimum basilicum* L.) on oxidative, anti-genotoxic and anti-inflammatory effects in human leukocytes cell cultures exposed to challenging agents. *Brazilian Journal of Pharmaceutical Sciences* 53: e15098.

Hosseini, A., Zare Mehrjerdi, M., and Aliniaeifard, S. (2018). Alteration of bioactive compounds in two varieties of basil (*Ocimum basilicum*) grown under different light spectra. *Journal of Essential Oil Bearing Plants* 21: 913–923.

Hussain, A.I., Anwar, F., Hussain Sherazi, S.T., and Przybylski, R. (2008). Chemical composition, antioxidant and antimicrobial activities of basil (*Ocimum basilicum*) essential oils depends on seasonal variations. *Food Chemistry* 108: 986–995.

Ismail, M. (2006). Central properties and chemical composition of *Ocimum basilicum*. Essential oil. *Pharmaceutical Biology* 44: 619–626.

Janssen, A.M., Scheffer, J.J., Ntezurubanza, L., and Baerheim Svendsen, A. (1989). Antimicrobial activities of some *Ocimum* species grown in Rwanda. *Journal of Ethnopharmacology* 26: 57–63.

Javanmardi, J., Khalighi, A., Kashi, A., Bais, H.P., and Vivanco, J.M. (2002). Chemical characterization of basil (*Ocimum basilicum* L.) found in local accessions and used in traditional medicines in Iran. *Journal of Agricultural and Food Chemistry* 50: 5878–5883.

Jayasinghe, C., Gotoh, N., Aoki, T., and Wada, S. (2003). Phenolics composition and antioxidant activity of sweet basil (*Ocimum basilicum* L.). *Journal of Agricultural and Food Chemistry* 51: 4442–4449.

Kamińska, M., Mogielnicki, A., Stankiewicz, A. et al. (2005). Angiotensin II via AT1 receptor accelerates arterial thrombosis in renovascular hypertensive rats. *Journal of Physiology and Pharmacology: An Official Journal of the Polish Physiological Society* 56: 571–585.

Kathirvel, P. and Ravi, S. (2012). Chemical composition of the essential oil from basil (*Ocimum basilicum* Linn.) and its in vitro cytotoxicity against HeLa and HEp-2 human cancer cell lines and NIH 3T3 mouse embryonic fibroblasts. *Natural Product Research* 26: 1112–1118.

Kaurinovic, B., Popovic, M., Vlaisavljevic, S., and Trivic, S. (2011). Antioxidant capacity of *Ocimum basilicum* L. and *Origanum vulgare* L. extracts. *Molecules (Basel, Switzerland)* 16: 7401–7414.

Kaya, I., Yigit, N., and Benli, M. (2008). Antimicrobial activity of various extracts of *Ocimum basilicum* L. and observation of the inhibition effect on bacterial cells by use of scanning electron microscopy. *African Journal of Traditional, Complementary, and Alternative Medicines: AJTCAM* 5: 363–369.

Kintzios, S., Kollias, H., Straitouris, E., and Makri, O. (2004). Scale-up micropropagation of sweet basil (Ocimum basilicum L.) in an airlift bioreactor and accumulation of rosmarinic acid. *Biotechnology Letters* 26 (6): 521-523. doi: 10.1023/B:BILE.0000019561.89044.30.

Kuerban, A., Almulaiky, Y.Q., Sheikh, R.A. et al. (2017). *In vivo* antithrombotic activity of ethanolic extract from *Ocimum basilicum* L. *Annual Research and Review in Biology* 20: 1–6.

Lee, S.-J., Umano, K., Shibamoto, T., and Lee, K.-G. (2005). Identification of volatile components in basil (*Ocimum basilicum* L.) and thyme leaves (*Thymus vulgaris* L.) and their antioxidant properties. *Food Chemistry* 91: 131–137.

Lis-Balchin, M., Deans, S.G., and Eaglesham, E. (1998). Relationship between bioactivity and chemical composition of commercial essential oils. *Flavour and Fragrance Journal* 13: 98–104.

Mahajan, N., Rawal, S., Verma, M., Poddar, M., and Alok, S. (2013). A phytopharmacological overview on Ocimum species with special emphasis on *Ocimum sanctum*. *Biomedicine and Preventive Nutrition* 3: 185–192.

Makri, O. and Kintzios, S. (2008). Ocimum sp. (Basil): botany, cultivation, pharmaceutical properties, and biotechnology. *Journal of Herbs, Spices & Medicinal Plants* 13: 123–150.

Maroon, J.C., Bost, J.W., and Maroon, A. (2010). Natural anti-inflammatory agents for pain relief. *Surgical Neurology International* 1: 80.

Marotti, M., Piccaglia, R., and Giovanelli, E. (1996). Differences in essential oil composition of basil (*Ocimum basilicum* L.) Italian cultivars related to morphological characteristics. *Journal of Agricultural and Food Chemistry* 44: 3926–3929.

Marwat, S.K., Khan, M.S., Ghulam, S., Anwar, N., Mustafa, G., and Usman, K. (2011). Phytochemical constituents and pharmacological activities of sweet Basil-*Ocimum basilicum* L. (Lamiaceae). *Asian Journal of Chemistry* 23: 3773–3782.

Mathews, S., Singhal, R.S., and Kulkarni, P.R. (1993). *Ocimum basilicum*: a new non-conventional source of fibre. *Food Chemistry* 47: 399–401.

Matos, M.J., Santana, L., Uriarte, E., Abreu, O.A., Molina, E., and Yordi, E.G. (2015). Coumarins-an important class of phytochemicals. In: *Phytochemicals-Isolation, Characterisation and Role in Human Health* (ed. A.V. Rao and L.G. Rao), 113–140. Rijeka, Croatia: IntechOpen.

Meera, R., Devi, P., Kameswari, B., Madhumitha, B., and Merlin, N.J. (2009). Antioxidant and hepatoprotective activities of *Ocimum basilicum* Linn. and *Trigonella foenum-graecum* Linn. against H2O2 and CCL4 induced hepatotoxicity in goat liver. *Indian Journal of Experimental Biology* 47: 584–590.

Mehring, A., Haffelder, J., Chodorski, J., Stiefelmaier, J., Strieth, D., and Ulber, R. (2020). Establishment and triterpenoid production of *Ocimum basilicum* cambial meristematic cells. *Plant Cell, Tissue and Organ Culture (PCTOC)* 143: 573–581.

Miraj, S. and Kiani, S. (2016). Study of pharmacological effect of *Ocimum basilicum*: a review. *Der Pharmacia Lettre* 8: 276–280.

Mueller, M., Hobiger, S., and Jungbauer, A. (2010). Anti-inflammatory activity of extracts from fruits, herbs and spices. *Food Chemistry* 122: 987–996.

Naidu, J.R., Ismail, R., Kumar, P., Jothy, S.L., Chen, Y., and Sasidharan, S. (2015). Antiplatelet activity and quantification of polyphenols content of methanol extracts of *Ocimum basilicum* and *Mentha spicata*. *Research Journal of Pharmaceutical, Biological and Chemical Sciences* 6: 1236–1244.

Nascimento, L.D.d., Moraes, A.A.B., Costa, K.S.D. et al. (2020). Bioactive natural compounds and antioxidant activity of essential oils from spice plants: new findings and potential applications. *Biomolecules* 10: 988.

Ntonga, P.A., Baldovini, N., Mouray, E., Mambu, L., Belong, P., and Grellier, P. (2014). Activity of *Ocimum basilicum, Ocimum canum*, and *Cymbopogon citratus* essential oils against *Plasmodium falciparum* and mature-stage larvae of *Anopheles funestus* s.s. *Parasite (Paris, France)* 21: 33.

Oliveira, J.S., Porto, L.A., Estevam, C.d.S. et al. (2009). Phytochemical screening and anticonvulsant property of Ocimum basilicum leaf essential oil. *Boletin Latinoamericano y del Caribe de Plantas Medicinales y Aromaticas* 8: 195–202.

Pandey, H., Pandey, P., Singh, S., Gupta, R., and Banerjee, S. (2015). Production of anti-cancer triterpene (betulinic acid) from callus cultures of different Ocimum species and its elicitation. *Protoplasma* 252 (2): 647-655. doi: 10.1007/s00709-014-0711-3.

Poprac, P., Jomova, K., Simunkova, M., Kollar, V., Rhodes, C.J., and Valko, M. (2017). Targeting free radicals in oxidative stress-related human diseases. *Trends in Pharmacological Sciences* 38: 592–607.

Prabhat, D., Sabuj, S., and Ranjan, S. (2010). Phytochemical and pharmacological investigation of the plant *Ocimum basilicum* with special reference to its anti ulceration property. *Journal of Pharmaceutical Research* 3: 63–66.

Pyne, R., Honig, J., Vaiciunas, J. et al. (2017). A first linkage map and downy mildew resistance QTL discovery for sweet basil (*Ocimum basilicum*) facilitated by double digestion restriction site associated DNA sequencing (ddRADseq). *PloS One* 12: e0184319.

Qamar, K.A., Dar, A., S Siddiqui, B. et al. (2010). Anticancer activity of *Ocimum basilicum* and the effect of ursolic acid on the cytoskeleton of MCF-7 human breast cancer cells. *Letters in Drug Design & Discovery* 7: 726–736.

Rahman, S., Islam, R., Kamruzzaman, A., and Jamal, A.H.M. (2011). *Ocimum sanctum* L.: a review of phytochemical and pharmacological profile. *American Journal of Drug Discovery and Development* 4: 1–15. ISSN 2150-427x, doi: 10.3923/ajdd.2011.

Raina, P., Deepak, M., Chandrasekaran, C., Agarwal, A., Wagh, N., and Kaul-Ghanekar, R. (2016). Comparative analysis of anti-inflammatory activity of aqueous and methanolic extracts of *Ocimum basilicum* (basil) in RAW264. 7, SW1353 and human primary chondrocytes in respect of the management of osteoarthritis. *Journal of Herbal Medicine* 6: 28–36.

Rezzoug, M., Bakchiche, B., Gherib, A. et al. (2019). Chemical composition and bioactivity of essential oils and Ethanolic extracts of *Ocimum basilicum* L. and *Thymus algeriensis* Boiss. & Reut. from the Algerian Saharan Atlas. *BMC Complementary and Alternative Medicine* 19: 146.

Rodríguez-González, Á., Álvarez-García, S., González-López, Ó., Silva, F.D., and Casquero, P.A. (2019). Insecticidal properties of Ocimum basilicum and Cymbopogon winterianus against Acanthoscelides obtectus, insect pest of the common bean (*Phaseolus vulgaris*, L.). *Insects* 10.

Saha, S., Mukhopadhyay, M. K., Ghosh, P. D., and Nath, D. (2012). Effect of methanolic leaf extract of Ocimum basilicum L. on benzene-induced hematotoxicity in mice. *Evidence-Based Complementary and Alternative Medicine* 2012: 176385. doi: 10.1155/2012/176385.

Santos, F.A. and Rao, V.S. (2000). Antiinflammatory and antinociceptive effects of 1,8-cineole a terpenoid oxide present in many plant essential oils. *Phytotherapy Research: PTR* 14: 240–244.

Siddiqui, B.S., Aslam, H., Ali, S.T., Begum, S., and Khatoon, N. (2007). Two new triterpenoids and a steroidal glycoside from the aerial parts of *Ocimum basilicum*. *Chemical & Pharmaceutical Bulletin* 55: 516–519.

Simon, J.E., Reiss-Bubenheim, D., Joly, R.J., and Charles, D.J. (1992). Water stress-induced alterations in essential oil content and composition of sweet basil. *Journal of Essential Oil Research* 4: 71–75.

Singh, S. (1999). Evaluation of gastric anti-ulcer activity of fixed oil of *Ocimum basilicum* Linn. and its possible mechanism of action. *Indian Journal of Experimental Biology* 37: 253–257.

Stanojevic, L.P., Marjanovic-Balaban, Z.R., Kalaba, V.D., Stanojevic, J.S., Cvetkovic, D.J., and Cakic, M.D. (2017). Chemical composition, antioxidant and antimicrobial activity of basil (*Ocimum basilicum* L.) essential oil. *Journal of Essential Oil-bearing Plants* 20: 1557–1569.

Szabadics, J. and Erdélyi, L. (2000). Pre- and postsynaptic effects of eugenol and related compounds on *Helix pomatia* L. neurons. *Acta biologica Hungarica* 51: 265–273.

Tamil Selvi, M., Thirugnanasampandan, R., and Sundarammal, S. (2015). Antioxidant and cytotoxic activities of essential oil of *Ocimum canum* Sims. from India. *Journal of Saudi Chemical Society* 19: 97–100.

Tohti, I., Tursun, M., Umar, A., Turdi, S., Imin, H., and Moore, N. (2006). Aqueous extracts of *Ocimum basilicum* L. (sweet basil) decrease platelet aggregation induced by ADP and thrombin *in vitro* and rats arterio-venous shunt thrombosis *in vivo*. *Thrombosis Research* 118: 733–739.

Touiss, I., Harnafi, M., Khatib, S. et al. (2019). Rosmarinic acid-rich extract from *Ocimum basilicum* L. decreases hyperlipidemia in high fat diet-induced hyperlipidemic mice and prevents plasma lipid oxidation. *Physiology and Pharmacology* 23: 197–207.

Touiss, I., Khatib, S., Bekkouch, O., Amrani, S., and Harnafi, H. (2017). Phenolic extract from *Ocimum basilicum* restores lipid metabolism in Triton WR-1339-induced hyperlipidemic mice and prevents lipoprotein-rich plasma oxidation. *Food Science and Human Wellness* 6: 28–33.

Tsai, K., Lin, B., Perng, D., Wei, J., Yu, Y., and Cherng, J.-M. (2011). Immunomodulatory effects of aqueous extract of *Ocimum basilicum* (Linn.) and some of its constituents on human immune cells. *Journal of Medicinal Plant Research* 5: 1873–1883.

Umar, A., Imam, G., Yimin, W. et al. (2010). Antihypertensive effects of *Ocimum basilicum* L. (OBL) on blood pressure in renovascular hypertensive rats. *Hypertension Research: Official Journal of the Japanese Society of Hypertension* 33: 727–730.

Umar, A., Zhou, W., Abdusalam, E. et al. (2014). Effect of *Ocimum basilicum* L. on cyclo-oxygenase isoforms and prostaglandins involved in thrombosis. *Journal of Ethnopharmacology* 152: 151–155.

Vani, S.R., Cheng, S.F., and Chuah, C.H. (2009). Comparative study of volatile compounds from genus Ocimum. *American Journal of Applied Sciences* 6: 523–528.

Weidinger, A. and Kozlov, A.V. (2015). Biological activities of reactive oxygen and nitrogen species: oxidative stress versus signal transduction. *Biomolecules* 5: 472–484.

Wie, M.B., Won, M.H., Lee, K.H. et al. (1997). Eugenol protects neuronal cells from excitotoxic and oxidative injury in primary cortical cultures. *Neuroscience Letters* 225: 93–96.

Yamasaki, K., Nakano, M., Kawahata, T. et al. (1998). Anti-HIV-1 activity of herbs in Labiatae. *Biological & Pharmaceutical Bulletin* 21: 829–833.

Yeşiloğlu, Y. and Sit, L. (2012). Antioxidant properties of various solvent extracts from purple basil. *Spectrochimica Acta. Part A, Molecular and Biomolecular Spectroscopy* 95: 100–106.

Zahran, E.M., Abdelmohsen, U.R., Khalil, H.E., Desoukey, S.Y., Fouad, M.A., and Kamel, M.S. (2020). Diversity, phytochemical and medicinal potential of the genus Ocimum L. (Lamiaceae). *Phytochemistry Reviews* 19: 907–953.

Zheljazkov, V.D., Cantrell, C.L., Tekwani, B., and Khan, S.I. (2008). Content, composition, and bioactivity of the essential oils of three basil genotypes as a function of harvesting. *Journal of Agricultural and Food Chemistry* 56: 380–385.

12

Himalayan Peony (*Paeonia emodi* Royle)

Enlightening Bioactive Compounds and Biological Applications towards Sustainable Use

Prabhakar Semwal[1,2,*], *Sakshi Painuli*[1,3], *Natália Cruz-Martins*[4,5,6,*], *and Ashish Thapliyal*[1]

[1]*Department of Biotechnology, Graphic Era University, Dehradun, Uttarakhand, India*
[2]*Uttarakhand State Council for Science and Technology, Dehradun, Uttarakhand, India*
[3]*Himalayan Environmental Studies and Conservation Organization, Dehradun, Uttarakhand, India*
[4]*Faculty of Medicine, University of Porto, Porto, Portugal*
[5]*Institute for Research and Innovation in Health (i3S), University of Porto, Porto – Portugal*
[6]*TOXRUN – Toxicology Research Unit, University Institute of Health Sciences, CESPU, CRL, 4585-116 Gandra, Portugal*
**Corresponding authors: semwal.prabhakar@gmail.com (P.S.);ncmartins@med.up.pt (N.C.-M.)*

12.1 Introduction

The *Paeonia* genus belongs to Paeoniaceae family and is an economically and ecologically important plant genus. According to the plant list database, it contains about 179 scientific names, among them 36 are accepted and 128 are synonyms (http://www.theplantlist.org/, accessed on 15 May 2020). Some species, including *Paeonia albiflora, Paeonia lactiflora, Paeonia ovate, Paeonia rubra*, and *Paeonia suffruticosa* have been used in Asian and European countries for the treatment of cardiovascular (CV) diseases, obesity, atherosclerosis, diabetes, and inflammatory conditions (Long et al. 2012; Mencherini et al. 2011; Mo et al. 2011; Wu et al. 2010).

Paeonia emodi Royle is an important member of this genus, also known as the Himalayan paeony. It is an endemic species of the Himalayan region that has been used for different purposes, due to its extraordinary ornamental and medicinal properties (Yan et al. 2019). To *P. emodi*, prominent biological activities have been attributed, including antioxidant, anti-inflammatory, cardioprotective, anti-epileptic, nephroprotective, hepatoprotective, lipoxygenase inhibitor, and antimicrobial effects (Badola and Negi 2017; Ibrar et al. 2018, 2019; Ilahi et al. 2016; Kishore et al. 2017; Raish et al. 2017; A. Sharma et al. 2018). Apart from this, it is also used against heart failure, hypertension, palpitations and atherosclerosis (Ghayur et al. 2008). It is the most popular wild edible plant in the Himalayas, being used by these communities for its renowned nutritional and medicinal value (Farooquee et al. 2004; Fayaz et al. 2019; Joshi et al. 2018; Jugran et al. 2016). However, due to their multiple applications and high demands, *P. emodi* is facing great pressure concerning its sustainability (K. Khan et al. 2019). Hence, this chapter aims to provide an updated overview on the Himalayan paeony ethnomedicinal uses, chemical composition and biological attributes of its derived extracts and isolated compounds.

Biomolecules from Natural Sources: Advances and Applications, First Edition. Edited by Vijai Kumar Gupta, Satyajit D. Sarker, Minaxi Sharma, María Elida Pirovani, Zeba Usmani, and Chelliah Jayabaskaran.

12.2 Methodology

A detailed literature search was conducted using different scientific search engines, such as PubMed, ScienceDirect, Scopus, Web of Science, SpringerLink, ACS publications, Google scholar, JSTOR, Taylor and Francis online, Wiley online library databases, using the keywords "*Paeonia emodi*", "Himalayan paeony" and "Chandra". Original articles, review articles, book chapters and conference proceedings published in English were selected to extract the required information from 1963 to 2020 (April).

12.3 Geographical Distribution, Taxonomy, and Nomenclature

P. emodi is indigenous to India, Pakistan, Nepal, Afghanistan, and other Asian countries, where it grows in the Himalayan region at altitudes ranging from 2300 to 2700 m (Haq et al. 2011; Samant and Joshi 2005).

The term *P. emodi* was first mentioned in the *Numerical List of Dried Specimens of Plants* under the *Herbarium of East India Company* in London (Wallich Herbarium at Kew). These samples were collected in 1831 under the supervison of Dr. Nathaniel Wallich (Calcutta Botanical Garden, India) and John Forbes Royle (a botanist), who validated this name by publishing a proper description of the taxon in 1834. Regarding taxonomy, *P. emodi* belongs to the Genus Paeonia, Family Paeoniaceae, Order Dilleniales, Class Magnoliopsida, Phylum Magnoliophyta and Kingdom Plantae. It is often known as the Himalayan paeony and Paeony rose in English; Chandayra, Pawin and Ud-salap in Hindi; Chandra in Sanskrit; Dhandharu in Garhwali; and Unsalib, Pamekh and ood-salib in Urdu (M. Ahmad et al. 2018).

12.4 Ethnomedicinal Uses

Ethnomedicinal applications of any product contribute to the development of evidence-based medicine, also markedly promoting drug discovery around the globe (Gyllenhaal et al. 2012). Traditionally, every plant part is able to be used to treat different diseases triggered by the presence of biologically active molecules. *P. emodi* is frequently used in folk medicine around the world for the treatment of dementia, epilepsy, diarrhea, vomiting, hemorrhoids, the nervous system, gynecological and uterine diseases, intestinal and muscular pain, among others. Detailed information on *P. emodi* ethnomedicinal uses is presented in Table 12.1.

12.5 Chemical Composition

Medicinal plants have been used for centuries in the treatment of multiple diseases, with effects being conferred by the presence of high-value bioactive compounds (Ghanbari et al. 2012). The literature-based screening of bioactive components revealed that *P. emodi* is a major source of secondary metabolites, among them alkaloids, flavonoids, glycosides,

Table 12.1 Ethnomedicinal uses of *P. emodi* in different regions.

Uses	Region	References
Roots		
Wound and weakness	Pakistan	(K. Khan et al. 2019)
Uterine disease and blood purifier	Pakistan	(Fayaz et al. 2019)
Whooping cough, diarrhea and intestinal spasm	India	(Negi and Maikhuri 2017)
Ulcer, diarrhea, eczema, whooping cough, intestinal spasm and cuts	India	(Negi et al. 2010)
Dementia and blood stagnation syndrome	China	(Okubo et al. 2000)
Biliousness, uterine diseases, dropsy and nervous affections	Pakistan	(Ghayur et al. 2008)
Unani medicine	Saudi Arabia	(Raish et al. 2016)
Gynecological disorders, rheumatic pain, diarrhea and vomiting	Pakistan	(M. Ahmad et al. 2014)
Epilepsy	Pakistan	(Ishtiaq et al. 2012)
Stomach problems	India	(Prakash 2014)
Anti-pyretic, epilepsy, backache	Pakistan	(Begum et al. 2014)
Muscular pain, rheumatism, backache	Pakistan	(Shah et al. 2016)
Gastrointestinal disorder	Pakistan	(Jamal et al. 2017)
Leaves		
Vegetable	Pakistan	(Fayaz et al. 2019)
Blood purifier, dysentery, colic, piles	India	(Semwal et al. 2010)
Dysentery, diarrhea	India	(Pandey et al. 2017)
Abdominal pain, vomiting.	India	(Malik et al. 2015)
Blood purifier, dysentery, colic	India	(Maikhuri et al. 2000)
Lean period consumption	India	(Misra et al. 2008)
Dyspepsia, dysentery	India	(Phondani et al. 2010)
Dysentery	India	(Nautiyal et al. 2005)
Cholera, whooping cough	Pakistan	(Adnan et al. 2014)
Vegetable	India	(Joshi et al. 2018), (Farooquee et al. 2004)
Flowers		
Cough	Pakistan	(Fayaz et al. 2019)
Intestinal pain, hemorrhoids, diarrhea, whooping cough	India	(Gaur 1999)
Hemorrhoid, anti-diarrheal, antispasmodic	India	(Ummara et al. 2013)
Halt diarrhea	India	(Rana and Datt 1997)
Tuber		
Epilepsy, uterine disease, nervous diseases, colic, hysteria, dropsy, obstructions, bilious	Pakistan	(Riaz et al. 2004)

(Continued)

Table 12.1 (Continued)

Uses	Region	References
Hepatoprotective	India	(Kshirsagar et al. 2011)
Skin disease, uterine disease, memory enhancer, blood purifier, dropsy, hysteria, colic, bilious	Pakistan	(Ismail et al. 2003)
Uterine disease	India	(Singh and Rawat 2011)
Blood purifier, epilepsy, rheumatism, uterine disease, dropsy, hysteria, cholera, stimulant, whooping cough	Pakistan	(Hussain and Ghani 2008)
Asthma	Pakistan	(Usmanghani et al. 1997)
Rhizomes		
Backache and used as tonic	Pakistan	(Ali et al. 2018)
Dropsy, epilepsy, headache, vomiting, backbone ache, abdominal spasms, hysteria, dizziness	Pakistan	(Shinwari et al. 2003)
Body pain	Pakistan	(Khalid et al. 2017)
Body weakness	Pakistan	(Sher et al. 2016)
Body tonic, abdominal pain, sexual tonic	Pakistan	(Aziz et al. 2017)
Diarrhea, blood purifier, body pain, epilepsy, uterine disease, bone pain	Pakistan	(Hamayun et al. 2007)
Seeds		
Emetic, purgative, cathartic	Pakistan	(Ghayur et al. 2008)
Emetic, backache, epilepsy, blood purifier, colic	Pakistan	(Ullah et al. 2011)
Rheumatism, backache	Pakistan	(S. M. Khan et al. 2013)
Whole plant		
Diarrhea, vomiting, blood purifier	Pakistan	(Gilani et al. 2017)

phenol, tannins, terpenoids, saponins, proteins, steroids, amino acids, fixed oils and fats, carbohydrates, among others (Kishore et al. 2017; Uddin et al. 2013; Zaidi et al. 2012; Zargar et al. 2014). Recently, Yang and co-workers (Yang et al. 2020) evaluated the roots of seven peonies for screening for bioactive compounds following chemical characterization by HPLC-DAD and HPLC-Q-TOF-MS. A total of 21 metabolites were identified, including paeonol (1), phenol (3), flavonoid (1), tannins (7), and mono-terpene glycosides (9), in all seven peonies species were studied, with both composite on and content of metabolites significantly varying among them. Ibrar and group (Ibrar et al. 2019) evaluated the ethyl acetate fraction of *P. emodi* for possible identification of active compounds. A total of 11 compounds were identified, mostly esculetin, stevioside, spirodecanedione, and methyl eugenol. *P. emodi* methanol seed extract and its fractions were also evaluated for screening and identification of chemical compounds by using GC-MS analysis. *P. emodi* methanol extract showed the presence of carbonyl compounds, while the fraction indicates 13-docosenamide, 9-octadecenamide, trans-13-doco-senamide, etc. (A. Sharma et al. 2018). The leaves of *P. emodi* were extracted with aqueous solution and protein, carbohydrates, proline, methionine, vitamin A, C & E, phytic acid and tannin were isolated from the species (Jugran et al. 2016). Verma and co-workers (Verma et al. 2015) evaluated the chemical composition of essential oils extracted from *P. emodi* root by GC-FID, GC-MS and NMR. In this study, a total of 24 compounds were identified, with the major constituents being salicylaldehyde (85.5%), *cis*-myrtanal (4.9%), myrtenal (1.8%), *trans*-myrtanol (1.6%) and nopinone (1.4%). A few other bioactive compounds have been isolated from these species by a number of researchers around the world (Asif et al. 1983; Haroon ur Rashid et al. 1972; Muhammad et al. 1999; Nawaz et al. 2000; Riaz, Anis, Aziz-ur-Rehman et al. 2003; Riaz, Anis, Malik et al. 2003; Riaz et al. 2004; Tantry et al. 2012). A few selected bioactive compounds are presented in Figure 12.1.

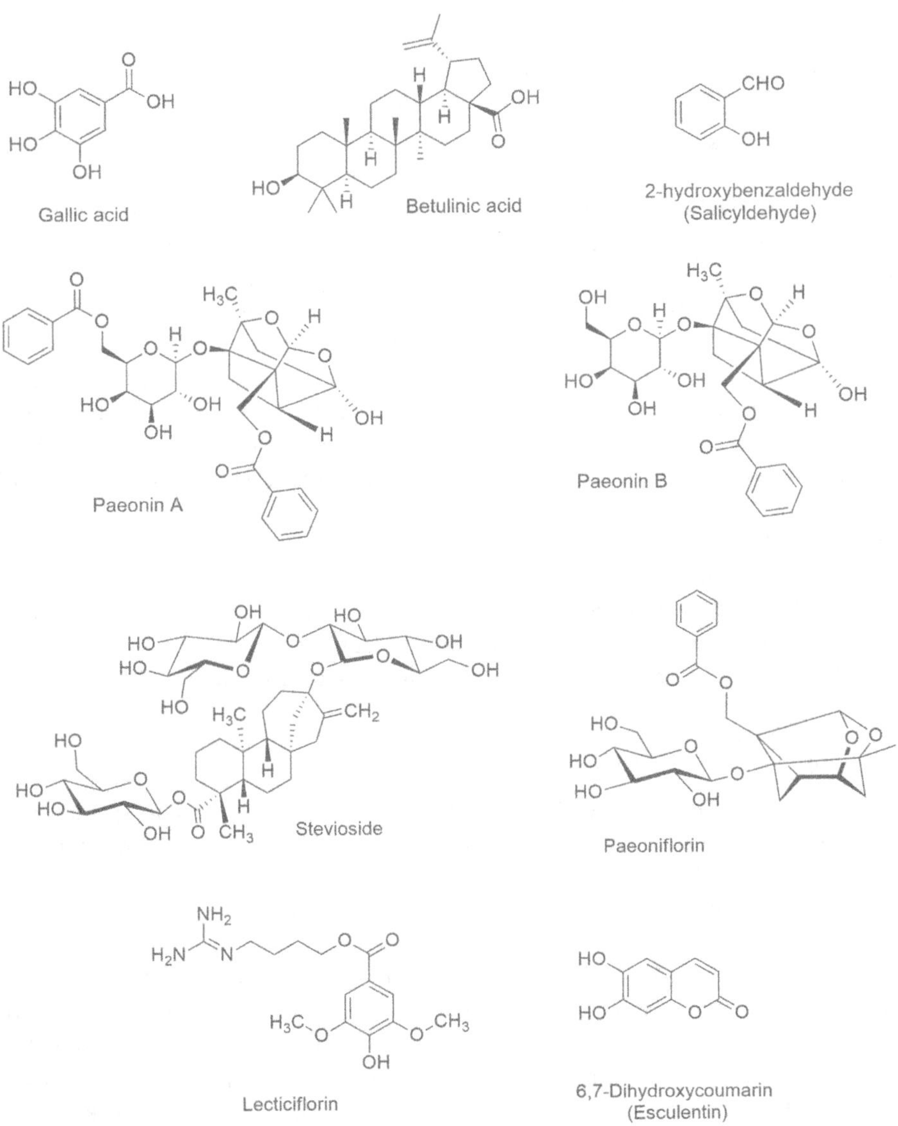

Figure 12.1 Most abundant bioactive compounds of *P. emodi*.

12.6 Bioactive Effects

P. emodi ethnomedicinal claims have been progressively validated through increasingly deep studies using its crude extracts/fractions/pure compounds and to search for new potentialities. Research findings have indicated that the different extracts/fractions of roots, leaves, and areal parts of *P. emodi* possess a variety of pharmacological activities, including antioxidant, anti-inflammatory, cardio-protective, anti-epileptic, nephroprotective, lipoxygenase inhibitory, hepatoprotective, and antimicrobial effects. These health potentialities are briefly discussed below and summarized in Table 12.2.

12.6.1 Cardioprotective Activity

Ibrar et al. (2019) assessed the cardio-protective effects of *P. emodi* rhizomes extracted with methanol and re-extracted with various solvents (n-hexane, ethyl acetate, aqueous and chloroform) in balb/C mice models. Serum levels of alanine amino transferase (ALT), aspartate amino transferase (AST), lactate dehydrogenase (LDH) and creatine phosphokinase (CPK) were evaluated. *P. emodi* ethyl acetate (Pe.EA) fraction showed a significant cardio-protective activity among all *P. emodi* samples. Pe.EA fraction significantly reduced the serum levels of ALT ($p<0.001$), AST ($p<0.001$), LDH ($p<0.001$), and CPK ($p<0.05$) at a dose of 300 mg kg^{-1} b.w. as compared to the standard group. Pe.EA fraction also led to strong membrane stabilization ($IC_{50} = 850$ μg mL^{-1}) at different concentrations (1000, 500, 250, 125 μg mL^{-1}), thus validating its cardio-protective activity (Ibrar et al. 2019).

Ibrar and co-workers (Ibrar et al. 2018) also evaluated the ethyl acetate fraction and sub-fractions of *P. emodi* for cardio-protective potential. The ethyl acetate sub-fraction (Pe.EA40) was the most active and used for the synthesis of gold nanoparticles (Pe.EA40-AuNPs). Both samples significantly reduced the serum level of ALT (66.07 IU L^{-1}), AST (77.08 IU L^{-1}), LDH (84.86 IU L^{-1}), and CPK (265.34 IU L^{-1}) at a dose of 80 mg kg^{-1} b.w. as compared to the standard group, while nanoparticles (Pe. EA 40-AuNPs) reduced ALT (60.74 IU L^{-1}), AST (75.47 IU L^{-1}), LDH (80.48 IU L^{-1}) and CPK (247.54 IU L^{-1}) levels, respectively. Research conducted by Ghayur and group (Ghayur et al. 2008) revealed that *P. emodi* roots ethanol extract exhibits a significant cardio-protective, vasodilatory, anti-platelet, and airway relaxant activities in guinea pig atria and rat aorta.

12.6.2 Antimicrobial Activities

Literature-based screenings have shown that *P. emodi* possesses a broad spectrum of antimicrobial potential against a large number of microbial strains. In a recent study, the shade dried leaves, roots and seeds of *P. emodi* were extracted with methanol and tested against six fungal strains, namely *Candida glabrata, C. albicans, C. tropicalis, C. parapsilosis, C. krusei* and *C keyfer*. The methanol seed extract showed promising antifungal activity against *C. albicans* (6.25 mg mL^{-1}), *C. glabrata* (3.12 mg mL^{-1}), and *C. parapsilosis* (12.50 mg mL^{-1}) when compared with the known standard drug (Amphotericin B). The minimal inhibitory concentration (MIC) values of the plant extract were recorded in the range 100–3.12 mg mL^{-1}. Moreover, the synergistic effects of methanol seed extract with known standard drugs were also addressed (A. Sharma et al. 2018). Research

Table 12.2 Other biological activities of *P. emodi*.

Extract/Fractions & Conc./Dose	Tested system	Activity (possible mechanism)	References
Rhizomes (Ethyl acetate) [200, 400, 800 μL]	Human blood	Ethyl acetate fraction showed significant thrombolytic activity [16.33–53.22%, $p<0.001$]	(Ibrar et al. 2019)
Seeds (Methanol)	Human embryonic kidney cells (HEK-293) and Human cervical cancer cells (HeLa)	Plant extract showed less significant cytotoxic effect against both cells [10%, 15%, MIC = 3.12 mg mL^{-1}]	(A. Sharma et al. 2018)
Roots (Alcoholic and hydro-alcoholic) [100, 200, 400 mg kg^{-1}]	Wister rats	Both extracts showed protective effects against diabetic nephropathy	(Kishore et al. 2017)
Leaves (Aqueous) [500 μg]	DNA nicking assay	Plant extract conferred significant protection to DNA and reduced oxidative stress	(Jugran et al. 2016)
Leaves (Aqueous) [100 μL]	Spectro-photometric assay	Plant extract exhibited significant trypsin inhibition activity	(Jugran et al. 2016)
Whole plant (Ethanol, ethyl-acetate, dichloromethane, n-hexane and aqueous) [0.2–0.8 mg mL^{-1}]	DPPH, ABTS, Superoxide anion, Nitric oxide, Hydrogen peroxide assay	Aqueous and ethyl-acetate fractions possessed significant antioxidant activity	(Uddin et al. 2013)
Roots (Ethanol) [300–600 mg kg^{-1}]	Swiss male albino mice	Plant extract significantly decreased oxidative injury in the mice brain tissue and acted as a potential candidate of anti-epileptic drug	(Zaidi et al. 2012)
Areal part (Ethanol) [3 mg mL^{-1}]	*B. subtilis, E. coli, S. typhi, S. aureus, P. aeruginosa, Shigella flexenari, T. longifusus, C. albicans, C. glabrata, A. flavus, M. canis, F. solani*	Plant extract did not show significant results for antibacterial and antifungal activity	(T. Khan, Ahmad, Khan et al. 2005)
Leaves (Ethanol) [400 μg]	*T. schoenleinii, P. boydii, C. albicans, M. canis, T. simii, F. solanivar Lycopersici,* and *M. phaseolina*	Plant extract showed significant antifungal activity [inhibition (20–55.5%) against all strains]	(Ismail et al. 2003)
Areal part (Methanol) [10 μL]	Spectro-photometric method	Plant extract showed significant lipoxygenase inhibitory activity [90% inhibition of lipoxygenase synthesis]	(T. Khan et al. 2009)

(Continued)

Table 12.2 (Continued)

Extract/Fractions & Conc./Dose	Tested system	Activity (possible mechanism)	References
Areal part (Ethanol, *n*-hexane, chloroform, ethyl acetate, and *n*-butanol)	DPPH assay	All fractions possessed good to significant antioxidant activity [60–83%]	(T. Khan, Ahmad, Nisar et al. 2005)
Areal part (Ethanol, *n*-hexane, chloroform, ethyl acetate, and *n*-butanol) [0.5 mg/mL]	Jack bean and *Bacillus pasteurii*	Plant extract had significant urease inhibitory activity against jack bean (74%) and *Bacillus pasteurii* (80%). Ethyl acetate, *n*-butanol and water fraction also revealed acceptable urease inhibitory effects	(T. Khan, Ahmad, Nisar et al. 2005)
Areal part (Ethanol, *n*-hexane, chloroform, ethyl acetate, and *n*-butanol)	-	Plant extract showed reasonable α-chymotrypsin inhibitory activity (54%). Ethyl acetate fraction possessed significant inhibitory activity (91%) against α-chymotrypsin	(T. Khan, Ahmad, Nisar et al. 2005)
Roots (Ethanol) [200–300 mg kg^{-1}]	Mice and rats	Plant extract showed significant anticonvulsant and anti-anxiety activity	(Nizami and Jafri 2005)
1β,3β,23-trihydroxyolean-12-en-28-oic acid (Roots, Methanol)	*Escherichia coli*	Isolated compound showed significant β-Glucuronidase inhibitory activity	(Riaz, Anis, Aziz-ur-Rehman et al. 2003)
Tuber (Alcoholic) [250 mg]	Rats, guinea pig, rabbit, dog and human uteri	Plant extract did not show significant uterotonic activity against any tested models	(M. Sharma and Mishra 1969)
Tuber	*In-vitro* and *in-vivo* models	Plant extract showed uterine stimulant activity	(Misra et al. 1968)

findings by Badola and Negi indicated that synthesized silver nanoparticles of *P. emodi* leaves possess significant antimicrobial effects against *Staphyloccus aureus, Bacillus subtilis, Salmonella typhi* and *Escherichia coli* (zone of inhibition (ZOI) between 9.8 and 10.5 mm) when compared with positive drug (Streptomycin, 12–15 mm) (Badola and Negi 2017).

The antibacterial potential of *P. emodi* methanolic root extract was evaluated against four bacterial strains, namely *E. coli, S. aureus, P. aeruginosa*, and *K. pneumonia*. The plant extract showed moderate antibacterial activity against all tested strains (ZOI between 8 to 25 mm) (El-Zawahry et al. 2013). At the same time, another research group assessed the antimicrobial profiling of *P. emodi* ethanol extract and its fractions against a series of bacterial and fungal strains. The ethyl acetate fraction revealed significant antibacterial and antifungal activity against *B. subtilis* (ZOI 27 mm) and

Fusarium solani (ZOI 35 mm) when compared to the standard drug (Miconazole, 35 mm) (Uddin et al. 2013).

P. emodi leaves extracted with n-hexane, ethyl acetate and chloroform were assessed for their antibacterial and antifungal activity against *E. coli, P. aeruginosa, K. pneumonia, S. aureus, S. epidermidis, S. typhi, Aspergillus flavus, A. fumigates, A. niger* and *Fusarium solani*. The plant extract showed significant antibacterial activity against all bacterial strains (ZOI in the range 8–20 mm), while chloroform and n-hexane fractions showed antifungal activity (Mufti et al. 2012). The antibacterial activity of *P. emodi* aerial parts extracted with hexane, ethanol and chloroform were studied using the disk diffusion method against *E. coli, P. aeruginosa, K. pneumonia, S. typhi, B. subtilis, B. cereus and C. albicans*. Most extracts evidenced acceptable antimicrobial activity against the tested strains (ZOI in the range 7–18 mm) (Fazal et al. 2012).

12.6.3 Anti-inflammatory Activity

Ahmad and co-workers evaluated the anti-inflammatory activity of *P. emodi* roots-extracted polysaccharide in carrageenan-induced paw edema. In this study, male albino rats were used and divided into three groups. Experimental rats were pre-treated with polysaccharides (100–200 mg kg^{-1} b.w.) and known standard drug (10 mg kg^{-1} b.w.), orally 1 h before carrageenan injection. The results indicated that *P. emodi* polysaccharide extract (200 mg kg^{-1}) prohibited the carrageenan-induced edema formation and possessed strong anti-inflammatory activity ($p<0.05$) (A. Ahmad et al. 2015).

12.6.4 Anti-hyperlipidemic Activity

In a recent study, *P. emodi* rhizomes were extracted with methanol, re-extracted with ethyl acetate and tested for its anti-hyperlipidemic effects in mice models. Experimental animals were divided into six groups, three groups were treated orally with ethyl acetate fraction (Pe.EA) at 75, 150, 300 mg kg^{-1} b.w. respectively. *P. emodi* ethyl acetate fraction effectively lowered lipid serum levels and demonstrated a dose-dependent anti-hyperlipidemic activity (Ibrar et al. 2019).

Ibrar and group examined the anti-hyperlipidemic activity of *P. emodi* ethyl acetate fraction (Pe.EA40) and its gold nanoparticles (Pe.EA40-AuNPs) at two different doses (80 and 40 mg kg^{-1} b.w.). Experimental mice were divided into four groups, two groups were treated with Pe.EA40 (80 mg kg^{-1} b.w.) and Pe.EA40-AuNPs (40 mg kg^{-1} b.w.) orally. Research findings indicated that both samples significantly reduced the serum level in plasma (Ibrar et al. 2018).

Shade-dried rhizomes of *P. emodi* were extracted with ethanol and its anti-hyperlipidemic effects were assessed in male Wistar rats, divided into six groups and treated orally with plant extract (200 mg kg^{-1} b.w.) and a known standard drug. The plant extract markedly reduced the level of TGs, TC, LDL-c, and AI ($p<0.05$). In addition, the plant extract was revealed to be safe and with potential application as an anti-hyperlipidemic in treated models (Zargar et al. 2014).

12.6.5 Hepatoprotective Activity

P. emodi roots extracted with ethyl acetate were assessed for its hepatoprotective effects. The plant extract significantly reduced the expression of CYP3A2 and CYP2C11 proteins compared to control (50%, 55.55%, $p<0.05$) (Raish et al. 2017). In addition, shade-dried rhizomes of *P. emodi* extracted with methanol and its various fractions (*n*-butanol, chloroform and aqueous) were also addressed for its hepatoprotective potential. The authors stated that the plant extract and its chloroform and butanol fractions significantly reduced the serum level of ALT, AST, and ALP in paracetamol-intoxicated mice (Ilahi et al. 2016).

12.6.6 Antioxidant Activity

Shade-dried *P. emodi* rhizomes ground and extracted with methanol and re-extracted with other solvents were tested for antioxidant activity using DPPH and ABTS assay (1000, 500, 250, 125 $\mu g\ mL^{-1}$). The most active fraction, ethyl acetate, exhibited significant antioxidant activity in terms of IC_{50} value (324 and 342 $\mu g\ mL^{-1}$) (Ibrar et al. 2019). Also, the alcoholic and hydro-alcoholic extracts from *P. emodi* roots were examined for the level of antioxidant enzymes in Wistar rats. Both extracts (100, 200, 400 $mg\ kg^{-1}$ b.w.) significantly reduced the level of GSH in kidney, liver, and pancreas in rats with diabetic neuropathy. In other words, an increased level of antioxidant enzymes was recorded in plant extract-treated rats when compared to control (Kishore et al. 2017).

A comparative study of the antioxidant potential of *P. emodi* aqueous extract (at three development stages) was performed by using DPPH, ABTS, FRAP and hydroxyl radical scavenging assay. Maximum DPPH scavenging activity (3.55 mM AAE/100 g) was recorded in boiled juvenile leaves, and maximum ABTS activity (71.13 mM AAE/100 g) in dried budding leaves. FRAP assay and hydroxyl radical scavenging assay demonstrated maximum activity in dried budding leaves (3.39 mM AAE/100 g) and dried mature leaves (1.69 mM AAE/100 g), respectively (Jugran et al. 2016). At same time, *P. emodi* rhizomes were extracted with methanol and re-extracted with different solvents (*n*-butanol, chloroform, and aqueous) to assess the antioxidant potential. The chloroform and butanol fractions revealed significant antioxidant activity (IC_{50} = 7.1, 6.5 $mg\ L^{-1}$) at various concentrations (Ilahi et al. 2016).

In another study, the antioxidant potential of *P. emodi* rhizomes was evaluated by using DPPH, reducing power, hydroxyl radical scavenging, superoxide radical and H_2O_2 scavenging activity. Research findings demonstrated that plant extracts (hydro-alcoholic and aqueous) exerted a marked antioxidant activity in each assay, except in H_2O_2 scavenging assay. *P. emodi* hydro-alcoholic and aqueous extract showed, respectively, 93% and 70% scavenging activity (100 $\mu g\ mL^{-1}$) in DPPH assay and 54.3% and 50.3% scavenging activity (50 $\mu g\ mL^{-1}$), respectively, in hydroxyl radical scavenging assay. Both extracts exhibited concentration-dependent effects in reducing power assay and superoxide radical scavenging assay, while poor activity was recorded in H_2O_2 assay (Zargar et al. 2014).

Graphical abstract

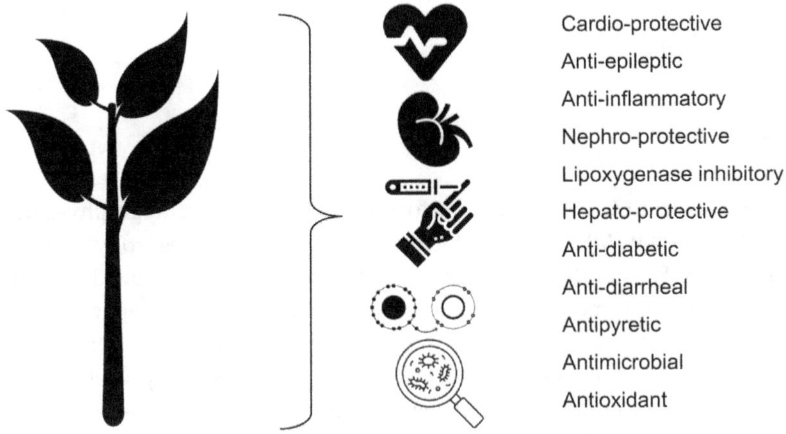

12.7 Conclusion

P. emodi traditional uses, chemical composition and pharmacological properties have been increasingly addressed and validated. The literature-based screening reveled that *P. emodi* is an important species of the Himalayan region and is used in different traditional systems. It is made clear in this chapter that every part of the plant is an important source of bioactive compounds but the root and its alcoholic fractions have been considered as the most effective part against different ailments. Although some aspects related to its chemical composition, characterization, and mechanism of action have been elucidated, some others remain unclear and, therefore, should be further investigated, even using advanced techniques of genomics, proteomics and metabolomics. Most studies are limited to the *in vitro* effects, and there are very few *in vivo* studies. Also, toxicological screening is important to understand the safety profile of any herbal product and these studies on *P. emodi* are limited. In addition, due to multiple applications and illegal trade, the *Paeonia* species is facing threats to its sustainability, so the healthy participation of local people, government bodies, and nongovernmental organizations regarding cultivation, conservation and sustainable utilization of the species will definitely be helpful in improving the population size of the species in nature.

Conflict of interest

The authors declare no conflict of interests

Acknowledgments

The authors thank the help and support received from the Graphic Era (Deemed to be University), Dehradun, India.

Contribution

All authors contributed equally

Abbreviations

P. emodi: *Paeonia emodi*, HPLC-DAD: High performance liquid chromatography-diode array detector, HPLC-Q-TOF-MS: High performance liquid chromatography coupled with high resolution TOF, GC-FID: Gas chromatography-flame ionization detection, GC-MS: Gas chromatography mass spectrophotometer, NMR:Nuclear magnetic resonance, ALT: Alanine amino transferase, AST: Aspartate amino transferase, LDH: Llactate dehydrogenase, CPK: Creatine phosphokinase, Pe.EA: Paeonia emodi ethyl acetate fraction, IC_{50}: Inhibitory concentration, b.w: Body weight, mg kg^{-1}: Milligram per kilogram, mg/mL: Milligram per milliliter, MIC: Minimum inhibitory concentration, ZOI: Zone of inhibition, mm: millimeter, μg/mL: microgram per milliliter, DPPH: 2,2- diphenyl-1-picrylhydrazyl, ABTS: 2,2- azinobis (3-ethyl-benzothiazoline-6-sulfonic acid), FRAP: Ferric reducing antioxidant power, H_2O_2: Hydrogen peroxide, GSH: Glutathione, mM: Milli molar, AAE/100 g: Ascorbic acid equivalents per 100 g.

References

Adnan, M., Bibi, R., Mussarat, S., Tariq, A., and Shinwari, Z.K. (2014). Ethnomedicinal and phytochemical review of Pakistani medicinal plants used as antibacterial agents against Escherichia coli. *Annals of Clinical Microbiology and Antimicrobials* 13 (1): 40.

Ahmad, A., Alkharfy, K.M., Wani, T.A., and Raish, M. (2015). Application of Box–Behnken design for ultrasonic-assisted extraction of polysaccharides from Paeonia emodi. *International Journal of Biological Macromolecules* 72: 990–997.

Ahmad, M., Malik, K., Tariq, A., Zhang, G., Yaseen, G., Rashid, N., ... Khan, M.P.Z. (2018). Botany, ethnomedicines, phytochemistry and pharmacology of Himalayan paeony (Paeonia emodi Royle.). *Journal of Ethnopharmacology* 220: 197–219.

Ahmad, M., Sultana, S., Fazl-i-Hadi, S., Ben Hadda, T., Rashid, S., Zafar, M., ... Yaseen, G. (2014). An Ethnobotanical study of Medicinal Plants in high mountainous region of Chail valley (District Swat-Pakistan). *Journal of Ethnobiology and Ethnomedicine* 10 (1): 36.

Ali, A., Badshah, L., and Hussain, F. (2018). Ethnobotanical appraisal and conservation status of medicinal plants in Hindukush Range, District Swat, Pakistan. *Journal of Herbs, Spices & Medicinal Plants* 24 (4): 332–355.

Asif, M., Ahmad, M.S., Mannan, A., Itohc, T., and Matsumoto, T. (1983). Analysis ofPaeonia emodi root oil. *Journal of the American Oil Chemists' Society* 60 (3): 581–583.

Aziz, M.A., Khan, A.H., Adnan, M., and Izatullah, I. (2017). Traditional uses of medicinal plants reported by the indigenous communities and local herbal practitioners of Bajaur Agency, Federally Administrated Tribal Areas, Pakistan. *Journal of Ethnopharmacology* 198: 268–281.

Badola, G. and Negi, D. (2017). Antibacterial and Photocatalytic activities of silver nanoparticles synthesized by Paeonia emodi leaves extract. *International Journal of Engineering, Science and Mathematics* 6 (8): 78–87.

Begum, S., AbdElslam, N.M., Adnan, M., Tariq, A., Yasmin, A., and Hameed, R. (2014). Ethnomedicines of highly utilized plants in the temperate Himalayan region. *African Journal of Traditional, Complementary and Alternative Medicines* 11 (3): 132–142.

El-Zawahry, Y., Reda, F., and Azazy, W. (2013). Synergistic effect of combination treatment by certain plant extracts and some antibiotics on the resistance of pathogenic bacteria to some common antibiotics. *Life Science Journal* 10 (4): 3477–3489.

Farooquee, N.A., Majila, B., and Kala, C. (2004). Indigenous knowledge systems and sustainable management of natural resources in a high altitude society in Kumaun Himalaya, India. *Journal of Human Ecology* 16 (1): 33–42.

Fayaz, M., Jain, A.K., Bhat, M.H., and Kumar, A. (2019). Ethnobotanical survey of Daksum forest range of Anantnag District, Jammu and Kashmir, India. *Journal of Herbs, Spices & Medicinal Plants* 25 (1): 55–67.

Fazal, H., Ahmad, N., Abbasi, B.H., and Abbass, N. (2012). Selected medicinal plants used in herbal industries; their toxicity against pathogenic microoraganisms. *Pakistan Journal of Botany* 44 (3): 1103–1109.

Gaur, R.D. (1999). Flora of the District Garhwal, North West Himalaya: With Ethnobotanical Notes. Srinagar, Uttar Pradesh, India: TransMedia.

Ghanbari, R., Anwar, F., Alkharfy, K.M., Gilani, A.-H., and Saari, N. (2012). Valuable nutrients and functional bioactives in different parts of olive (Olea europaea L.)—a review. *International Journal of Molecular Sciences* 13 (3): 3291–3340.

Ghayur, M.N., Gilani, A.H., Rasheed, H., Khan, A., Iqbal, Z., Ismail, M., ... Janssen, L.J. (2008). Cardiovascular and airway relaxant activities of peony root extract. *Canadian Journal of Physiology and Pharmacology* 86 (11): 793–803.

Gilani, S.N.Z., Bhat, B.A., Rafiqi, F.A., and Rahi, S. (2017). Nusrat, Samira Khan and Intizar Ahmad (2017): awareness among university students on forests and traditional use of medicinal plants in kashmir valley: a sample survey. *International Journal of Innovative Research and Review* 5 (2): 11–27.

Gyllenhaal, C., Kadushin, M., Southavong, B., Sydara, K., Bouamanivong, S., Xaiveu, M., ... Loc, P. (2012). Ethnobotanical approach versus random approach in the search for new bioactive compounds: support of a hypothesis. *Pharmaceutical Biology* 50 (1): 30–41.

Hamayun, M., Khan, M.A., Chudhary, M.F., and Ahmad, H. (2007). Studies on traditional knowledge of medicinal herbs of Swat Kohistan, District Swat, Pakistan. *Journal of Herbs, Spices & Medicinal Plants* 12 (4): 11–28.

Haq, F., Ahmad, H., and Alam, M. (2011). Traditional uses of medicinal plants of Nandiar Khuwarr catchment (District Battagram), Pakistan. *Journal of Medicinal Plants Research* 5 (1): 39–48.

Haroon ur Rashid, M., Begum, S., and Malik, M. (1972). Chemical investigations on the roots of Paeonia emodi Wall. *Pakistan Journal of Forestery* 22 (2): 109–111.

Hussain, M. and Ghani, A. (2008). Herbal remidies used for gastrointestinal disorders in Kaghan valley, NWFP, Pakistan. *Pakistan Journal of Weed Science Research* 14 (3–4): 169–200.

Ibrar, M., Khan, M.A., and Imran, M. (2018). Evaluation of Paeonia emodi and its gold nanoparticles for cardioprotective and antihyperlipidemic potentials. *Journal of Photochemistry and Photobiology. B, Biology* 189: 5–13.

Ibrar, M., Khan, M.A., Nisar, M., and Khan, M. (2019). Evaluation of Paeonia emodi for its cardioprotective potentials: an investigative study towards possible mechanism. *Journal of Ethnopharmacology* 231: 57–65.

Ilahi, I., Khan, J., Ghaffar, R., Hussain, A., Rahman, K., Wahab, S., ... Zeb, L. (2016). In vitro antioxidant and hepatoprotective activities of Paeonia emodi (Wall.) rhizome methanol extract and its phenolic compounds rich fractions. *Pakistan Journal of Pharmaceutical Sciences* 29 (5): 1787–1794.

Ishtiaq, M., Mumtaz, A.S., Hussain, T., and Ghani, A. (2012). Medicinal plant diversity in the flora of Leepa Valley, Muzaffarabad (AJK), Pakistan. *African Journal of Biotechnology* 11 (13): 3087–3098.

Ismail, M., Iqbal, Z., Ahmad, B., Zakir, S., and Niaz, U. (2003). Biological and pharmacological properties of two indigenous medicinal plants, Rheum emodi and Paeonia emodi. *Pakistan Journal of Biological Sciences* 6 (9): 984–986.

Jamal, Z., Pervez, A., Hussain, M., Shah, G.M., Shah, S.H., and Ahmed, M. (2017). Ethnomedicinal plants used for gastrointestinal ailments by the rural communities of Kaghan Valley, Mansehra, Pakistan. *Journal of Applied Environmental and Biological Sciences* 7 (12): 41–48.

Joshi, S., Ballabh, B., Negi, P., and Dwivedi, S. (2018). Diversity, distribution, use pattern and evaluation of wild edible plants of Uttarakhand, India. *Defence Life Science Journal* 3 (2): 126–135.

Jugran, A.K., Chaudhary, W.Y., Bahukhandi, A., Bhatt, I.D., Rawal, R.S., and Dhyani, P.P. (2016). Effect of processing and storage methods on the nutritional, anti-nutritional, and anti-oxidant properties of paeonia emodi, wall. ex. royle. *Applied Biochemistry and Biotechnology* 180 (2): 322–337.

Khalid, M., Bilal, M., Hassani, D., Zaman, S., and Huang, D. (2017). Characterization of ethno-medicinal plant resources of karamar valley Swabi, Pakistan. *Journal of Radiation Research and Applied Sciences* 10 (2): 152–163.

Khan, K., Rahman, I.U., Soares, E.C., Ali, N., and Ijaz, F. (2019). Ethnoveterinary therapeutic practices and conservation status of the medicinal flora of Chamla Valley, Khyber Pakhtukhwa, Pakistan. *Frontiers in Veterinary Science* 6: 122.

Khan, S.M., Page, S., Ahmad, H., Shaheen, H., Ullah, Z., Ahmad, M., and Harper, D.M. (2013). Medicinal flora and ethnoecological knowledge in the Naran Valley, Western Himalaya, Pakistan. *Journal of Ethnobiology and Ethnomedicine* 9 (1): 4.

Khan, T., Ahmad, M., Ahmad, W., Saqib, Q.N.u., and Choudhary, M.I. (2009). Preliminary evaluation of the antispasmodic and lipoxygenase inhibitory effects of some selected medicinal plants. *Pharmaceutical Biology* 47 (12): 1137–1141.

Khan, T., Ahmad, M., Khan, H., and Khan, M.A. (2005). Biological activities of aerial parts of Paeonia emodi Wall. *African Journal of Biotechnology* 4 (11): 1313–1316.

Khan, T., Ahmad, M., Nisar, M., Ahmad, M., Arif Lodhi, M., and Choudhary, M.I. (2005). Enzyme inhibition and radical scavenging activities of aerial parts of Paeonia emodi Wall. (Paeoniaceae). *Journal of Enzyme Inhibition and Medicinal Chemistry* 20 (3): 245–249.

Kishore, L., Kaur, N., and Singh, R. (2017). Nephroprotective effect of Paeonia emodi via inhibition of advanced glycation end products and oxidative stress in streptozotocin-nicotinamide induced diabetic nephropathy. *Journal of Food and Drug Analysis* 25 (3): 576–588.

Kshirsagar, A., Mohite, R., Aggrawal, A., and Suralkar, U. (2011). Hepatoprotective medicinal plants of Ayurveda-A review. *Asian Journal of Pharmaceutical and Clinical Research* 4 (3): 1–8.

Long, J., Gao, M., Kong, Y., Shen, X., Du, X., Son, Y.-O., ... Mo, X. (2012). Cardioprotective effect of total paeony glycosides against isoprenaline-induced myocardial ischemia in rats. *Phytomedicine* 19 (8–9): 672–676.

Maikhuri, R., Nautiyal, S., Rao, K., and Semwal, R. (2000). Indigenous knowledge of medicinal plants and wild edibles among three tribal sub communities of the central Himalayas, India. *Indigenous Knowledge and Development Monitor* 8 (2): 7–13.

Malik, Z.A., Bhat, J.A., Ballabha, R., Bussmann, R.W., and Bhatt, A. (2015). Ethnomedicinal plants traditionally used in health care practices by inhabitants of Western Himalaya. *Journal of Ethnopharmacology* 172: 133–144.

Misra, M.B., Dikshit, B.B., Mishra, S.S., and Misra, R.K. (1968). A preliminary pharmacology of Paeonia emodi Wall. *Indian Journal of Medical Sciences* 22 (7): 463–465.

Mencherini, T., Picerno, P., Festa, M., Russo, P., Capasso, A., and Aquino, R. (2011). Triterpenoid constituents from the roots of Paeonia rockii ssp. rockii. *Journal of Natural Products* 74 (10): 2116–2121.

Misra, S., Maikhuri, R., Kala, C., Rao, K., and Saxena, K. (2008). Wild leafy vegetables: a study of their subsistence dietetic support to the inhabitants of Nanda Devi Biosphere Reserve, India. *Journal of Ethnobiology and Ethnomedicine* 4 (1): 15.

Mo, X., Zhao, N., Du, X., Bai, L., and Liu, J. (2011). The protective effect of peony extract on acute myocardial infarction in rats. *Phytomedicine* 18 (6): 451–457.

Mufti, F.U.D., Ullah, H., Bangash, A., Khan, N., Hussain, S., Ullah, F., ... Jabeen, M. (2012). Antimicrobial activities of Aerva javanica and Paeonia emodi plants. *Pakistan Journal of Pharmaceutical Sciences* 25 (3): 565–569.

Muhammad, P., Ahmad, S., Rubnawaz, H., Ullah, N., and Malik, A. (1999). New monoterpene glycosides from Paeonia emodi. *Zeitschrift für Naturforschung B* 54 (4): 544–548.

Nautiyal, S., Rajan, K., and Shibasaki, R. (2005). Interaction of biodiversity and economic welfare- a case study from the Himalayas of India. *Journal of Environmental Informatics* 6 (2): 111–119.

Nawaz, H.R., Malik, A., Khan, P.M., Shujaat, S., and Rahman, A.-u. (2000). A novel β-glucuronidase inhibiting triterpenoid from Paeonia emodi. *Chemical & Pharmaceutical Bulletin* 48 (11): 1771–1773.

Negi, V.S. and Maikhuri, R. (2017). Forest resources consumption pattern in Govind wildlife sanctuary, western Himalaya, India. *Journal of Environmental Planning and Management* 60 (7): 1235–1252.

Negi, V.S., Maikhuri, R., Phondani, P., and Rawat, L. (2010). An inventory of indigenous knowledge and cultivation practices of medicinal plants in Govind Pashu Vihar Wildlife Sanctuary, Central Himalaya, India. *International Journal of Biodiversity Science, Ecosystem Services & Management* 6 (3–4): 96–105.

Nizami, Q. and Jafri, M. (2005). Anticonvulsant, antianxiety and memory enhancing activities of delphinium denudatum and Paeonia emodi. Paper presented at the European Neuropsychopharmacology.

Okubo, T., Nagai, F., Seto, T., Satoh, K., Ushiyama, K., and Kano, I. (2000). The inhibition of phenylhydroquinone-induced oxidative DNA cleavage by constituents of Moutan Cortex and Paeoniae Radix. *Biological & Pharmaceutical Bulletin* 23 (2): 199–203.

Pandey, N.C., Bhatt, D., Arya, D., Chopra, N., Upreti, B.M., Joshi, G., and Tewari, L.M. (2017). Diversity of ethno-medicinal plant: a case study of Bageshwar district Uttarakhand. *Journal of Medicinal Plants Studies* 5 (2): 11–24.

Phondani, P.C., Maikhuri, R.K., Rawat, L.S., Farooquee, N.A., Kala, C.P., Vishvakarma, S.R., ... Saxena, K. (2010). Ethnobotanical uses of plants among the Bhotiya tribal communities of Niti Valley in Central Himalaya, India. *Ethnobotany Research and Applications* 8: 233–244.

Prakash, R. (2014). Traditional uses of medicinal plants in Uttarakhand Himalayan region. *Scholars Academic Journal of Biosciences* 2 (5): 345–353.

Raish, M., Ahmad, A., Alkharfy, K.M., Al-Jenoobi, F.I., Al-Mohizea, A.M., Mohsin, K., ... Shakeel, F. (2016). Antioxidant potential and in situ analysis of major and trace element determination of Ood-saleeb, a known unani herbal medicine by ICP-MS. *Biological Trace Element Research* 172 (2): 521–527.

Raish, M., Ahmad, A., Alkharfy, K.M., Jan, B.L., Mohsin, K., Ahad, A., ... Al-Mohizea, A.M. (2017). Effects of Paeonia emodi on hepatic cytochrome P450 (CYP3A2 and CYP2C11) expression and pharmacokinetics of carbamazepine in rats. *Biomedicine and Pharmacotherapy* 90: 694–698.

Rana, T. and Datt, B. (1997). Ethnobotanical observation among Jaunsaris of Jaunsar-Bawar, Dehra Dun (UP), India. *International Journal of Pharmacognosy* 35 (5): 371–374.

Riaz, N., Anis, I., Aziz-ur-Rehman, Malik, A., Ahmed, Z., Muhammad, P., ... Atta-ur-Rahman. (2003). Emodinol, ß-glucuronidase inhibiting triterpene from Paeonia emodi. *Natural Product Research* 17 (4): 247–251.

Riaz, N., Anis, I., Malik, A., Ahmed, Z., Muhammad, P., Nawaz, S.A., and Choudhary, M.I. (2003). Paeonins A and B, lipoxygenase inhibiting monoterpene galactosides from Paeonia emodi. *Chemical and Pharmaceutical Bulletin* 51 (3): 252–254.

Riaz, N., Malik, A., Rehman, A.-u., Ahmed, Z., Muhammad, P., Nawaz, S.A., ... Choudhary, M.I. (2004). Lipoxygenase inhibiting and antioxidant oligostilbene and monoterpene galactoside from Paeonia emodi. *Phytochemistry* 65 (8): 1129–1135.

Samant, S.S. and Joshi, H.C. (2005). Plant diversity and conservation status of Nanda Devi National Park and comparison with highland National Parks of the Indian Himalayan Region. *The International Journal of Biodiversity Science and Management* 1 (1): 65–73.

Semwal, D., Saradhi, P.P., Kala, C., and Sajwan, B. (2010). Medicinal plants used by local Vaidyas in Ukhimath block, Uttarakhand. *Indian Journal of Traditional Knowledge* 9 (3): 480–485.

Shah, S.A., Shah, N.A., Ullah, S., Alam, M.M., Badshah, H., Ullah, S., and Mumtaz, A.S. (2016). Documenting the indigenous knowledge on medicinal flora from communities residing near Swat River (Suvastu) and in high mountainous areas in Swat-Pakistan. *Journal of Ethnopharmacology* 182: 67–79.

Sharma, A., Singh, S., Tewari, R., Bhatt, V., Sharma, J., and Maurya, I. (2018). Phytochemical analysis and mode of action against Candida glabrata of Paeonia emodi extracts. *Journal de Mycologie Medicale* 28 (3): 443–451.

Sharma, M. and Mishra, S. (1969). A pharmacological study of some abortifacient plants. *Indian Journal of Physiology and Pharmacology* 13 (3): 139–141.

Sher, H., Bussmann, R.W., Hart, R., and de Boer, H.J. (2016). Traditional use of medicinal plants among Kalasha, Ismaeli and Sunni groups in Chitral District, Khyber Pakhtunkhwa province, Pakistan. *Journal of Ethnopharmacology* 188: 57–69.

Shinwari, Z., Khan, A., Nakaike, T., and Kyōkai, N. (2003). Medicinal and other useful plants of District Swat, Pakistan. *Al-Aziz Communications, Peshawar* 4: 78–83.

Singh, G. and Rawat, G. (2011). Ethnomedicinal survey of Kedarnath wildlife sanctuary in Western Himalaya, India. *Indian Journal of Fundamental and Applied Life Sciences* 1 (1): 35–46.

Tantry, M.A., Dar, J.A., Khuroo, M.A., and Shawl, A.S. (2012). Nortriterpenoids from the roots of Paeonia emodi. *Phytochemistry Letters* 5 (2): 253–257.

Uddin, G., Sadat, A., and Siddiqui, B.S. (2013). Phytochemical screening, in vitro antioxidant and antimicrobial activities of the crude fractions of Paeonia emodi wall. Ex Royle. *Middle-East Journal of Scientific Research* 17 (3): 367–373.

Ullah, S., Abbasi, M., Raza, M., Khan, S., Aziz-ur-Rehman, Muhammad, B., and Mughal, M. (2011). Antibacterial activity of some selected plants of Swat valley. *Bioscience Research* 8 (1): 15–18.

Ummara, U., Bokhari, T.Z., Altaf, A., Younis, U., and Dasti, A. (2013). Pharmacological study of Shogran valley flora, Pakistan. *International Journal of Scientific Engineering and Research* 4 (9): 1–9.

Usmanghani, K., Saeed, A., and Alam, T. (1997). Indusyunic Medicine. Karachi, Pakistan: University of Karachi Press. 601.

Verma, R.S., Padalia, R.C., Chauhan, A., Yadav, A., and Chanotiya, C.S. (2015). Essential oil composition of Himalayan Peony (Paeonia emodi Royle). *Journal of Essential Oil Research* 27 (6): 477–480.

Wu, S.H., Wu, D.G., and Chen, Y.W. (2010). Chemical constituents and bioactivities of plants from the genus Paeonia. *Chemistry & Biodiversity* 7 (1): 90–104.

Yan, Z.G., Xie, L.H., Wang, N., Sun, D.Y., Bai, Z.Z., Niu, L.X., ... Ji, X.T. (2019). Phenotypic characteristics and fatty acid composition of seeds from different herbaceous peony species native to China. *Chemistry & Biodiversity* 16 (3): e1800589.

Yang, Y., Li, S.-s., da Silva, J.A.T., Yu, X.-n., and Wang, L.-s. (2020). Characterization of phytochemicals in the roots of wild herbaceous peonies from China and screening for medicinal resources. *Phytochemistry* 174: 112331.

Zaidi, S.A., Pathan, S.A., Singh, S., Ahmad, F., Jamil, S., and Khar, R. (2012). Effect of repeated administration of Paeonia emodi wall root extract in experimental models of epilepsy and behaviour. *Journal of Pharmacology and Toxicology* 7 (2): 64–77.

Zargar, B.A., Masoodi, M.H., Ahmed, B., and Ganie, S.A. (2014). Antihyperlipidemic and antioxidant potential of Paeonia emodi Royle against high-fat diet induced oxidative stress. *ISRN Pharmacology* 2014: Article ID 182362.

13

Health Properties of Dietary Monoterpenes

Rafael Chelala Moreira[1], Kele A.C. Vespermann[2], Gustavo Molina[2], Juliano Lemos Bicas[1], and Mario Roberto Marostica Junior[1]

[1]Department of Food Science and Nutrition, School of Food Engineering, University of Campinas (Unicamp), Campinas – SP, Brazil.
[2]Institute of Science and Technology, Federal University of Jequitinhonha and Mucury Valleys (UFVJM), Diamantina – MG, Brazil.
**Corresponding author: mmarosti@unicamp.br*

13.1 Introduction

Terpenes make up the largest class of chemical compounds known in living organisms, being also the most diverse from the chemical, structural and functional points of view, with more than 80000 compounds already known (Christianson 2017). Mostly found in plants, they attract a lot of interest from the food, pharmaceutical, cosmetic, and agricultural industries (Zhou and Pichersky 2020).

Terpenes are synthesized, in most organisms, by the mevalonate pathway from isoprene units, C5 components, sequentially joined as structures $(C5)_n$ to form several classes. Terpene classes can be divided into hemiterpenes (C5), monoterpenes (C10), sesquiterpenes (C15), diterpenes (C20), sesterterpenes (C25), triterpenes (C30), tetraterpenes (C40), and polyterpenes ($(C5)_n$; $n > 8$). Among the important roles played by terpenes are defense, growth, pigmentation, communication, and hormone production (Christianson 2017; Silvestre and Gandini 2008; Zhou and Pichersky 2020).

Monoterpenes ($C_{10}H_{16}$), are formed by two units of isoprene and present various chemical structures and biological activities. Some monoterpenes are classified as Generally Recognized as Safe (GRAS) by the Food and Drugs Administration (FDA) and have aroused great interest in the formulation of fear drugs due to their effects against some diseases (Iraji et al. 2020). The main biological effects of monoterpenes reported in the literature are anticancer, antioxidants, antidiabetics, antimicrobials, antivirals, analgesics, anti-inflammatories, and antihypertensives (Bakkali et al. 2008). In this chapter, anticancer, anti-inflammatory, antidiabetic, antioxidant activities and the cardiovascular and antihypertensive effects of monoterpenes will be treated in more detail.

Monoterpenes being the main constituents of several essential oils from medicinal aromatic plants, this review focuses on the main and most current scientific information on dietary monoterpene's effects *in vivo* and their role in health.

Biomolecules from Natural Sources: Advances and Applications, First Edition. Edited by Vijai Kumar Gupta, Satyajit D. Sarker, Minaxi Sharma, María Elida Pirovani, Zeba Usmani, and Chelliah Jayabaskaran.

13.2 Monoterpenes: Definition, Food Sources, and Dietary Intake

Monoterpenes are units composed of two isoprene molecules, with the general formula $C_{10}H_{16}$. They are the most fragrant class of terpenoids, constituting 90% of the essential oils. They are secondary metabolites of plants, which, although classified as non-essential for viability, mediate important interactions between plants and their environment. Despite being mostly found in plants, specialized classes occur in some animals and microorganisms (Tchimene et al. 2013).

Monoterpenes are a class of varied compounds, not only because of the diversity of their basic skeletons but also because there is the possibility of the formation of a large number of stereoisomers, due to the presence of stereogenic centers in each skeleton. Thus there is a possibility of the formation of several chemical classes, such as alcohols, aldehydes, ketones, and carboxylic acids. An important feature is that basic skeletons can easily interconvert to give rise to any monoterpene (or its derivatives) from any abundant precursor (Silvestre and Gandini 2008).

Monoterpenes are divided into three subgroups: acyclic (such as myrcene, linalool, geraniol), monocyclic (such as α-terpineol and terpinolene), and bicyclic (such as α-pinene, thujone, camphor, fenchone). Within these subgroups there are other classifications: unsaturated hydrocarbons (limonene), alcohols (menthol), aldehydes and ketones (mirtenal, carvone), lactones (monoterpenic lactones are called iridoids, such as nepetalactone), and tropolones (γ-tujaplicin) (Santos et al. 2011). Currently, more than 3000 different natural monoterpenes have been identified.

Monoterpenes are non-nutritive dietary compounds found in the essential oils of citrus fruits, cherry, mint, and herbs. Examples of dietary monoterpenes include R-(+)-limonene in orange and other citrus peel oils, caraway and dill; perillyl alcohol in cherry and spearmint; carvone in caraway and spearmint; and geraniol in lemongrass oil. Due to their aroma, monoterpenes, have been used widely for flavor in food and beverage industries in products like ice-creams, candies, baked goods, pudding, gelatin, chewing gums, non-alcoholic beverages, etc. (Ravichandran et al. 2018; Wagner and Elmadfa 2003).

Limonene, one of the most common monoterpenes, is found in the essential oils of natural fruit like grapefruit (95%), tangerine (94%), orange (91%), mandarine (72%), lemon (65%), and elemi (50%) (González-Mas et al. 2019). The average daily dietary intake of R-(+)-limonene, the principal active form of limonene in the USA, has been estimated to be about 0.3 mg kg^{-1} (Flavor and Extract Manufacturers' Association 1991). Table 13.1 summarizes data for daily consumption and Table 13.2 summarizes industrial use of some monoterpenes.

13.3 Anticancer Activity

Cancer is a large group of diseases that can affect any organ or tissue in the body. It is characterized by the uncontrollable growth of abnormal cells, which go beyond their usual limits to invade adjacent parts of the body and/or spread to other organs. Other

Table 13.1 Daily consumption data for some monoterpenes.

Monoterpene	Individual consumption ((mg kg^{-1}) per day)	Principal food sources	Reference
Limonene	0.3	Citrus oil and citrus peels	(Flavor and Extract Manufacturers' Association 1991)
	0.1132		(Burdock 2010)
Perillyl alcohol	0.00000014	Cherries, ginger grass essential oil, bergamot, berries, and peppermint	(Burdock 2010)
Geraniol	0.001694	Gingergrass oil, lemongrass oil, apple juice, citrus peel oil, citrus juice, and berries	(Burdock 2010)
Menthol	0.1694	Peppermint and other mint oils	(Burdock 2010)
Carvone	0.1624	Spearmint oil, mint oils, caraway, dill, citrus oil, citrus juice (lemon, lime, and orange), and celery seed	(Burdock 2010)
Linalool	0.01822	Bergamot, apple, citrus peel oils and juices, berries, grapes, and guava.	(Burdock 2010)
α-pinene	0.01949	Citrus juices and peel oils, bilberry, cranberry, various spices, mint essential oils, and carrot.	(Burdock 2010)
α-terpineol	0.01723	Citrus peel oils, cardamom, cinnamon, clary sage, various spice, and berries.	(Burdock 2010)

common names for cancer are neoplasia and malignant tumor. Cancer is the second leading cause of death worldwide, accounting for about 9.6 million deaths, or one in six deaths, in 2018. In men, the most common types are lung, prostate, colorectal, stomach, and liver. In women, breast cancer, colorectal, lung, cervical, and thyroid cancer are the most common (World Health Organization 2021a).

Cancer treatment usually includes radiation therapy, chemotherapy, and/or surgery. As each type of cancer needs a specific treatment regimen, an early and correct diagnosis of the disease is essential (World Health Organization 2021a). Conventional chemotherapy has a number of serious toxic side effects, such as hair loss, nausea and vomiting, diarrhea, loss of appetite, decreased white blood cells, and tiredness, leading to considerable interruption of treatments. In order to reduce the toxic effects caused by the treatment of cancer, phytochemicals, such as monoterpenes, appear as naturally abundant molecules, of low cost and as low or even non-toxic with various pharmacological properties, including anticancer activity (Choudhari et al. 2020).

The investigation of natural compounds has been successful in the discovery of new drugs used in anticancer therapy. Currently, more than two-thirds of the medicines used

Table 13.2 Industrial use data for some monoterpenes (ppm).

Food category	Limonene		Perillyl alcohol		Geraniol		Menthol	
	Usual	Maximum	Usual	Maximum	Usual	Maximum	Usual	Maximum
Alcoholic beverages	792,70	812,30	-	-	0,93	2,76	17,59	60,32
Baked goods	342,40	500,50	2,00	8,00	17,00	25,08	65,87	93,06
Chewing gum	343,00	3024,00	-	-	33,02	43,15	464,70	2274,00
Frozen dairy	339,20	436,80	1,05	2,25	8,58	12,00	40,44	72,27
Gelatins, puddings	346,70	451,80	1,00	2,00	2,11	4,85	17,66	25,01
Hard candy	1043,00	1390,00	-	-	12,60	42,24	1407,00	2083,00
Meat products	70,36	79,74	-	-	3,99	8,90	-	-
Non-alcoholic beverages	178,60	198,80	0,55	1,75	3,49	5,86	6,95	9,56
Soft Candy	388,70	511,80	2,68	5,69	15,92	24,44	130,00	181,90
Food category	**Carvone**		**Linalool**		**α-pinene**		**α-terpineol**	
	Usual	**Maximum**	**Usual**	**Maximum**	**Usual**	**Maximum**	**Usual**	**Maximum**
Alcoholic beverages	125,90	144,60	0,24	0,42	5,52	10,55	0,58	2,12
Baked goods	93,67	115,40	11,90	17,64	35,06	110,70	15,93	19,52
Chewing gum	92,16	349,50	40,74	61,00	159,90	279,60	13,53	83,47
Frozen dairy	150,10	197,40	6,47	9,57	13,05	38,94	9,23	14,00
Gelatins, puddings	87,00	90,00	7,31	9,76	6,14	12,13	8,56	12,64
Hard candy	155,80	369,40	3,13	14,89	252,50	265,30	10,67	17,39
Meat products	0,10	0,10	18,72	45,65	10,00	20,00	5,27	10,54
Non-alcoholic beverages	34,38	41,00	3,57	6,87	9,42	22,89	3,40	5,41
Soft Candy	193,80	225,60	6,38	10,42	31,01	62,72	13,51	16,50

Adapted from Burdock, G. A. (2010).

to treat cancer come from natural products or are developed using the knowledge acquired with the activity of its ingredients (Effert 2010). Vieira et al. (2016) reported from a meta-analysis that an increase in intake of up to approximately 400 g per day of varied fruits and vegetables can reduce the risk of lung cancer by 27%. This result corroborates the World Health Organization (2021a) who recommen eating a healthy diet, including fruit and vegetables to prevent cancer. However, the anticancer effects of fruits varied in different cancer types (Wang et al. 2021).

In vitro and *in vivo* studies have increasingly shown the antitumor activity of monoterpenes in cancer cells by promoting cell cycle arrest, apoptosis, autophagy, and/or senescence. The main anticancer mechanisms reported for these monoterpenic compounds include the inhibition of proteins involved in cell survival and proliferation (Ras, ERK, Akt), modulation of oxidative stress and endoplasmic reticulum stress and depolarization of the mitochondrial membrane potential. Increased efficacy of antineoplastic drugs combined with monoterpenes has also been reported, suggesting their potential as new anticancer molecules also used as sensitizing compounds (Mączka et al. 2020; Rodenak-Kladniew et al. 2020a, 2020b; Ye et al. 2020; Yu et al. 2018).

Limonene is known to be chemopreventive against many types of cancer. Its tumor-inhibiting action occurs both in the initiation phase and in the promotion phase. There is scientific evidence that limonene interferes with apoptosis, cell cycle/proliferation, angiogenesis, and DNA damage repair (Araújo-Filho et al. 2021; Hafidh et al. 2018). Limonene, despite having good tolerability and favorable toxicity, is rapidly distributed and metabolized in active monoterpenes, by cells, has the limitation of being poorly soluble in water and being rapidly metabolized *in vivo*. For this reason, in order to improve its pharmacological properties, the encapsulation of limonene in various wall materials, such as biopolymers, proteins and lipid nanoparticles, is currently being studied (Angellotti et al. 2020; Badawy et al. 2020; Bidyarani and Kumar 2019; Miller et al. 2013; Mukhtar et al. 2018).

Perillyl alcohol is a monoterpene that has aroused great interest for its potent antitumor activity. It has chemopreventive and chemotherapeutic activities against several tumors (such as skin, lung, stomach, leukemia), including glioblastomas (GBM), the most common and malignant type of brain tumor. It also stands out for its antitumor activity against pancreatic carcinomas in non-toxic doses, without affecting the rate of DNA synthesis, and for its effectiveness against prostate tumor cells, through the inhibition of telomerase activity. Its anticancer effects are presumably related to its Na^+ / K^+-ATPase inhibitory properties, as well as its activity as an intracellular acidifier, similar to cardiac glycosides (Garcia et al. 2015; Marson et al. 2019; Shojaei et al. 2014; Sobral et al. 2014; Stark et al. 1995).

Carvacrol is a phenolic monoterpene found in essential oils of oregano, thyme, peppermint, and wild bergamot. It has several medicinal potentials for presenting antiviral, antifungal, anti-inflammatory, and anticancer effects. It has a strong cytotoxic, genotoxic and pro-apoptotic activity against cancer cells, in a dose-dependent manner (Sharifi-Rad et al. 2018). *In vitro* studies were performed against liver, stomach and colon cancer (Sobral et al. 2014;). However, its highlight is in combating the proliferation of breast cancer cells. Recent studies have shown that carvacrol suppresses breast cancer cells by regulating the cell cycle (Li et al. 2021). In addition to anticancer activities, carvacrol can also act to reduce the adverse effects of chemotherapy, as it has a gastroprotective and anti-inflammatory effect (Alvarenga et al. 2016).

Another monoterpene that stands out for its anticancer effects is thymoquinone, which transforms the signaling pathways of cancer progression and causes cancer cell apoptosis and death (antiproliferative and proapoptotic activities). Not only does it improve the anticancer activity of chemotherapy drugs, but it also mitigates its side effects (Gupta et al. 2016; Sobral et al. 2014).

These studies demonstrate that monoterpenes can be used as biopharmaceuticals in the treatment and prevention of cancer. However, more research is needed in order to understand the mechanisms that lead to anticarcinogenic effects and to improve the pharmacological effects of these compounds.

13.4 Anti-inflammatory Activity

Inflammation is a complex and non-specific biological immune response of tissues against aggressive agents such as pathogens, irritants, or damaged cells. It can be categorized as acute or chronic and involves a cascade of biochemical events that encompass the local vascular system, the immune system, and different types of cells found in the injured tissue. Acute inflammation is the initial response and is characterized by the increased movement of plasma and cells of the innate immune system, such as neutrophils and macrophages, from the blood to the injured tissues. Chronic inflammation refers to a progressive change in the type of cells present at the site of the inflammatory reaction and is characterized by the simultaneous destruction and healing of the injured tissue (Baumgart and Carding 2007; da Silveira e Sá et al. 2013; Ferrero-Miliani et al. 2007).

One type of inflammatory disease that affects millions of people around the world is inflammatory bowel disease (IBD), characterized by a group of chronic intestinal diseases, the most significant manifestations of which are ulcerative colitis (UC) and Crohn's disease (CD). The inflammation of the intestinal mucosa in IBD is due to an exacerbated immune response, which presents interleaved periods of accentuated symptoms and remission. Currently, there is still no cure for this disease (Baumgart and Carding 2007).

Monoterpenes have been increasingly explored to combat inflammation because of their potential biological activities, including beneficial effects in animal models for colitis. Some results of anticancer activity, as limonene and perillyl alcohol, also suggest anti-inflammatory action. In addition, there is the benefit of these compounds being abundant in nature and being present in herbs, spices, and fruits that are part of human food (da Silveira e Sá et al. 2013; Riella et al. 2012;).

The inhibitory effect of 1,8-cineole on the production of mediators stimulated by lipopolysaccharide (LPS) and interleukin (IL) -1β (IL-1β) by human monocytes *in vitro*, showing a decrease in the production of tumor necrosis factor α (TNF-α), IL-1β, leukotriene B4 (LTB4) and thromboxane B2, which suggests the usefulness of 1,8-cineole against airway inflammation in bronchial asthma and other steroid-sensitive disorders was reported in the 1990s. Subsequently, a clinically relevant anti-inflammatory action of 1,8-cineole in bronchial asthma was reported (da Silveira e Sá et al. 2013; Juergens et al. 2003).

Geraniol, popularly used to treat diarrhea, dermatitis, and intestinal inflammation in East Asia, and nerol also have an anti-inflammatory effect against colitis, as they reduce the inflammatory cytokines interleukin-1β (IL-1β) and TNF-α, respectively. Geraniol also had an antioxidant effect, evidenced by its ability to reduce lipid peroxidation and increase the total antioxidant capacity in colonic tissue (González-Ramírez et al. 2016; Soubh et al. 2015).

A recent study by our research group investigated the effect of a diet supplemented with limonene-1,2-diol, R-(+)- and S-(-)- α-terpineol on TNBS-induced colitis in male Wistar rats. An anti-inflammatory effect was observed, as the consumption of monoterpenoids reduced the levels of tumor necrosis factor-alpha (TNF-α) up to five times in most groups. In addition, diets supplemented, mainly with S-(-)- α-terpineol and limonene-1,2-diol, promoted a significant increase in the gene expression of the anti-inflammatory cytokine IL-10. In general, no significant differences were observed in the effects of the different enantiomers of α-terpineol, which indicates that its mechanism of anti-inflammatory action does not depend on the spatial conformation (Alexandrino et al. 2020).

In order to characterize the anti-inflammatory effect of monoterpenes present in mint species, Sousa et al. (2020) measured the production of nitric oxide (NO) induced by bacterial lipopolysaccharide (LPS) in murine macrophages. Nine compounds decreased significantly by more than 30%. As a result, it was found that the order of potency (S) – (+) – carvone > (R)-(-)-carvone > (+)-dihydrocarveol > (S)-8-hydroxycarvotanacetone > (R)-8 -hydroxycarvotanacetone> (+) – dihydrocarvone> (-) – carveol> (-) – dihydrocarveol> (S) – (-) – pulegone. The most potent compound (S)-(+)-carvone, significantly decreased the expression of NOS2 and IL-1β in macrophages and in a cell model of osteoarthritis using primary human chondrocytes. This result suggests that (S) – (+) – carvone may be effective in preventing diseases related to inflammation, such as osteoarthritis.

Based on the above, the anti-inflammatory potential of some bioactive monoterpenes was demonstrated in experimental models. The data presented suggest the therapeutic potential of this chemical class as a source for the development of new anti-inflammatory agents.

13.5 Antidiabetic Activity

Diabetes mellitus, according to the World Health Organization (2021b), is a chronic disease that occurs when the pancreas does not produce enough insulin or the body cannot effectively use the insulin it produces. Insulin is a hormone that regulates blood sugar. The most common effect of uncontrolled diabetes is hyperglycemia (increased blood sugar). Over time, it causes serious damage to many systems in the body, especially the nerves, vision, and blood vessels (World Health Organization 2021b).

There are three types of diabetes mellitus. Type 1 (formerly known as insulin-dependent, juvenile, or childhood-onset) is characterized by deficient insulin production. Its most common treatment is the daily administration of insulin. Its cause is unknown and medicine has no way of preventing it. Type 2 (formerly called non-insulin-dependent or adult-onset) is a result of the body's ineffective use of insulin, which produces the hormone correctly. It is the type of disease that mostly affects the population around the world, as it is the result of excess body weight and physical inactivity, problems that often occur in a large part of the global population. The third type, gestational diabetes, occurs in pregnant women and is characterized by an excessive fetal size leading to serious risks of complications in childbirth. Symptoms may be similar to those of type 1 diabetes but are generally less pronounced. Diabetes treatment involves food planning and physical activity, along with lowering blood glucose and levels of other risk factors, such as smoking tobacco, which damage blood vessels. As there is still no cure for diabetes, much rigor is needed to control the disease (World Health Organization 2021b).

Although insulin is essential for the treatment of diabetes, other compounds are also used oral medications to control the disease, such as α-glucosidase inhibitors, which target the digestion of carbohydrates in the intestine, thus limiting the availability of glucose captured by the blood and biguanides (the compounds similar to metformin) that suppress the release/production of glucose in the liver. However, these treatments are responsible for side effects, such as the acquisition of resistance over time and gastrointestinal complications, in addition to having a high cost (Stein et al. 2013; Upadhyay et al. 2018). Due to the above, more alternatives are sought for the control of diabetes, such as the use of herbal medicines and known bioactive compounds, such as monoterpenes (Murali and Saravanan 2012).

Tan et al. (2016) studied the *in vitro* effect of 12 monoterpenes (geraniol, nerol, citral, (R) – (-) – linalool, (R) – (+) – limonene, (S) – (-) – perillyl alcohol, (R) – (+) – β-citronelol, (S) – (-) – β-citronelol, α-terpineol, l-menthol, γ-terpinene and terpinolene) against antidiabetic and antiobesogenic activities. Monoterpenes exhibited low radical scavenging properties (DPPH and ABTS), even at high concentrations. Some monoterpenes inhibited α-amylase and α-glucosidase activity and stimulated glucose uptake and lipolysis. Monoterpenes such as (R) – (+) – limonene stimulated glucose uptake and lipolysis. In this study, it is worth mentioning that mRNA expression of glucose transporter 1 (GLUT1) was up-regulated, but glucose transporter 4 (GLUT4) was not affected and adipose triglyceride lipase (ATGL) was suppressed.

Madhuri and Naik (2017) studied the effect of borneol against hyperglycemia, in streptozotocin-induced diabetic Wistar rats. This study revealed that borneol positively modulates hyperglycemia against STZ-induced diabetic rats in a dose-dependent manner.

One of the well-defined mechanisms for testing antidiabetic effects is the quantification of insulin secretion. An agent with antidiabetic effects promotes increased secretion of this hormone. A number of monoterpenes were tested for insulin secretion in beta-pancreatic cells. While carvacrol had no significant effect on the serum insulin level at the dose tested, carvone, citronellol, geniposide, geraniol, and myrtenal were shown to increase the insulin level in STZ-induced diabetes (Babukumar et al. 2017; Bayramoglu et al. 2014; Habtemariam 2017; Muruganathan and Srinivasan 2016; Rathinam and Pari 2016; Srinivasan and Muruganathan 2016; Zhang et al. 2015).

In recent years, a series of studies have been carried out with the aim of demonstrating the antidiabetic potential of monoterpenes *in vitro* and *in vivo*. In this sense, Table 13.3 summarizes some results. However, despite studies showing that monoterpenes can be used as biopharmaceuticals in the treatment and prevention of diabetes, crucial evidence that is still lacking is the performance of clinical studies in humans (Habtemariam 2017).

Table 13.3 Antidiabetic effects of some monoterpenes.

Monoterpene	Experimental model	Biological effect	References
(R)-(+)-limonene	*In* vitro, 3T3-L1 cell culture; α-amylase and α-glucosidase enzymes	Increase GLUT1 expression at mRNA level	(Tan et al. 2016)
(R)-(+)-limonene	*In vivo*, STZ-induced diabetic rats – 50, 100 and 200 mg $kg,^{-1}$ p.o. for 45 days.	Reverse the following diabetic effect: increased blood glucose and glycosylated hemoglobin levels, increased activity of gluconeogenic enzymes (glucose 6-phosphatase and fructose 1,6-bisphosphatase) and decreased activity of glycolytic enzyme, glucokinase and liver glycogen.	(Murali, Saravanan 2012)
Cymene	*In vitro*, Advanced glycation end products (AGEs).	100 μM – Inhibit AGE formation; prevent glycation mediated transition of α-helix to β-pleated sheet structure transition	(Joglekar et al. 2014)
Cymene	*In vivo*, STZ-induced diabetic rats – 20 mg kg^{-1}, p.o. for 60 days	Improve HbA1c and nephropathic parameters	(Joglekar, Panaska, Arvindekar 2014)
Borneol	*In vivo*, STZ-induced diabetic rats – 25 or 50 mg kg^{-1}, p.o. for 30 day.	Lower blood glucose and HbA1c; increase blood insulin; reverse the diabetes-induced increase in the levels of TC, TGs LDL-C, VLDL-C; restore body weight loss; increase liver glycogen level.	(Madhuri and Naik 2017)
Carvacrol	*In vivo*, STZ-induced diabetes in rats – 25 and 50 mg kg^{-1}, p.o. for 7 days.	No effect on serum insulin levels	(Bayramoglu et al. 2014)
Carvone	*In vivo*, STZ-induced diabetic rats – 10 mg kg^{-1}, i.p. for 14 days.	Reduce plasma glucose, HbA1c; improve the levels of hemoglobin and insulin	(Muruganathan and Srinivasan 2016)
Citronellol	*In vivo*, STZ-induced diabetic rats – 25, 50, and 100 mg kg^{-1}, p.o. for 30 days.	Improve the levels of insulin, hemoglobin and hepatic glycogen with significant decrease in glucose and HbA1c levels.	(Srinivasan and Muruganathan 2016)

(Continued)

Table 13.3 (Continued)

Monoterpene	Experimental model	Biological effect	References
Geniposide	*In vivo*, STZ-induced diabetic rats – 800 mg kg^{-1} per day, p.o. for 46 days.	Improve insulin and blood glucose	Zhang et al. 2015)
Geniposide	In vitro. Pancreatic β-cells – cultured primary cells of rats origin	10 μM – Potentiate insulin secretion via activating the glucagon-like-1 receptor (GLP-1R) as well as the adenylyl cyclase (AC)/cAMP signaling pathway	Zhang et al. 2016)
Geraniol	*In vivo*, STZ-induced diabetic rats – 100, 200 and 400 mg kg^{-1}, p.o. for 45 days	Improve the levels of insulin, hemoglobin and hepatic glycogen content; decrease plasma glucose,; preserve the normal histological appearance of hepatic cells and pancreatic β-cells.	(Babukumar et al. 2017)
Myrtenal	*In vivo*, STZ-induced diabetic rats – 80 mg kg^{-1}, p.o. for 28 days	Decrease plasma glucose; increase plasma insulin levels; up-regulate IRS2, Akt and GLUT2 in liver; increase IRS2, Akt and GLUT4 protein expression in skeletal muscle.	(Rathinam and Pari 2016)
Paeoniflorin	*In vitro*, 3T3-L1 adipocytes treated with tumour necrosis factor (TNF)-α	50 μg/mL – Increase insulin-stimulated glucose; promote serine phosphorylation of IRS-1 and insulin-stimulated phosphorylation of AKT	(Kong et al. 2013)
Thymol	*In vivo*, HFD- induced type 2 diabetes in C57BL/6J mice – 40 mg kg^{-1}, intragastric for 5 weeks.	Improved glucose homeostasis, decreased kidney weight, biochemical parameters in serum and urine and restored the TGF-β and VEGF proteins in HFD-induced diabetic mice	(Saravanan and Pari 2016).

13.6 Antioxidant Activity

Antioxidants are molecules capable of inhibiting, delaying, or removing oxidative damage, acting as free radical scavengers, in the inactivation of peroxides and other oxygen species, as methal chelator, and in the extinction of secondary lipid oxidation products, even at low concentrations (Néri-Numa et al. 2019). This oxidative stress is a metabolic response of the body to an imbalance in the proportion of antioxidants and oxidants, due to adverse conditions, the deficiency of antioxidants or an increase in the amount of oxidant species, such as reactive oxygen species (ROS) and reactive nitrogen species (RNS) (Azat Aziz et al. 2019; Gulcin 2020).

The ROS and RNS have both beneficial and deleterious effects on the organism, varying according to their concentration, in the first case, ROS occurs as by-products of metabolism, normal respiration, and the autoxidation of xenobiotics (Gulcin 2020). On the other hand, in the presence of an excess of ROS or NOS the biological system can develop different pathologies, like diabetes, cancer, coronary, and neurodegenerative diseases (Francenia Santos-Sánchez et al. 2019; Néri-Numa et al. 2019). This imbalance can affect redox balance resulting in cell damage and the endogenous antioxidant defense system, such as superoxide dismutase (SOD), catalase (CAT), glutathione peroxidase, or non-enzymatic compounds, such as bilirubin and albumin, may be compromised and, consequently, it fails to guarantee complete protection of the organism. In order to compensate for this deficit, exogenous antioxidants can be used through the ingestion of food, medicinal plants, and pharmaceuticals (Francenia Santos-Sánchez et al. 2019).

Medicinal plants are an inexhaustible source of new bioactive compounds, and many drugs currently used are products derived from them (Bouyahya et al. 2019). Among these bioactive compounds, essential oils and their volatile compounds are of great interest, due to several reports describing their biological properties, such as antioxidant, antimicrobial, and antidiabetic (Bouyahya et al. 2019). This natural antioxidant property of many essential oils is especially interesting for promoting human health efficiently and with less adverse effects than conventional treatments and can be useful to prevent disorders caused by oxidative stress, atherosclerosis, neurodegenerative, autoimmune, and cancer diseases (Arantes et al. 2019).

A special example, the *Mentha* species, especially mint (*Mentha canadensis* L), spearmint (*Mentha spicata* L.), and peppermint (*Mentha piperita* L.), are sources of essential oils of great commercial relevance, due to their fragrance and profile of volatile compounds such as pulegone, menthol, limonene, menthone, piperitone, and carvone (Asghari et al. 2018). Both the plants themselves and their essential oils are widely used in the preparation of teas, in popular medicine, condiments, food additives, cosmetics, and pharmaceuticals. Furthermore, some studies reported their antioxidant activity, as free radical scavenging, metal chelating activity, β-carotene bleaching, and reducing power, probably associated with the presence of oxygenated monoterpenes, such as 1,8-cineole, rotundifolone, menthol, pulegone, and menthofurane (Asghari et al. 2018; Benabdallah et al. 2018).

Some studies have reported that genetic factors, climatic conditions, the part of the plant used, geographical situation, and phenological stage can affect the essential oil composition, and, as a result, also the antioxidant activity (Bouyahya et al. 2019; Farhadi et al. 2020; Leporini et al. 2020; Simirgiotis et al. 2020). As an example, the essential oil of post-flowering *Centaurium erythraea* was richer in oxygenated monoterpene than the vegetative and flowering stage, on the other hand, the essential oil of the flowering stage presents most antioxidant activity and antityrosinase effect (Bouyahya et al. 2019). From that point of view, the chemical composition of the essential oil needs to be standardized for a commercial application, in order to achieve the proposed purpose, e.g., as an antioxidant or food preservative (Dammak et al. 2019).

In this sense, the antioxidant activity of essential oils could be associated with multiple systems, as they possess a chemical mixture with many functional groups, polarities, and chemical behavior, in which major or minor constituents or a synergic action of both can be responsible for their better antioxidant activity (Benabdallah et al. 2018). It is worth mentioning that it was suggested that the major component, terpinen-4-ol is responsible for the antioxidant activity of *Melaleuca alternifolia* essential oil in a silver catfish (*Rhamdia quelen*) model (Souza et al. 2018). On the other hand, the monoterpene fraction did not contribute to the antioxidant activity of *Nigella sativa* essential oil, since these compounds were lost in the non-volatile fraction, and phenolic compounds and thymoquinone can be responsible for the oil antioxidant activity (A.F.C. Silva et al. 2020).

More specifically, the antioxidant activity could be correlated to oxygenated monoterpenes and mixture of mono- and sesquiterpene hydrocarbons composition (Leporini et al. 2020) and it was also suggested that the high reducing potential of some essential oils can be linked to the presence of hydrogen donating components, such as 1,8-cineole, linalool, and carvacrol (Asghari et al. 2018; Ranjbaran et al. 2019). As an example, several essential oils rich in an oxygenated monoterpene, 1,8-cineole, proved to be an excellent free radical scavenging, such as *Laurus nobilis, Calamintha nepeta, Thymus mastichina, Mentha aquatica* and, *Mentha arvensis* (Arantes et al. 2019; Benabdallah et al. 2018; Dammak et al. 2019).

In another study, the authors also suggest that the antioxidant potential of the essential oil from *Origanum virens* could be associated with a high content of oxygenated and hydrocarbon monoterpenes and the presence of hydrocarbon sesquiterpene, indicating a synergetic effect (Arantes et al. 2019). In this sense, in an evaluation of antioxidant activity of several monoterpenes, greater free radical scavenging activities for some oxygenated monoterpenes (geraniol and thymol) and a hydrocarbon monoterpene (myrcene) were verified (Badawy et al. 2019)

Oxidative stress may lead to liver damage, metabolic syndrome, cardiovascular, and neurodegenerative diseases in humans, for example. Therefore, compounds that inhibit oxidative stress can prevent the onset of stress-related diseases (Amini et al. 2020). Some studies have demonstrated the potential of commercial standard monoterpenes, myrtenol, limonene, citral, and α-pinene to protect biological systems against oxidative damage by the restoration of detoxifying enzymes levels *in vivo* models of myocardial ischemia-reperfusion injury, chronic immobilization, N-nitrosodiethylamine-induced hepatocellular carcinoma, and isoproterenol-induced myocardial infarction, respectively (Amini et al. 2020; Britto et al. 2018; Krishnan et al. 2020; Zhang et al. 2020). Also, dietary intake (0.25–0.5 M) of menthol stimulates the antioxidant system and suppresses oxidative stress in rainbow trout (*Oncorhynchus mykiss*) without negative effects on fish growth (Hoseini et al. 2020) and safranal can be used to prevent photoaging or premature aging, as a result of its free radical scavenging capability and photoprotective activity demonstrated in *in vitro* assay (Madan and Nanda 2018).

In another important field, essential oils or their fractions could be an excellent alternative to synthetic food additives to prevent oxidation, improving stability, preventing loss of their sensory and nutritional qualities (Wang et al. 2019). As an example, rosemary essential oil or its distillate fractions proved to be an excellent alternative natural antioxidant for sunflower oil, as a result of its high capacity to capture free radicals and lowest peroxide values, conjugated dienes, and trienes (Mezza et al. 2018).

These results suggest that essential oils and/or monoterpenes may find application in several industrial sectors, such as cosmetics and food preservatives, as substitutes for synthetic antioxidants, or as biopharmaceuticals in the treatment and prevention of degenerative diseases. In this sense, Table 13.4 summarizes some results concerning the antioxidant and biological effects of some monoterpenes and essential oils in different studies using *in vitro* assays and animal models.

13.7 Cardiovascular and Antihypertensive Effects of Monoterpenes

Cardiovascular diseases (CVDs) are a series of disorders that affect the heart and blood vessels, leading to heart attacks and strokes, which are due to a combination of risk factors that include tobacco use, unhealthy diet and obesity, physical inactivity and harmful use of alcohol, hypertension, diabetes, and hyperlipidaemia (World Health Organization 2017). Ischaemic heart disease and strokes are the leading cause of death recorded by the World Health Organization, approaching 15.2 million cases in 2019 or 27% of all deaths across the world, an increase of about 2.8 million cases compared to 2000 (World Health Organization 2020b). Another important factor linked to heart attacks and strokes is hypertension or elevated blood pressure, which is associated with excessive salt consumption and the other risk factors mentioned above for CVDs (World Health Organization 2019).

CVDs represent a major social and public health problem worldwide, given the fact that approximately 75% of deaths occurred in low- and middle-income countries (World Health Organization 2020a). In addition, the high cost of treatments that directly impacts the budget of public health systems and the adverse effects of the drugs currently used makes relevant the discovery of new therapies (de Andrade et al. 2017; Saljoughian et al. 2018).

In this way, plants and their derivatives, such as plant extracts, essential oils or their active compounds, have been used for centuries in popular medicine, and some of them are already traditionally used in the prevention and treatment of CVDs (Patrignani et al. 2021). In this sense, essential oils and their monoterpenes appear as promising drugs for the treatment and/or prevention of these diseases, since they can lead to vasorelaxation, hypotension, promote the cardiovascular system, and decrease heart rate (Saljoughian et al. 2018; M. R. V. Santos et al. 2011). However, the discovery of a new drug in medicinal plants requires an investigation of the therapeutic potential of its components through pharmacological screening (Menezes et al. 2010). Despite that, there are reports of patent filing dealing with the use of terpenes in preclinical tests in the control of hypertension, diabetes, myocardial ischemia, and vasorelaxant effects, demonstrating a state of transition between research and potential application in humans, of these isolated compounds (E.A.P. Silva et al. 2019).

Table 13.4 The antioxidant capacity and related metabolic effects of monoterpenes and essential oils by *in vitro* and *in vivo* assays.

Monoterpene	Monoterpene source	Method and Biological Effect	References
Myrtenol	Commercial standard	Oral administration in male Wistar rats (50 mg kg^{-1}) for seven consecutive days ↑ Detoxifying enzymes SOD, CAT, GPx, GR Prevent the overgeneration of ROS and hidroperoxide formation induced by myocardial ischemia-reperfusion injury	(Britto et al. 2018)
α-pinene	Commercial standard	Oral administration in male Wistar albino rats (50 mg kg^{-1}) for seven consecutive days ↑ Detoxifying enzymes SOD, CAT, and GPx ↓ TBARS and LOOH formation in isoproterenol-induced myocardial infarction model	(Zhang et al. 2020)
Menthol	Commercial standard	Dietary intake (0.1–0.5 M) in rainbow trout (*Oncorhynchus mykiss*) for 60 days Total antioxidant capacity - ↑ Detoxifying enzymes SOD and CAT ↓ MDA levels	(Hoseini et al. 2020)
Limonene	Commercial standard	Oral gavage in rats (10 mg kg-1) during 21 days ↑ Detoxifying enzyme GSH and ↓ MDA levels ↓ expression of pro-inflammatory TNF-α, IL-1β, IL-6, and NF-κB mRNA induced in rats exposed to chronic immobilization	(Amini et al. 2020)
Citral	Commercial standard	Oral gavage (100 mg kg-1) during 6 weeks in male Wistar albino with N-nitrosodiethylamine-induced hepatocellular carcinoma ↑ Detoxifying enzyme SOD, CAT, GPx, GSH, and GR ↓ TBARS, LOOH, CD, and PC formation	(Krishnan et al. 2020)
Terpinolene α-phellandrene	Commercial standard	NO scavenging activity IC_{50} (μM) – 409.4; 216.9 ABTS IC_{50} (μM) – 497.4; 367.7 FRAP IC_{50} (μM) – 1325.7; 1619.6	(Scherer et al. 2019)
Safranal	Commercial standard	DPPH IC_{50} (μg/mL) – 22.7 Protective effect of 6.6	(Madan and Nanda 2018)
Terpinen-4-ol (41.98%) γ-terpinene (20.15%) α-terpinene (9.85%)	*Melaleuca alternifolia* essential oil Terpinen-4-ol commercial standard	Silver catfish (*Rhamdia quelen*) model ↓ TBARS, LOOH levels ↑ Detoxifying enzyme GSH	(Souza et al. 2018)

(Continued)

Table 13.4 (Continued)

Monoterpene	Monoterpene source	Method and Biological Effect	References
Myrcene, α-pinene, camphor, menthone, citronellol, geraniol, linalool, menthol, α-terpineol, thymol	Commercial standard	Free radical scavenging activity – DMPD IC_{50} (mg/mL), respectively: 22.1; 880.74; 1101.3; 1217.47; 289.67; 19.15; 530.68; 1047.71; 480.56; 31.43	(Badawy et al. 2019)
1,8-cineole (21.9%) Camphor (25%) α-thujone (27.5%)	*Salvia officinalis* essential oils	DPPH IC_{50} (mg/mL) – 11.12	(Dammak et al. 2019)
1,8-cineole (35%) Camphor (32%)	*Lavandula dentate* essential oil	DPPH IC_{50} (mg/mL) –14.03	(Dammak et al. 2019)
1,8-cineole (43.2%) β-terpynil acetate (13.7%) Eugenol methyl ether (11.7%)	*Laurus nobilis* essential oil	DPPH IC_{50} (mg/mL) –0.35	(Dammak et al. 2019)
Cymene Thymoquinone	*Nigella sativa essential* oil	DPPH IC_{50} (μg/mL) – 6.7	(A.F.C. Silva et al. 2020)
1,8-cineole (16.98–21.89%) Camphor (7.27–11.08%) α-pinene (10.37–10.96%) Trans-caryophyllene (10.58-8.62%)	*Salvia rosmarinus* essential oil	DPPH IC_{50} (μg/mL) – 29.84–33.21% ABTS IC_{50} (μg/mL) – 22.61–35.43 β-carotene bleaching IC_{50} (μg mL^{-1}) – 33.18–45.21 FRAP (μM Fe (II)/g) – 2.95–5.59	(Leporini et al. 2020)
1,8-cineole (27.9%) Menthone (22%) Menthol (16.3%)	*Calamintha nepeta* essential oil	DPPH 0.1 mg quercetin Eq./mL essential oil Reducing power 0.2 mg quercetin Eq./mL essential oil β-carotene/linoleic acid 11 mg quercetin Eq./mL essential oil	(Arantes et al. 2019)
p-Cymene (12.2%) γ-Terpinene (20.2%) Thymol (19.4%)	*Origanum virens* essential oil	DPPH 0.6 mg quercetin Eq./mL essential oil Reducing power 1.7 mg quercetin Eq./mL essential oil β-carotene/linoleic acid 113 mg quercetin Eq./mL essential oil	(Arantes et al. 2019)

(Continued)

Table 13.4 (Continued)

Monoterpene	Monoterpene source	Method and Biological Effect	References
1,8-cineole (71.2%) α-terpineol (9.7%)	*Thymus mastichina* essential oil	DPPH 0.5 mg quercetin Eq./mL essential oil Reducing power 0.3 mg quercetin Eq./mL essential oil β-carotene/linoleic acid 16 mg quercetin Eq./mL essential oil	(Arantes et al. 2019)
1,8-cineole (33.5%) Linalool (15.1%) Menthone (12.9%) Trans-piperitone oxide(12.6%)	*Mentha longifolia* var. *calliantha* essential oil	DPPH – 5.8 mmol TEs g^{-1} oil ABTS – 186.63 mmol TEs g^{-1} oil FRAP – 102 mmol TEs g^{-1} oil CUPRAC – 337 mmol TEs g^{-1} oil Metal chelating – 17.40 mg EDTAEs g^{-1} oil Molybdenum total antioxidant – 9.17 mmol TEs g^{-1} oil	(Asghari et al. 2018)
γ-terpinene (10.6%) Z-sabinene hydrate (13.4%) Thymol (15.4%)	*Origanum vulgare* essential oil	DPPH IC_{50} μg mL^{-1} – 4750 ABTS IC_{50} μg mL^{-1} – 1252.74 FRAP IC_{50} μg mL^{-1} – 274.53	(Simirgiotis et al. 2020)
1,8-cineole (10.26%) Menthofurane (73.38%)	*Mentha aquatica* essential oil	DPPH IC_{50} – 0.69 mg mL^{-1} β-carotene bleaching IC_{50} – 0.16 mg mL^{-1} Chelating ability IC_{50} – 1.73 mg mL^{-1}	(Benabdallah et al. 2018)
1,8-cineole (18.16%) Menthofurane (29.69%)	*Mentha arvensis* essential oil	DPPH IC_{50} – 0.76 mg mL^{-1} β-carotene bleaching IC_{50} – 0.22 mg mL^{-1} Chelating ability IC_{50} – 1.72 mg mL^{-1}	
1,8-cineole (6.73%) Menthol (49.89%) p-Menthone (20.84%) Cis-carene (4.99%)	*Mentha piperita* essential oil	DPPH IC_{50} – 4.75 mg mL^{-1} β-carotene bleaching IC_{50} – 1.25 mg mL^{-1} Chelating ability IC_{50} – 2.39 mg mL^{-1}	
Pulegone(59.12%) Neomenthol (20.76%)	*Mentha pulegium* essential oil	DPPH IC_{50} – 0.97 mg mL^{-1} β-carotene bleaching IC_{50} – 0.625 mg mL^{-1} Chelating ability IC_{50} – 2.64 mg mL^{-1}	
Rotundifolone (65.99%)	*Mentha rotundifolia* essential oil	DPPH IC_{50} – 1.50 mg mL^{-1} β-carotene bleaching IC_{50} – 0.54 mg mL^{-1} Chelating ability IC_{50} –4.03 mg mL^{-1}	
Rotundifolone (52.17%)	*Mentha villosa* essential oil	DPPH IC_{50} – 1.86 mg mL^{-1} β-carotene bleaching IC_{50} – 0.62 mg mL^{-1} Chelating ability IC_{50} –1.72 mg mL^{-1}	

(Continued)

Table 13.4 (Continued)

Monoterpene	Monoterpene source	Method and Biological Effect	References
1,8-cineole (29.24%) Camphor (20.33%) α-pinene (15.77%)	*Rosmarinus officinalis* essential oil	IC_{50} μg/mL – 4.39 ↓ Peroxide values, conjugated dienes and conjugated trienes during storage of sunflower oil	(Mezza et al. 2018)
Menthol (9.46–20.82%) Carvacrol (8.73–25.61%)	*Centaurium erythraea* essential oil from vegetative, flowering and, post-flowering stage	DPPH IC_{50} μg mL^{-1}– 61.53; 47.18; 69.25 ABTS IC_{50} μg mL^{-1} – 134.81; 65.34; 101.62 FRAP IC_{50} μg mL^{-1}– 74.42; 53.25; 82.06 Inhibition of tyrosinase activity IC_{50} μg mL^{-1} – 103.6; 41.86; 49.18	(Bouyahya et al. 2019)
1,8-cineole (34.51; 32.73; 21.28%) Camphor (7.27; 5.98; 14.07%) Borneol (6.91; 6.17; 3.37%)	*Achillea millefolium* essential oil from vegetative, flowering and, fruit set stages	DPPH IC_{50} mg mL^{-1}– 25.54; 25.87; 27.39, respectively	(Farhadi et al. 2020)
α-pinene Limonene	Some wine terpenoids	DPPH IC_{50} mg mL^{-1}– 12.57; 13.35 Reducing power – 213.7 and 133.48 μg L-ascorbic acid mL^{-1}	(Wang et al. 2019)

ABTS, 2,2′azinobis(3-ethylbenzothiazoline)-6-sulfonic acid radical cátion; **CUPRAC**, cupric ion reducing capacity; **DMPD**, *N,N*-dimethyl-1,4-phenylenediamine; **DPPH**, 2,2-diphenyl-1-picrylhydrazyl; **FRAP**, Ferric reducing antioxidant power; **MDA**, malondialdehyde; **GPx**, glutathione peroxidase; **CAT**, catalase; **GSH**, glutathione; **GR**, glutathione reductase; **POX**, peroxidase; **SOD**, superoxide dismutase; **ROS**, reactive oxygen species; **TBARS**, thiobarbituric acid reactive substances; **LOOH**, lipid hydroperoxides; **CD**, conjugated dienes; **PC**, protein carbonyls; **NO**, nitric oxide.

One of the main risk factor linked to heart attacks and strokes is hypertension or elevated blood pressure, which is associated with excessive salt consumption, a diet high in saturated fat and trans fats, low intake of fruits and vegetables, physical inactivity, consumption of tobacco and alcohol, and being overweight or obese (World Health Organization 2019). The most commonly used methods in the treatment of hypertension include the use of angiotensin-converting enzyme inhibitors (ACE-I), diuretics, calcium channel blockers, beta blockers, cholesterol lowering drugs, anti-platelet and anti-coagulants to decrease platelet aggregation, and anti-arrhythmic treatments (Saljoughian et al. 2018). In this sense, this section provides some reports of the use of monoterpenes in the prevention and/or treatment of CVDs.

In general, the hypotensive activity of alcohol monoterpenes is higher than hydrocarbons monoterpenes, in addition the position of hydroxyl in the benzene ring may also influence the effectiveness of hypotensive activity of these compounds (Menezes et al. 2010; Saljoughian et al. 2018). As an example, α-pinene, β-pinene, citronellol and linalool induziram hypotension and tachycardia in non-anaesthetized normotensive rats (Menezes et al. 2010). In another study, D-limonene (20–40 mg kg^{-1}) also produce intense hypotension in rats, however, in this case is associated with bradycardia (Nascimento et al. 2019).

(−)-Carveol presents a vasorelaxant activity on human umbilical arteries (HUAs), where it is suggested that this vasorelaxation is linked to the inhibition of L-type voltage-operated calcium channels (VOCCs) and the partial modulation of Ca^{2+} activated K^+ channels (BKCa), and the blockade of these channels, provided by carveol (R.E.R. da Silva et al. 2020). In another model, it was demonstrate that (-)-borneol has a vasorelaxant effect in rat aortic rings, which is dependent on the presence of vascular endothelium with NO-pathway. On the other hand, their relaxing endothelium-independent effect is depend on the K_{ATP} channel (S. E. Santos et al. 2019). Similarly, isoespintanol exerts a vasorelaxant effect on thoracic aortic rings through NO-dependent pathways, and has also suggest that an inhibition of calcium influx could be the involved mechanism (Rinaldi et al. 2019).

High levels of cholesterol and triglycerides in plasma are associated with with the development of atherosclerosis and coronary heart disease. Camphene is able to reduce the levels of cholesterol and triglycerides (TG) in plasma hyperlipidemic rats (Vallianou and Hadzopoulou-Cladaras 2016). Similar results were found with perillyl alcohol at 100 and 200 mg kg^{-1} doses, where there was a reduction in the serum triglyceride, LDL, VLDL, total cholesterol and increase HDL levels (Hassan et al. 2019)

In the same way, citronellal improves endothelial dysfunction and prevents the growth of atherosclerosis in rats by reducing oxidative stress in a dose dependent-manner, moreover at 150 mg kg^{-1} per day. Atherosclerotic plaque size reduction size was equivalent to lovastatin at a regular dose (100 mg kg^{-1}) (Lu et al. 2019).

Carvacrol has cardioprotective effects, such as myocardial ischemic damage, resulting from multiple mechanisms involving its antioxidant and antiapoptotic activities, which are mediated by MAPK/ERK and Akt/eNOS signaling pathways (Patrignani et al. 2021). In this same context, it was suggested that D-Limonene ameliorates myocardial infarction injury in a murine model, antioxidant and antiapoptotic activities, resulting in a reduction in the the size of the infarcted area, promoted cardioprotection, prevented histological alterations, completely abolished oxidative stress damage, restored superoxide dismutase activity, and suppressed pro-apoptotic enzymes (Durço et al. 2019)

These results suggest that monoterpenes may find application as biopharmaceuticals in the treatment and prevention of CVDs. In this sense, Table 13.5 summarizes some results concerning the cardiovascular and antihypertensive effects of some monoterpenes in different studies using *in vitro* assays and animal models.

Table 13.5 Cardiovascular effects of monoterpenes.

Monoterpene	Monoterpene source	Experimental model	Biological Effect	References
α-pinene	Commercial standard	*In vivo* Non-anaesthetized normotensive rats	Hypotension and tachycardia	(Menezes et al. 2010)
β-pinene	Commercial standard	*In vivo* Non-anaesthetized normotensive rats	Hypotension and tachycardia	(Menezes et al. 2010)
Citronellol	Commercial standard	*In vivo* Non-anaesthetized normotensive rats	Hypotension and tachycardia	(Menezes et al. 2010)
Linalool	Commercial standard	*In vivo* Non-anaesthetized normotensive rats	Hypotension and tachycardia	(Menezes et al. 2010)
D-Limonene	Commercial standard	*In vivo* Wistar rats	Hypotension and bradycardia	(Nascimento et al. 2019)
		In vivo Murine model	Ameliorates myocardial infarction injury	(Durço et al. 2019)
(−)-Carveol	Commercial standard	*In vitro* Human umbilical arteries (HUAs)	Vasorelaxant effect	(R.E.R. da Silva et al. 2020)
Borneol	Commercial standard	*In vitro* Isolated rat aorta	Vasorelaxant effect	(S. E. Santos et al. 2019)
Isoespintanol	Commercial standard	*In vitro* Thoracic aortic rings isolated from rats	Vasorelaxant effect	(Rinaldi et al. 2019)
Camphene	Commercial standard	*In vivo* Hyperlipidemic rats	Hypocholesterolemic and hypotriglyceridemic effects	(Vallianou and Hadzopoulou-Cladaras 2016)
Perillyl alcohol	Commercial standard	*In vivo* Hyperlipidemic rats	Hypocholesterolemic and hypotriglyceridemic effects	(Hassan et al. 2019)
Citronellal	Commercial standard	*In vivo* Atherosclerotic model of rats	Prevents the formation of carotid atherosclerotic plaque	(Lu et al. 2019)

13.8 Conclusion

Based on the above, regular consumption of monoterpenes has great potential to promote positive biological activities, such as antioxidant, antidiabetic, cardioprotective, and anti-inflammatory effects. Also highlighted was the great potential of this class of compounds against cancer and its role in alleviating symptoms of traditional chemotherapy treatments, such as gastroprotective effects. Other properties that also arouse interest are antimicrobial, anesthetic, analgesic and antiallergic effects.

Considering the non-toxic character and the benefits associated with its consumption, research related to the discovery and proof of its biological effects has been increasingly encouraged, reinforcing the need to use natural compounds as health promoters.

Acknowledgements

This study was financed in part by the Coordination for the Improvement of Higher Education Personnel (CAPES) (Finance Code 001 and process number 88887.470149/2019-00); the National Council of Technological and Scientific Development (CNPq) (process numbers 403328/2016-0 and 301496/2019-6) and the São Paulo Research Foundation (FAPESP) (process numbers 2015/50333-1, 2018/11069-5, 2015/13320-9, 2019/13465-8, and 2020/06814-3). MRMJ acknowledges Red Iberomericana de Alimentos Autoctonos Subutilizados (ALSUB-CYTED, 118RT0543).

References

Alexandrino, T.D., Moya, A.M.T.M., de Medeiros, T.D.M., Morari, J., Velloso, L.A., Leal, R.F., Maróstica, M.R., Pastore, G.M., Cazarin, C.B.B., and Bicas, J.L. (2020). Anti-inflammatory effects of monoterpenoids in rats with TNBS-induced colitis. *PharmaNutrition* 14: 100240. doi: 10.1016/j.phanu.2020.100240.

Alvarenga, E.M., Souza, L.K., Araújo, T.S., Nogueira, K.M., Sousa, F.B., Araújo, A.R., Martins, C.S., Pacífico, D.M., de C Brito, G.A., Souza, E.P., Sousa, D.P., and Medeiros, J.V. (2016). Carvacrol reduces irinotecan-induced intestinal mucositis through inhibition of inflammation and oxidative damage via TRPA1 receptor activation. *Chemico-Biological Interactions* 260: 129–140.

Amini, R., Asle-Rousta, M., and Aghazadeh, S. (2020). Hepatoprotective effect of limonene against chronic immobilization induced liver damage in rats. *Naunyn-Schmiedeberg's Archives of Pharmacology* 393 (11): 2053–2059. doi: 10.1007/s00210-020-01915-0.

Angellotti, G., Murgia, D., Presentato, A., D'Oca, M.C., Scarpaci, A.G., Alduina, R., Raimondi, M.V., and De Caro, V. (2020). Antibacterial PEGylated solid lipid microparticles for Cosmeceutical purpose: formulation, characterization, and efficacy evaluation. *Materials* 13 (9): 2073. doi: 10.3390/ma13092073.

Arantes, S.M., Piçarra, A., Guerreiro, M., Salvador, C., Candeias, F., Caldeira, A.T., and Martins, M.R. (2019). Toxicological and pharmacological properties of essential oils of

Calamintha nepeta, Origanum virens and *Thymus mastichina* of Alentejo (Portugal). *Food and Chemical Toxicology* 133 (November): 110747. doi: 10.1016/j.fct.2019.110747.

Araújo-Filho, H.G. de, dos Santos, J.F., Carvalho, M.T.B., Picot, L., Fruitier-Arnaudin, I., Groult, H., Quintans-Júnior, L.J., and Quintans, J.S.S. (2021). Anticancer activity of limonene: a systematic review of target signaling pathways. *Phytotherapy Research* n/a: 1–14. doi: 10.1002/ptr.7125.

Asghari, B., Zengin, G., Bahadori, M.B., Abbas-Mohammadi, M., and Dinparast, L. (2018). Amylase, glucosidase, tyrosinase, and cholinesterases inhibitory, antioxidant effects, and GC-MS analysis of wild mint (*Mentha longifolia* var. *calliantha*) essential oil: a natural remedy. *European Journal of Integrative Medicine* 22 (September): 44–49. doi: 10.1016/j.eujim.2018.08.004.

Azat Aziz, M., Diab, A.S., and Mohammed, A.A. (2019). Antioxidant categories and mode of action. In: Antioxidants. IntechOpen. doi: 10.5772/intechopen.83544.

Babukumar, S., Vinothkumar, V., Sankaranarayanan, C., and Srinivasan, S. (2017). Geraniol, a natural monoterpene, ameliorates hyperglycemia by attenuating the key enzymes of carbohydrate metabolism in streptozotocin-induced diabetic rats. *Pharmaceutical Biology* 55: 1442–1449. doi: 10.1080/13880209.2017.1301494.

Badawy, M.E.I., Lotfy, T.M.R., and Shawir, S.M.S. (2020). Facile synthesis and characterizations of antibacterial and antioxidant of chitosan monoterpene nanoparticles and their applications in preserving minced meat. *International Journal of Biological Macromolecules* 156: 127–136. doi: 10.1016/j.ijbiomac.2020.04.044.

Badawy, M.E.I., Marei, G.I.Kh., Rabea, E.I., and Taktak, N.E.M. (2019). Antimicrobial and antioxidant activities of hydrocarbon and oxygenated monoterpenes against some foodborne pathogens through *in vitro* and *in silico* studies. *Pesticide Biochemistry and Physiology* 158 (May). Elsevier: 185–200. doi: 10.1016/j.pestbp.2019.05.008.

Bakkali, F., Averbeck, S., Averbeck, D., and Idaomar, M. (2008). Biological effects of essential oils – a review. *Food and Chemical Toxicology* 46: 446–475. doi: 10.1016/j.fct.2007.09.106.

Baumgart, D.C. and Carding, S.R. (2007). Inflammatory bowel disease: cause and immunobiology. *The Lancet* 369: 1627–1640. doi: 10.1016/S0140-6736(07)60750-8.

Bayramoglu, G., Senturk, H., Bayramoglu, A., Uyanoglu, M., Colak, S., Ozmen, A., and Kolankaya, D. (2014). Carvacrol partially reverses symptoms of diabetes in STZ-induced diabetic rats. *Cytotechnology* 66: 251–257. doi: 10.1007/s10616-013-9563-5.

Benabdallah, A., Boumendjel, M., Aissi, O., Rahmoune, C., Boussaid, M., and Messaoud, C. (2018). Chemical composition, antioxidant activity and acetylcholinesterase inhibitory of wild *Mentha* species from northeastern Algeria. *South African Journal of Botany* 116: May 131–139. doi: 10.1016/j.sajb.2018.03.002.

Bidyarani, N. and Kumar, U. (2019). Study of limonene loaded Zein nanoparticles for sustainable agricu lture. In: Advances in Spectroscopy: Molecules to Materials (ed. D.K. Singh, S. Das, and A. Materny), Vol. 236, 257–267. Singapore: Springer Singapore. doi: 10.1007/978-981-15-0202-6_19.

Bouyahya, A., Belmehdi, O., El Jemli, M., Marmouzi, I., Bourais, I., Abrini, J., Faouzi, My El A., Dakka, N., and Bakri, Y. (2019). Chemical variability of *Centaurium erythraea* essential oils at three developmental stages and investigation of their *in vitro* antioxidant, antidiabetic, dermatoprotective and antibacterial activities. *Industrial Crops and Products* 132 (June): 111–117. doi: 10.1016/j.indcrop.2019.01.042.

Britto, R.M.de, da Silva-Neto, J.A., Mesquita, T.R.R., de Vasconcelos, C.M.L., de Almeida, G.K.M., de Jesus, I.C.G., dos Santos, P.H. et al. (2018). Myrtenol protects against myocardial ischemia-reperfusion injury through antioxidant and anti-apoptotic dependent mechanisms. *Food and Chemical Toxicology* 111 (January): 557–566. doi: 10.1016/j.fct.2017.12.003.

Burdock, G.A. (2010). Fenaroli's Handbook of Flavor Ingredients. 6e. Boca Raton: CRC Press. doi: 10.1016/S0015-6264(76).80260–X.

Choudhari, A.S., Mandave, P.C., Deshpande, M., Ranjekar, P., and Prakash, O. (2020). Phytochemicals in cancer treatment: from preclinical studies to clinical practice. *Frontiers in Pharmacology* 10:1614. doi: 10.3389/fphar.2019.01614.

Christianson, D.W. (2017). Structural and chemical biology of terpenoid cyclases. *Chemical Reviews* 117: 11570–11648. doi: 10.1021/acs.chemrev.7b00287.

da Silveira e Sá, R., Andrade, L.N., and de Sousa, D.P. (2013). A review on anti-inflammatory activity of monoterpenes. *Molecules* 18: 1227–1254. doi: 10.3390/molecules18011227.

Dammak, I., Hamdi, Z., El Euch, S.K., Zemni, H., Mliki, A., Hassouna, M., and Lasram, S. (2019). Evaluation of antifungal and anti-ochratoxigenic activities of *Salvia officinalis, Lavandula dentata* and *Laurus nobilis* essential oils and a major monoterpene constituent 1,8-cineole against *Aspergillus carbonarius. Industrial Crops and Products* 128 (February): 85–93. doi: 10.1016/j.indcrop.2018.11.006.

de Andrade, T., Brasil, G., Endringer, D., da Nóbrega, F., and de Sousa, D. (2017). Cardiovascular activity of the chemical constituents of essential oils. *Molecules* 22 (9): 1539. doi: 10.3390/molecules22091539.

Durço, A.O., de Souza, D.S., Heimfarth, L., Miguel-dos-Santos, R., Rabelo, T.K., de Oliveira Barreto, T., Rhana, P. et al. (2019). D-limonene ameliorates myocardial infarction injury by reducing reactive oxygen species and cell apoptosis in a murine model. *Journal of Natural Products* 82 (11): 3010–3019. doi: 10.1021/acs.jnatprod.9b00523.

Efferth, T. (2010). Editorial: cancer therapy with natural products and medicinal plants. *Planta Medica*. doi: 10.1055/s-0030-1250062.

Farhadi, N., Babaei, K., Farsaraei, S., Moghaddam, M., and Pirbalouti, A.G. (2020). Changes in essential oil compositions, total phenol, flavonoids and antioxidant capacity of *Achillea millefolium*at different growth stages. *Industrial Crops and Products* 152 (January). Elsevier: 112570. doi: 10.1016/j.indcrop.2020.112570.

Ferrero-Miliani, L., Nielsen, O.H., Andersen, P.S., and Girardin, S.E. (2007). Chronic inflammation: importance of NOD2 and NALP3 in interleukin-1β generation. *Clinical and Experimental Immunology* 147: 227–235. doi: 10.1111/j.1365-2249.2006.03261.x.

Flavor and Extract Manufacturers' Association. (1991). D-Limonene Monograph, 1-4. Washington, DC: Flavor and Extract Manufacturers' Association.

Francenia Santos-Sánchez, N., Salas-Coronado, R., Villanueva-Cañongo, C., and Hernández-Carlos, B. (2019). Antioxidant compounds and their antioxidant mechanism. In: Antioxidants. IntechOpen. doi: 10.5772/intechopen.85270.

Garcia, D.G., Castro-Faria-Neto, H.C. de, Silva, C.I. da, Souza e Souza, K.F.C. de, Gonçalves-de-Albuquerque, C.F., Silva, A.R., Amorim, L.M. da F. de, Freire, A.S., Santelli, R.E., Diniz, L.P., Gomes, F.C.A., Faria, M.V. de C., and Burth, P. (2015). Na/K-ATPase as a target for anticancer drugs: studies with perillyl alcohol. *Molecular Cancer* 14: 105. doi: 10.1186/s12943-015-0374-5.

González-Mas, M.C., Rambla, J.L., López-Gresa, M.P., Blázquez, M.A., and Granell, A. (2019). Volatile compounds in citrus essential oils: a comprehensive review. *Frontiers in Plant Science* 10: 12. doi: 10.3389/fpls.2019.00012.

González-Ramírez, A.E., González-Trujano, M.E., Orozco-Suárez, S.A., Alvarado-Vásquez, N., and López-Muñoz, F.J. (2016). Nerol alleviates pathologic markers in the oxazolone-induced colitis model. *European Journal of Pharmacology* 776: 81–89. doi: 10.1016/j.ejphar.2016.02.036.

Gulcin, İ. (2020). Antioxidants and antioxidant methods: an updated overview. *Archives of Toxicology* 94 (3): 651–715. doi: 10.1007/s00204-020-02689-3.

Gupta, B., Ghosh, K.K., and Gupta, R.C. (2016). Chapter 39 – thymoquinone. In: Nutraceuticals (ed. R.C. Gupta), 541–550. Boston, MA: Academic Press. doi: 10.1016/B978-0-12-802147-7.00039-5.

Habtemariam, S. (2017). Antidiabetic potential of monoterpenes: a case of small molecules punching above their weight. *International Journal of Molecular Sciences* 19: 4. doi: 10.3390/ijms19010004.

Hafidh, R.R., Hussein, S.Z., MalAllah, M.Q., Abdulamir, A.S., and Abu Bakar, F. (2018). A high-throughput quantitative expression analysis of cancer-related genes in human HepG2 cells in response to limonene, a potential anticancer agent. *Current Cancer Drug Targets* 18 (8): 807–815. doi: 10.2174/1568009617666171114144236.

Hassan, T., Elanchezhiyan, C., Naseer, I., and Veerakumar, D. (2019). Protective effect of perillyl alcohol (POH) a monoterpene: on high fat diet induced hyperlipidemia in albino Wistar rats a preliminary study. *International Journal of Pharmaceutical Sciences and Research* 10 (7): 3395–3398. doi: 10.13040/IJPSR.0975-8232.10(7).3395-98.

Hoseini, S.M., Mirghaed, A.T., Paray, B.A., Hoseinifar, S.H., and Van Doan, H. (2020). Effects of dietary menthol on growth performance and antioxidant, immunological and biochemical responses of rainbow trout (*Oncorhynchus mykiss*). *Aquaculture* 524 (July): 735260. doi: 10.1016/j.aquaculture.2020.735260.

Iraji, A., Yazdanpanah, S., Alizadeh, F., Mirzamohammadi, S., Ghasemi, Y., Pakshir, K., Yang, Y., and Zomorodian, K. (2020). Screening the antifungal activities of monoterpenes and their isomers against Candida species. *Journal of Applied Microbiology* 129: 1541–1551. doi: 10.1111/jam.14740.

Joglekar, M.M., Panaskar, S.N., and Arvindekar, A.U. (2014). Inhibition of advanced glycation end product formation by cymene – a common food constituent. *Journal of Functional Foods* 6: 107–115. doi: 10.1016/j.jff.2013.09.024.

Juergens, U.R., Dethlefsen, U., Steinkamp, G., Gillissen, A., Repges, R., and Vetter, H. (2003). Anti-inflammatory activity of 1.8-cineol (eucalyptol) in bronchial asthma: a double-blind placebo-controlled trial. *Respiratory Medicine* 97: 250–256. doi: 10.1053/rmed.2003.1432.

Kong, P., Chi, R., Zhang, L., Wang, N., and Lu, Y. (2013). Effects of paeoniflorin on tumor necrosis factor-α-induced insulin resistance and changes of adipokines in 3T3-L1 adipocytes. *Fitoterapia* 91: 44–50. doi: 10.1016/j.fitote.2013.08.010.

Krishnan, P., Sundaram, J., Salam, S., Subramaniam, N., Mari, A., Balaraman, G., and Thiruvengadam, D. (2020). Citral inhibits N-nitrosodiethylamine-induced hepatocellular carcinoma via modulation of antioxidants and xenobiotic-metabolizing enzymes. *Environmental Toxicology* 35 (9): 971–981. doi: 10.1002/tox.22933.

Leporini, M., Bonesi, M., Loizzo, M.R., Passalacqua, N.G., and Tundis, R. (2020). The essential oil of *Salvia rosmarinus* Spenn. from Italy as a source of health-promoting compounds: chemical profile and antioxidant and cholinesterase inhibitory activity. *Plants* 9 (6): 798. doi: 10.3390/plants9060798.

Li, L., He, L., Wu, Y., and Zhang, Y. (2021). Carvacrol affects breast cancer cells through TRPM7 mediated cell cycle regulation. *Life Sciences* 266: 118894. doi: 10.1016/j.lfs.2020.118894.

Lu, J.-X., Guo, C., Ou, W.-S., Jing, Y., Niu, H.-F., Song, P., Li, Q.-Z. et al. (2019). Citronellal prevents endothelial dysfunction and atherosclerosis in rats. *Journal of Cellular Biochemistry* 120 (3): 3790–3800. doi: 10.1002/jcb.27660.

Mączka, W., Wińska, K., and Grabarczyk, M. (2020). One hundred faces of geraniol. *Molecules* 25. doi: 10.3390/molecules25143303.

Madan, K. and Nanda, S. (2018). *In-vitro* evaluation of antioxidant, anti-elastase, anti-collagenase, anti-hyaluronidase activities of safranal and determination of its sun protection factor in skin photoaging. *Bioorganic Chemistry* 77: (April): 159–167. doi: 10.1016/j.bioorg.2017.12.030.

Madhuri, K. and Naik, P.R. (2017). Ameliorative effect of borneol, a natural bicyclic monoterpene against hyperglycemia, hyperlipidemia and oxidative stress in streptozotocin-induced diabetic Wistar rats. *Biomedicine & Pharmacotherapy* 96: 336–347. doi: 10.1016/j.biopha.2017.09.122.

Marson, B.M., Picheth, G., Kuczera Pereira, T., Kita, D.H., Valdameri, G., Rocha Alencar Menezes, L., Gioppo Nunes, G., Alves de Freitas, R., and Pontarolo, R. (2019). Effect of different tensoactives on the morphology and release kinetics of PLA-b-PEG microcapsules loaded with the natural anticancer compound perillyl alcohol. *Journal of Pharmaceutical Sciences* 108: 860–869. doi: 10.1016/j.xphs.2018.09.003.

Menezes, I.A.C., Barreto, C.M.N., Antoniolli, Â.R., Santos, M.R.V., and de Sousa, D.P. (2010). Hypotensive activity of terpenes found in essential oils. *Zeitschrift Für Naturforschung C* 65 (9–10): 562–566. doi: 10.1515/znc-2010-9-1005.

Mezza, G.N., Borgarello, A.V., Grosso, N.R., Fernandez, H., Pramparo, M.C., and Gayol, M.F. (2018). Antioxidant activity of rosemary essential oil fractions obtained by molecular distillation and their effect on oxidative stability of sunflower oil. *Food Chemistry* 242 (March): 9–15. doi: 10.1016/j.foodchem.2017.09.042.

Miller, J.A., Lang, J.E., Ley, M., Nagle, R., Hsu, C.-H., Thompson, P.A., and Chow, H.-H.S. (2013). Human breast tissue disposition and bioactivity of limonene in women with early-stage breast cancer. *Cancer Prevention Research* 6 (6): 577–584. doi: 10.1158/1940-6207.

Mukhtar, Y.M., Adu-Frimpong, M., Xu, X., and Yu, J. (2018). Biochemical significance of limonene and its metabolites: future prospects for designing and developing highly potent anticancer drugs. *Bioscience Reports* 38 (6): BSR20181253. doi: 10.1042/BSR20181253.

Murali, R. and Saravanan, R. (2012). Antidiabetic effect of d-limonene, a monoterpene in streptozotocin-induced diabetic rats. *Biomedicine & Preventive Nutrition* 2: 269–275. doi: 10.1016/j.bionut.2012.08.008.

Muruganathan, U. and Srinivasan, S. (2016). Beneficial effect of carvone, a dietary monoterpene ameliorates hyperglycemia by regulating the key enzymes activities of carbohydrate metabolism in streptozotocin-induced diabetic rats. *Biomedicine & Pharmacotherapy* 84: 1558–1567. doi: 10.1016/j.biopha.2016.11.025.

Nascimento, G.A. do, de Souza, S.D., Lima, B.S., de Vasconcelos, C.M.L., de Souza Araújo, A.A., Durço, A.O., Quintans-Junior, L.J. et al. (2019). Bradycardic and antiarrhythmic effects of the D-limonene in rats. *Arquivos Brasileiros de Cardiologia* 113 (5): 925–932. doi: 10.5935/abc.20190173.

Néri-Numa, I., Vespermann, K.A.C., Carvalho, C.H., Paulino, B.N., Macedo, M.J., Pastore, G.M., Bicas, J.L., and Molina, G. (2019). Bioactive potential of sesquiterpenes. *Food Aroma Evolution*: 695–718.

Patrignani, F., Prasad, S., Novakovic, M., Marin, P.D., and Bukvicki, D. (2021). Lamiaceae in the treatment of cardiovascular diseases. *Frontiers in Bioscience (Landmark Edition)* 26 (February): 612–643. http://www.ncbi.nlm.nih.gov/pubmed/33049685.

Ranjbaran, A., Kavoosi, G., Mojallal-Tabatabaei, Z., and Ardestani, S.K. (2019). The antioxidant activity of *Trachyspermum ammi* essential oil and thymol in murine macrophages. *Biocatalysis and Agricultural Biotechnology* 20 (July): 101220. doi: 10.1016/j.bcab.2019.101220.

Rathinam, A. and Pari, L. (2016). Myrtenal ameliorates hyperglycemia by enhancing GLUT2 through Akt in the skeletal muscle and liver of diabetic rats. *Chemico-Biological Interactions* 256: 161–166. doi: 10.1016/j.cbi.2016.07.009.

Ravichandran, C., Badgujar, P.C., Gundev, P., and Upadhyay, A. (2018). Review of toxicological assessment of d-limonene, a food and cosmetics additive. *Food and Chemical Toxicology* 120: 668–680. doi: 10.1016/j.fct.2018.07.052.

Riella, K.R., Marinho, R.R., Santos, J.S., Pereira-Filho, R.N., Cardoso, J.C., Albuquerque-Junior, R.L.C., and Thomazzi, S.M. (2012). Anti-inflammatory and cicatrizing activities of thymol, a monoterpene of the essential oil from Lippia gracilis, in rodents. *Journal of Ethnopharmacology* 143: 656–663. doi: 10.1016/j.jep.2012.07.028.

Rinaldi, G.J., Rojano, B., Schinella, G., and Mosca, S.M. (2019). Participation of NO in the vasodilatory action of isoespintanol. *Vitae* 26 (2): 78–83. doi: 10.17533/udea.vitae.v26n2a03.

Rodenak-Kladniew, B., Castro, A., Stärkel, P., Galle, M., and Crespo, R. (2020a). 1,8-Cineole promotes G0/G1 cell cycle arrest and oxidative stress-induced senescence in HepG2 cells and sensitizes cells to anti-senescence drugs. *Life Sciences* 243: 117271. doi: 10.1016/j.lfs.2020.117271.

Rodenak-Kladniew, B., Castro, M.A., Crespo, R., Galle, M., and García de Bravo, M. (2020b). Anti-cancer mechanisms of linalool and 1,8-cineole in non-small cell lung cancer A549 cells. *Heliyon* 6: e05639. doi: 10.1016/j.heliyon.2020.e05639.

Saljoughian, S., Roohinejad, S., Bekhit, A.E.-D.A., Greiner, R., Omidizadeh, A., Nikmaram, N., and Mousavi Khaneghah, A. (2018). The effects of food essential oils on cardiovascular diseases: a review. *Critical Reviews in Food Science and Nutrition* 58 (10): 1688–1705. doi: 10.1080/10408398.2017.1279121.

Santos, M.R.V., Moreira, F.V., Fraga, B.P., de Sousa, D.P., Bonjardim, L.R., and Junior Quintans, L. (2011). Cardiovascular effects of monoterpenes: a review. *Brazilian Journal of Pharmacognosy* 21 (4): 764–771. doi: 10.1590/S0102-695X2011005000119.

Santos, S.E., Ribeiro, F.P.R.A., Menezes, P.M.N., Duarte-Filho, L.A.M., Quintans, J.S.S., Quintans-Junior, L.J., Silva, F.S., and Ribeiro, L.A.A. (2019). New insights on relaxant effects of (-)-borneol monoterpene in rat aortic rings. *Fundamental & Clinical Pharmacology* 33 (2): 148–158. doi: 10.1111/fcp.12417.

Saravanan, S. and Pari, L. (2016). Protective effect of thymol on high fat diet induced diabetic nephropathy in C57BL/6J mice. *Chemico-Biological Interactions* 245: 1–11. doi: 10.1016/j.cbi.2015.11.033.

Scherer, M.M. de C., Marques, F.M., Figueira, M.M., Peisino, M.C.O., Schmitt, E.F.P., Kondratyuk, T.P., Endringer, D.C., Scherer, R., and Fronza, M. (2019). Wound healing activity of terpinolene and α-phellandrene by attenuating inflammation and oxidative stress *in vitro*. *Journal of Tissue Viability* 28 (2): 94–99. doi: 10.1016/j.jtv.2019.02.003.

Sharifi-Rad, M., Varoni, E.M., Iriti, M., Martorell, M., Setzer, W.N., del Mar Contreras, M., Salehi, B., Soltani-Nejad, A., Rajabi, S., Tajbakhsh, M., and Sharifi-Rad, J. (2018). Carvacrol and human health: a comprehensive review. *Phytotherapy Research* 32: 1675–1687. doi: 10.1002/ptr.6103.

Shojaei, S., Kiumarsi, A., Moghadam, A.R., Alizadeh, J., Marzban, H., and Ghavami, S. (2014). Chapter two – perillyl alcohol (monoterpene alcohol), limonene. In: The Enzymes, (ed. S.Z. Bathaie and F. Tamanoi), 7–32. Academic Press. doi: 10.1016/B978-0-12-802215-3.00002-1.

Silva, A.F.C., Haris, P.I., Serralheiro, M.L., and Pacheco, R. (2020). Mechanism of action and the biological activities of *Nigella sativa* oil components. *Food Bioscience* 38 (December): 100783. doi: 10.1016/j.fbio.2020.100783.

Silva, E.A.P., Carvalho, J.S., Guimarães, A.G., Barreto, R. de S.S., Santos, M.R.V., Barreto, A.S., and Quintans-Júnior, L.J. (2019). The use of terpenes and derivatives as a new perspective for cardiovascular disease treatment: a patent review (2008–2018). *Expert Opinion on Therapeutic Patents* 29 (1): 43–53. doi: 10.1080/13543776.2019.1558211.

Silva, R.E.R. da, de Alencar Silva, A., Pereira-de-Morais, L., de Sousa Almeida, N., Iriti, M., Kerntopf, M.R., Menezes, I.R.A. de, Coutinho, H.D.M., and Barbosa, R. (2020). Relaxant effect of monoterpene (−)-carveol on isolated human umbilical cord arteries and the involvement of ion channels. *Molecules* 25 (11): 2681. doi: 10.3390/molecules25112681.

Silvestre, A.J.D. and Gandini, A. (2008). Chapter 2 – terpenes: major sources, properties and applications. In: Monomers, Polymers and Composites from Renewable Resources (ed. M.N. Belgacem and A. Gandini), 17–38. Amsterdam: Elsevier. doi: 10.1016/B978-0-08-045316-3.00002-8.

Simirgiotis, M.J., Burton, D., Parra, F., López, J., Muñoz, P., Escobar, H., and Parra, C. (2020). Antioxidant and antibacterial capacities of *origanum vulgare* l. Essential oil from the arid andean region of chile and its chemical characterization by gc-ms. *Metabolites* 10 (10): 1–12. doi: 10.3390/metabo10100414.

Sobral, M.V., Xavier, A.L., Lima, T.C., and de Sousa, D.P. (2014). Antitumor activity of monoterpenes found in essential oils. *The Scientific World Journal* 2014: 953451. doi: 10.1155/2014/953451.

Soubh, A.A., Abdallah, D.M., and El-Abhar, H.S. (2015). Geraniol ameliorates TNBS-induced colitis: involvement of Wnt/β-catenin, p38MAPK, NFκB, and PPARγ signaling pathways. *Life Sciences* 136: 142–150. doi: 10.1016/j.lfs.2015.07.004.

Sousa, C., Leitão, A.J., Neves, B.M., Judas, F., Cavaleiro, C., and Mendes, A.F. (2020). Standardised comparison of limonene-derived monoterpenes identifies structural determinants of anti-inflammatory activity. *Scientific Reports* 10: 7199. doi: 10.1038/s41598-020-64032-1.

Souza, C.F., Baldissera, M.D., Silva, L. de L., Geihs, M.A., and Baldisserotto, B. (2018). Is monoterpene terpinen-4-ol the compound responsible for the anesthetic and antioxidant activity of *Melaleuca alternifolia* essential oil (tea tree oil) in silver catfish? *Aquaculture* 486 (February): 217–223. doi: 10.1016/j.aquaculture.2017.12.025.

Srinivasan, S. and Muruganathan, U. (2016). Antidiabetic efficacy of citronellol, a citrus monoterpene by ameliorating the hepatic key enzymes of carbohydrate metabolism in streptozotocin-induced diabetic rats. *Chemico-Biological Interactions* 250: 38–46. doi: 10.1016/j.cbi.2016.02.020.

Stark, M.J., Burke, Y.D., McKinzie, J.H., Ayoubi, A.S., and Crowell, P.L. (1995). Chemotherapy of pancreatic cancer with the monoterpene perillyl alcohol. *Cancer Letters* 96: 15–21. doi: 10.1016/0304-3835(95)03912-G.

Stein, S.A., Lamos, E.M., and Davis, S.N. (2013). A review of the efficacy and safety of oral antidiabetic drugs. *Expert Opinion on Drug Safety* 12 (2): 153–175. doi: 10.1517/14740338.2013.752813.

Tan, X.C., Chua, K.H., Ravishankar Ram, M., and Kuppusamy, U.R. (2016). Monoterpenes: novel insights into their biological effects and roles on glucose uptake and lipid metabolism in 3T3-L1 adipocytes. *Food Chemistry* 196: 242–250. doi: 10.1016/j.foodchem.2015.09.042.

Tchimene, M.K., Okunji, C.O., Iwu, M.M., and Kuete, V. (2013). 1 – monoterpenes and related compounds from the medicinal plants of Africa. In: Medicinal Plant Research in Africa (ed. V. Kuete), 1–32. Oxford: Elsevier. doi: 10.1016/B978-0-12-405927-6.00001-1.

Upadhyay, J., Polyzos, S.A., Perakakis, N., Thakkar, B., Paschou, S.A., Katsiki, N., Underwood, P., Park, K.-H., Seufert, J., Kang, E.S., Sternthal, E., Karagiannis, A., and Mantzoros, C.S. (2018). Pharmacotherapy of type 2 diabetes: an update. *Metabolism* 78: 13–42. doi: 10.1016/j.metabol.2017.08.010.

Vallianou, I. and Hadzopoulou-Cladaras, M. (2016). Camphene, a plant derived monoterpene, exerts its hypolipidemic action by affecting SREBP-1 and MTP expression. *PLoS ONE* 11 (1): 1–21. doi: 10.1371/journal.pone.0147117.

Vieira, A.R., Abar, L., Vingeliene, S., Chan, D.S.M., Aune, D., Navarro-Rosenblatt, D., Stevens, C., Greenwood, D., and Norat, T. (2016). Fruits, vegetables and lung cancer risk: a systematic review and meta-analysis. *Annals of Oncology* 27: 81–96. doi: 10.1093/annonc/mdv381.

Wagner, K.-H. and Elmadfa, I. (2003). Overview focusing on mono-, di- and tetraterpenes. *Annals of Nutrition & Metabolism* 47: 95–106.

Wang, C.-Y., Chen, Y.-W., and Hou, C.-Y. (2019). Antioxidant and antibacterial activity of seven predominant terpenoids. *International Journal of Food Properties* 22 (1): 230–238. doi: 10.1080/10942912.2019.1582541.

Wang, J., Gao, J., Xu, H., Qian, Y., Xie, L., Yu, H., and Qian, B. (2021). Citrus fruit intake and lung cancer risk: a meta-analysis of observational studies. *Pharmacological Research* 166: 105430. doi: 10.1016/j.phrs.2021.105430.

World Health Organization (2017). Cardiovascular diseases (CVDs) – key facts. https://www.who.int/news-room/fact-sheets/detail/cardiovascular-diseases-(cvds).

World Health Organization (2019). Hypertension – key facts. https://www.who.int/news-room/fact-sheets/detail/hypertension.

World Health Organization (2020a). Cardiovascular diseases (CVDs) – treatment. https://www.who.int/health-topics/cardiovascular-diseases/#tab=tab_1.

World Health Organization (2020b). Leading causes of death and disability – visual summary of global and regional trends 2000–2019. https://www.who.int/data/stories/leading-causes-of-death-and-disability-2000-2019-a-visual-summary.

World Health Organization (2021a). Cancer. https://www.who.int/news-room/fact-sheets/detail/cancer

World Health Organization (2021b). Diabetes. https://www.who.int/news-room/factsheets/detail/diabetes

Ye, Z., Liang, Z., Mi, Q., and Guo, Y. (2020). Limonene terpenoid obstructs human bladder cancer cell (T24 cell line) growth by inducing cellular apoptosis, caspase activation, G2/M phase cell cycle arrest and stops cancer metastasis. *Journal of the Balkan Union of Oncology* 25: 280–285.

Yu, X., Lin, H., Wang, Y., Lv, W., Zhang, S., Qian, Y., Deng, X., Feng, N., Yu, H., and Qian, B. (2018). D-limonene exhibits antitumor activity by inducing autophagy and apoptosis in lung cancer. *Onco Targets and therapy* 11: 1833–1847. doi: 10.2147/OTT.S155716.

Zhang, B., Wang, H., Yang, Z., Cao, M., Wang, K., Wang, G., and Zhao, Y. (2020). Protective effect of alpha-pinene against isoproterenol-induced myocardial infarction through NF-KB signaling pathway. *Human and Experimental Toxicology* 39 (12): 1596–1606. doi: 10.1177/0960327120934537.

Zhang, L., Yang, B., and Yu, B. (2015). Paeoniflorin protects against nonalcoholic fatty liver disease induced by a high-fat diet in mice. *Biological and Pharmaceutical Bulletin* 38: 1005–1011. doi: 10.1248/bpb.b14-00892.

Zhang, Y., Ding, Y., Zhong, X., Guo, Q., Wang, H., Gao, J., Bai, T., Ren, L., Guo, Y., Jiao, X., and Liu, Y. (2016). Geniposide acutely stimulates insulin secretion in pancreatic β-cells by regulating GLP-1 receptor/cAMP signaling and ion channels. *Molecular and Cellular Endocrinology* 430: 89–96. doi: 10.1016/j.mce.2016.04.020.

Zhou, F. and Pichersky, E. (2020). More is better: the diversity of terpene metabolism in plants. *Current Opinion in Plant Biology* 55: 1–10. doi: 10.1016/j.pbi.2020.01.005.

14

Biomolecules Derived from Whey

Strategies for Production and Biological Properties

M. C. Perotti, C. I. Vénica, I. V. Wolf, M. A. Vélez, G. H. Peralta, A. Quiberoni, and C. V. Bergamini*

Instituto de Lactología Industrial (INLAIN), Universidad Nacional del Litoral (UNL), Consejo Nacional de Investigaciones Científicas y Técnicas (CONICET), Facultad de Ingeniería Química, Santiago del Estero 2829, 3000 Santa Fe, Argentina
**Corresponding authors: cperotti@fiq.unl.edu.ar*

14.1 Introduction

The dairy industry is an indispensable part of the food industry; it plays an important role in the global economy as well as in human nutrition.

According to the Food and Agriculture Organization (FAO) of the United Nations, world milk production has continued to rise until it reached an estimated level of 860 million tons in 2020, despite market disruptions caused by the COVID-19 pandemic (www.fao.org/3/cb1993en/cb1993en.pdf). The dairy industry processes raw milk into numerous dairy products with the subsequent generation of different by-products; whey is the most abundant by-product (Jelen 2011).

Before the 1970s, whey was a polluting waste product because it has a lot of organic matter resulting in high values of biological and chemical oxygen demand (BOD and COD, > 30 and > 60 g O_2 L^{-1}, respectively). It was drained into water courses if not used as animal feed or applied to fields as liquid fertilizer. The great expansion and industrialization that the dairy industry experienced in the twentieth century, together with the stringent environmental regulations implemented worldwide, lobbied industry to stop dumping whey into streams and municipal sewage systems. These legislative restrictions on whey disposal prompted the search for ways to take advantage of the whey, encouraging a deeper exploration of its composition in order to make profit from it valuable constituents for different purposes (Carvalho et al. 2013; Dullius et al. 2018; Jelen 2011; Prazeres et al. 2012; Smithers 2008, 2015). So, the transformation of whey into valuable products constitutes a logical use to return whey to the human food chain in the light of global food shortage (Mann et al. 2019).

Whey represents about 85–95% of the milk volume processed from different dairy products (cheese, casein, cream, etc.) and retains 55% of the milk nutrients, including lactose, whey proteins, water-soluble vitamins, and mineral salts. Thus, whey is one of the biggest reservoirs of food protein available today with elevated nutritional value, and is a very important source of carbohydrate for the world. Worldwide whey production is estimated

Biomolecules from Natural Sources: Advances and Applications, First Edition. Edited by Vijai Kumar Gupta, Satyajit D. Sarker, Minaxi Sharma, María Elida Pirovani, Zeba Usmani, and Chelliah Jayabaskaran.

at around 190 million tons/year (half of which is produced in the United States), which represents approximately 1.5 million tons of increasingly high-value protein and 8.6 million tons of lactose. This situation has created opportunities for the better management of whey, which has led to the development of modern technologies like centrifugation, membrane filtration, spray drying, chromatography, fermentation, among others. A variety of valuable products mainly derived from bovine milk whey have been obtained and commercialized: whey protein concentrates and isolates (WPC and WPI, respectively), whey protein fractions (β-lactoglobulin – β-LG -, α-lactalbumin –α-LA -, casein glycomacropeptide – GMP -, lactoferrin, and lactoperoxidase), protein hydrolysates (WPH) and peptides, and non-protein products such as oligosaccharides, some of which are highly accepted as ingredients in food, pharmaceuticals, and other applications (Abd El-Salam et al. 2009; Jelen 2011; Tetra Pak 2015; Prazeres et al. 2012; Singh 2014). On the other hand, commercialized products from non-bovine whey are not common in the market and information describing such products is limited.

In this chapter, we explore the physicochemical composition of whey from bovine milk, the most representative technologies used for the production of added-value products and bioactive compounds or biomolecules from industrialization of whey such as whey protein fractions, peptides, and oligosaccharides, beneficial properties and food applications of these products. In addition, the use of whey as the encapsulating material of bioactive compound-loaded liposomes, and the presence of bacteriophages in whey derived products, are also discussed.

14.2 Physicochemical Composition of Whey

There are two types of whey depending on the way it is obtained. *Acid whey* (AW) is produced from the manufacture of casein (or cottage cheese); milk is coagulated by direct addition of acid (organic or mineral) and the casein precipitates at its isoelectric point (pH 4.6 at room temperature). *Sweet whey* (SW) (or rennet whey) is obtained from the manufacture of cheese in which rennet treatment is used for milk coagulation. The production of SW accounts for 95% of the total whey production (Jelen 2011). The typical composition of AW and SW is shown in Table 14.1.

The main components of both whey types after water (93–95%) are lactose (β-D-Gal-($1\rightarrow4$)-D-Glu) (approximately 70% of the total solids), whey proteins (8–10%), and minerals (12–15%). The major protein fractions are β-LG, α-LA, lactoferrin and lactoperoxidase, bovine serum albumin (BSA), and immunoglobulins. Minor differences in the mineral content, acidity, and composition of the protein fraction between both types of whey are mentioned; however, they have a deep effect on the functional and nutritional properties of whey. In particular, AW has a higher ash and lower protein content than SW; demineralization of casein micelles during acid milk coagulation produces higher mineral content in AW and a substantially increased acidity.

During rennet clotting of milk a fragment of the κ-casein is released, termed glycomacropeptide (GMP), which ends up in the cheese whey. Thus, GMP constitutes approximately 20% of the SW protein fraction, but it is not found in AW (Abd El-Salam et al. 2009; Anand et al. 2013; Jelen 2011).

Table 14.1 Physicochemical composition of whey (%, w/v).

Parameters	Sweet whey (SW)	Acid whey (AW)
pH	5.8–6.6	3.9–4.6
Total solids	6.3–7.0	6.3–7.0
Protein	0.6–1.0	0.5–0.8
Lactose	4.60–5.20	4.40–4.60
Fat	0.50	0.04
Ash	0.5	0.8
Titratable acidity	0.05–0.20	0.40–0.60

14.3 Processing of Whey and Derived Products

The production and utilization of whey derived products has experienced growth, which has been stimulated by an increase in the production of cheese around the world (www.fao.org/3/ca8861en/Dairy.pdf; Prazeres et al. 2012). Also, the advent of powerful industrial technologies for the separation or isolation of lactose, proteins and other components has driven the dairy industry to produce products of tailored composition and functionality.

The arrival of membrane filtration technology allowed the separation and fractionation of whey components; in general, this is followed by spray drying in order to obtain a powder product (< 5% moisture). It is a molecular sieving technique that employs a 150 μm thick semipermeable surface supported by a more porous layer of similar material on a reinforcing base. Permeate (soluble compounds of low molecular weight) flows through ther membrane, while passage of the retentate (other materials) is blocked (Carter et al. 2018; Henning et al. 2006; Kumar et al. 2013). Five types of membrane filtration: microfiltration (MF), ultrafiltration (UF), nanofiltration (NF), electrodialysis (ED), and reverse osmosis (RO), are used (Tunick 2008).

Typical operations for whey powder production consist of evaporation in multistage vacuum evaporators, precrystallization of lactose to minimize the problem of hygroscopicity, followed by spray drying (Jelen 2011). Demineralization of whey is an important pretreatment process achieved by ED, ion exchange resins and NF, or their combinations, to make a suitable ingredient for certain foods (Gernigon et al. 2011; Henning et al. 2006). For infant formula, a reduction of 90–95% of minerals is needed in order to mimic the mineral composition of human milk. In ice cream production, a reduction of 50–70% of minerals is enough (Gernigon et al. 2011; Jelen 2011).

Different protein fractions can be removed selectively or totally from whey and concentrated by the membrane. The concentration and state of proteins are important factors that differentiate these products. In the case of protein concentrates (WPC), the total protein content varies between 34% and 80% resulting in different products: WPC34, WPC50, WPC60, WPC75, and WPC80, which are obtained by UF/diafiltration (DF), evaporation and spray drying (Abd El-Salam et al. 2009); WPC34 and

WPC80 are the most common and widely used products (Anand et al. 2013). WPI is the highest protein-containing whey product without lactose; it is manufactured by DF, ion exchange, NF, or their combinations, concentrated and spray dried (Abd El-Salam et al. 2009; Jelen 2011; Tunick 2008). Both spray drying and especially evaporation can cause denaturation of globular proteins resulting in loss of solubility and other functionality defects, so these operations should be controlled carefully (Abd El-Salam et al. 2009; Carter et al. 2018). Methods for obtaining fractions rich in β-LG and α-LA are based on differential solubility at different pH, temperature, and ionic strength. UF and chromatographic techniques are the most widespread methods used to separate GMP from SW (Abd El-Salam et al. 2009). On the other hand, hydrolysis processes can be applied to modify functional properties and improve biological properties of whey proteins (Abd El-Salam and El-Shibiny 2018). However, whey proteins are not easily broken down with proteases because of their globular structure. Some researchers have enhanced the enzymatic hydrolysis of whey proteins applying pretreatments with advanced technologies (such as sonication and ohmic treatment, among others). Also, these treatments increase the biological properties and functionality of hydrolysates (Alizadeh and Aliakbarlu 2020; Wu et al. 2018). As well, it is essential to carefully control the hydrolysis process to avoid the production of bitter-tasting peptides which may limit the application of whey protein hydrolysates in foods (Le Maux et al. 2016).

UF permeate from whey is rich in lactose and minerals but almost devoid of protein. It can be dried by spray drying to obtain powder lactose. However, this product has low commercial value and, during storage, has problems with loss of product by crystallization, as part of the amorphous lactose becomes crystalline lactose. In addition, lactose is little used in the food industry because of its low sweetening power and low solubility. Nonetheless, lactose can be used for the production of added value ingredients. Lactose is the precursor for a number of bioactive compounds or biomolecules produced by chemical, physical, or enzymatic conversion. These derived products have an established and expanding place in food and pharmaceutical industries, such as galactooligosaccharides (GOS), lactulose, lactitol, lactobionic acid, lactosucrose, among others (Ganzle 2011; Singh 2014; Vera et al. 2021).

The diagram in Figure 14.1 summarizes some of the processes for converting whey in derived products. Table 14.2 shows the typical chemical composition of some dried products derived from whey (Tetra Pak 2015).

14.4 Functional and Nutritional Aspects and Applications of Whey Derived Products

Whey derived products enriched in proteins are widely used in many food (baked, dairy and meat products, and sports drinks) and pharmaceutical applications because of their broad range of functionality, nutritive value, and health benefits (Henning et al. 2006; Kumar et al. 2013; Soares de Castro et al. 2017). In addition, the demand for whey proteins has been boosted by the consumer request for high-protein foods and supplements.

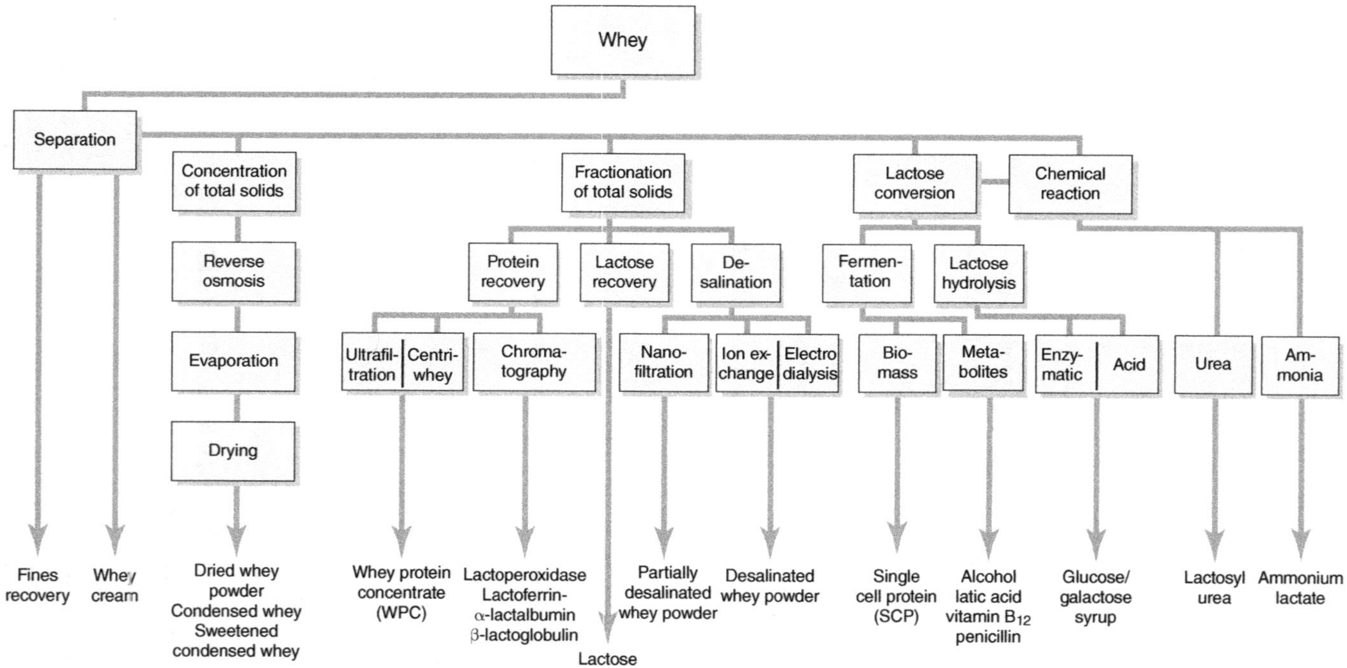

Figure 14.1 Processing of whey and derived products. (from *Handbook of Dairy Processing*, Tetra Pak 2015).

The functionality of food proteins refers to their physicochemical properties; in general, these properties are classified as hydrodynamic or hydration-related properties (water absorption, solubility, viscosity, and gelation) and surface-active properties (emulsification, foaming, and film formation) (Soares de Castro et al. 2015).

Whey protein foams well, remains soluble at a wide range of pH (from 2 to 10), binds large amounts of water, and allows the formation of a three-dimensional network due to gelation property (Abd El-Salam et al. 2009). The nutritional value of the whey proteins is superior to casein on the basis of essential and sulfur-containing amino acids (Anand et al. 2013; Madureira et al. 2007). In particular, β-LG is a rich source of cysteine and branched-chain amino acids (BCAA: leucine, isoleucine, valine). It is denatured into gel, a property that has been exploited in meat, fish, beverages, and formulated foods (Anand et al. 2013; Tunick 2008). α-LA is rich in cysteine, methionine and BCAA, and is a calcium-binding protein. A purified bovine α-LA is widely used in infant formula, due to its similarity to human α-LA. It is also added to sports drinks as BCAA is used by muscles for energy and protein synthesis (Anand et al. 2013; Geng et al. 2015; Marella et al. 2011). Lactoferrin and lactoperoxidase are minor protein fractions but commercially important. Lactoferrin is an iron-binding glycoprotein. It has antibacterial, antiviral, and antioxidant properties, and modulates iron metabolism and immune functions. Lactoperoxidase has antimicrobial properties (Anand et al. 2013; Tunick 2008).

Enzymatic hydrolysis of whey proteins and formation of carbohydrate-whey protein conjugates (glycation) are well-known methods for increasing their aggregate value. Functional properties (solubility, viscosity, thermal stability, and emulsifying and foaming properties) and more importantly, biological properties are modified and improved (Abd El-Salam and El-Shibiny 2018; Soares de Castro et al. 2017). A wide variety of applications are recognized for whey protein hydrolyzed (WPH) as they provide a number of benefits compared to products containing unhydrolyzed protein. WPH is a source of bioactive peptides that are inactive in the intact protein but become active after hydrolysis. Depending on the amino acid content, sequences, and structures, the peptides generated

Table 14.2 Typical chemical composition of powder products derived from whey (%, w/w). (Based on Tetra Pak 2015.)

Product	Total protein	Lactose	Fat	Ash
Regular whey powder	11.0–14.0	73.5		8.5
Demineralized (70%) whey powder	11.0–15.0	70.0–83.0	0.5–1.8	3.5
Demineralized (90%) whey powder				1.0
Ultrafiltration permeate powder	1.0	90.0		9.0
WPC	34.0–82.0	4.0–55.0	1.0–10.0	2.5–8.0
WPI	80.0–92.0	< 1.0	0.2–1.0	2.0–4.2
α-LA rich product	78.0	1.5	12.0	3.0
β-LG rich product	76.3–88.0	0.1–13.4	0.2–10.5	3.0–1.2
GMP	83.0–87.0	< 2.0	< 1.0	7.0–9.0

can have different biological functions (Mann et al. 2019; Meisel 2005). The available information about the biological activities of hydrolysates/peptides derived from whey is abundant (Corrochano et al. 2018; Mann et al. 2019; Zhao and Ashaolu 2020). Commercial products with different degrees of hydrolysis (4–30%) and thus different peptide profiles can be obtained (Abd El-Salam et al. 2009; Gauthier and Pouliot 2003). Generally, hydrolysates designed for sport drinks, nutritional bars, and hypoallergenic infant formulas have a moderate–high degree of hydrolysis. Other applications are beverages, yogurts and smoothies, pet food, and weight management products (Pihlanto-Lepälä 2001; Pinto Coelho Silvestre et al. 2012).

On the other hand, one of the currently active research topics for commercial exploitation is the conversion of the lactose present in whey to biomolecules or bioactive compounds. Among them, oligosaccharides of galactose or GOS are an important group of biomolecules that have gained notoriety from the knowledge of their beneficial effects on health, mainly in a prebiotic role. They are obtained at industrial scale to be used as functional ingredients for multiple applications. Lactosucrose is another biomolecule that is under study as an "emerging prebiotic" and with promising potential applications (Ganzle 2011; Hanau et al. 2020; Kruschitza and Nidetzkya 2020).

Specifically, below, we will focus on the structure, characteristics, technology of production, biological properties, and applications of some peptides such as antihypertensive, antimicrobial and antioxidant, GMP, GOS and lactosucrose.

14.5 Bioactive Peptides Derived from Whey Protein

As mentioned above, the high nutritional value of whey, in addition to lactose, minerals and vitamins, are related to proteins (Anand et al. 2013). It is well documented that whey proteins are a source of essential amino acids, but it has also been recognized that these proteins exert many other functionalities through the formation of biologically active peptides. These peptides have been defined by Kitts and Weiler (2003) as specific protein fragments that have a positive impact on body functions or conditions and may finally promote health. Such peptides are inactive within the sequence of the precursor protein and can be released *in vivo* by digestive enzymes during gastrointestinal transit, by fermentation with starter culture during manufacture of fermented product like yogurt or cheese (*in situ* production) or by hydrolysis with specific enzymes/microorganisms on specific substrate (Gobbetti et al. 2002; Korhonen 2009; Korhonen and Pihlanto 2006; Mann et al. 2019; Moughan et al. 2014). Biological activities ascribed to peptides including antimicrobial, angiotensin-converting enzyme (ACE inhibitory), antioxidant, antithrombotic, cholesterol-lowering ability, enhancement of mineral absorption/bioavailability, immunomodulatory, and opioid. Moreover, some peptides are multifunctional and can exert more than one of these effects (Hartmann and Meisel 2007). The biological activities are largely influenced by the type, number, sequence, properties, and configuration of amino acids present in the peptide.

The potential of peptides to promote human health by reducing the risk of diseases or enhancing natural immune protection has attracted huge interest from researchers, manufacturers, and consumers alike over the past two decades (Agyie et al. 2016;

Hartmann and Meisel 2007). The discovery or production of peptides from whey owing health-promoting benefits to the immune, cardiovascular, nervous, endocrine and digestive systems, constitute an innovative approach which can provide added value to whey (Ballatore et al. 2020). In that sense, there is an increasing interest in incorporating these peptides as ingredients in functional foods (Dullius et al. 2018; Hartmann and Meisel 2007; Vavrusova et al. 2015). Dullius et al. (2018) summarized some commercially sold functional food ingredients based on whey derived bioactive peptides.

Although there are many laboratory-scale studies on the production of peptides, *in vitro* and *in vivo* evaluation of the bioactivity, the successful development large-scale production has its limitations either because it is difficult to scale up, high in cost, time consuming, or bioactive activities affected (Agyei et al. 2016).

14.5.1 Antihypertensive Peptides

Worldwide, a great percentage (approximately 40% in 2008) of people older than 25 years are affected by hypertension, which is one of the main causes of cardiovascular diseases (WHO 2013). Several regulators of blood pressure exist in the organism, but one of the main is the renin-angiotensin-aldosterone system, in which the angiotensin-converting enzyme (ACE) plays an essential role. ACE (peptidyl dipeptide hydrolase, EC 3.4.15.1) catalyzes the conversion of angiotensin I (produced from angiotensinogen by renin) to angiotensin II, a potent vasoconstrictor. In addition, ACE degrades a vasodilatador peptide: bradykinin, and stimulates the release of aldosterone in the adrenal cortex. So, ACE leads to a rise in blood pressure. Many of the pharmaceuticals used for the management and the reduction of blood pressure in hypertensive patients are ACE inhibitors; these compounds are thought to be competitive substrates for ACE (Hayes et al. 2007; Martin et al. 2015; Pihlanto-Leppälä 2000; WHO 2013). Food-derived peptides are natural sources of ACE inhibitors, which are more attractive alternatives than synthetic drugs because the latter can have side effects (Hussein et al. 2020). ACE-inhibitory peptides derived from whey proteins are called lactokinins (FitzGerald and Meisel 1999); they have been widely studied and the information has been reviewed by numerous researchers (Madureira et al. 2010; Pihlanto-Leppälä 2000; Pihlanto-Leppälä et al. 2000; Tavares and Malcata 2013; Yamamoto and Takano 1999).

Different treatments, individual or combined, have been efficient to produce bioactive peptides derived from whey protein with ACE-inhibitory activity. These peptides commonly contain up to 10 amino acids (Hayes et al. 2007). In general, trypsin has been the enzyme that has led to the production of hydrolysates with the higher activity; however, other enzymes have also been effective (Madureira et al. 2010). Abubakar et al. (1998) reported 56–96% of ACE-inhibition in bovine whey protein hydrolysates after treatment with different proteases (pepsin, trypsin, chymotrypsin, proteinase K, actinase E, thermolysin, and papain). In this work, the highest activity was observed for the hydrolysates obtained with thermolysin and proteinase K; the peptide with the highest ACE-inhibitory activity in the hydrolysate with proteinase K was β-LG f(78–80). Pihlanto-Leppälä et al. (2000) also assayed different enzymes, but ACE-inhibitory peptides from bovine α-LA and β-LG were only obtained after hydrolysis with trypsin alone, and with an enzyme combination containing pepsin, trypsin, and chymotrypsin. The ACE-inhibitory peptides identified were α-LA fractions (50–52), (99–108), and (104–108) and β-LG fractions (22–25), (32–40), (81–83), (94–100), (106–111), and (142–146).

ACE-inhibitory peptides were also obtained from the whey protein of other species in addition to bovine. In this way, Hernández-Ledesma et al. (2002) studied the production of ACE-inhibitory peptides from caprine and ovine β-LG using digestive (trypsin and chymotrypsin) and microbial (proteinase K and thermolysin) enzymes. They found that bacterial enzymes led to the production of hydrolysates with higher activities. In addition, they identified four novel ACE-inhibitory peptides from caprine β-LG hydrolysates with thermolysin: f(46–53) and f(122–125) (both with potent activity), and f(58–61) and f(103–105) (both with lower activity).

Besides the production of antihypertensive peptides with digestive and microbial enzymes, other treatments have been also assayed. In this way, Gammoh et al. (2020) demonstrated that both ultrasonication, and *Lactobacillus* fermentation of whey protein from camel milk increased the inhibitory activity against ACE and antioxidant activity. In another study, the pretreatment of whey protein with ultrasound enhanced the enzymolysis with alcalase was reviewed; this fact led to a higher ACE-inhibitory activity of hydrolysates (Wu et al. 2018).

The inhibition of the ACE activity in *in vitro* assays is indicative of the potential of a given peptide to act as a hypotensive agent *in vivo*. To exert its activity, bioactive peptides must reach the target organ intact; however, they may be susceptible to further degradation after oral ingestion (FitzGerald et al. 2004). The stability of these peptides to different proteolytic enzymes has been evaluated in *in vitro* studies. Mullally et al. (1997) showed that the ACE-inhibitory peptide β-LG f(142–148), obtained by hydrolysis with trypsin, was resistant to further digestion with pepsin, while it was hydrolyzed to a very low level by chymotrypsin. However, Walsh et al. (2004) observed that this peptide was degraded when it was exposed to digestive enzymes (pepsin + corolase PP) or incubated with human serum (which contains proteinases/peptidases). Thus, this ACE-inhibitory peptide is not adequate to elicit a hypotensive response in humans.

In addition to *in vitro* assays, several *in vivo* studies have been done, which allow the hypotensive effect to be reliably assessed. Based on these works, it is possible to confirm that some hydrolysates may serve as innovative natural ingredients for functional foods with the goal of attenuating hypertension. In this way, Martin et al. (2015) demonstrated that the peptide IW (Ile-Trp) and a hydrolysate containing this peptide IW (obtained with a combination of elastase and trypsin) decreased the blood pressure in spontaneously hypertensive rats (SHR). Hussein et al. (2020) demonstrated the efficacy of a hydrolysate, obtained by treatment with alcalase, to decrease the blood pressure in SHR. In another study, a significant diminution of the blood pressure in SHR was observed for whey protein hydrolysates obtained with trypsin, proteinase K and actinase E (Abubakar et al. 1998). These authors found contradictory results between *in vitro* and *in vivo* studies. In effect, they observed that the hydrolysates with thermolysin showed the highest ACE-inhibitory activity without antihypertensive effect in SHR, while hydrolysates with actinase E showed low *in vitro* activity but a strong *in vivo* effect on the blood pressure.

Finally, human studies related to the hypotensive effects of peptides are scarce, above all for peptides derived from whey protein. A commercial β-LG hydrolysate, which claims to reduce blood pressure, named BioZate® (Davisco Foods International Inc.), is commercially available (Madureira et al. 2010). The ingestion of this hydrolysate by 30 males and females, unmedicated, nonsmoking, hypercholesterolemic, borderline hypertensives over a six week period led to a significant diminution of the diastolic and systolic blood pressure (FitzGerald et al. 2004).

Besides the ACE-inhibitory peptides, there are other peptides derived from whey protein that could decrease blood pressure by other mechanisms. Thus, the hypotensive effect of milk protein hydrolysates may be caused by several peptides with different mechanism of action (FitzGerald et al. 2004). Sipola et al. (2002) reported that the antihypertensive effect of peptides β-lactorphin (Tyr-Leu-Leu-Phe), derived from β-LG, and α-lactorphin (Tyr-Gly-Leu-Phe), derived from α-LA, was mediated via opioid receptors.

14.5.2 Antimicrobial Peptides

Several antimicrobial peptides have been identified in hydrolysates of whey protein. Their mechanism of action has been barely studied. However, it is accepted that all antimicrobial peptides primarily act on the plasma membrane through electrostatic interaction. This fact results in (a) the formation of transient transmembrane channels that alter the membrane permeability and/or energy generation, or (b) the disruption of the plasma membrane with a consequent cell lysis (Benkerroum 2010). It is difficult to determine what factor contributes most to the antibacterial activity: amino acid composition, isoelectric point, net charge, number of charges or charge distribution along the peptide sequence (Demers-Mathieu et al. 2013).

Bactericidal properties, mainly against Gram-positive bacteria, were reported in hydrolysates of α-LA with endopeptidases and hydrolysates of β-LG with trypsin (Pellegrini et al. 1999, 2001). In particular, four peptides derived from β-LG (f(15–20), f(25–40), f(78–83), and f(92–100)) showed bactericidal activity (Pellegrini et al. 2001). Regarding peptides derived from hydrolysis of α-LA with trypsin or chymotrypsin, three antibacterial peptides (f(1–5), f(17–31)S-S(109–114), and f(61–68)S-S(75–80)) were identified. Two of these peptides are composed of two polypeptide chains held together by a disulfide bridge. The peptides produced by digestion of α-LA with pepsin did not have antibacterial activity (Pellegrini et al. 1999). Demers-Mathieu et al. (2013) demonstrated that an anionic-peptide-enriched fraction obtained from trypsin-hydrolyzed whey protein and concentrated by NF inhibited the growth of two Gram-positive pathogenic bacteria (*Listeria monocytogenes* and *Staphylococcus aureus*). In this work, two anionic peptides derived from β-LG: f(84–91) and f(125–135) showed antibacterial activity. In another study, Pihlanto-Leppälä et al. (1999) observed antibacterial properties against *Escherichia coli* JM103 of hydrolysates of α-LA and β-LG, which were enhanced after UF. Théolier et al. (2013) found antibacterial activity in WPI hydrolyzed with pepsin, but not in hydrolysates with trypsin or chymotrypsin; these last results differed from other studies. Six peptides with antimicrobial activity were derived from β-LG: f(14–18), f(123–125), f(50–54), f(143–146), f(134–136), and f(147–149), while one peptide was derived from α-LA: f(117–121).

The iron-binding glycoprotein lactoferrin possesses antimicrobial and antifungal activities against different bacteria and yeast (Orsi 2004). In addition, lactoferricin, an antimicrobial peptide derived from this protein also displays a broad spectrum of activity against bacteria, fungi, protozoa, and viruses. Lactoferricin was reported for the first time by Bellamy et al. (1992a, 1992b) and several authors have since studied their antimicrobial activity against different microorganisms. Bovine lactoferricin corresponds to the fraction f(17–41) of the lactoferrin (Madureira et al. 2010).

Although few antimicrobial peptides derived from milk proteins have been discovered until now, these proteins have great potential to provide more peptides having applications in the clinical area as health-promoting food supplements and in the food area as biopreservatives (Mohanty et al. 2016).

14.5.3 Antioxidant Peptides

The oxidative decomposition of unsaturated fatty acids and proteins, and subsequent chain reactions with the production of reactive oxygen species (ROS) occurring during manufacture and storage of food, are some of the main deterioration processes that affect food quality (flavor, texture, color) (Elias et al. 2008; Lund and Baron 2010). Therefore, the use of antioxidant compounds, especially from natural sources, is of great interest in the food industry in order to delay oxidative processes (Pihlanto-Leppälä 2006).

On the other hand, ROS are produced inside the body as a by-product of normal oxygen metabolism. Consequently, cellular systems are exposed to permanent oxidative stress produced by ROS. Fortunately, the organism has its own defense system against reactive species composed by antioxidant enzymes (catalase, superoxide dismutase, and glutathione peroxidase) and endogenous antioxidants (i.e., glutathione) (Pihlanto-Leppälä 2006; Valko et al. 2007; Virtanen et al. 2007). However, when there is an imbalance between the ROS generated and the capacity of the biological system to eliminate them, for example when the antioxidant defense system loses its capacity of response (e.g., elderly people), oxidative stress can occur causing damage to vital cellular components. This stress has been associated with a great number of degenerative and age specific diseases such as cardiovascular disease, cancer, diabetes, cataracts, neurodegenerative disorders, and aging (Scicchitano et al. 2018). Therefore, enhancement of the body's antioxidant defenses through diet would be a practical approach for reducing oxidative stress. In that sense, some peptides have been demonstrated to be powerful compounds with antioxidant ability (Kitts and Weiler 2003; Sarmadi and Ismail 2010).

Antioxidant compounds develop this activity through different pathways: by preventing the formation of free radicals, by scavenging free radicals and ROS, or by chelating transition metal ions which act as pro-oxidants. However, the defined mechanism relating to the antioxidant activity of peptides has not been fully understood to date (Mann et al. 2019). Antioxidant properties of the peptides are related to their composition in amino acids, structure, sequence, and hydrophobicity. Tyr, Trp, Met, Lys, Cys, and His are examples of amino acids that have antioxidant activity (Mann et al. 2019).

Some studies have described the antioxidant potential *in vitro* of different whey derived products such as WPC or WPI (Corrochano et al. 2019; Önay-Ucar et al. 2014). In an effort to increase the antioxidant activity of these products, different processing methods have been assayed, such as enzymatic hydrolysis (Hernández-Ledesma et al. 2005; Mann et al. 2015; Peña-Ramos et al. 2001; Peng et al. 2010; Zhidong et al. 2013), fractionation (Hernández-Ledesma et al. 2005), thermal treatment, pressure treatment, and polymerization (Corrochano et al. 2019). The hydrolysis and fractionation methods seem to be very promising (Elias et al. 2008), so we will focus on these below. Different factors have been tested: concentration of protein, type of enzyme, hydrolysis conditions (temperature, time, pH), method of isolating, and fractionation of peptides.

Peng et al. (2010) investigated the effect of the concentration of WPI and time of hydrolysis with alcalase on the antioxidant activity. Radical scavenging ability was enhanced with increasing hydrolysis time up to five hours and with the increase in protein concentration. Although unhydrolyzed WPI displayed antioxidant activity, it was far less potent than the hydrolyzed product. Hydrolyzed WPI was fractionated by Sephadex G-10 gel filtration chromatography yielding four fractions (I, II, III and IV) composed of peptides of >40k, 2.8–40k, 0.1–2.8k, and <0.1k, respectively. Fraction III composed of short peptides, exhibited the strongest radical scavenging effect (Peng et al. 2009). Hernández-Ledesma et al. (2005) reported for the first time the antioxidant properties of peptides obtained from bovine α-LA and β-LG employing commercial digestive and microbial proteases (pepsin, trypsin, chymotrypsin, thermolysin, and corolase PP). Higher radical scavenging activity was obtained for both hydrolysates with corolase PP and permeates (3 kDa) produced from both hydrolysates had 71% and 85% higher activity than their corresponding hydrolysates. A total of 42 peptide fragments were identified by HPLC-MS/MS in the β-LG hydrolysate by corolase PP. Of special interest was the peptide Trp-Tyr-Ser-Leu-Ala-Met-Ala-Ala-SerAsp-Ile, since it had a much stronger activity than that of synthetic antioxidant BHA (butylated hydroxyanisole). Athira et al. (2015) performed hydrolysis of UF retentate of whey using alcalase; hydrolysis conditions (pH, temperature, and time) were optimized for maximum antioxidant activity using response surface methodology. Maximum antioxidant activity was found for WPH obtained at pH 9.00, 55 °C and eight hours of reaction. The antioxidant peptides from hydrolysates were further enriched by ultrafiltration through 3 kDa membrane and seven peptides (β-Lg f(123–131), β-Lg f(122–131), β-Lg f(124–131), β-Lg f(123–134), β-Lg f(122–131), β-Lg f(96–100), and β-Lg f(94–100)) were identified by LC-MS/MS. Furthermore, hydrolysate successfully mitigated the increase in the concentration of oxidative biomarkers (glutathione pyruvate transaminase, alkaline phosphatase, and creatinine) of mice in which oxidative stress was induced with an overdose of paracetamol (Athira et al. 2013). Mann et al. (2015) reported that a WPH prepared with corolase showed maximum antioxidant activity as compared to WPH prepared with flavourzyme and alcalase. Peptide fragments derived from β-LG and α-LA were identified in the hydrolysates by LC-MS/MS. O'Keeffe et al. (2017) transferred at semi-pilot-scale previous results obtained at laboratory-scale. The study consisted of the hydrolysis of WPC with corolase, ultrafiltration of hydrolysates using membrane having nominal cut-off of 5 kDa, fractionation using laboratory-scale ultrafiltration and SP RP-HPLC, and analysis of antioxidant activity by *in vitro* method. A 0.65 kDa permeate obtained by UF from 5 kDa permeate demonstrated antioxidant activity. Further, this permeate resulted in a fraction (SP_F3) with a 4.4-fold increase in activity. Peptides within SP_F3 fraction were identified using UPLC-ESI-MS/MS as Met-Pro-Ile. In the same way, Power-Grant et al. (2015) informed that enzymatic hydrolysis of WPC significantly enhanced its antioxidant activity compared to intact product. Bamdad et al. (2017) identified eight peptides in the fraction of β-LG hydrolyzed by alcalase having the most antioxidant activity. According to Corrochano et al. (2019), hydrolysis of WPI with commercially available proteases: alcalase, bromelain, and neutrase, did not enhance antioxidant activity during simulated gut transit. This implies that the proteolytic conditions of the gut are capable of generating equivalent antioxidant capacity during simulated gastric digestion of WPI. Of the four major whey proteins tested, α-LA fraction exerted the best antioxidant property, which increased after simulated gastric digestion, suggesting that α-LA is the preferred whey protein candidate to supplement foods. Ballatore et al. (2020) optimized the hydrolysis of WPC with tripsin using surface response methodology. A strong antioxidant activity was detected for the hydrolyzed product and for a fraction ≤ 3 kDa Additionally, both products showed a high cytoprotection effect against a stress agent in epithelial cells of the rat ileum. In contrast, few encouraging results were obtained by Tavares et al. (2011) since none of hydrolysates from WPC, α-LA and GMP with *Cynara cardunculus* containing two aspartic proteases, exhibited a relevant antioxidant activity.

On the other hand, some human intervention studies reported positive results of foods supplemented with WPH. Lollo et al. (2014) conducted a study with elite Brazilian soccer players who were supplemented with a WPH (0.5 g protein/kg of body mass per day) for 12 weeks. The supplementation led to a decrease in *in vivo* markers of oxidative stress and tissue damage. Increased levels of antioxidant biomarkers in plasma, such as glutathione were detected in elderly patients with ischemic stroke fed with an enteral formula containing WPH (de Aguilar-Nascimento et al. 2011). Power-Grant et al. (2016) provided preliminary evidence of an association between the change in the antioxidant activity *in vitro* of a milk-based protein matrix (MPM) formulated with milk protein concentrate (MPC) and WPH and the change in antioxidant activity in plasma of healthy women (50–70 years) after ingestion. The authors indicated that further investigations are required to identify the bioactive peptides/amino acids within the MPM that elicit the *in vitro* and *in vivo* antioxidant effect.

14.6 Glycomacropeptide

14.6.1 Definition, Structure, and Technologies

Glycomacropeptide, also called caseinomacropeptide (CMP), is a group of C-terminal hydrophilic peptides derived from κ-casein. During cheese making, milk is coagulated with chymosin (rennet) and κ-casein is cleaved in Phe105-Met106 bond into two fractions: para-κ-casein (residues 1 to 105), and GMP (residues 106 to 169), which remains soluble in whey (Abd El-Salam 2006; Manso and Lopez-Fandiño 2004; Thomä-Worringer et al. 2006).

GMP exists naturally in several forms as a result of extensive post-translational glycosylation phosphorylation and genetic variants.

The sequence of 64 amino acid residues is very particular. BCAA prevails in its structure; nevertheless, aromatic amino acids such as phenylalanine (Phe), tryptophan and tyrosine are absent. The existence of genetic variants for κ-casein determines two main genetic variants for GMP named A and B, which differ in two amino acids.

The content of carbohydrates in the molecule varies from 0 to 15% (Fakuda et al. 2004). Thus, from a glycosylation viewpoint, GMP consists of two fractions: glycosylated and phosphorylated fraction (g-GMP) and non-glycosylated but phosphorylated fraction (a-GMP). g-GMP contains different types of carbohydrate chains attached to specific sites of the structure. Commonly, g-GMP presents up to five carbohydrate chains composed of sialic acid, galactose, and N-acetyl galactosamine, which are bonded to threonine and serine residues in different positions (121, 131, 133, 136 (var A); 142, and 165) (Neelima et al. 2013). On the other hand, depending on the κ-casein variant that forms GMP, it may contain phosphorus up to 0.4% (Rojas and Torres 2013).

SW is considered the main source for production of GMP at industrial scale (Abd El-Salam 2006). GMP is the third most abundant protein in whey, after β-Lac and α-Lac and makes up 15–25% of the total whey proteins. The levels of GMP reported in SW are around 1.2–1.5 g L^{-1} (Thomä-Worringer et al. 2006). Derived whey products such as WPC and WPI are enriched in GMP (Bonnaillie et al. 2014). GMP can be also produced by hydrolyzing reconstituted κ-casein or caseinates with chymosin.

The different methods developed for the separation of GMP from whey or derived products can be grouped as follows:

- *Methods based on thermal treatments*

This method is based on heat precipitation of whey proteins. GMP isolation and recovery by means of thermal treatment of whey has been evaluated in some descriptive studies. The most appropriate conditions that could allow industrial production have been established. Assays have been performed by thermal treatment (90 °C) of SW; peptides attributed to olygomers of GMP were recovered at levels of 34% (Rojas and Torres 2013). On the other hand, thermal treatments are usually employed as a first step of other techniques (Berrocal and Neeser 1991; Nielsen and Tromholt 1994; Saito et al. 1991). Some of these processes have been patented.*Methods based on selective precipitation*

Tricloroacetic acid, ethanol, and acetic acid are the most common reactives used in the selective precipitation of GMP and its separation from other proteins. The type of reactive and concentration influence the carbohydrate profile and the yield of GMP. However, the evaluation of low cost procedures is required for the industrial application of this method (Li and Mine 2004; Rojas et al. 2009).

- *Methods based on membrane filtration*

The UF process has been reported to be the most effective for recovering GMP from whey. It is based on the concept that at pH 3.5 GMP permeates through membranes with a molecular weight cut-off (MWCO) ranging from 20 to 50 kDa, whereas the majority of whey proteins (β-LG, α- LA, immunoglobulins, etc.) are retained (Neelima et al. 2013). The first patented method was made by Tanimoto et al. (1991). They established that the pH control of whey is an essential step. Whey must be adjusted to 3.5, as at pH 4 or higher the GMP tends to aggregate making it difficult to pass through membrane. At pH three or less, the sialic acid bond to peptide chain became unstable, decreasing the beneficial effects of the obtained GMP. Then, whey is ultrafiltred at 50 °C using membranes of 20 kDa. Additionally, a step of diafiltration is necessary to increase the recovery of GMP in permeate. Finally, the pH of permeate is adjusted to seven, where GMP forms non-covalent linked polymers, and can be ultrafiltered or concentrated using the same or another membrane.

In the last decades, different modifications of this standard process have been proposed (Kawasaki et al. 1993; Li and Mine 2004; Martin-Diana et al. 2002; Tolkach and Kulozik 2005). The most recent innovation in UF technologies is the use of membranes with surface modified containing positively charged polymer brushes. By placing a strong positive charge on a 1000 kDa membrane and adjusting the pH of whey close to the isoelectric point of GMP, a 14-fold increase in selectivity was observed compared to procedure with unmodified membrane (Arunkumar and Etzel 2018).

- *Chromatographic methods*

Chromatography is the most common method for the recovery of GMP from whey; different types such as ion exchange chromatography (Doultani et al. 2003; Tanimoto et al. 1992; Tek et al. 2005), gel chromatography (Nakano and Ozimek 2000), affinity chromatography (Saito et al. 1991), and hydrophobic interaction chromatography (Silva-Hernandez et al., 2002) have been reported. These procedures have allowed variable levels of GMP to be recovered. However, a series of limitations such as a low throughput caused by a low flow rate, low whey loading volume, low recovery, and the cost of resins make most of these applications impractical at industrial scale (Thomä-Worringer et al. 2006). A commercially available GMP ingredient isolated by ion exchange chromatography is not pure enough for some food applications.

- *Combined methods*

As mentioned earlier, in most large-scale technologies to isolate GMP from whey ion exchange chromatography or UF is used. Li and Mine (2004) concluded that the combination of UF and anionic chromatography might be a suitable and practical approach for industrial scale. Combined processes such as cation exchange chromatography, UF/DF, and lyophilization have been assayed at pilot scale in order to purify GMP and decrease Phe levels (LaClair et al. 2009). In some cases, the costs of purification makes the process impractical at a commercial scale.

14.6.2 Biological Properties

Significant biological functions and health benefits have been attibuted to GMP or peptides derived from it. Some studies suggest that is not only the peptide but also the sugar moieties, especially sialic acids, that are responsible for various biological properties of GMP.

Brody (2000) summarizes the literature reports of biological properties of bovine milk GMP which include: ability to bind cholera and *Escherichia coli* enterotoxins, inhibition of bacterial and viral adhesion, suppression of gastric secretions, promotion of beneficial intestinal bacteria (bifidobacteria), and modulation of immune responses. It has also been suggested that GMP can affect gastrointestinal function through the inhibition of gastric secretion and that this action could be mediated by peptide hormones, such as cholecystokinin (CCK) released from intestinal cells, which is a potent activator of satiety signals, regulating energy and food intake. The CCK-like activity of GMP was confirmed by subsequent studies (Sukkar et al. 2013).

On the other hand, in the last decades a great deal of evidence has been accumulated in relation to the role of GMP in the mineral transport and anticariogenic activity. These studies demonstrate that GMP is able to: enhance zinc absorption (Kelleher et al. 2003), exert inhibitory activity against enamel demineralization, and promote tooth enamel remineralization (Aimutis 2004; Elgamily et al. 2019), reduce the erosive effect of acidic drinks (Setareh Nejad et al. 2009, 2010), prevent adhesion of cariogenic bacteria to tooth surfaces (Schüpbach et al. 1996), inhibit bone loss in ovariectomized rats by increasing calcium bioavailability (Neeser et al. 2000), enhance the content of calcium in fermurs of mice, especially during the period of a full body development (Burns et al. 2015).

14.6.3 Applications

In the last decades, the use of GMP is growing. Because of its biological properties, GMP is considered to be a potential ingredient for functional foods and nutraceuticals (Li and Mine 2004). Several food and ingredient suppliers (Davisco Foods International Inc., Agropur Ingredients, Right Food, Erie Foods and Arla Foods) produce isolates of GMP for use as functional ingredient. In addition, other ingredients derived from SW, such as: powder whey, WPC, WPI, among others, contain mostly β-LG, α-LA and GMP. Therefore, the incorporation of these ingredients in food formulations is an indirect way of adding GMP and improving the nutritional and functional value. They also increase the protein levels and provide bioavailable calcium and other minerals. Yogurt, infant formula, and some types of cheese are good examples of foods where these ingredients are employed in the formulation. However, few studies have reported on the nutritional and functional evaluation of food supplemented with GMP.

Human milk is very rich in sialic acids whereas bovine milk provides lower levels of this compounds. Thus, infant formulas formulated from bovine milk must be supplemented with sialic acids, either free or bound to GMP in order to emulate human milk (Claumarchirant et al. 2016; Wang et al. 2001).

On the other hand, the lack of Phe in amino acids sequence of GMP, as mentioned above, makes it a potential ingredient for foods intended for patients suffering phenylketonuria (PKU). Individuals with PKU cannot metabolize Phe and must follow a low-Phe diet in which most dietary protein is provided by a Phe-free amino acid formula (LaClair et al. 2009). Some studies have demonstrated the acceptability of products made with GMP (Lim et al. 2007). GMP has been modified and adapted for PKU to provide an alternative protein source through supplementation with rate-limiting amino acids (GMP-AA).

14.7 Non-protein Whey Products

14.7.1 Galactooligosaccharides

14.7.1.1 Definition, Structure and Technologies

GOS are non-digestible carbohydrates, consisting of a chain of a variable number of galactose units and usually linked to a terminal glucose. The general structure can be represented as (galactose)n-glucose, where n is 2–10. They are produced enzymatically in a reaction of lactose transgalactosylation catalyzed by β-galactosidases (β-D-galactoside galacto-hydrolase) (EC 3.2.1.23) that occurs simultaneously with lactose hydrolysis reaction (Xavier et al. 2018). The mechanism of these reactions can be described as follows. The β-galactosidase is coupled to a molecule of lactose. A glucosyl moiety is released and the enzyme and galactosyl moiety form a covalent bond. At this point, hydrolysis or transgalactosylation can occur depending on the available galactosyl acceptor. The hydrolysis takes place if the galactosyl acceptor is water and so a galactose molecule is released. In turn, if the acceptor is a saccharide (glucose, galactose, lactose, or GOS), transgalactosylation reaction occurs obtaining a mixture of GOS (Sangwan et al. 2011). As the GOS yields attainable are low (rarely exceeding 40%), the product obtained contains, in addition to GOS, significant amounts of unreacted lactose and monosaccharides (glucose and galactose). The composition of the GOS fraction is heterogeneous and complex, as different types of GOS in terms of degree of polymerization (2 to 10 monomer units), types of glycosidic bonds between units (β-(1→2), β-(1→3), β-(1→4), β-(1→6)) and molecular weight are obtained (Ackerman et al. 2017; Kittibunchakul et al. 2020). The yield in the GOS production and the composition of the resulting mixture depends on several factors: enzyme origin (yeast, fungi, bacteria), degree of purity (crude cell extract, purified enzyme), free or immobilized enzyme, concentration of the lactose substrate, source of substrate (lactose solution or whey), and reaction conditions (pH, temperature and time, ion concentration). In relation to these aspects, information published is abundant (Bosso Tomal et al. 2019; Mano et al. 2019; Martínez-Villaluenga et al. 2008; Neri et al. 2009; Palai et al. 2012; Park and Oh 2010).

In our institute, different variables were assayed for the production of GOS: (i) lactose concentration (5–25% w/v) from different matrices: purified lactose solution, whey processed by UF and NF and mixtures of milk-whey, (ii) enzyme doses (0.16–3.00 g L^{-1}), and (iii) commercial preparations of high purity β-galactosidase from different origin (*K. lactis, Sacharomyces marxianus* and *Bifidobacterium bifidum*). In general, we observed that the GOS yield increased with the lactose concentration, and the higher yields were obtained in less reaction time with the higher enzyme dose. A slight degradation of the GOS was also observed after the maximum was reached; this fact was more noticeable for the assay with low initial lactose concentration (Vénica et al. 2015, 2017). From purified lactose solution, maximum GOS yields were approximately 8, 15 and 26%, for 5, 10 and 20% w/v of initial lactose concentration, respectively, and with 0.25 and 0.40 g L^{-1} of β-galactosidase from *K. lactis* (Vénica et al. 2015). From milk-whey mixture, response surface method was applied to model the behavior of the variables studied: lactose concentration and dose of β-galactosidase from *K. lactis*; this statistical tool showed to be suitable for predicting the concentration and maximum yield of GOS and the reaction time required to reach the maximum. The total GOS concentrations obtained ranged from 0.65 to 2.78% w/v, corresponding to yields of 10 to 18%, respectively (Vénica et al. 2017). In relation to GOS production from whey, whey was ultrafiltered by polyethersulfone membrane (10 000 Da) and the permeate obtained was concentrated by NF with polyamide (200 Da) membrane up to 25% w/v of lactose. After that, the incubation of the lactose solution with β-galactosidases (1 g L^{-1}) from *K. lactis* and *S. marxianus* demonstrated similar performance in the production of GOS; maximum yields of 15% were obtained between 90 and 135 min. While using β-galactosidase from *B. bifidum* (1 to 3 g L^{-1}) lower transgalactosylase capacity was observed, reaching a maximum yield of only 6% at 120 min (results not yet published). Similar results were reported by many authors (Martínez-Villaluenga et al. 2008; Neri et al. 2009; Palai et al. 2012; Urrutia et al. 2013).

Commercial GOS preparations are composed of a mixture of several types of GOS in liquid or powder form and contains generally between 32 to 75% of total GOS. The main GOS producing companies are Friesland Foods Domo (Netherlands), Yakult Honsha (Japan), Nisin Sugar Manufacturing Co (Japan), Ingredion Inc. (South Korea), and Clasado Bioscience (Malta, UK). The lowest content of total GOS are found in the Bimuno Travelaid with 32% w/w, the highest in VivinalGOS Easy Drying Syrup with 72% on dry matter, while other preparations have intermediate values (for example, Bimuno Powder with 50% w/w, Oligomate 55NP powder with 55% w/w, Vivinal GOS Syrup with 59% on dry matter) (Vera et al. 2016). Products with higher levels of GOS are very scarce; one example is TOS-100, a purified powder containing 99% oligosaccharides (Martins et al. 2019). Therefore, purification processes continue to be a key issue in the GOS production, as the main secondary products of enzymatic reaction, unreacted lactose, and glucose and galactose, must be removed (Córdova et al. 2017; Goulas et al. 2002; Vera et al. 2016). To this aim, several methods of purification have been proposed.

- *Methods based on membrane filtration*

NF technology appears to be a potential industrially scalable method for purification and concentration of oligosaccharide mixtures (Goulas et al. 2002). The reported antecedents indicate good performance, although the effectiveness of the process depends on the type of membrane (MWCO, pore size distribution, morphology and material), conditions of process (pressure, temperature, flow), and the application or not of DF. The high osmotic pressure, high viscosity of concentrated carbohydrate solutions, and the similar molecular weight of compounds to be separated, are the main constraints of the separation and purification processes (Vera et al. 2016).

Ren et al. (2015) reported the use of a NF separation system with DF, coupled to a continuous enzymatic production of GOS. The membrane material was composed of a three-layer composite film (polyethylene piperazine, polysulfone, and polyester) of 400–600 Da; they obtained a purity of 57% of GOS. According to Michelon et al. (2014), PES membranes with MWCO of 1000 and 400 Da, are characterized by low rejection coefficients, high permeate fluxes, and high selectivity, while aromatic polyamide membranes (150 to 300 Da) have high rejection coefficients, low permeate fluxes, and low selectivity. They obtained a purification factor of 1.68 and 61% of GOS recovery with PES membrane of 400 Da. In addition, a GOS purity of 77% was achieved with a polyamide thin film composite membrane with a microporous polysulfone support layer (400 Da), after three DF stages (Das et al. 2011). Likewise, performances of different membranes (cellulose acetate 500–600 Da, cellulose acetate 800–1000 Da from two different brands and polyamide 150–300 Da) were evaluated for the purification of a commercial GOS mixture (Feng et al. 2009). The best result was obtained for one of the cellulose acetate membranes (800–1000 Da) working with a constant DF volume, 50 °C and 6 bar, as a reduction of 90% monosaccharides and 52% lactose was achieved.

- *Methods based on selective fermentation (bioconversion)*

Bioconversion with yeast strains (from the genera *Kluyveromyces* and *Saccharomyces*) and thermophilic lactobacilli (*L. helveticus*) have been reported as a plausible technological alternative for GOS purification. The basis of this method is the selective removal of the metabolizable sugars (glucose, galactose, and lactose) from GOS mixture by fermentation. The effectiveness of the process depends on the strains used and the carbohydrate composition of the mixture (Sangwan et al. 2014). Even though high purity is attainable, this methodology has some important drawbacks, such as the high ratio of biomass to carbohydrate mass required and the production of metabolites like ethanol, acetic acid, and glycerol (Vera et al. 2016); biomass could be removed by centrifugation (Sangwan et al. 2014).

S. cerevisiae was used to eliminate monosaccharides by anaerobic glycolysis, transforming them into ethanol and CO_2. The studies carried out showed that glucose was completely metabolized and galactose was partially or completely metabolized (Aburto et al. 2016; Goulas et al. 2007; Li et al. 2008; Yoon et al. 2003). However, lactose content remained unaltered so the purity obtained with *S. cerevisiae* was modest (Goulas et al. 2007). *Kluyveromyces* cells have the advantage of metabolizing monosaccharides and also lactose (Vera et al. 2016). High levels of purity of GOS (> 95%) have been reported with *K. marxianus* and *K. lactis* as the consumption of monosaccharides and lactose were almost complete (Cheng et al. 2006; Guerrero et al. 2014; Li et al. 2008). However, the fermentation process has to be exhaustively controlled since GOS consumption was observed after 30 hours (Cheng et al. 2006). Similar to *Kluyveromyces* cells, *L. helveticus* can metabolize monosacharides and also lactose. Sangwan et al. (2014) purified a GOS mixture applying successive fermentation steps with *S. cerevisiae, K. lactis* and *L. helveticus*, respectively; they achieved a level of GOS purification of 91.7%.

- *Chromatographic and selective precipitation with ethanol methods*

Simulated moving bed (SMB) chromatography has been employed for carbohydrate fractionation and GOS purification. This method allowed us to get a purity of GOS of 99.9% with high yield as an extract containing exclusively lactose and monosaccharides was obtained. This technology allows both high product purity and high process yield. However, flow velocities ranging from 0.5 to 4 cm min^{-1} are necessary so quite big chromatographic columns are required which involves high cost (Wisniewski et al. 2013).

The selective adsorption with activated charcoal is another method reported. Different concentrations of ethanolic solutions 1 to 15% (v/v) were assayed to remove mono- and disaccharides, whereas GOS of higher polymerization degrees were recovered using ethanol 50% (v/v). With 1% (v/v) ethanol, monosaccharides were only eliminated, whereas the other carbohydrates were mostly recovered. As the ethanol concentration increased up to 15% (v/v), most of the disaccharides were eliminated, but also part of the trisaccharides. Ethanol treatments were not selective enough and substantial amounts of trisaccharides were lost. As a consequence, a compromise between purity and amount of recovered GOS should be considered before selecting the most suitable conditions of treatment. The main disadvantages of this treatment are the incomplete removal of disaccharides, as well as the loss of trisaccharides (Hernandez et al. 2009).

On the other hand, selective precipitation with ethanol is another method employed for GOS purification. Sen et al. (2011) applied this approach to aqueous solutions with different total concentration of saccharides (28, 60 and 81 g L^{-1}) from commercial preparation GOS; they studied different concentrations of ethanol (70, 85 and 90% v/v) and working temperatures (10, 25 and 40 ºC). The higher GOS enrichment occurred for the lower total saccharide concentration and higher ethanol concentration, while the temperature had less impact on the precipitation. GOS was enriched 2.3 fold in the precipitate formed with ethanol 90% (v/v) and with 28 g L^{-1} of total saccharide at 40 ºC. The purification efficacy can be increased by performing sequential precipitations with ethanol, but the recovery of the initial GOS is significantly reduced.

14.7.1.2 Biological Properties

The most outstanding aspect is its prebiotic role. Prebiotics represent one of the substances most used to maintain a healthy microbiome. Recently, the importance of the intestinal microbiota in health has become increasingly prominent (Farias et al. 2019). The concept of prebiotic was first introduced in 1995 by Gibson and Roberfroid (1995), which was updated and now prebiotics are defined as: "a selectively fermented ingredient that allows specific changes, both in the composition and/or activity of the gastrointestinal microflora that confers benefits upon host well-being and health". In general, prebiotics are carbohydrate compounds; however, not all carbohydrates can be considered prebiotics (Gibson et al. 2004; Glenn et al. 2004).

Prebiotics are resistant to salivary, pancreatic and intestinal enzymes, and the stomach acid medium. They are fermented by the beneficial microflora of the large intestine inhibiting the growth of pathogenic and putrefactive bacteria and thereby contribute to reduction of toxic metabolites, prevention of diarrhea, constipation relief, and improvement of lactose tolerance (Bosso Tomal et al. 2019). Metabolism of GOS leads to production of short chain fatty acids (acetate, propionate, butyrate) which is related to the improvement of several physiological effects including intestinal function, minerals absorption, regulation of lipid and glucose metabolism, as well as reducing the risk of colon cancer, among other things (Ackerman et al. 2017; Bosso Tomal et al. 2019; Farias et al. 2019; Sangwan et al. 2011; Whisner et al. 2013). Different authors have worked on identifying the individual structure of GOS, which requires the use of sophisticated analytical techniques; this aspect is important since the biological properties depend on the linkage type and polymerization degree. However, there is no in-depth understanding how these individual GOS structures differ functionally (Böger et al. 2019a; Coulier et al. 2009). A recent article evidenced that the prebiotic effect of GOS is mainly associated to trisaccharide (mostly 4' and 6'-galactosyl-lactose) and tetrasaccharides (mostly 6'-digalactosyl-lactose) (Aburto et al. 2016). However, it has also been reported that disaccharides containing β-(1→6) bonds (allolactose and galactobiose), present bifidogenic properties similar to trisaccharides (Rodriguez-Colinas et al. 2013).

14.7.1.3 Applications

GOS have a safe history of use in food and infant nutrition. The first ingredient containing GOS was launched in Japan in the late 1980s (Bosso Tomal et al. 2019).

Due to the recognized beneficial effects, GOS ingredients are added to the formulation of infant formulas in order to mimic breast milk as closely as possible (Otieno 2010). Breast milk naturally contains a high concentration of human milk oligosaccharide (HMO) (0.5–2% w/w) forming part of the complex chemical structures which constitute the third most abundant component (Crisà 2013). Conversely, in milks from ruminant mammals (cows, goats, sheep) the levels naturally detected of these compounds are very low (Korhonen 2010; Pandya and Haenlein 2010; Playne 2003;

Vandenplas 2002). The synthesis of HMO-like compounds is expensive and laborious; in contrast, the industrial production of GOS ingredient is relatively easy at industrial scale. In addition, HMO and GOS have similar effects on the growth of infant colonic microbiota (Böger et al. 2019b).

GOS have the advantage of being very soluble and tolerate high temperature and low pH, unlike other prebiotics like inulin and fructo-oligosaccharides (FOS) (Sangwan et al. 2011). These technological properties are important as they allow the incorporation of GOS ingredients in food matrices of different natures: fruit juices (Klewicki 2007), ice cream (Balthazar et al. 2015), yogurt (Prasad et al. 2013), cheese (Belsito et al. 2017), confectionary (Lans et al. 2018), bakery products, beverages, and snack bars (Sangwan et al. 2011).

14.7.2 Lactosucrose

14.7.2.1 Definition, Structure, and Production

Lactosucrose (O-β-D-galactopyranosyl-(1,4)-O-α-D-glucopyranosyl-(1,2)-β-D-fructofuranoside) is a non-digestible carbohydrate. It is a rare trisaccharide composed of D-galactose, D-glucose, and D-fructose (Han et al. 2009). Raffinose (O-α-D-galactopyranosyl-(1,6)-O-α-D-glucopyranosyl-(1,2)-β-D fructofuranoside), naturally found in some vegetables, is an isomer of lactosucrase (Hanau et al. 2020; Silvério et al. 2015).

Lactosucrose scarcely exists in nature and is difficult to synthesize chemically (Mu et al. 2013). Although, it can be synthesized from sucrose and lactose involving levansucrose (EC 2.4.1.10) and β-fructofuranosidase (EC 3.2.1.26) (also known as saccharase or invertase), which are able to catalyze both hydrolysis and transglycosylation reactions. In the transglycosylation reaction, the fructosyl moiety of sucrose is transferred to the hydroxyl group of lactose producing lactosucrose plus glucose. However, other compounds other than lactose can act as acceptors, such as water or sucrose; when water is the acceptor, the hydrolysis occurs and when sucrose is the acceptor, the synthesis of FOS takes place (Silvério et al. 2015). An alternative route considers the use of β-galactosidase, where the galactosyl moiety of lactose is transferred to the hydroxyl group of sucrose; in this case, lactose can act both as donor and acceptor of the galactosyl group so that a mixture of lactosucrose and GOS is produced (Vera et al. 2021). As well, hydrolysis of the formed product can also occur, compromising the lactosucrose yield production (Silvério et al. 2015).

The first report describing the synthesis of this compound using levansucrase from *Aerobacter levanicum* was performed by Avigad and co-workers and date from 1957. In the 1990s, a process of production of lactosucrose using a β-fructofuranosidase from *Arthrobacter* K-1, was patented (Fujita et al. 1990). In 1992, Ensuiko sugar refining Co., Ltd., a Japanese-based company, started the commercial production of lactosucrose, which is still the only producing company (Long et al. 2020).

Whole cells of seven bacteria (*Bacillus amyloliquefaciens, B. subtilis, G. stearothermophilus, Paenibacillus polymyxa, Pseudomonas syringae, Rahnella aquatilis*, and *Sterigmatomyces elviae*) containing levansucrase activity, were evaluated for the production of lactosucrose. The highest lactosucrose concentration was found for *B. subtilis* (Park et al. 2005). Although, β-fructofuranosidase is cheap and widely available in bacteria, fungi, and plants, *Arthrobacter* sp. K-1 and *Bacillus* sp. No. 417 are the only two bacteria that can produce β-fructofuranosidase with excellent transfructosylation activity (Xiao et al. 2019). In addition, according to Xiao et al. (2019), β-galactosidase from *Bacillus circulans* is the only member of β-galactosidase that can produce lactosucrose.

Several variables influence the production yields and the composition of the reaction mixture obtained at the end of the enzymatic process, such as reaction conditions (pH, temperature, concentration of substrates, ratio enzyme/substrate), type and dose of biocatalyst employed. Mixtures may contain variable amounts of unreacted substrates (lactose and sucrose), glucose, fructose, and galactose formed from hydrolysis of sucrose or lactose, and unknown saccharides resulting from the transfer activity. Silvério et al. (2015) summarized the results obtained from several works in which free enzyme, whole cells, recombinant enzyme and immobilized enzyme of levansucrose, β-fructofuranosidase and β-galactosidase were used.

In general, the presence of monosaccharide by-products inhibits the transfructosylation activity, thereby causing low yields of lactosucrose synthesis (about 25%). In order to improve the synthesis of lactosucrose by increasing yield and productivity, promising enzymes with enhanced catalytic activity, genetic modifications of enzymes, use of several enzymes and immobilization strategies have been tested (Vera et al. 2021). Multiple enzyme systems composed of glucose oxidase in combination with β-fructofuranosidase (Jong and Seung 1993) and with levansucrose (Han et al. 2009), have been assessed to eliminate the glucose, converting it to gluconic acid. Arakawa et al. (2002) reported the use of an invertase-deficient yeast together with β-fructofuranosidase from *Arthrobacter* sp., to remove residual glucose. The reaction is finished by heating and the reaction mixture is purified by decoloration, carbonation, filtration, desalination, ultrafiltration, and concentration. This procedure is carried out at industrial scale in a batch process, allowed three kinds of commercially available products to be obtained, composed of a saccharide mixture with 40 and 55% of lactosucrose (Nyuka-oligo LS-40 L, LS-55 L, and LS-55 P). Another patented methodology for industrial application performed by Okabe et al. (2008), achieved higher lactosucrose content (at least 70%) and low amounts of by-product (less than 3%). This invention combined a purified β-fructofuranosidase from *Bacillus* sp. V230 and a sucrose unassimilable yeast (*Saccharomyces cerevisiae*) which is able to assimilate monosaccharides but it cannot consume or hydrolyze oligosaccharides, including disaccharides. Finally, they removed unreacted substrates by crystallization using temperature control and addition of organic solvents (methanol, ethanol, and acetone). Long et al. (2020) highlighted that the co-immobilization of bi-enzymes: β-fructofuranosidase and glucose oxidase improved the lactosucrose production with low cost. The immobilization enhanced the thermal stability and pH tolerance of two enzymes in comparison to free enzymes. The use of immobilized enzymes reduces the process cost because enzymes can be reused several times (Garcia-Galan et al. 2011).

On the other hand, Li et al. (2009) showed for the first time, in a very exhaustive study, the effectiveness of a commercial β-galactosidase from *Bacillus circulans* for the production of lactosucrose. The chemical structure of the compounds formed was confirmed by mass spectroscopy and nuclear magnetic resonance (NMR) analyses after their isolation from the reaction mixture by using activated charcoal and Celite column; tri- and tetrasaccharides were synthesized in addition to lactosucrose. The composition of the mixture changed considerably with the reaction conditions (pH, temperature, time, substrate concentration, molar ratio donor/acceptor, enzyme dose), which were optimized to obtain the maximum lactosucrose productivity. More recently, Silva Duarte et al. (2017) evaluated a new bioprocess for the synthesis of lactosucrose using β-galactosidase from *B. circulans* immobilized on chitosan. They evaluated the effect of temperature and pH on the production of lactosucrose and other oligosaccharides. Changes of the reaction conditions modified the qualitative profile of the final product without changes in the total content of oligosaccharides produced; at pH7 and 30°C, 79 g L^{-1} of lactosucrose, 37 g L^{-1} of GOS and 250 g L^{-1} of total oligosaccharides and at pH5 and 64°C, 40, 62 and 250 g L^{-1}, respectively, were obtained. It was clearly demonstrated that lactosucrose was produced at low temperatures and high pH, while GOS were greater at high temperature and low pH.

14.7.2.2 Biological Properties

Lactosucrose belongs to the group of non-digestible carbohydrate substances classified as "emerging prebiotic", together with xylooligosaccharides, isomaltooligosaccharides and chitooligosaccharides, to differentiate them from well-established prebiotics such as GOS, lactulose, inulin, and FOS. This classification takes into account the development state and regulatory status. Mu et al. (2013) summarized the studies on the beneficial functions of lactosucrose, such as the intestinal microflora maintenance and intestinal protection, calcium absorption-promoting activity, among others. More recently, Li et al. (2015) evaluated the production of short chain fatty acid (SCFA) by *in vitro* fermentation of human intestinal microbiota of lactosucrose and two analogues (4′-galactosyl-lactosucrose and iso-lactosucrose). It is well known that the fermentation in the colon depends on factors related to the gut environment (different parts of the large intestine) as well as the chemical structure of the available substrates (type of monosaccharide moieties, degree of polymerization, conformation of links between the monosaccharides). The results showed that the two analogues of lactosucrose seem to present potential prebiotic activity even higher than lactosucrose.

14.7.2.3 Applications

The incorporation of lactosucrose in the formulation of functional foods and beverages has increased in the last two decades in Japan. Lactosucrose was certified as a Foods for Specific Health Uses (FOSHU) ingredient in Japan (Mu et al. 2013). In addition, it is a low-digestive and low-cariogenic sweetener. As a sweetener as well as an "emerging prebiotic" ingredient, lactosucrose has been included in confectionaries, desserts, bakery products, yoghurts, ice creams, infant formula, coffee, tea, chocolates, chewing gum, juices, soups, and mineral water (Silvério et al. 2015). However, its use is largely confined to Japan.

14.8 Application of Whey Proteins as Coating Material for Bioactive Compound-Loaded Liposomes

Increased knowledge about the functionality of whey proteins has led to new and interesting applications. Their use as encapsulating materials for the protection and delivery of bioactive components is one of these new applications (Smithers 2008). Thus, liposome technology has appeared as an emerging area of interest.

Having various areas of applicability, liposomes have been the subject of many studies since the 1960s. Liposomes are vesicles that form spontaneously by the self-assembly of phospholipids in aqueous solutions. They consist of phospholipid bilayers that contain an aqueous core, which enables the encapsulation of both fat-soluble and water-soluble compounds. The encapsulation efficiency of the compound of interest, stability and size of the liposomes are key parameters in the formulation of these structures, and depend on different variables involved during their preparation: type and concentration of phospholipids, nature of the compound to be encapsulated, preparation method, inclusion of other compounds to improve stability of liposomes: cholesterol, ionic lipoids, among others (Mozafari et al. 2008; Shashi et al. 2012; Takeuchi et al. 2005). In the recent years, this technology has become a promising target for food ingredient research (Liu et al. 2020; Vélez et al. 2017). Its use in the food industry is focused on four fundamental objectives: (a) improve bioactive compound stability during food storage increasing shelf life and reducing the appearance of off-flavors, since several bioactive compounds oxidize, volatilize, or hydrolyze easily (Jimenez et al. 2008; Marsanasco et al. 2011); (b) increase the bioavailability of the compounds in the gastrointestinal tract (Chaudhry et al. 2008; Mozafari et al. 2008; Sekhon 2010); (c) improve the sensory characteristics of food (Laloy et al. 1998); (d) release control of compounds so as to produce a delay or sustained release (Mozafari et al. 2008; Rao et al. 1994).

Unfortunately, liposomes still maintain several drawbacks, such as degradation, aggregation, fusion, and leakage of core materials, due to their high sensitivity to the external environment. This could be an obstacle for the use of liposomes as carriers of bioactive compounds in food due to the interaction of the vesicles with compounds from the food matrix (mainly sugars and salts) (Laye et al. 2008), as well as oxidation, hydrolysis or enzymatic degradation over storage, hydrolysis under low pH and enzymatic conditions during gastric digestion (Liu et al. 2016). Several studies have shown that coating of liposomes is an effective method to reduce the lipid membrane damage and the leakage of encapsulated compounds. Coatings of different chemical nature, e.g., chitosan, sodium alginate (Liu et al. 2013, 2016), Ɛ poly L-lysine (Alemán et al. 2016), have been assayed. However, there is little information about the application of milk proteins for liposome coatings. Vélez et al. (2015) formulated liposomes containing cis-5,8,11,15,17-eicosapentaenoic acid (EPA) and cis-4,7,10,13,16,19-docosahexaenoic acid (DHA) by the film hydration technique employing phosphatidylcholine and cholesterol, obtaining high structural integrity after coating the vesicles with whey proteins from milks of different origins (cows, sheep, and goats). WPI have been employed recently to coat liposomes successfully due to the ionic interactions between liposomes surface and the proteins (Pan et al. 2020). Owing to a great quantity of branched chain amino acids, WPI is high in nutritive value and it could be advantageous compared to chitosan due to its less fishy aftertaste (Frenzel et al. 2015). Varying pH enables WPI to coat anionic or cationic liposomes, e.g., when WPI is dissolved in acid solution, it is attracted to the surface of negatively charged liposomes because of its oppositely charged amino groups; WPI is generally positive at pH 4.5 as the pH is below the isoelectric point of β-LG (pI = 5.2), the major protein in WPI (Gomaa et al. 2017). Frenzel and Steffen-Heins (2015) reported that physical stability of WPI coated liposomes increased compared to uncoated liposomes, as indicated by prolonged shelf-life, deletion of osmotic effects in the presence of salts or sugars and a lower sensitivity towards low pH values during *in vitro* gastric digestion. Pan et al. (2020) showed that astaxanthin (3,3'-dihydroxy-β,'-carotene-4,4'-dione), one of ketocarotenoid pigments found naturally in the freshwater microalgae, was successfully encapsulated in WPI coated liposome: the physical stability under heating and light was significantly improved as compared with uncoated liposomes. Gomaa et al. (2017) demonstrated the efficiency of dual coating (pectin and WPI) on the protection of liposomes loaded with antimicrobial peptides against gastrointestinal digestion.

So far, there have been few comprehensive studies on liposomes functionalizing food matrices with regard to the effect on food manufacture and storage stability, *in vitro* digestion or masking of undesired flavors. Vélez et al. (2020) tested the incorporation of liposomal conjugated linoleic acid (CLA) during yogurt making, and studied the effect of thermal treatment and the addition of antioxidants. They showed that liposomal CLA did not modify the fermentation process, it could be added before and after the thermal treatment and high CLA recoveries were found. The addition of antioxidants did not influence CLA content and high oxidation stability was found. Ghorbezade et al. (2017) assayed the addition of omega-3 fatty acids (EPA and DHA) through liposomes to yogurt; they found that encapsulated fish oil presented higher recovery than addition of the free oil. Rasti et al. (2017) studied addition of omega-3 liposomes and omega-3 free oil in milk, finding similar and higher recovery values in both cases at the beginning of storage, but recovery after three days of storage was higher in milks with liposomes. Wechtersbach et al. (2012) included ascorbic acid loaded liposomes into apple juice; the stability of vitamin C increased. Also, ascorbic acid and alpha-tocopherol loaded liposomes were mixed successfully into orange juice (Marsanasco et al. 2011) and liposomes containing vitamin E and folic acid were tested in chocolate milk finding good acceptability (Marsanasco et al. 2015). Frenzel et al. (2015) tested WPI coated and spray-dried liposomes against those of uncoated liposomes applied as transporters of flavonoid quercetin in a whey drink. Vesicles were effective in masking flavonoid bitterness. Relative to uncoated liposomes, the WPI coating improved liposome application into a dairy drink, regarding the physical stability of liposomes towards osmotically active food compounds (sugar or salt), compared to different storage conditions and simulated gastric fluid stress. These stabilizing effects were due to the reduction of semi-permeability of the membrane as the WPI coating formed a barrier on the liposomal surface.

14.9 Bacteriophages in Whey Derived Products: From Threat to Reality

At present, and despite decades dedicated to minimizing their occurrence, bacteriophage (phage) infections continue to be the major threat to the fermentative dairy industry (Pujato et al. 2019). Phages are ubiquitous in the industrial environment and their total eradication is just unachievable, so efforts are focused on maintaining their concentration below dangerous levels for dairy fermentations. As is well known, the lactic acid bacteria (LAB) constituent of the starter cultures used in these technological processes can be infected by phages, resulting in a partial or total inhibition of their growth and activity, obtaining products with low organoleptic and/or microbiological quality. In these cases, the economic losses for the dairy industry can become very important. It is estimated that the capacity of dairy plants is reduced by 10% due to these infections and that, approximately, 2% of world cheese production is affected by the diminished acidification activity of starters (Murphy et al. 2017). Several technological strategies had been designed to control the level of phages in manufacturing environments, including the adequate design of the processing plants, correct sanitization and ventilation, modification of the processes, use of phage inhibitor culture media, rotation of the starter cultures, and use of strains with improved phage resistance (Fernández et al. 2017).

The main source of phage entry to the industrial environment is raw milk (Briggiler Marcó et al. 2019). Several studies have shown that 37% of raw milk destined for the production of yogurt carried phages able to infect different LAB of industrial relevance (del Rio et al. 2007; Madera et al. 2004). Other sources of phage contamination include supplies and the manufacturing environment itself. As already mentioned, the addition of whey derived products as ingredients in the manufacture of fermented milk (such as yogurt), some types of cheeses and other fermented foods, has become widespread in recent years, in order to increase yield and nutritional value, as well as improve the texture and functional properties of the final product (Wagner et al. 2017a). Unfortunately, the addition of these ingredients to fermentation processes has caused an unexpected problem: these whey derivatives could contain high concentrations of phages (specific of several genera and species of LAB), able to infect the starter cultures. As acknowledged, raw milk used for cheese making is normally subjected to a pasteurization process (72° C – 15 s) prior to use, in order to ensure its microbiological quality. However, a large number of phages are still infective after the heat treatment, being able to spread during the process and reach very high concentrations in cheese whey (> 10^9 Plaques Forming Units/ml – PFU/ml) (Atamer et al. 2013; Briggiler Marcó et al. 2019; Wagner et al. 2017a).

The treatment of cheese whey prior to its processing for derivate products begins with clarification, skimming and a heat treatment, being the last focused to inactivate contaminants and LAB of the starters that are still viable. This treatment cannot be as rigorous because it is necessary to avoid protein denaturation and, although many phages are inactivated at this stage, there are more and more frequent reports of LAB phages that remain infective in whey and its derivatives (Atamer et al. 2009, 2011; Briggiler Marcó et al. 2019; Capra et al. 2013; Pujato et al. 2014; Wagner et al. 2017b, 2018). Subsequently, and to obtain the corresponding WPC, the whey proteins are concentrated by means of UF and finally subjected to spray drying. UF processes not only retain the desired proteins, but also the phages potentially present in the whey, due to the fact that the phage particles have a size equal to or greater than that of the whey protein particles. Regarding spray drying, although the phages are subjected to the stress factors present (thermal, osmotic and oxidative) in this process, some phage particles remain infective. Recent reports, focused on the analysis of whey derived products, demonstrated a great dissemination of phages infective of homofermentative LAB, both from mesophilic and thermophilic starter cultures (*Lactococcus lactis* and *Streptococcus thermophilus*, respectively), and of heterofermentative LAB, especially belonging to *Leuconostoc pseudomesenteroids* and *Leuconostoc mesententeroides* species. These studies also found a high biodiversity of phages, many of which exhibited alarmingly high thermal resistance (Samtlebe et al. 2015, 2017a, 2017b; Wagner et al. 2017a, 2017b).

As detailed, the use of derivatives of cheese whey as ingredients in food is increasing. However, beyond those countries with high quality standards in the food industry, this manufacturing sector is, in general, unaware of the magnitude of the risk involved in its use in new fermentation processes. Recently, and due to the specific requirement of the dairy industries and suppliers of the sector, studies aimed at evidencing the presence of phages in WPC samples obtained from local suppliers has been started in Argentina (results not yet published). The first results demonstrated that most of the samples studied (80%) were contaminated with phages that were lytic on one or more commercial strains of *S. thermophilus*, currently used by the local fermentative dairy industry. It is noteworthy that some samples were especially problematic because, from them, it was possible to isolate phages that infected a large number of strains from different commercial starter cultures. Another important aspect to highlight is that the phage concentration in many samples (58%) was around levels considered highly dangerous ($> 10^4$ PFU/g) (Atamer et al. 2013). Concerning the genetic characterization of the isolated phages, a very high diversity among them was confirmed, a situation that reinforces the idea of the danger in the use of these whey derived products without any type of prior evaluation in this regard. All the existing information related to this problem highlights that the presence of infective phages in WPC samples is a significant threat when they are used as ingredients in new fermentation processes, intensified by the high concentration evidenced, in some cases, and the great diversity among them. A broad characterization of these phages (phenotypic and genetic) is essential in order to implement control strategies that ensure correct activity of the starter cultures and the desired quality of final products. Likewise, a phage monitoring protocol in these derivatives should be implemented in the dairy industry in order to minimize the risks caused by these valuable products.

14.10 Conclusion

The market for functional ingredients is constantly rising due to growing health awareness among consumers and changes in eating patterns around the world, driving the growing need for healthy foods. This fact encourages the development and discovery of new ingredients and new and innovative applications to be of current research interest.

Whey is an abundant and highly polluting by-product of the dairy industry. However, in the last decades, the perception of whey as a pollutant has changed with the discovery of valuable constituents with bioactive properties or the conversion of these constituents in valuable biomolecules or high added derived products. A variety of these, mainly derived from bovine milk whey, are obtained at industrial scale and commercialized for multiple applications (food, pharmaceutical, among other).

In this chapter, we presented the typical physicochemical composition of whey, an overview of the technologies employed for converting whey in ingredients, their structures, nutritional and functional aspects, biological activities, purification or isolating methods, and food applications. The scientific interest of this topic is demonstrated through the large number of research papers available to date. The results published are widely variable and in some cases are controversial. Regarding the purity of the ingredient in the compound or biomolecule of interest, for some cases the technological methods are still too expensive to be applied on an industrial scale. Furthermore, the analytical tools that are necessary to verify or perform a thorough and complete characterization of the ingredient are also expensive and not routine. On the other hand, in relation to health benefits, further research is needed particularly in humans to fully substantiate the role of antihypertensive, antimicrobial and antioxidant peptides derived from whey protein, GMP, GOS and lactosucrose.

References

Abd El-Salam, M. and El-Shibiny, S. (2018). Glycation of whey proteins: technological and nutritional implications. *International Journal of Biological Macromolecules* 112: 83–92.

Abd El-Salam, M.H. (2006). Separation of casein glycomacropeptide from whey: methods of potential industrial application. *International Journal of Dairy Science* 1: 93–99.

Abd El-Salam, M., El-Shibiny, S., and Salem, A. (2009). Factors affecting the functional properties of whey protein products: a review. *Food Research International* 25: 251–270.

Abubakar, A, Saito, T, Kitazawa, H, Kawai, Y, Itoh, T (1998). Structural analysis of new antihypertensive peptides derived from cheese whey protein by proteinase K digestion. *Journal of Dairy Science* 81: 3131–3138.

Aburto, C., Guerrero, C., Vera, C., Wilson, L., and Illanes, A. (2016). Simultaneous synthesis and purification (SSP) of galacto-oligosaccharides in batch operation. *LWT - Food Science and Technology* 72: 81–89.

Ackerman, D.L., Craft, K.M., and Townsend, S.D. (2017). Infant food applications of complex carbohydrates: structure, synthesis, and function. *Carbohydrate Research* 437: 16–27.

Agyei, D., Ongkudon, C.M., Wei, C.Y., Chan, A.S., and Danquah, M.K. (2016). Bioprocess challenges to the isolation and purification of bioactive peptides. *Food and Bioproducts Processing* 98: 244–256.

Aimutis, W. (2004). Bioactive properties of milk proteins with particular focus on anticariogenesis. *The Journal of Nutrition* 134 (4): 989S–995S.

Alemán, A., Mastrogiacomo, I., López-Caballero, M.E., Ferrari, B., Montero, M.P., and Gómez-Guillén, M.C. (2016). A novel functional wrapping design by complexation of ε-polylysine with liposomes entrapping bioactive peptides. *Food and Bioprocess Technology* 9 (7): 1113–1124.

Alizadeh, O. and Aliakbarlu, J. (2020). Effects of ultrasound and ohmic heating pretreatments on hydrolysis, antioxidant and antibacterial activities of whey protein concentrate and its fractions. *LWT - Food Science and Technology* 131: 109913.

Anand, S., Nath, K.S., and Chenchaiah, M. (2013). Chapter 22 Whey and whey products. In: Milk and Dairy Products in Human Nutrition: Production, Composition and Health (ed. Y. Park and G. Haenlein), 477–497. Oxford: John Wiley & Sons Ltd.

Arakawa, K., Aoyama, Y., Ikeda, H., Mikuni, K., Fijita, K., and Hara, K. (2002). The development of lactosucrose production and its applications in foods for specified health use. *Journal of Applied Glycoscience* 49 (1): 63–72.

Arunkumar, A. and Etzel, M. (2018). Fractionation of glycomacropeptide from whey using positively charged ultrafiltration membranes. *Foods* 7 (166): 1–10.

Atamer, Z., Ali, Y., Neve, H., Heller, K.J., and Hinrichs, J. (2011). Thermal resistance of bacteriophages attacking flavour-producing dairy *Leuconostoc* starter cultures. *International Dairy Journal* 21: 327–334.

Atamer, Z., Dietrich, J., Müller-Merbach, M., Neve, H., Heller, K.J., and Hinrichs, J. (2009). Screening for and characterization of *Lactococcus lactis* bacteriophages with high thermal resistance. *International Dairy Journal* 19: 228–235.

Atamer, Z., Samtlebe, M., Heller, K.J., and Hinrichs, J. (2013). Review: elimination of bacteriophages in whey and whey products. *Frontiers in Microbiology* 4: 1–9.

Athira, S., Mann, B., Sharma, R., and Kumar, R. (2013). Ameliorative potential of whey protein hydrolysate against paracetamol-induced oxidative stress. *Journal of Dairy Science* 96: 1431–1437.

Athira, S., Mann, B., Sharma, R., Kumar, R., Saini, P., and Singh, A.K. (2015). Production and characterization of whey protein hydrolysate having antioxidant activity from cheese whey. *Journal of the Science of Food and Agriculture* 95. doi: 10.1002/jsfa.7032.

Ballatore, M., Bettiol, M., Vanden Braber, N., Aminahuel, C., Rossi, V., Petroselli, G., Erra-Balsells, R., Cavaglieri, L., and Montenegro, M. (2020). Antioxidant and cytoprotective effect of peptides produced by hydrolysis of whey protein concentrate with trypsin. *Food Chemistry* 319. doi: 10.1016/j.foodchem.2020.126472.

Balthazar, C.F., Silva, H.L.A., Celeguini, R.M.S., Santos, R., Pastore, G.M., Conte Junior, C.A., Freitas, M.Q., Nogueira, L.C., Silva, M.C., and Cruz, A.G. (2015). Effect of galactooligosaccharide addition on the physical, optical, and sensory acceptance of vanilla ice cream. *Journal of Dairy Science* 98: 4266–4272.

Bamdad, F., Bark, S., Kwon, C.H., Suh, J.W., and Sunwoo, H. (2017). Anti-inflammatory and antioxidant properties of peptides released from beta-lactoglobulin by high hydrostatic pressure-assisted enzymatic hydrolysis. *Molecules* 22. doi: 10.3390/molecules22060949.

Bellamy, W., Takase, M., Wakabayashi, H., Kawase, K., and Tomita, M. (1992a). Antibacterial spectrum of lactoferricin B, a potent bactericidal peptide derived from the N-terminal region of bovine lactoferrin. *Journal of Applied Microbiology* 73: 472–479.

Bellamy, W., Takase, M., Yamauchki, K., Wakabayashi, H., Kawase, K., and Tomita, M. (1992b). Identification of the bactericidal domain of lactoferrin. *Biochimica et Biophysica Acta* 1121: 130–136.

Belsito, P.C., Ferreira, M.V.S., Cappato, L.P., Cavalcanti, R.N., Vidal, V.A.S., Pimentel, T.C., Esmerino, E., Balthazar, C.F., Neto, R.P.C., Tavares, M.I.B., Zacarchenco, P.B., Freitas, M.Q., Silva, M.C., Raices, R.S.L., Pastore, G.M., Pollonio, M.A.R., and Cruz, A.G. (2017). Manufacture of Requeijao cremoso processed cheese with galactooligosaccharide. *Carbohydrate Polymers* 174: 869–875.

Benkerroum, N. (2010). Antimicrobial peptides generated from milk proteins: a survey and prospects for application in the food industry. A review. *International Journal of Dairy Technology* 63 (3): 320–338.

Berrocal, R. and Neeser, J.R. (1991) Process for producing kappa-casein-glycomacropeptide. European Patent Office, EPO453782A1.

Böger, M., Hekelaar, J., van Leeuwen, S.S., Dijkhuizen, L., and Lammerts van Bueren, A. (2019a). Structural and functional characterization of a family GH53 β-1,4-galactanase from *Bacteroides thetaiotaomicron* that facilitates degradation of prebiotic galactooligosaccharides. *Journal of Structural Biology* 205: 1–10.

Böger, M., van Leeuwen, S.S., van Bueren, A.L., and Dijkhuizen, L. (2019b). Structural identity of galactooligosaccharide molecules selectively utilized by single cultures of probiotic bacterial strains. *Journal of Agricultural and Food Chemistry* 67: 13969–13977.

Bonnaillie, L., Qi, P., Wickham, E., and Tomasula, P. (2014). Enrichment and purification of casein glycomacropeptide from whey protein isolate using supercritical carbon dioxide processing and membrane ultrafiltration. *Foods* 3: 94–109.

Bosso Tomal, A.A., Farinazzo, F.S., Bachega, A., Bosso, A., da Silva, J.B., and Suguimoto, H. (2019). Galactooligosacharides and human health implications. *Nut. Food Sci. Int. J.* 9 (1). doi: 10.19080/NFSIJ.2019.09.555754

Briggiler Marcó, M., Suárez, V., Quiberoni, A., and Pujato, S. (2019). Inactivation of dairy bacteriophages by thermal and chemical treatments. *Viruses* 11: 1–28.

Brody, E.P. (2000). Biological activities of bovine glycomacropeptide. *British Journal of Nutrition* 84: 39–46.

Burns, P., Binetti, A., Torti, P., Kulozik, U., Forzani, L., Renzulli, P., Vinderola, G., and Reinheimer, J. (2015). Administration of caseinomacropeptide-enriched extract to mice enhances the calcium content of femur in a low-calcium diet. *International Dairy Journal* 44: 15–20.

Capra, M.L., Neve, H., Sorati, P.C., Atamer, Z., Hinrichs, J., Heller, K.J., and Quiberoni, A. (2013). Extreme thermal resistance of phages isolated from dairy samples: updating traditional phage detection methodologies. *International Dairy Journal* 30: 59–63.

Carter, B., Patel, H., Barbano, D., and Drake, M. (2018). The effect of spray drying on the difference in flavor and functional properties of liquid and dried whey proteins, milk proteins, and micellar casein concentrates. *Journal of Dairy Science* 101: 1–10.

Carvalho, F., Prazeres, A., and Rivas, J. (2013). Cheese whey wastewater: characterization and treatment. *Science of the Total Environment* 445–446: 385–396.

Chaudhry, Q., Scotter, M., Blackburn, J., Ross, B., Boxall, A., Castle, L., Aitken, R., and Watkins, R. (2008). Applications and implications of nanotechnologies for the food sector. *Food Additives & Contaminants. Part A, Chemistry, Analysis, Control, Exposure & Risk Assessment* 25 (3): 241–258.

Cheng, C.-C., Yu, M.-C., Cheng, T.-C., Sheu, D.-C., Duan, K.-J., and Tai, W.-L. (2006). Production of high-content galacto-oligosaccharide by enzyme catalysis and fermentation with *Kluyveromyces marxianus*. *Biotechnology Letters* 28: 793–797.

Claumarchirant, L., Sanchez-Siles, L.M., Matencio, E., Alegría, A., and Lagarda, M.J. (2016). Evaluation of sialic acid in infant feedings: contents and bioavailability. *Journal of Agricultural and Food Chemistry* 64: 8333–8342.

Córdova, A., Astudillo, C., Santibañez, L., Cassano, A., Ruby-Figueroa, R., and Illanes, A. (2017). Purification of galacto-oligosaccharides (GOS) by three-stage serial nanofiltration units. *Chemical Engineering Research & Design* 117: 488–499.

Corrochano, A., Buckin, V., Kelly, P., and Giblin, L. (2018). Whey proteins as antioxidants and promoters of cellular antioxidant pathways. *Journal of Dairy Science* 101: 1–15.

Corrochano, A., Sariçay, Y., Arranz, E., Kelly, P., Buckin, V., Giblin, L. (2019). Comparison of antioxidant activities of bovine whey proteins before and after simulated gastrointestinal digestion. *Journal of Dairy Science* 102: 1–14.

Coulier, L., Timmermans, J., Bas, R., Van Den Dool, R., Haaksman, I., Klarenbeek, B., Slaghek, T., and Van Dongen, W. (2009). In-depth characterization of prebiotic galactooligosaccharides by a combination of analytical techniques. *Journal of Agricultural and Food Chemistry* 57: 8488–8495.

Crisà, A. (2013). Chapter 7 Milk carbohydrates and oligosaccharides. In: Milk and Dairy Products in Human Nutrition. Production, Composition and Health (ed. Y.W. Park and G.F.W. Haenlein), 129–141. Oxford: Wiley-Blackwell.

Das, R., Sen, D., Sarkar, A., Bhattacharyya, S., and Bhattacharjee, C. (2011). A comparative study on the production of galacto-oligosaccharide from whey permeate in recycle membrane reactor and in enzymatic batch reactor. *Industrial & Engineering Chemistry Research* 50 (2): 806–816.

de Aguilar-Nascimento, J.E., Regis Prado Silveira, B., and Borges Dock-Nascimento, D. (2011). Early enteral nutrition with whey protein or casein in elderly patients with acute ischemic stroke: a double-blind randomized trial. *Nutrition* 27 (4): 440–444.

del Rio, B., Binetti, A., Martín, M., Fernández, M., Magadán, A., and Alvarez, M. (2007). Multiplex PCR for the detection and identification of dairy bacteriophages in milk. *Food Microbiology* 24 (1): 75–78.

Demers-Mathieu, V., Gauthier, S.F., Britten, M., Fliss, I., Robitaille, G., and Jean, J. (2013). Antibacterial activity of peptides extracted from tryptic hydrolyzate of whey protein by nanofiltration. *International Dairy Journal* 28: 94–101.

Doultani, S., Turhan, K., and Etzel, M. (2003). Whey protein isolate and glycomacropeptide recovery from whey using ion exchange chromatography. *Journal of Food Science* 68 (4): 1389–1395.

Dullius, A., Goettert, M., and Volken de Souza, C. (2018). Whey protein hydrolysates as a source of bioactive peptides for functional foods – biotechnological facilitation of industrial scale-up. *Journal of Functional Foods* 42: 58–74.

Elgamily, H., Safwat, E., Soliman, Z., Salama, H., El-Sayed, H., and Anwar, M. (2019). Antibacterial and remineralization efficacy of casein phosphopeptide, glycomacropeptide nanocomplex, and probiotics in experimental toothpastes: an in vitro comparative study. *European Journal of Dental* 13 (3): 391–398.

Elias, R., Kellerby, S., and Decker, E. (2008). Antioxidant activity of proteins and peptides. *Critical Reviews in Food Science and Nutrition* 48: 430–441.

Fakuda, S., Roig, S., and Prata, L. (2004). Correlation between acidic ninhydrin and HPLC methods to evaluate fraudulent addition of whey in milk. *Lait* 84: 501–512.

Farias, D.P., Fernandes de Araújo, F., Neri-Numa, I.A., and Pastore, G.M. (2019). Prebiotics: trends in food, health and technological applications. *Trends in Food Science and Technology* 93: 23–35.

Feng, Y.M., Chang, X.L., Wang, W.H., and Ma, R.Y. (2009). Separation of galacto-oligosaccharides mixture by nanofiltration. *Journal of the Taiwan Institute of Chemical Engineers* 40 (3): 326–332.

Fernández, L., Escobedo, S., Gutierrez, D., Portilla, S., Martinez, B., Garcia, P., and Rodriguez, A. (2017). Bacteriophages in the dairy environment: from enemies to allies. *Antibiotics* 6: 1–14.

FitzGerald, R.J. and Meisel, H. (1999). Lactokinins: whey protein derived ACE inhibitory peptides. *Nahrung* 43: 165–167.

FitzGerald, R.J., Murray, B.A., and Walsh, D.J. (2004). Hypotensive peptides from milk proteins. *The Journal of Nutrition* 134: 980S–988S.

Frenzel, M., Krolak, E., Wagner, A.E., and Steffen-Heins, A. (2015). Physicochemical properties of WPI coated liposomes serving as stable transporters in a real food matrix. *LWT - Food Science and Technology* 63 (1): 527–534.

Frenzel, M. and Steffen-Heins, A. (2015). Whey protein coating increases bilayer rigidity and stability of liposomes in food-like matrices. *Food Chemistry* 173: 1090–1099.

Fujita, K., Hara, K., Hashimoto, H., and Kitahara, S. (1990). Transfructosylation catalyzed by β-fructofuranosidase I from Arthrobacter sp. K-1. *Agricultural and Biological Chemistry* 54: 2655–2661.

Gammoh, S., Alu'datt, M.H., Tranchant, C.C., Al-U'datt, D.G., Alhamad, M.N., Rababah, T., Kubow, S., Haddadin, M.S.Y., Ammari, Z., Maghaydah, S., and Banat, H. (2020). Modification of the functional and bioactive properties of camel milk casein and whey proteins by ultrasonication and fermentation with *Lactobacillus delbrueckii* subsp. *lactis*. *LWT - Food Science and Technology* 129: 109501. doi: 10.1016/j.lwt.2020.109501.

Ganzle, M. (2011). Lactose: galacto-oligosaccharides. In: Encyclopedia of Dairy Science (ed. J.W. Fuquay, P.L.H. McSweeney and P.F. Fox), 2195–2202. London: Elsevier Ltd.

Garcia-Galan, C., Berenguer-Murcia, A., Fernandez-Lafuente, R., and Rodrigues, R.C. (2011). Potential of different enzyme immobilization strategies to improve enzyme performance. *Advanced Synthesis & Catalysis* 353: 2885–2904.

Gauthier, S. and Pouliot, Y. (2003). Functional and biological properties of peptides obtained by enzymatic hydrolysis of whey proteins. *Journal of Dairy Science* 86: E78–E87.

Geng, X., Tolkach, A., Otte, J., and Ipsen, R. (2015). Pilot-scale purification of α-lactalbumin from enriched whey protein concentrate by anion-exchange chromatography and ultrafiltration. *Dairy Science & Technology* 95: 353–368.

Gernigon, G., Schuck, P., Jeantet, R., and Burling, H. (2011). Demineralization. In: Encyclopedia of Dairy Science (ed. J.W. Fuquay, P.L.H. McSweeney and P.F. Fox), 738–743. London: Elsevier Ltd.

Ghorbanzade, T., Jafari, S.M., Akhavan, S., and Hadavi, R. (2017). Nano-encapsulation of fish oil in nano-liposomes and its application in fortification of yogurt. *Food Chemistry* 216: 146–152.

Gibson, G.R., Probert, H.M., Loo, J.V., Rastall, R.A., and Roberfroid, M.B. (2004). Dietary modulation of the human colonic microbiota: updating the concept of prebiotics. *Nutrition Research Reviews* 17 (2): 259–275.

Gibson, G.R. and Roberfroid, M.B. (1995). Dietary modulation of the human colonic microbiota: introducing the concept of prebiotics. *The Journal of Nutrition* 125: 1401–1412.

Glenn, R., Gibson, G.R., Probert, H.M., Loo, J.V., Rastall, R.A., and Roberfroid, M.B. (2004). Dietary modulation of the human colonic microbiota: updating the concept of prebiotics. *Nutrition Research Reviews* 17 (2): 259–275.

Gobbetti, M., Stepaniak, L., De Angelis, M., Corsetti, A., and Di Cagno, R. (2002). Latent bioactive peptides in milk proteins: proteolytic activation and significance in dairy processing. *Critical Reviews in Food Science and Nutrition* 42 (3): 223–239.

Gomaa, A.I., Martinent, C., Hammami, R., Fliss, I., and Subirade, M. (2017). Dual coating of liposomes as encapsulating matrix of antimicrobial peptides: development and characterization. *Frontiers in Chemistry* 5 (103). doi: 10.3389/fchem.2017.00103.

Goulas, A., Kapasakalidis, P.G., Sinclair, H.R., Rastall, R.A., and Grandison, A.S. (2002). Purification of oligosaccharides by nanofiltration. *Journal of Membrane Science* 209 (1): 321–335.

Goulas, A., Tzortzis, G., and Gibson, G.R. (2007). Development of a process for the production and purification of α- and β-galactooligosaccharides from *Bifidobacterium bifidum* NCIMB 41171. *International Dairy Journal* 17 (6): 648–656.

Guerrero, C., Vera, C., Novoa, C., Dumont, J., Acevedo, F., and Illanes, A. (2014). Purification of highly concentrated galacto-oligosaccharide preparations by selective fermentation with yeasts. *International Dairy Journal* 39: 78–88.

Han, W.-C., Byun, S.-H., Kim, M.-H., Sohn, E.H., Lim, J.D., Um, B.H., Kim, C.H., Kang, S.A., and Jang, K.H. (2009). Production of lactosucrose from sucrose and lactose by a levansucrase from Zymomonas mobilis. *Journal of Microbiology and Biotechnology* 19: 1153–1160.

Hanau, S., Almugadam, S.H., Sapienza, E., Cacciari, B., Manfrinato, M.C., Trentini, A., and Kennedy, J.F. (2020). Schematic overview of oligosaccharides, with survey on their major physiological effects and a focus on milk ones. *Carbohydrate Polymer Technologies and Applications* 1. doi: 10.1016/j.carpta.2020.100013.

Hartmann, R. and Meisel, H. (2007). Food-derived peptides with biological activity: from research to food applications. *Current Opinion in Biotechnology* 18: 163–169.

Hayes, M., Stanton, C., Fitzgerald, G.F., and Ross, R.P. (2007). Putting microbes to work: dairy fermentation, cell factories and bioactive peptides. Part II: bioactive peptide functions. *Biotechnology Journal* 2: 435–449.

Henning, D., Baer, R., Hassan, A., and Dave, R. (2006). Major advances in concentrated and dry milk products, cheese, and milk fat-based spreads. *Journal of Dairy Science* 89 (11): 1179–1188.

Hernandez, O., Ruiz Matute, A.I., Olano, A., Moreno, F.J., and Sanz, M.L. (2009). Comparison of fractionation techniques to obtain prebiotic galactooligosaccharides. *International Dairy Journal* 19 (9): 531–536.

Hernández-Ledesma, B., Davalos, A., Bartolome, B., and Amigo, L. (2005). Preparation of antioxidant enzymatic hydrolysates from alpha-lactalbumin and beta-lactoglobulin. Identification of active peptides by HPLC-MS/MS. *Journal of Agricultural and Food Chemistry* 53: 588–593.

Hernández-Ledesma, B., Recio, I., Ramos, M., and Amigo, L. (2002). Preparation of ovine and caprine β-lactoglobulin hydrolysates with ACE-inhibitory activity. Identification of active peptides from caprine β-lactoglobulin hydrolysed with thermolysin. *International Dairy Journal* 12: 805–812.

Hussein, F.A., Chay, S.Y., Ghanisma, S.B.M., Zarei, M., Auwal, S.M., Hamid, A.A., Ibadullah, W.Z.W., and Saari, N. (2020). Toxicity study and blood pressure–lowering efficacy of whey protein concentrate hydrolysate in rat models, plus peptide characterization. *Journal of Dairy Science* 103: 2053–2064.

Jelen, P. (2011). Whey processing. In: Encyclopedia of Dairy Science (ed. J.W. Fuquay, P.L.H. McSweeney and P.F. Fox), 3839–3845. London: Elsevier Ltd.

Jimenez, M., Garcia, H.S., and Beristain, C.I. (2008). Sensory evaluation of dairy products supplemented with microencapsulated conjugated linoleic acid (CLA). *LWT - Food Science and Technology* 41 (6): 1047–1052.

Jong, W.Y. and Seung, K.S. (1993). The production of high-content fructo-oligosaccharides from sucrose by the mixed-enzyme system of fructosyltransferase and glucose oxidase. *Biotechnology Letters* 15 (6): 573–576.

Kawasaki, Y., Kawakami, H., Tanimoto, M., Dosako, S., Tomizawa, A., Kotake, M., and Nakajima, I. (1993). pH-Dependent molecular weight changes of casein glycomacropeptide and its preparation by ultrafiltration. *Milchwissenschaft* 48: 191–196.

Kelleher, S.L., Chatterton, D., Nielsen, K., and Leonnerdal, B. (2003). Glycomacropeptide and α-lactalbumin supplementation of infant formula affects growth and nutritional status in infant rhesus monkeys. *American Journal of Clinical Nutrition* 77: 1261–1268.

Kittibunchakul, S., van Leeuwen, S.S., Dijkhuizen, L., Haltrich, D., and Nguyen, T.-H. (2020). Structural comparison of different galacto-oligosaccharide mixtures formed by β–galactosidases from lactic acid bacteria and Bifidobacteria. *Journal of Agricultural and Food Chemistry* 68: 4437–4446.

Kitts, D. and Weiler, K. (2003). Bioactive proteins and peptides from food sources. Applications of bioprocesses used in isolation and recovery. *Current Pharmaceutical Design* 9 (16): 1309–1323.

Klewicki, R. (2007). The stability of gal-polyols and oligosaccharides during pasteurization at a low pH. *LWT - Food Science and Technology* 40: 1259–1265.

Korhonen, H. (2009). Milk-derived bioactive peptides: from science to applications. *Journal of Functional Foods* 1 (2): 177–187.

Korhonen, H. and Pihlanto, A. (2006). Bioactive peptides: production and functionality. *International Dairy Journal* 16: 945–960.

Korhonen, H.J. (2010). Chapter 2. Bioactive components in bovine milk. In: Bioactive Components in Milks and Dairy Products (ed. Y.W. Park), 13–42. State Avenue: Wiley-Blackwell.

Kruschitza, A. and Nidetzkya, B. (2020). Downstream processing technologies in the biocatalytic production of oligosaccharides. *Biotechnology Advances* 43. doi: 10.1016/j.biotechadv.2020.107568.

Kumar, P., Sharma, N., Ranjan, R., Kumar, S., Bhat, Z., and Jeong, D. (2013). Perspective of membrane technology in dairy industry: a review. *Asian-Australasian Journal of Animal Sciences* 26 (9): 1347–1358.

LaClair, C.E., Ney, D.M., MacLeod, E.L., and Etzel, M.R. (2009). Purification and use of glycomacropeptide for nutritional management of phenylketonuria. *Journal of Food Science* 74 (4): E199–E206.

Laloy, E., Vuillemard, J.-C., and Simard, R. (1998). Characterization of liposomes and their effect on the properties of Cheddar cheese during ripening. *Lait* 78 (4): 401–412.

Lans, A.M., Frelka, J.C., Paluri, S., and Vodovotz, Y. (2018). Physical properties and sensory analysis of galacto-oligosaccharide glassy confections. *LWT - Food Science and Technology* 96: 499–506.

Laye, C., McClements, D.J., and Weiss, J. (2008). Formation of biopolymer-coated liposomes by electrostatic deposition of chitosan. *Journal of Food Science* 73 (5): N7–N15.

Le Maux, S., Nongonierma, A., Barre, C., and FitzGerald, R. (2016). Impact of enzyme inactivation conditions during the generation of whey protein hydrolysates on their physicochemical and bioactive properties. *Food Chemistry* 199: 246–251.

Li, E. and Mine, Y. (2004). Technical note: comparison of chromatographic profile of glycomacropeptide from cheese whey isolated using different methods. *Journal of Dairy Science* 87: 174–177.

Li, W., Wang, K, Sun, Y., Ye, H., Hu, B., Zeng, X. (2015). Lactosucrose and its analogues derived from lactose and sucrose: Influence of structure on human intestinal microbiota *in vitro*. *Journal of Functional Foods* 17: 73–82.

Li, W., Xiang, X., Tang, S., Hu, B., Tian, L., Sun, Y., Ye, H., and Zeng, X. (2009). Effective enzymatic synthesis of lactosucrose and its analogues by beta-D-galactosidase from Bacillus circulans. *Journal of Agricultural and Food Chemistry* 57 (9): 3927–3933.

Li, Z., Xiao, M., Lu, L., and Li, Y. (2008). Production of non-monosaccharide and high purity galactooligosaccharides by immobilized enzyme catalysis and fermentation with immobilized yeast cells. *Process Biochemistry* 43: 896–899.

Lim, K., Van Calcar, S., Nelson, K., Gleason, S., and Ney, D. (2007). Acceptable low-phenylalanine foods and beverages can be made with glycomacropeptide from cheese whey for individuals with PKU. *Molecular Genetics and Metabolism* 92: 176–178.

Liu, W., Hou, Y., Jin, Y., Wang, Y., Xu, X., and Han, J. (2020). Research progress on liposomes: application in food, digestion behavior and absorption mechanism. *Trends in Food Science and Technology* 104: 177–189.

Liu, W., Liu, J., Li, T., Liu, C., and Liu, W. (2013). Improved physical and in vitro digestion stability of a polyelectrolyte delivery system based on layer-by-layer self-assembly alginate-chitosan-coated nanoliposomes. *Journal of Agricultural and Food Chemistry* 61 (17): 4133–4144.

Liu, W., Liu, W., Ye, A., Peng, S., Wei, F., Liu, C., and Han, J. (2016). Environmental stress stability of microencapsules based on liposomes decorated with chitosan and sodium alginate. *Food Chemistry* 196: 396–404.

Lollo, P.C.B., Amaya-Farfan, J., Faria, I.C., Salgado, J.V.V., Chacon-Mikahil, M.P.T., Cruz, A.G., Oliveira, C.A.F., Montagner, P.C., and Arruda, M. (2014). Hydrolysed whey protein reduces muscle damage markers in Brazilian elite soccer players compared with whey protein and maltodextrin. A twelve-week in-championship intervention. *International Dairy Journal* 34: 19–24.

Long, J., Pan, T., Xie, Z., Xu, X., and Jin, Z. (2020). Co-immobilization of β-fructofuranosidase and glucose oxidase improves the stability of Bi-enzymes and the production of lactosucrose. *LWT - Food Science and Technology* 128. doi: 10.1016/j.lwt.2020.109460.

Lund, M. and Baron, C. (2010). Chapter 2 Chemical deterioration and physical instability of food and beverages. In: Protein Oxidation in Foods and Food Quality (ed. L.H. Skibsted, J. Risbo, and M.L. Andersen), 33–69. Cambridge: Woodhead Publishing Ltd.

Madera, C., Monjardin, C., and Suárez, J.E. (2004). Milk contamination and resistance to processing conditions determine the fate of *Lactococcus lactis* bacteriophages in dairies. *Applied and Environmental Microbiology* 70: 7365–7371.

Madureira, A.R., Pereira, C.I., Gomes, A.M.P., Pintado, M.E., and Malcata, F.X. (2007). Bovine whey proteins. Overview of their main biological properties. *Food Research International* 40: 1197–1211.

Madureira, A.R., Tavares, T., Gomes, A.M.P., Pintado, M.E., and Malcata, F.X. (2010). Invited review: physiological properties of bioactive peptides obtained from whey proteins. *Journal of Dairy Science* 93: 437–455.

Mann, B., Athira, S., Sharma, R., Kumar, R., and Sarkar, P. (2019). Chapter 14 Bioactive peptides from whey proteins. In: Whey Proteins. From Milk to Medicine (ed. H. Deeth and N. Bansal), 519–548. Queensland: Academic Press. Elsevier Inc.

Mann, B., Kumari, A., Kumar, R., Sharma, R., Prajapati, K., Mahboob, S., and Athira, S. (2015). Antioxidant activity of whey protein hydrolysates in milk beverage system. *Journal of Food Science and Technology* 52: 3235–3241.

Mano, M.C.R., Paulino, B.N., and Pastore, G.M. (2019). Whey permeate as the raw material in galacto-oligosaccharide synthesis using commercial enzymes. *Food Research International* 124: 78–85.

Manso, M.A. and Lopez-Fandiño, R. (2004). k-casein macropeptides from cheese whey: physicochemical, biological, nutritional, and technological features for possible use. *Food Reviews International* 20: 329–355.

Marella, C., Muthukumarappan, K., and Metzge, L. (2011). Evaluation of commercially available, wide-pore ultrafiltration membranes for production of α-lactalbumin–enriched whey protein concentrate. *Journal of Dairy Science* 94: 1165–1175.

Marsanasco, M., Márquez, A.L., Wagner, J.R., Chiaramoni, N.S., and Alonso, S. Del V. (2015). Bioactive compounds as functional food ingredients: characterization in model system and sensory evaluation in chocolate milk. *Journal of Food Engineering* 166: 55–63.

Marsanasco, M., Márquez, A.L., Wagner, J.R., Del, S., Alonso, V., and Chiaramoni, N.S. (2011). Liposomes as vehicles for vitamins E and C: an alternative to fortify orange juice and offer vitamin C protection after heat treatment. *Food Research International* 44 (9): 3039–3046.

Martin, M., Kopaliani, I., Jannasch, A., Mund, C., Todorov, V., Henle, T., and Deussen, A. (2015). Antihypertensive and cardioprotective effects of the dipeptide isoleucine–tryptophan and whey protein hydrolysate. *Acta Physiologica* 215: 167–176.

Martin-Diana, A.B., Fraga, M., and Fontecha, J. (2002). Isolation and characterization of caseinomacropeptide from bovine, ovine and caprine cheese whey. *European Food Research and Technology* 214: 282–286.

Martínez-Villaluenga, C., Cardelle-Cobas, A., Corzo, N., Olano, A., and Villamiel, M. (2008). Optimization of conditions for galactooligosaccharide synthesis during lactose hydrolysis by β-galactosidase from *Kluyveromyces lactis* (Lactozym 3000 L HP G). *Food Chemistry* 107: 258–264.

Martins, G.N., Ureta, M.M., Tymczyszyn, E.E., Castilho, P.C., and Gomez-Zavaglia, A. (2019). Technological aspects of the production of fructo and galacto-oligosaccharides. Enzymatic synthesis and hydrolysis. *Frontiers in Nutrition* 6 (78): 1–24.

Meisel, H. (2005). Biochemical properties of peptides encrypted in bovine milk proteins. *Current Medicinal Chemistry* 12: 1905–1919.

Michelon, M., Manera, A.P., Carvalho, A.L., and Maugeri Filho, F. (2014). Concentration and purification of galacto-oligosaccharides using nanofiltration membranes. *International Journal of Food Science & Technology* 49 (8): 1953–1961.

Mohanty, D., Jena, R., Choudhury, P.K., Pattnaik, R., Mohapatra, S., and Saini, M.R. (2016). Milk derived antimicrobial bioactive peptides: a review. *International Journal of Food Properties* 19: 837–846.

Moughan, P.J., Rutherfurd, S.M., Montoya, C.A., and Dave, L.A. (2014). Food-derived bioactive peptides – a new paradigm. *Nutrition Research Reviews* 27: 16–20.

Mozafari, M.R., Johnson, C., Hatziantoniou, S., and Demetzos, C. (2008). Nanoliposomes and their applications in food nanotechnology. *Journal of Liposome Research* 18 (4): 309–327.

Mu, W., Chen, Q., Xiao, W., Zhang, T., and Jiang, B. (2013). Current studies on physiological functions and biological production of lactosucrose. *Applied Microbiology and Biotechnology* 97: 7073–7080.

Mullally, M.M., Meisel, H., and FitzGerald, R.J. (1997). Identification of a novel angiotensin-I-converting enzyme inhibitory peptide corresponding to a tryptic fragment of bovine β-lactoglobulin. *FEBS Letters* 402: 99–101.

Murphy, J., Mahony, J., Fitzgerald, G.F., and van Sinderen, D. (2017). Bacteriophages infecting lactic acid bacteria. In: Cheese: Chemistry, Physics and Microbiology (ed. P. McSweeney, P. Fox, P. Cotter, and D. Everett), 249–272. Cambridge: Academic Press.

Nakano, T. and Ozimek, L. (2000). Purification of glycomacropeptide from caseinate hydrolysate by gel chromatography and treatment with acidic solution. *Journal of Food Science* 65 (4): 588–590.

Neelima, R.S., Rajput, Y.S., and Mann, B. (2013). Chemical and functional properties of glycomacropeptide (GMP) and its role in the detection of cheese whey adulteration in milk: a review. *Dairy Science & Technology* 93: 21–43.

Neeser, J.R., Offord Cavin, E., Felix, R., Tullberg-Reinert, H., Ginty, F., and Barclay, D. (2000) Milk protein hydrolysate for addressing a bone or dental disorder. International Patent. WO 0049885.

Neri, D., Balcão, V., Costa, R., Rocha, I., Ferreira, E., Torres, D., Rodrigues, L., Carvalho, L., and Teixeira, J. (2009). Galacto-oligosaccharides production during lactose hydrolysis by free *Aspergillus orizae* β-galactosidase and immobilized on magnetic polysiloxanepolyvinyl alcohol. *Food Chemistry* 115: 92–99.

Nielsen, P.M. and Tromholt, N. (1994) Method for production of a kappa-casein glycomacropeptide and use of a kappa-casein glycomacropeptide. WO1994015952A1.

O'Keeffe, M.B., Conesa, C., and FitzGerald, R.J. (2017). Identification of angiotensin converting enzyme inhibitory and antioxidant peptides in a whey protein concentrate hydrolysate produced at semi-pilot scale. *International Journal of Food Science & Technology* 52: 1751–1759.

Okabe, H., Aga, H., Kubota, M., and Miyabe, T. (2008) Lactosucrose high content saccharide, its preparation and uses. US Patent No. 20080027027.

Önay-Ucar, E., Arda, N., Pekmez, M., Yilmaz, A.M., Boke-Sarikahya, N., Kirmizigul, S., and Yalcin, A.S. (2014). Comparison of antioxidant capacity, protein profile and carbohydrate content of whey protein fractions. *Food Chemistry* 150: 34–40.

Orsi, N. (2004). The antimicrobial activity of lactoferrin: current status and perspectives. *BioMetals* 17: 189–196.

Otieno, D.O. (2010). Synthesis of β-galactooligosaccharides from lactose using microbial β-galactosidases. *Comprehensive Reviews in Food Science and Food Safety* 9: 471–482.

Palai, T., Mitra, S., and Bhattacharya, P.K. (2012). Kinetics and design relation for enzymatic conversion of lactose into galacto-oligosaccharides using commercial grade β-galactosidase. *Journal of Bioscience and Bioengineering* 114 (4): 418–423.

Pan, L., Zhang, X., Fan, X., Li, H., Xu, B., and Li, X. (2020). Whey protein isolate coated liposomes as novel carrier systems for astaxanthin. *European Journal of Lipid Science and Technology* 122 (4). doi.org/10.1002/ejlt.201900325.

Pandya, A.J. and Haenlein, G.F.W. (2010). Chapter 5 Bioactive components in buffalo milk. In: Bioactive Components in Milks and Dairy Products (ed. Y.W. Park), 105–158. State Avenue: Wiley-Blackwell.

Park, A.-R. and Oh, D.-K. (2010). Galacto-oligosaccharide production using microbial β-galactosidase: current state and perspectives. *Applied Microbiology and Biotechnology* 85: 1279–1286.

Park, N.H., Choi, H.J., and Oh, D.K. (2005). Lactosucrose production by various microorganisms harboring levansucrase activity. *Biotechnology Letters* 27 (7): 495–497.

Pellegrini, A., Dettling, C., Thomas, U., and Hunziker, P. (2001). Isolation and characterization of four bactericidal domains in the bovine β-lactoglobulin. *Biochimica et Biophysica Acta* 1526: 131–140.

Pellegrini, A., Thomas, U., Bramaz, N., Hunziker, P., and Fellenberg, R. (1999). Isolation and identification of three bactericidal domains in the bovine α-lactalbumin molecule. *Biochimica et Biophysica Acta* 1426: 439–448.

Peña-Ramos, E.A. and Xiong, Y.L. (2001). Antioxidative activity of whey protein hydrolyzates in a liposomal system. *Journal of Dairy Science* 84: 2577–2583.

Peng, X., Kong, B., Xia, X., and Liu, Q. (2010). Reducing and radical-scavenging activities of whey protein hydrolysates prepared with Alcalase. *International Dairy Journal* 20 (5): 360–365.

Peng, X., Xiong, Y.L., and Kong, B. (2009). Antioxidant activity of peptide fractions from whey protein hydrolysates as measured by electron spin resonance. *Food Chemistry* 113 (1): 196–201.

Pihlanto-Lepälä, A. (2001). Bioactive peptides derived from bovine whey proteins: opioid and ace-inhibitory peptides. *Trends in Food Science and Technology* 11: 347–356.

Pihlanto-Leppälä, A. (2006). Antioxidative peptides derived from milk proteins. *International Dairy Journal* 16: 1306–1314.

Pihlanto-Leppälä, A. (2000). Bioactive peptides derived from bovine whey proteins: opioid and ace-inhibitory peptides. *Trends in Food Science and Technology* 11: 347–356.

Pihlanto-Leppälä, A., Koskinen, P., Phlola, K., Tupasela, T., and Korhonen, H. (2000). Angiotensin I-converting enzyme inhibitory properties of whey protein digests: concentration and characterization of active peptides. *The Journal of Dairy Research* 67 (1): 53–64.

Pihlanto-Leppälä, A., Marnila, P., Hubert, L., Rokka, T., Korhonen, H.J., and Karp, M. (1999). The effect of alpha-lactalbumin and beta-lactoglobulin hydrolysates on the metabolic activity of Escherichia coli JM103. *Journal of Applied Microbiology* 87: 540–545.

Pinto Coelho Silvestre, M., Ramalho Silva, M., Dias Medeiros Silva, V., Santana de Souza, M., de Oliveira Lopes Junior, C., and de Oliveira Afonso, W. (2012). Analysis of whey protein hydrolysates: peptide profile and ACE inhibitory activity. *Brazilian Journal of Pharmaceutical Sciences* 48 (4): 747–757.

Playne, M. (2003). Galacto-oligosaccharides. In: Encyclopedia of Dairy Science (ed. H. Roginski, J. Fuquay, and P. Fox), 1151–1158. London: Academic Press.

Power-Grant, O., Bruen, C., Brennan, L., Giblin, L., Jakeman, P., and FitzGerald, R.J. (2015). In vitro bioactive properties of intact and enzymatically hydrolysed whey protein: targeting the enteroinsular axis. *Food & Function* 6: 972–980.

Power-Grant, O., McCormack, W.G., Ramia De Cap, M., Amigo Benavent, M., Fitzgerald, R.J., and Jakeman, P. (2016). Evaluation of the antioxidant capacity of a milk protein matrix in vitro and in vivo in women aged 50–70 years. *International Journal of Food Sciences and Nutrition* 67: 325–334.

Prasad, L.N., Sherkat, F., and Shah, N.P. (2013). Influence of galactooligosaccharides and modified waxy maize starch on some attributes of yogurt. *Journal of Food Science* 78: M77–M83.

Prazeres, A., Carvalho, F., and Rivas, F. (2012). Cheese whey management: a review. *Journal of Environmental Management* 110: 48–68.

Pujato, S., Quiberoni, A., and Mercanti, D. (2019). Bacteriophages on dairy foods. *Journal of Applied Microbiology* 126: 14–30.

Pujato, S.A., Guglielmotti, D.M., Ackermann, H.W., Patrignani, F., Lanciotti, R., Reinheimer, J.A., and Quiberoni, A. (2014). *Leuconostoc* bacteriophages from blue cheese manufacture: long-term survival, resistance to thermal treatments, high pressure homogenization and chemical biocides of industrial application. *International Journal of Food Microbiology* 177: 81–88.

Rao, D.R., Chawan, C.B., and Veeramachaneni, R. (1994). Liposomal encapsulation of b-galactosidase: comparison of two methods of encapsulation and in vitro lactose digestibility. *Journal of Food Biochemistry* 18 (4): 239–251.

Rasti, B., Erfanian, A., and Selamat, J. (2017). Novel nanoliposomal encapsulated omega-3 fatty acids and their applications in food. *Food Chemistry* 230: 690–696.

Ren, H., Fei, J., Shi, X., Zhao, T., Cheng, H., Zhao, N., Chen, Y., and Ying, H. (2015). Continuous ultrafiltration membrane reactor coupled with nanofiltration for the enzymatic synthesis and purification of galactosyl-oligosaccharides. *Separation and Purification Technology* 144: 70–79.

Rodriguez-Colinas, B., Kolida, S., Baran, M., Ballesteros, A.O., Rastall, R.A., and Plou, F.J. (2013). Analysis of fermentation selectivity of purified galacto-oligosaccharides by in vitro human faecal fermentation. *Applied Microbiology and Biotechnology* 97: 5743–5752.

Rojas, E. and Torres, G. (2013). Isolation and recovery of glycomacropeptide from milk whey by means of thermal treatment. *Food Science and Technology Campinas* 33 (1): 14–20.

Rojas, E., Valbuena, E., Torres, G., García, A., Piñero, M., and Mila Galindo, L. (2009). Aislamiento y rendimiento del GMP mediante precipitación del lactosuero con ácido tricloroacético. *Revista Científica* (Maracaibo) 19 (3): 295–302.

Saito, T., Yamaji, A., and Itoh, T. (1991). A new isolation method of caseinoglycopeptide from sweet cheese whey. *Journal of Dairy Science* 74: 2831–2837.

Samtlebe, M., Wagner, N., Brinks, E., Neve, H., Heller, K.J., Hinrichs, J., and Atamer, Z. (2017a). Production of phage free cheese whey: design of a tubular laboratory membrane filtration system and assessment of a feasibility study. *International Dairy Journal* 71: 17–23.

Samtlebe, M., Wagner, N., Neve, H., Heller, K.J., Hinrichs, J., and Atamer, Z. (2015). Application of a membrane technology to remove bacteriophages from whey. *International Dairy Journal* 48: 38–45.

Samtlebe, M., Wagner, N., Neve, H., Heller, K.J., Hinrichs, J., and Atamer, Z. (2017b). Reduction of *Lactococcus lactis* phage contamination in whey by means of membrane filtration: impact of phage morphology and of bacterial host cells functioning as "phage fishing tool". *International Dairy Journal* 68: 88–94.

Sangwan, V., Tomar, S., Singh, R.R.B., Singh, A.K., and Ali, B. (2011). Galactooligosaccharides: novel components of designer foods. *Journal of Food Science* 76: 103–111.

Sangwan, V., Tomar, S.K., Ali, B., Singh, R.R.B., Singh, A.K., and Mandal, S. (2014). Galactooligosaccharides purification using microbial fermentation and assessment of its prebiotic potential by in vitro method. *International Journal of Current Microbiology and Applied Sciences* 3 (4): 573–585.

Sarmadi, B.H. and Ismail, A. (2010). Antioxidative peptides from food proteins: a review. *Peptides* 31: 1949–1956.

Schüpbach, R., Neeser, R., Golliard, M., Rouvet, M., and Guggenheim, B. (1996). Incorporation of caseinoglycomacropeptide and caseinophosphopeptide into the salivary pellicle inhibits adherence of mutans streptococci. *Journal of Dental Research* 75: 1779–1788.

Scicchitano, B.M., Pelosi, I., Sica, G., and Musarò, A. (2018). The physiopathologic role of oxidative stress in skeletal muscle. *Mechanisms of Ageing and Development* 170: 37–44.

Sekhon, B.S. (2010). Food nanotechnology – an overview. *Nanotechnology, Science and Applications* 3: 1–15.

Sen, D., Gosling, A., Stevens, G.W., Bhattacharya, P.K., Barber, A.R., Kentish, S.E., Bhattacharjee, C., and Gras, S.L. (2011). Galactosyl oligosaccharide purification by ethanol precipitation. *Food Chemistry* 128: 773–777.

Setareh Nejad, A., Kanekanian, A., and Tatham, A. (2009). The inhibitory effect of glycomacropeptide on dental erosion. *Dairy Science & Technology* 89: 233–239.

Setareh Nejad, A., Kanekanian, A., Tatham, A., and Abedi, A. (2010). The protective effect of caseinomacropeptide against dental erosion using hydroxyapatite as a model system. *International Dairy Journal* 20: 652–656.

Shashi, K., Satinder, K., and Bharat, P.A. (2012). A complete review on liposomes. *International Research Journal of Pharmacy* 3: 10–16.

Silva Duarte, L., Schöffer, J. de N., Soibelmann Glock Lorenzoni, A., Rodrigues, E., and Hertz, P. (2017). A new bioprocess for the production of prebiotic lactosucrose by animmobilized β-galactosidase. *Process Biochemistry* 55: 96–103.

Silva-Hernández, E., Nakano, T., and Ozimek, L. (2002). Isolation and analysis of κ-casein glycomacropeptide from goat sweet whey. *Journal of Agricultural and Food Chemistry* 50 (7): 2034–2038.

Silvério, S.C., Macedo, E.A., Teixeira, J.A., and Rodrigues, L. (2015). Perspectives on the biotechnological production and potential applications of lactosucrose: a review. *Journal of Functional Foods* 19: 74–90.

Singh, R.S. (2014). Chapter 24 Biotechnological approaches for valorization of whey. In: Advances in Industrial Biotechnology (ed. R.S. Singh, A. Pandey, and C. Larroche), 443–478. New Delhi: International Publishing House Pvt. Ltd.

Sipola, M., Finckenberg, P., Vapaatalo, H., Pihlanto-Leppälä, A., Korhonen, H., Korpela, R., and Nurminen, M.-L. (2002). α-lactorphin and β-lactorphin improve arterial function in spontaneously hypertensive rats. *Life Sciences* 71 (11): 1245–1253.

Smithers, G. (2008). Whey and whey proteins-from 'gutter-to-gold'. *International Dairy Journal* 18: 695–704.

Smithers, G. (2015). Whey-ing up the options - Yesterday, today and tomorrow. *International Dairy Journal* 48: 2–14.

Soares de Castro, R., Pavan Bagagli, M., and Sato, H.H. (2015). Improving the functional properties of milk proteins: focus on the specificities of proteolytic enzymes. *Current Opinion in Food Science* 1: 64–69.

Soares de Castro, R., Fontenele Domingues, M.A., Ohara, A., Okuro, P.K., Gonçalves dos Santos, J., Peres Brexó, R., and Sato, H.H. (2017). Whey protein as a key component in food systems: physicochemical properties, production technologies and applications. *Food Structure*. doi: 10.1016/j.foostr.2017.05.004.

Sukkar, S.G., Vaccaro, A., Ravera, G.B., Borrini, C., Gradaschi, R., Sacchi-Nemours, A.M., Cordera, R., and Andraghetti, G. (2013). Appetite control and gastrointestinal hormonal behavior (CCK, GLP-1, PYY 1–36) following low doses of a whey protein-rich nutraceutic. *Mediterranean Journal of Nutrition and Metabolism* 6: 259–266.

Takeuchi, H., Matsui, Y., Sugihara, H., Yamamoto, H., and Kawashima, Y. (2005). Effectiveness of submicron-sized, chitosan-coated liposomes in oral administration of peptide drugs. *International Journal of Pharmaceutics* 303 (1–2): 160–170.

Tanimoto, M., Kawasaki, Y., Dosako, S., Ahiko, K., and Nakajima, I. (1992). Large-scale preparation of κ-casein glycomacropeptide from rennet casein whey. *Bioscience, Biotechnology, and Biochemistry* 56: 140–141.

Tanimoto, M., Kawasaki, Y., Shinmoto, H., Dosako, S., and Tomizawa, A. (1991) Process for producing kappa-casein glycomacropeptides. US Patent 5075424.

Tavares, T.G., Contreras, M.M., Amorim, M., Martín-Alvarez, P.J., Pintado, M.E., and Recio, I. (2011). Optimisation, by response surface methodology, of degree of hydrolysis and antioxidant and ACE-inhibitory activities of whey protein hydrolysates obtained with cardoon extract. *International Dairy Journal* 21: 926–933.

Tavares, T.G. and Malcata, F.X. (2013). Whey proteins as source of bioactive peptides against hypertension. In: Bioactive Food Peptides in Health and Diseases, (ed. B. Hernández-Ledesma and C.C. Hsieh), 75–95. Rijeka: Intech.

Tek, H.N., Turhan, K.N., and Etzel, M.R. (2005). Effect of conductivity, pH and elution buffer salinity on glycomacropeptide recovery from whey using anion-exchange chromatography. *Journal of Food Science* 70: E295–E300.

Tetra Pak (2015). Whey processing. Chapter 15, Dairy Processing Handbook. https://dairyprocessinghandbook.tetrapak.com/chapter/whey-processing.

Théolier, J., Hammami, R., Labelle, P., Fliss, I., and Jean, J. (2013). Isolation and identification of antimiobial peptides derived by peptic cleavage of whey protein isolate. *Journal of Functional Foods* 5 (2): 706–714.

Thomä-Worringer, C., López-Fandiño, R., and Sørensen, J. (2006). Health effects and technological features of caseinomacropeptide. *International Dairy Journal* 16: 1324–1333.

Tolkach, A. and Kulozik, U. (2005). Fractionation of whey proteins and caseinomacropeptide by means of enzymatic crosslinking and membrane separation techniques. *Journal of Food Engineering* 67 (1–2): 13–20.

Tunick, M.H. (2008). Chapter 1 Whey protein production and utilization: a brief history. In: Whey Processing, Functionality, and Health Benefits (ed. C. Onwulata and P. Huth), 1–14. Ames: John Wiley & Sons, Inc.

Urrutia, P., Rodriguez-Colinas, B., Fernandez-Arrojo, L., Ballesteros, A.O., Wilson, L., and Illanes, A. (2013). Detailed analysis of galactoologosaccharides synthesis with β-galactosidase from *Aspergillus orizae*. *Journal of Agricultural and Food Chemistry* 61: 1081–1087.

Valko, M., Leibfritz, D., Moncol, J., Cronin, M.T.D., Mazur, M., and Telser, J. (2007). Free radicals and antioxidants in normal physiological functions and human disease. *The International Journal of Biochemistry & Cell Biology* 39 (1): 44–84.

Vandenplas, Y. (2002). Oligosaccharides in infant formula. *British Journal of Nutrition* 87 (2): S293–S296.

Vavrusova, M., Pindstrup, H., Johansen, L., Andersen, M., Andersen, H., and Skibsted, L. (2015). Characterisation of a whey protein hydrolysate as antioxidant. *International Dairy Journal* 47: 86–93.

Vélez, M.A., Perotti, M.C., Hynes, E.R., and Martinez Ferez, A. (2015). Formulation and characterization of liposomes containing fatty acids omega-3 for food applications. Proc. Fifth Euro-Global Summit Expo Food Beverages, *Journal of Food Processing & Technology* 6: 114.

Vélez, M.A., Perotti, M.C., Santiago, L., Gennaro, A.M., and Hynes, E.R. (2017). Bioactive compounds delivery using nanotechnology: design and applications in dairy food. In: Nutrient Delivery (ed. A.M. Grumezescu), 221–250. Netherlands: Elsevier.

Vélez, M.A., Zeiter, A., Capra, M.L., Pozza, L., Hynes, E.R., and Perotti, M.C. (2020). Developing yogurts containing CLA isomers-loaded liposomes and free CLA oil. *International Dairy Journal* accepted for publication. doi: 10.1016/j.idairyj.2020.104956.

Vénica, C.I., Bergamini, C.V., and Perotti, M.C. (2017). Response surface methodology as a tool for modeling the galactooligosaccharides production. *The Journal of Dairy Research* 84 (4): 464–470.

Vénica, C.I., Bergamini, C.V., Rebechi, S.R., and Perotti, M.C. (2015). Galacto-oligosaccharides formation during manufacture of different varieties of yogurt. Stability through storage. *LWT - Food Science and Technology* 63 (1): 198–205.

Vera, C., Córdova, A., Aburto, C., Guerrero, C., Suárez, S., and Illanes, A. (2016). Synthesis and purification of galacto-oligosaccharides: state of the art. *World Journal of Microbiology & Biotechnology* 32: 2–20.

Vera, C., Illanes, A., and Guerrero, C. (2021). Enzymatic production of prebiotic oligosaccharides. *Current Opinion in Food Science* 37: 160–170.

Virtanen, T., Pihlanto, A., Akkanen, S., and Korhonen, H. (2007). Development of antioxidant activity in milk whey during fermentation with lactic acid bacteria. *Journal of Applied Microbiology* 102 (1): 106–115.

Wagner, N., Brinks, E., Samtlebe, M., Hinrichs, J., Atamer, Z., Kot, W., Franz, C.M.A.P., and Neve, H. (2017a). Whey powders are a rich source and excellent storage matrix for dairy bacteriophages. *International Journal of Food Microbiology* 241: 308–317.

Wagner, N., Samtlebe, M., Franz, C.M.A.P., Neve, H., Heller, K.J., Hinrichs, J., and Atamer, Z. (2017b). Dairy bacteriophages isolated from whey powder: thermal inactivation and kinetic characterisation. *International Dairy Journal* 68: 95–104.

Wagner, N., Matzen, S., Walte, H.-G., Neve, H., Franz, C.M.A.P., Heller, K.J., and Hammer, P. (2018). Extreme thermal stability of *Lactococcus lactis* bacteriophages: evaluation of phage inactivation in a pilot-plant pasteurizer. *LWT - Food Science and Technology* 92: 412–415.

Walsh, D.J., Bernard, H., Murray, B.A., MacDonald, J., Pentzien, A.-K., Wright, G.A., Wal, J.-M., Struthers, A.D., Meisel, H., and FitzGerald, R.J. (2004). In vitro generation and stability of the lactokinin beta-lactoglobulin fragment (142–148). *Journal of Dairy Science* 87 (11): 3845–3857.

Wang, B., Brand-Miller, J., McVeagh, P., and Petocz, P. (2001). Concentration and distribution of sialic acid in human milk and infant formula. *American Journal of Clinical Nutrition* 74 (4): 510–515.

Wechtersbach, L., Ulrih, N.P., and Cigić, B. (2012). Liposomal stabilization of ascorbic acid in model systems and in food matrices. *LWT - Food Science and Technology* 45 (1): 43–49.

Whisner, C.M., Martin, B.R., Schoterman, M.H.C., Nakatsu, C.H., McCabe, L.D., McCabe, G.P., Wastney, M.E., van den Heuvel, E.G.H.M., and Weaver, C.M. (2013). Galacto-oligosaccharides increase calcium absorption and gut bifidobacteria in young girls: a double-blind cross-over trial. *British Journal of Nutrition* 110 (7): 1292–1303.

WHO (2013). A global brief on hypertension: silent killer, global public health crisis: world health day 2013. World Health Organization, Geneva.

Wisniewski, L., Antosova, M., and Polakovic, M. (2013). Simulated moving bed chromatography separation of galacto-oligosaccharides. *Acta Chim Slovaca* 6: 206–210.

Wu, Q., Zhang, X., Jia, J., Kuang, C., and Yang, H. (2018). Effect of ultrasonic pretreatment on whey protein hydrolysis by alcalase: thermodynamic parameters, physicochemical properties and bioactivities. *Process Biochemistry* 67: 46–54.

Xavier, J.R., Ramana, K.V., and Sharma, R.K. (2018). β-galactosidase: biotechnological applications in food processing. *Journal of Food Biochemistry* 42: 1–15.

Xiao, Y., Chen, Q., Guang, C., Zhang, W., and Mu, W. (2019). An overview on biological production of functional lactose derivatives. *Applied Microbiology and Biotechnology* 103 (9): 3683–3691.

Yamamoto, N. and Takano, T. (1999). Antihypertensive peptides derived from milk proteins. *Nahrung* 43: 159–164.

Yoon, S.H., Mukerjea, R., and Robyt, J.F. (2003). Specificity of yeast (*Saccharomyces cerevisiae*) in removing carbohydrates by fermentation. *Carbohydrate Research* 338: 1127–1132.

Zhao, C. and Ashaolu, T.J. (2020) Bioactivity and safety of whey peptides. *LWT - Food Science and Technology* 134. doi: 10.1016/j.lwt.2020.109935.

Zhidong, L., Benheng, G., Xuezhong, C., Zhenmin, L., Yun, D., Hongliang, H., and Wen, R. (2013). Optimisation of hydrolysis conditions for antioxidant hydrolysate production from whey protein isolates using response surface methodology. *Irish Journal of Agricultural and Food Research* 52: 53–65.

15

EPS from Lactobacilli and Bifidobacteria

Microbial Metabolites with Both Technological and Health-Promoting Properties

*Elisa C. Ale, Melisa A. Puntillo, María F. Rojas, and Ana G. Binetti**

Instituto de Lactología Industrial (CONICET-UNL), Facultad de Ingeniería Química (UNL), Santa Fe (3000), Argentina
Corresponding author: anabinetti@fiq.unl.edu.ar

15.1 Introduction

Exopolysaccharides are complex carbohydrates commonly produced by lactic acid bacteria (LAB) and bifidobacteria, which can be released to the medium (EPS) or remained attached to the cell surface (capsular polysaccharide or CPS) (Castro-Bravo et al. 2018b). Although their natural role has not been fully understood yet, it is believed that they may play a beneficial role for the producing bacteria by acting as a physical barrier under stressing environmental conditions, participating in cell interactions, and modifying different ecological niches (Castro-Bravo et al. 2018a). On the other hand, the benefits EPS exert on the host have been widely studied. They proved to have immunomodulatory effects (Ale et al. 2016a; Liu et al. 2017a), regulate the gut microbiota (Bengoa et al. 2020; Yan et al. 2020), protect the host against pathogenic microorganisms (Ale et al. 2016a; Paik et al. 2018), present antioxidant properties (Xu et al. 2021; Zhang et al. 2013), cholesterol-lowering effects (Korcz et al. 2018), and anti-cancer activity *in vitro* (Xiao et al. 2020a), among others. Moreover, some EPS present interesting rheological properties due to their thickening effects and water-holding capacity within different food matrices (cheese, yogurt, bread, etc.), when produced *in situ* (Tang et al. 2018; Yilmaz et al. 2015) or added as a food ingredient (Ale et al. 2016b). This last strategy is the least applied by the food industry since the EPS yields obtained from both lactobacilli and bifidobacteria are low in comparison with bacteria of other genera, and the costs of extraction/purification processes are generally high. A summary of the technological and health promoting effects attributed to EPS up to date is shown in Figure 15.1. It is important to highlight that all these properties strongly depend on the chemical nature of EPS, as it will determine not only their behavior in the host (Castro-Bravo et al. 2018b), but also in different food matrices (Zhou et al. 2019).

Considering this double techno-functional role, these polymers have been of great interest to the food industry. In the light of these potential applications of EPS and their producing bacteria, the present chapter intends to summarize the most recent information regarding the functional, technological, and chemical properties of EPS from bifidobacteria and lactobacilli, with the aim of providing scientific evidence that supports their application, highlighting the importance of deepening the current knowledge about these interesting molecules.

Biomolecules from Natural Sources: Advances and Applications, First Edition. Edited by Vijai Kumar Gupta, Satyajit D. Sarker, Minaxi Sharma, María Elida Pirovani, Zeba Usmani, and Chelliah Jayabaskaran.

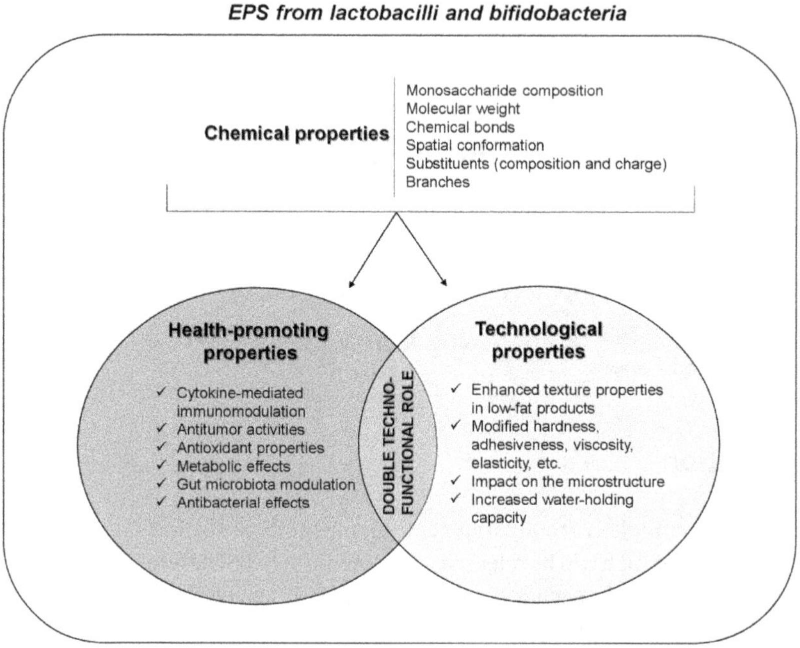

Figure 15.1 Chemical, technological, and functional properties of EPS from lactobacilli and bifidobacteria. The impact their chemical characteristics have on their properties (and possible applications) is shown.

15.2 Chemical and Structural Characterization

The chemical composition and structure of EPS depend on the producing LAB or bifidobacteria strain (Riaz Rajoka et al. 2018), and the fermentation conditions, such as composition of the culture medium and pH seem to influence the characteristics of the EPS produced (Ale et al. 2020; Dertli et al. 2013; Knoshaug et al. 2007; Lal et al. 2019; Polak-Berecka et al. 2015). These biopolymers are unique due to the nature/number/sequence of the monosaccharide residues, anomeric configurations (D or L), ring size (pyranose or furanose), type of glycosidic linkages, type/number/location of the non-carbohydrate substituents and the consequent three-dimensional structure, parameters that will impact on the rheological properties of these molecules (Gerwig 2019).

Based on their chemical composition, EPS are classified in homopolysaccharides (HoPS) and heteropolysaccharides (HePS). HoPS are either linear or branched polymers composed of only one type of neutral monosaccharide, mainly D-glucose (α-D-glucans and β-D-glucans, according to their linkages) or D-fructose (β-fructans) (Lynch et al. 2018b). α-D-glucans are sub-classified based on their bounds into dextrans [α-D-Glc(1→4)], alternans [α-D-Glc(1→6)/α-D-Glc(1→3)], mutans [α-D-Glc(1→3)], and reuterans [α-D-Glc(1→4)/α-D-Glc(1→6)], the last presenting branching points. In addition, the linkages that could be present in β-D-glucans are β-D-Glc(1→3) with (1,2) side chains. On the other hand, β-fructans are classified into inulin type [β-(2→1) bounds] and levan type [β-(2→6) bounds] (Torino et al. 2015). There is also another type of HoPS known as polygalactans, composed of repeating units of galactose with different linkages (Fagunwa et al. 2019). Regarding HePS, they are typically branched polymers which consist of repeating units of more than one type of monosaccharide at different ratios, mostly D-glucose, D-galactose and L-rhamnose, linked by α and β-linkages. Besides, other carbohydrates (e.g., mannose, L-fucose, N-acetylglucosides) and non-carbohydrates constituents (e.g., amino acids), or even charged groups (e.g., acetate, pyruvate, phosphate), can be present. The number of monosaccharides that form the HePS repeating unit generally range from two to eight (Lynch et al. 2018b).

Other characteristics that differentiate HoPS from HePS are the synthetic mechanisms and yields obtained under controlled fermentation conditions. Regarding their biosynthetic pathways, HoPS are extracellularly synthesized through trans glycosylation reactions, catalyzed by glucoside hydrolases (GH family) which use sucrose as precursor. Glucansucrases (GH family 70) and fructansucrases (GH family 68) cleave glycosidic bonds and transfer the resulting glucose or fructose residues to build α-glucans and β-fructans, respectively. Unlike α-glucans, β-glucans are intracellularly synthetized from UDP-glucose by a glycosyltransferase, resembling the mechanism of HePS biosynthesis. Although HePS synthesis has not been fully elucidated, it seems to begin in the cytoplasm, where glycosyltransferases use activated sugar-nucleotides (e.g., UDP-glucose) as precursors. The first step is mediated by the priming glycosyltransferase (pGT), a membrane-anchored enzyme, that catalyzes the binding of the first precursor to a phosphorylated lipid carrier. Then, after a series of transglycosylation steps, a flippase translocate the repeating unit from the cytoplasmic face of the membrane to the periplasmic face, where the repeating units are polymerized and released (Ale et al. 2019b; Oleksy and Klewicka 2018; Zhou et al. 2019). In general, higher yields are obtained for HoPS (g/L) than HePS (mg/L), but this mainly depends on the composition of the medium (especially the C and N sources), and the growth conditions (temperature, pH, time, oxygen; Ale et al. 2020).

HePS are produced by different LAB genera such as *Lactobacillus*, *Lactococcus*, and *Streptococcus*, while HoPS are generally produced by *Lactobacillus*, *Streptococcus*, *Leuconostoc*, *Oenococcus* and *Weisella* (Dimopoulou et al. 2016; van Hijum et al. 2006). Although genes involved in HePS synthesis have been widely found in most *Bifidobacterium* species, HoPS production has not been reported in this genus (Hidalgo-Cantabrana et al. 2014b; Inturri et al. 2017). Table 15.1 shows some examples of EPS from both *Lactobacillus* and *Bifidobacterium*, which have been recently characterized, together with some potential applications.

Since the chemical composition of EPS is quite complex and diverse, their characterization requires the complementation of different techniques. The first step is to obtain EPS from the culture broth (preferably under optimized conditions) by, for example, precipitation with absolute ethanol and, after harvesting the EPS-rich extract, a dialysis step against ultrapure/deionized water is performed. Then, the obtained extract is freeze-dried, and further purification steps can be considered, including treatment with DNAse I, Pronase E and precipitation with trichloroacetic acid to eliminate the remaining proteins and nucleic acids (Ale et al. 2016b). To determine the monosaccharide composition, purified EPS is degraded to their constituents by total acid hydrolysis (sulfuric or trifluoroacetic acid at 120 °C) followed by high-performance

Table 15.1 EPS produced by bifidobacteria (shaded in grey) and lactobacilli (not shaded). The producing strain, chemical structure, and characteristics together with their potential applications, are shown. MW: molecular weight; HePS: heteropolysaccharide; HoPS: homopolysaccharide; Ara: arabinose; Man: mannose; Glc: glucose; Gal: galactose; Fru: fructose; Rha: rhamnose; *p*: pyranose; *f*: furanose.

Strain	Chemical structure and molecular weight	Potential properties/ applications	Reference
B. animalis RH	Branched HePS formed by (1→4)-linked Glc, (1→3,4)-linked Man, (1→4)-linked Rha, and (1→4)-linked Gal as backbone and Gal distributed in branches. The molar ratio of monosaccharides was 0.4 Rha: 0.3 Ara: 1.6 Gal, 0.8 Glc and 1.2 Man.	Further studies are required	Shang et al. (2013)
B. longum W11	D-Glc and D-Gal, with <4% of Rha and galacturonic acid. Two repeating units: one linear chain [→6)-β-Gal*f*-(1→3)-α-Gal*p*-(1→] (MW = 2.81×10^5 Da) and the second unit had (80%) the same linear backbone but the β-Gal*f* is 5-substituted by β-Glc*p* unit (MW = 9.81×10^3 Da)	Antioxidant capability	Inturri et al. (2017)
B. animalis ssp. *lactis* Bb12 *B. breve* 11815 *B. longum* 11818	Unknown structure	EPS from *B. animalis* ssp. *lactis* Bb12 can be used for reducing syneresis and improving texture and viscosity of food products	Lal et al. (2019)
B. breve lw01	Rha, Ara, Gal, Glc and Man in a molar ratio of 0.35:0.44:1.38:0.67:1.65	Anticancer activity against head and neck squamous cell carcinoma cell line	Wang et al. (2019b)
L. plantarum subsp. *plantarum* KX041	HePS (Ara, Man, Glc and Gal in a molar ratio of 0.95:12.94:7.26:3.31). MW = 3.87×10^4 Da	Potential natural antioxidant for the food industry	Wang et al. (2017)
L. plantarum subsp. *plantarum* SKT109	HePS (Fru and Glc in a molar ratio of 3:1). MW = 2.1×10^6 Da	The use of this strain improved the rheology of fermented milk without causing post acidification during storage, and increased the concentration of characteristic flavor compounds	Wang et al. (2015a)

(Continued)

Table 15.1 (Continued)

Strain	Chemical structure and molecular weight	Potential properties/ applications	Reference
L. plantarum subsp. *plantarum* YW32	HePS (Man, Fru, Gal and Glc in a molar ratio of 8.2:1:4.1:4.2). MW = 1.03x10^5 Da	Application in functional foods: improvement of rheological properties, antioxidant activity, antibiofilm activity on several pathogenic bacteria and antitumor activity against human colon cancer HT-29 cells	Wang et al. (2015b)
L. fermentum Lf2	Three EPS were identified: a β-glucan with high molecular mass (1.23x10^6 g/mol) and two HePS of medium molecular mass (average of 8.8 × 10^4 g/mol). One HePS has a main chain of β-1,6-linked galactofuranoses which is non-stoichiometrically 2-*O*-glucosylated. The degree of substitution depends on the fermentation conditions. The second HePS is composed of Glc and Gal and has four monosaccharides in the repeating unit	Increase of both consistency and hardness of low-fat yogurt, without sensory defects. These EPS have also presented functional properties *in vivo* and *in vitro*, such as immunomodulatory effects, prevention of *Salmonella* and synbiotic effects when combined with a probiotic strain	Ahmed et al. (2020); Ale et al. (2016a, 2016b, 2019a), Vitlic et al. (2019)
L. mucosae VG1	D-galactan MW = 1.51 × 10^4 g/mol	Further studies are required	Fagunwa et al. (2019)
L. helveticus LZ–R–5	EPS composed of D-Gal and D-Glc, formed by linear repeating units: [→6)-β-D-Gal*p*-(1→3)-β-DGlc*p*-(1→3)-β-D-Glc*p*-(1→3)-β-D-Glc*p*-(1→3)-β-D-Glc*p*-(1→]. MW = 5.41 × 10^5 Da	Immunostimulatory activity	You et al. (2020)
L. reuteri DSM17938 and *L. reuteri* L26 Bioceol™	Both EPS were D-glucans: EPS-DSM17938 was relatively linear with (1→4) and (1→6) glycosidic linkages and MW = 6.5X10^5 Da. EPS-L26 was branched with (1→3) and (1→6) glycosidic linkages and MW = 8.2X10^5 Da	EPS-DSM17938 increased IL-1β expression and EPS-L26 up-regulated NF-κB, TNF-α and IL-6 mRNA. Potential role as prophylactic agent for gastrointestinal infections	Kšonžeková et al. (2016)
L. rhamnosus KF5	Two fractions of HePS: fraction S1 with 1.36 × 10^4 Da, contained Glc, Ara, glucosamine, galactosamine and Gal in a molar ratio of 2.03:1.29:1.25:0.72:0.61; fraction S2 with 1.23 × 10^6 Da, formed by Rha, Glc, Gal in a molar ratio of 1.73:1.47: 1.00	Immunomodulatory activity (stimulation of splenocyte proliferation *in vitro*)	Shao et al. (2014)

(Continued)

Table 15.1 (Continued)

Strain	Chemical structure and molecular weight	Potential properties/ applications	Reference
L. casei SB27	Two fractions of high MW. The main chain of LW1was formed by (1→4)-linked Gal*p* and (1→4)-linked Glc*p* with side chains (1→4,6)-linked Gal*p* through the O-6 position. LW1 MW = 2.51×10^4 Da. LW2 was composed of a main chain which involved (1→4)-linked Gal*p* and (1→4)-linked Glc*p* with its side chains being (1→3)-Gal*p* through the O-6 position of (1→3,6)-Gal*p* linked to the main chain. LW2 MW = 1.23×10^4 Da	Potential functional food ingredient which could be used as a source of natural antitumour drugs (*in vitro* antitumour activities against H-29 human colon cancer cells)	Di et al. (2017)
L. johnsonii FI9785	EPS-1 branched dextran with every backbone residue substituted with 2-linked Glc and sites with substituent groups: 1-phosphoglycerol and *O*-acetyl groups. EPS-2 was an HePS with a repeating unit: -6)-α-Glc*p*-(1–3)-β-Glc*p*-(1–5)-β-Gal*f*-(1– 6)-α-Glc*p*-(1–4)-β-Gal*p*-(1– 4)-β-Glc*p*-(1- and a single *O*-acetyl group	Further studies are required	Dertli et al. (2013)
L. sakei MN1	HoPS α -glucan: α-D-Glc*p*b1→6). MW = 1.7×10^8 Da	Antiviral agent in aquaculture	Nácher-Vázquez et al. (2015); Zarour et al. (2017)

anion exchange chromatography (Ale et al. 2016b; Kant Lakra et al. 2020) or gas chromatography (Abid et al. 2018; Wang et al. 2017). There are several methods to determine the polymer molecular weight (MW), but it is typically measured by high performance size exclusion chromatography (HP-SEC) coupled with refractive index detector (RI) or, more recently, a multi-angle laser light-scattering detector (MALLS) (Inturri et al. 2017). Structural information such as type of monosaccharides, ring size, anomeric configuration, positions of glycosidic linkages and presence of functional groups can be obtained using some methods/techniques such as ^{1}H and ^{13}C NMR spectroscopy, Fourier transform infrared (FTIR) spectroscopy, X-ray diffraction (Dahech et al. 2013; Kant Lakra et al. 2020).

Up-to-date information about microbial EPS can be found in some databases such as GlyTouCan repository, Bacterial Polysaccharide Gene Database, PolysacDB, Carbohydrate Structure Database, and Polysac3DB. In particular, a well-organized EPS database (EPS-DB) has been recently created, where details about structure, chemical composition, taxonomic and genetic information, bacterial strain, growth conditions, bibliography, and functional properties of bacterial EPS are shown (Birch et al. 2019).

The high diversity in chemical composition and structure of EPS results in a variety of functional and technological properties, and studies on the structure/function relationship might be important to exploit their potential applications (Gerwig 2019; Mende et al. 2016). For instance, Chen et al. (2019) suggested that galactose, rhamnose and glucosamine content of EPS produced by *Limosilactobacillus reuteri* Mh-001 enhanced the anti-inflammatory activity in RAW246.7 macrophages, whereas higher glucose content might suppress this effect. Additionally, the EPS composition has proved to influence on the rheology of different food matrices as well, for example, Chen et al. (2016) compared the effects of dextrans (*Weissella cibaria* 10M) and reuterans (*L. reuteri* TMW 1.656) on the quality of wheat sourdough bread. Enzymatically produced reuterans with different α-(1→4) linkage content and dextrans were added to bread dough and their effects on loaf volume and crumb hardness were evaluated. They found that reuterans with increased α-(1→4) linkage (25–51%) obtained from mutant reuteransucrases enhanced wheat bread volume compared with wild type reuteran (14%). Moreover, reuterans with 39% α-(1→4) linkage and the lowest molecular weight (12.2 MDa) had comparable effects in improvement bread quality than dextrans with ≤ 2% of α-(1→4) linkage and 82.9 MDa. Xiao et al. (2020b) reported that chemical modifications (acetylation, carboxymethylation, sulfation, and phosphorylation) of EPS-2 produced by *Lactobacillus helveticus* MB2-1 had great effects on the structure, physicochemical, and antioxidant properties of this polymer. Briefly, the structure of the granules of EPS-2, thermal stability, and gelatinization properties changed in all cases and antioxidant capacities were improved, mainly with the introduction of sulfate groups. In the case of EPS from *Bifidobacterium*, some controversial results have been obtained. In general, it is hypothesized that neutral EPS with high MW can inhibit the production of pro-inflammatory cytokines, whereas smaller-sized EPS, or acidic ones, have immune stimulating properties (López et al. 2012). These authors found that EPS with a high MW fraction produced by *Bifidobacterium animalis* A1dOxR reduced immune response in human peripheral blood mononuclear cells (PBMC), while the lower MW-EPS produced by *B. animalis* A1 enhanced this response. In contrast, Hidalgo-Cantabrana et al. (2014a) found that the EPS from *B. animalis* A1dOxR had a mild pro-inflammatory (Th1) response on the murine PBMC (but not its purified high MW fraction), while the low-MW EPS obtained from *B. animalis* A1 presented no effects. These contradictions may be due to the different models applied, the first report described the effects on human cells, whereas the second used murine cells. According to Xu et al. (2019), the size and charge of EPS are the factors exerting the greatest impact on the immune properties. EPS with low MW have better immune effects possibly because of their high solubility in water and the feasibility to bind more easily to cell receptors. However, more parameters should be considered when studying the relationship between the chemical and functional/technological properties of these molecules, such as monosaccharide composition, linkage patterns and microstructure, paying special attention to the model applied.

15.3 Rheological Properties

EPS, especially those produced by LAB, can provide different food matrices with enhanced texture and organoleptic characteristics (Ale et al. 2016b; Lynch et al. 2018a). As mentioned above, it has been demonstrated that their technological properties depend on their chemical structure and composition. In this direction, combination of large polymerization degree and hydroxyl groups might be the dominant factors (Zhou et al. 2019). In general, most EPS consist of highly energetic chemical bonds and high polymerization and branching degree, characteristics that make these molecules stiff and rigid. Furthermore, hydroxyl groups strengthen their intermolecular interaction by the formation of hydrogen bonds, enhancing the intrinsic viscosity and rheology of different matrices. This effect can also be produced by carboxyl groups, which reinforce intermolecular hydrogen bonds and subsequently exert high viscosity (Zhou et al. 2019). On the other hand, acetylated groups make chains more elastic and flexible, as observed in gellan gum (Khan et al. 2007). It is widely known that the molecular backbone structure, as well as the chain length, play a crucial role. For example, (1→4) linkages are more rigid than (1→2) and (1→3), and α-linkages are more flexible than β-linkages (Laws et al. 2001).

The variety found in the chemical properties of these carbohydrates made it interesting to explore their application to diverse food matrices of different nature. For example, EPS producing cultures proved to exert promising textural effects on low-fat cheeses, providing rheological properties that resemble those found in the original whole-fat products. In this direction, Ayyash et al. (2018) used a commercial ropy culture and two EPS^+ strains of the species *Lactiplantibacillus plantarum* mixed at a ratio 1:1 (both isolated from camel milk) to make low-fat akawi cheese. This Middle Eastern type of cheese is widely consumed in North Africa, Middle Eastern, and Gulf Cooperation Council countries. The authors found that the magnitude of elastic G' and complex G* moduli of non-EPS cheese were higher than EPS^+ cheeses. In addition, cheeses without EPS exhibited more hardness and adhesiveness than treated ones at 0 and 7 days of storage, characteristics that may be attributed to the higher moisture contents found in the EPS^+ cheeses. The microstructure was also analyzed, and control cheeses presented large pores compared with those made with EPS-producing cultures at day 0, and had wider protein strands which might have provided resistance to compression. On the other hand, treated cheeses showed increased adhesiveness during storage, probably due to stronger interactions between proteins and water. Recently, EPS produced by one of these strains was chemically characterized (Ayyash et al. 2020). The average molecular weight was 3.8×10^5 Da and it was composed of arabinose, mannose, glucose, and galactose as the major monosaccharide constituents.

Similar results regarding the microstructure were described by Şanli et al. (2013), who studied the texture properties of low-fat and reduced fat Turkish cheeses made with EPS^+ starter culture. The protein matrix found in these cheeses was more open and presented a sponge-like structure, together with larger cavities (probably containing water) than the 10% fat control cheese. In addition, Hamet et al. (2015) isolated five strains of *Lacticaseibacillus paracasei* from kefir grains which were able to produce EPS of high MW. The authors found that the viscosity of the acid milk gels obtained did not depend on polysaccharide production but on the MW distribution of these molecules.

Most studies have evaluated the production of EPS *in situ* because of the low yield obtained, but some strains presented relatively high EPS yields making it feasible to apply these molecules as ingredients in different food matrices. In this sense, the EPS extract produced by *Limosilactobacillus fermentum* Lf2, a strain isolated as a non-starter culture of Cremoso cheese, demonstrated to have positive effects on the texture of yogurts when used as an additive. The EPS provided a creamy consistency and increased hardness, without the development of unwanted sensory defects (Ale et al. 2016b). The EPS produced was further analyzed and proved to have immunomodulatory effects (*in vivo* and *in vitro*) (Ale et al. 2016a; Vitlic et al. 2019) and to positively modulate the intestinal microbiota (Ale et al. 2019a). In addition, the chemical structure was recently elucidated (Ahmed et al. 2020) indicating that the purified extract was a complex mixture of a high molecular mass β-glucan (weight average mass of 1.23×10^6 g mol^{-1}) and two medium molecular mass HePS (weight average mass of 8.8×10^4 g mol^{-1}). Similarly, Ikeda et al. (2019) studied the rheological properties of the EPS secreted by *L. fermentum* MTCC 25067. They found that EPS molecules formed three-dimensional networks in an aqueous solution and that the gel-like dynamic viscoelasticity could be attributed to supramolecular assemblies. These assemblies may have resulted from intermolecular associations of the molecules in solution. In this case, the absolute molar mass of the EPS was determined to be 1.85×10^6 g mol^{-1} and, according to previous studies, it is a neutral HePS (Leo et al. 2007).

The effect of EPS^+ strains was also evaluated in non-dairy food matrices. For example, İspirli et al. (2020) studied the *in situ* α-glucan production of *L. reuteri* E81 in sourdough. The strain provided higher elasticity, and bread containing E81 sourdough presented decreased hardness in comparison with the control group at day 0 of storage. Images of the microstructure of lyophilized sourdough showed associations between glucan and starch granules forming a film-like structure. Furthermore, Li et al. (2014a) used *L. plantarum* 70810 to make fermented soymilk and found that this strain increased the water holding capacity, apparent viscosity and EPS amount in comparison with other starter cultures. Observation of microstructure indicated that the EPS-protein networks might have improved the texture.

On the other hand, the impact of EPS-producing bifidobacteria on the texture of fermented products has been analyzed to a lesser extent, probably because of the low yield obtained compared with EPS from LAB. Patrignani et al. (2018) studied seven strains belonging to the genus *Bifidobacterium aesculapii* and their potential to be applied as fermentation starters in soybean milk. Some of them showed good growth potential and provided the final products with good firmness and viscosity indexes. In addition, five strains proved to produce EPS *in situ*. Furthermore, Prasanna et al. (2012b) performed a screening of *Bifidobacterium* strains in low-fat UHT milk to evaluate their growth, acidification properties, EPS production, and ability to increase the viscosity of fermented milk. Among the 22 strains, EPS was produced by strains of the species *B. bifidum*, *B. breve*, *B. longum* and *B. infantis* at concentrations ranging from 25 to 140 mg L^{-1}. Further studies demonstrated that the EPS from some of these strains presented interesting emulsifying and rheological properties (Prasanna et al. 2012a). In another work from the same group (Prasanna et al. 2013), two of these strains were used to produce low-fat yogurt, and those made with the EPS-producing *Bifidobacterium* strains exhibited lower syneresis than the control group. The lowest was shown by *B. longum* subsp. *infantis* CCUG 52486, which also showed the highest storage modulus and firmness. The microstructure indicated that the observed parameters were due to EPS-milk proteins interactions. In summary, this strain was the most promising one, since it can be used with starter cultures to make low-fat set yogurts with both probiotic (You and Yaqoob 2012) and rheological properties.

Overall, this information suggests that EPS-producing strains of both genera, *Bifidobacterium* and *Lactobacillus*, are worth studying for the development of new fermented products with enhanced rheology and, in some cases, with health-promoting benefits as well. In general, the EPS production *in situ* is the most studied strategy to incorporate these molecules into food matrices, since the low yields obtained make it difficult to add them as purified ingredients on a large scale. The main reasons are the time-demanding purification steps to obtain EPS, together with the cost of the extraction process (ethanol precipitation, dialysis, centrifugation steps, enzymatic treatments, etc.).

15.4 Health-Promoting Effects

Lactobacilli and bifidobacteria have shown several and diverse health benefits in different models (*in vitro*, *in vivo*, and within many clinical trials) which were mainly related to their probiotic properties. Fortunately, these effects have been deeply studied seeking the mechanisms involved, and EPS seem to play a crucial role in the functionality of the producing strain. The health-promoting properties these molecules exert are described below, which are also summarized in Table 15.2, focusing on the most recent reports about this subject.

15.4.1 Antitumor Activity

EPS from many LAB and bifidobacteria have demonstrated antitumor or anticancer activity, and they seem to be potential antitumor agents due to their low cytotoxicity and side effects (Rahbar Saadat et al. 2019). Although there are many reports about this property, most studies were done using *in vitro* models. For example, EPS from some lactobacilli strains (*Lactobacillus helveticus* MB2-1, *L. plantarum* 70810, *Lacticaseibacillus casei* SB27, and *Lentilactobacillus kéfiri* MSR101) inhibited the proliferation of cancerous cell lines, such as colon carcinoma cell line HT-29 (Di et al. 2017; Li et al. 2015; Riaz Rajoka et al. 2019; Wang et al. 2014). The first two were effective against HepG-2 and BGC-823, while the others presented anticancer activity by the regulation of apoptosis-associated genes *Caspase-3* and *Caspase-8*. In addition, EPS from *Lactobacillus acidophilus* DSMZ 20079 and *Lactobacillus rhamnosus* exhibited antitumor activity against human colon cancer cells, Caco-2 cells (El-Deeb et al. 2018; Riaz Rajoka et al. 2018). Furthermore, cytotoxicity studies showed that HePS from *L. plantarum* RJF4 was non-toxic to normal cell lines and, at the same time, presented antiproliferative effect on MIA PaCa-2 pancreatic cancer cell line (Dilna et al. 2015). Another strain of the same species, *L. plantarum* NCU116, has also suppressed the proliferation of mouse epithelial colorectal cell line CT26 via caspase-mediated pathways (Zhou et al. 2017). In this direction, EPS from *Lactobacillus gasseri* G10 and H15, both strains isolated from healthy human vagina, presented antiproliferative activity against a human cervical carcinoma (HeLa) cell line (Sungur et al. 2017). On the other hand, the anticancer effects of EPS produced by bifidobacteria were studied to a lesser extent. For instance, Wang et al. (2019b) found that EPS isolated from *B. breve* lw01 presented anticancer activity against head and neck Squamous cell carcinoma cell line, involving the regulation of both the cell cycle and apoptosis mechanisms.

Table 15.2 Summary of the different bifidobacteria (shaded in grey) and lactobacilli (not shaded) EPS-producing strains. Their origins and functional properties attributed to EPS are shown (except for *B. animalis* IPLA R1, whose properties were attributed to the EPS-producing strain).

EPS-producing strain	Origin	Functional properties	Reference
B. animalis IPLA R1	Dairy products	Beneficial effects on metabolic disorders associated with obesity.	Salazar et al. (2019)
B. animalis RH	Feces from centenarians	Enhanced proliferation and phagocytosis activity in RAW264.7 cells.	Liu et al. (2017a)
B. breve lw01	Infant feces	Anticancer potentiality against head and neck squamous cell.	Wang et al. (2019b)
B. longum W11	Commercial strain	*In vitro* antioxidant activity on human lung fibroblast (MRC-5 cell line).	Inturri et al. (2017)
L. casei SB27	Yak milk	*In vitro* antitumor activities, inhibited proliferation of HT-29 colorectal cancer cells by regulation apoptosis-related gene expression.	Di et al. (2017)
L. casei WXD030	Chinese dairy products	Modulation of the immune response *in vitro* (RAW264.7) and potential adjuvant for foot-and-mouth disease (FMD) vaccine in a mouse model.	Xiu et al. (2018)
L. paracasei M7	Human breast milk	Antibiofilm activity against pathogens.	Bhat et al. (2019)
L. rhamnosus SHA111	Human breast milk	Antitumor activity against human colon cancer Caco-2 cells.	Riaz Rajoka et al. (2018)
L. plantarum subsp. *plantarum* BR2	Jack fruit	Antidiabetic potential.	Sasikumar et al. (2017)
L. plantarum subsp. *plantarum* H31	Pickled cabbage	Decreased activity of pancreas α-amylase and regulation of genes expression (*GLUT-4* and *AMPK*).	Huang et al. (2020)
L. plantarum subsp. *plantarum* KX041	Traditional Chinese pickle juice	Strong *in vitro* antioxidant activity.	Wang et al. (2017)
L. plantarum subsp. *plantarum* JLAU103	Hurood, Chinese traditional fermented dairy food	Strong *in vitro* antioxidant activity and stimulated proliferation and phagocytic activity on RAW264.7.	Min et al. (2019); Wang et al. (2020)
L. plantarum subsp. *plantarum* NCU116	Pickled vegetables	Anticancer activity on mouse intestinal epithelial cancer cells CT26.	Zhou et al. (2017)
B. bifidum WBIN03 and *L. plantarum* subsp. *plantarum* R315	Human feces and breast milk, respectively	Strong scavenging activity against DPPH, hydroxyl, and superoxide radicals. Antibacterial activity against pathogens.	Li et al. (2014b)
L. acidophilus DSMZ 20079	Human	*In vitro* anticancer potentiality against human colon cancer (CaCo-2).	El-Deeb et al. (2018)

(Continued)

Table 15.2 (Continued)

EPS-producing strain	Origin	Functional properties	Reference
Lactobacillus sp. Ca6	Gastrointestinal tract of indigenous poultry (Tunisia)	*In vitro* antibacterial activity against pathogens.	Trabelsi et al. (2017)
L. delbureckii subsp. *bulgaricus* (wild type and mutant)	Culture collection from the University of Ibadan	Strong *in vitro* antioxidant activity.	Adebayo-Tayo and Fashogbon (2020)
L. gasseri G10 and H15	Healthy human vagina	Inhibited proliferation of cervical cancer cell (HeLa) and modulation of the immune response.	Sungur et al. (2017)
L. helveticus MB2-1	Traditional Sayram ropy fermented milk	*In vitro* antitumor activity and inhibited proliferation of cancer cell lines (HT-29, HepG-2, BGC-823).	Li et al. (2015)
L. kéfiri MSR101	Chinese kefir grains	*In vitro* antitumor activities and inhibited proliferation of HT-29 colorectal cancer cells. Regulation through apoptosis-related gene expression.	Riaz Rajoka et al. (2019)
L. fermentum Lf2	Argentine cheese	Protection *in vivo* against *Salmonella* Typhimurium, immunomodulatory effects *in vivo* and *in vitro*, synbiotic effects when combined with a probiotic strain, and modulation of the gut microbiota.	Ale et al. (2016a, 2019a)
L. plantarum subsp. *plantarum* RJF4	Rotten jack fruit	Anticancer activity against MiaPaCa2-pancreatic cancer cell line.	Dilna et al. (2015)

To sum up, there is enough evidence about the effects these molecules have on different cancer cell lines. However, further *in vivo* studies, and/or clinical trials, are still missing. More importantly, research on the anticancer activities of purified EPS (instead of the producing strains) using mouse models should be addressed in the future.

15.4.2 Antioxidant Properties

High levels of free radicals are detrimental for our body as they can cause damage to biological macromolecules and tissues. Besides, oxidative stress plays an essential role in several health conditions such as cancer, liver cirrhosis, and fatty liver (Dilna et al. 2015). For this reason, the search for new natural antioxidant molecules is a field of interest nowadays, and EPS from LAB and bifidobacteria seem to be interesting options to consider. The antioxidant effect of EPS has been widely studied (mainly *in vitro*) as it is a topic of interest for the development of new functional foods. For example, EPS from *L. plantarum* JLAU103 and KX041 exhibited strong scavenging activities *in vitro* against hydroxyl, ABTS (2,2′-azinobis-3-ethylbenzothiazoline-6-sulfonic acid) and DPPH (1,1′-diphenyl-2-picrylhydrazyl) radicals (Min et al. 2019; Wang et al. 2017). In addition,

Adebayo-Tayo and Fashogbon (2020) studied the antioxidant properties of the EPS produced by wild-type and mutant *Lactobacillus delbureckii* subsp. *bulgaricus* EPSMLD and EPSWLD, respectively, and found that the DPPH scavenging activity, total antioxidant activity, H_2O_2 scavenging activity, and ferric reducing antioxidant power were higher than the control group (ascorbic acid). Furthermore, the purified EPS from the commercialized probiotic *B. longum* W11 proved to diminish intracellular ROS (reactive oxygen species) and RSH (thiols) levels in human fibroblasts MRC-5 at different concentrations (Inturri et al. 2017). The EPS isolated from another *Bifidobacterium* strain, *B. animalis* RH, has also shown antioxidant activity *in vitro* and radical-scavenging activity against hydroxyl and superoxide radicals (Xu et al. 2011). In this direction, Wang et al. (2019a) found that the CPS of *L. fermentum* S1 presented significant antioxidant activities *in vitro* and *in vivo* (*Caenorhabditis elegans*). This last work highlights the possibility to apply not only EPS (released to the medium), but also polysaccharides which are strongly attached to the cell surface. Nevertheless, the extraction method and the low yields obtained may make it difficult to use them as food ingredients, at least from an economic point of view.

15.4.3 Functional Properties Associated with Carbohydrate and Lipid Metabolisms

Despite the need for further research on the mechanisms involved, EPS from lactobacilli and bifidobacteria were widely associated with a variety of health benefits such as prevention of diabetes, obesity, and cholesterol-related conditions. For example, the antidiabetic potential of EPS obtained from *L. plantarum* H31-2, a strain isolated from pickled cabbage, was recently evaluated *in vitro* (Huang et al. 2020). The authors found that the pancreatic α-amylase was inhibited by both the crude and purified EPS fractions. In addition, the supernatant glucose concentration of insulin-resistant HepG2 cells was reduced by the purified EPS, and the expressions of *GLUT-4*, *Akt-2*, and *AMPK* were upregulated. In this direction, Sasikumar et al. (2017) proved that a high MW glucomannan produced by *L. plantarum* BR2 was able to inhibit *in vitro* both α- amylase and α-glucosidase activities.

On the other hand, some studies about the impact of EPS-producing strains were also addressed using *in vivo* models. For example, Salazar et al. (2019) evaluated the effects of *B. animalis* IPLA R1 (EPS^+) using a mouse model of induced obesity. They found that, after the treatment, mice presented reduced serum insulin and hepatic triglyceride levels. Besides, the treatment downregulated the expression of genes involved in the hepatic synthesis of fatty acids, while upregulated the expression of genes involved in fatty acid oxidation.

Additionally, it is known that elevated cholesterol and high blood pressure are among the risk factors of cardiovascular disease (CVD) (Yildiz and Karatas 2018), and EPS synthesized by LAB have demonstrated cholesterol-lowering activities through, probably, increased secretion of bile acids (Korcz et al. 2018). For instance, a high MW glucomannan from *L. plantarum* BR2 presented elevated antioxidant and radical scavenging activities, together with α-amylase and α-glucosidase inhibition (antidiabetic properties) and reduced cholesterol levels (*in vitro*). More interestingly, the sample did not reveal any cytotoxicity against H9C2 normal cells, suggesting its potential application as a food additive (Sasikumar et al. 2017). In addition, the EPS produced by *L. plantarum* RJF4 displayed these three properties as well: cholesterol-lowering ability, α-amylase inhibition, and antioxidant effects *in vitro* (Dilna et al. 2015). More studies regarding the impact of purified EPS from bifidobacteria on cholesterol metabolism should be addressed in the future, since most of them were focused on EPS produced by lactobacilli.

15.4.4 Microbiota Modulation

There is plenty of information about the microbiota modulation exerted by probiotics, but the interaction of EPS from lactobacilli or bifidobacteria with other gut microbial members has been studied to a lesser extent. Among these interactions, the effect against pathogens and the capacity of these molecules to be metabolized as substrates by other enteric bacteria (prebiotic property), are two phenomena that were studied in some detail (Castro-Bravo et al. 2018b). It should be noticed that, while a probiotic is "a live microorganism that, when administered in adequate amounts, confer a health benefit on the host" (Hill et al. 2014), a prebiotic is defined as "a substrate that is selectively utilized by host microorganisms conferring a health benefit" (Gibson et al. 2017). Furthermore, when prebiotics and probiotics are combined, a synbiotic effect can take place, giving better results than those obtained from the individual components. In other words, synbiotic was recently defined as "a mixture comprising live microorganisms and substrate(s) selectively utilized by host microorganisms that confers a health benefit on the host" (Swanson et al. 2020). As previously mentioned, their prebiotic effects strongly depend on the chemical characteristics of these molecules, since the nature of the linkages, branches, spatial distribution, MW, and chemical composition, among other parameters, will determine their resistance to gastrointestinal digestion. This way, the polysaccharides can reach the gut where those microorganisms with the necessary enzymatic pool can degrade them. Then, the monosaccharides released could influence on the microbiota composition through cross-feeding mechanisms (Rios-Covian et al. 2016). As described above, Salazar et al. (2019) studied the effects of EPS$^+$ *B. animalis* IPLA R1 using an obesity-induced mouse model and, apart from finding that this strain promoted beneficial changes in lipids and glucose homeostasis, they found it had a bifidogenic effect and increased the levels of *Bacteroides-Prevotella*. In this direction, Ale et al. (2019a) evaluated the health benefits of EPS from *L. fermentum* Lf2, solely or combined with a probiotic bifidobacteria, using a mouse model. Those mice receiving yogurt with EPS presented increased concentrations of total short chain fatty acids (SCFA), metabolites that positively modulate colonic function, and multiple inflammatory and metabolic processes (Gill et al. 2018). Besides, the levels of the *Clostridium coccoides* cluster (SCFA-producing group) increased during the treatment. A possible bifidogenic role was observed for the group that received both EPS and the bifidobacteria, since the levels of *Bifidobacterium* increased, suggesting a potential synbiotic effect. Recently, the capacity of EPS to be metabolized *in vitro* by infant fecal microbiota was analyzed (Bengoa et al. 2020). More specifically, these authors studied two EPS (EPS8339 and EPS83124) produced by *L. paracasei* strains which were isolated from kefir grains. They reported that EPS8339 produced an increment of propionate and butyrate together with an augmented proportion of the genera *Victivallis*, *Acidaminococcus* and *Comamonas* and a reduction in the proportion of enterobacteria, while EPS83124 caused an increment of butyrate levels and a marked reduction of *Enterobacteriaceae* with a significant increase in the genus *Comamonas*. The authors hypothesized that the observed responses were probably due to differences in their structural characteristics.

In another work (Yan et al. 2019), the impact of a ropy strain, *B. longum* subsp. *longum* YS108R, using a mouse model with DSS (dextran sulfate sodium)-induced colitis, was studied. The authors found that the strain could decrease the relative abundance of *Enterobacteriaceae* and *Peptostreptococcaceae* and increase the relative abundance of the *Bacteroidales* family S24-7. In summary, this strain could alleviate DSS-induced colitis by modulating the pro-inflammatory response and gut microbiota. Furthermore, a recent work of Püngel et al. (2020) evaluated the microbiota modulation ability of *B. breve* UCC2003 (EPS^+) and its corresponding isogenic EPS-deletion mutant using an *in vitro* infant colon system with feces from healthy, full-term breast-fed infants. The vessels were inoculated with *B. breve* and both microbiome dynamics (16S rRNA sequencing) and metabolomics (NMR) monitored. The authors found that the EPS^+ vessel had increased levels of *Tyzzerella*, *Faecalibacterium* and SCFA, with further taxa-metabolites correlations which were not found for the EPS-negative vessel. In nutshell, this information suggests that EPS, as well as EPS-producing bifidobacteria and lactobacilli, can positively modulate the gut microbiota, property that was successfully demonstrated using both *in vitro* and *in vivo* systems.

15.4.5 Immunomodulatory Effects

EPS have widely shown the ability to modulate the immune system, both innate and adaptive responses, mainly through the regulation of cytokines (Singh and Saini 2017). Apparently, EPS are able to stimulate antigen presenting cells (APC), such as dendritic cells (DC), through Toll-like receptors (TLR), especially TLR2 and TLR4. Cell surface pattern recognition receptors (PRR) may also function as EPS receptors as they can modify transcription and signaling factors, inducing cytokine production (Rahbar Saadat et al. 2019). There are many works on the immunomodulatory effects of these molecules using different models, for example, Wang et al. (2020) showed that the EPS from *L. plantarum* JLAU103 (EPS103) was able to promote the proliferation and improve the phagocytic activity of RAW264.7 macrophages. Besides, it promoted the release of IL-6, TNF-α, and nitric oxide in this cell line, a situation that was reverted in the presence of LPS (lipopolysaccharide). Apparently, these immunomodulatory effects were produced via the nuclear factor-kappa B (NF-κB) signaling pathway. Similarly, Xiu et al. (2018) found that the EPS derived from the probiotic strain *L. casei* WXD030 presented immunostimulant effects in both *in vitro* and *in vivo* assays. More interestingly, the authors indicated that the purified EPS could be an efficacious adjuvant to be used for the development of vaccines. Polysaccharides produced by bidifobacteria have also proved to have immunomodulatory effects *in vitro*, as shown for the EPS from *B. animalis* RH (Liu et al. 2017a). This study suggested that TLR4 was involved in this response. An *in vivo* assay with the same EPS demonstrated that these molecules were able to enhance the mucosal immunity in the small intestine together with the IFN-γ level in serum, balancing the Th1/Th2 responses (Xu et al. 2017).

A recent study from our group (Ale et al. 2019a) about the EPS extract from *L. fermentum* Lf2 showed that the purified EPS was able to increase the levels of the proinflammatory cytokine TNF-α, while this extract in its purified and crude forms enhanced the levels of the regulatory cytokine IL-10. Furthermore, and as described above, the EPS exerted a prebiotic role alone or combined with the probiotic strain *B. animalis* subsp. *lactis* INL1. It is worth noticing that, in this case, the crude EPS extract used produced health-promoting effects on mice when added to yogurt, a strategy that is probably more feasible to apply in the food industry than its purified form, as no further purification steps are required. In another work, the crude EPS fraction was able to act as a moderate immunomodulator, altering s-IgA and IL-6 levels in the small intestine when added to yogurt and milk, respectively. Additionally, this extract significantly prevented mice from *Salmonella* infection (Ale et al. 2016a).

Regarding allergic responses, Domingos-Lopes et al. (2017) suggested that EPS from *Leuconostoc citreum* L3C1E7 can ameliorate them and suppresses IgE synthesis. Similarly, EPS from *B. longum* subsp. *longum* 35624™ was proposed as a potential molecule able to reduce the severity of chronic eosinophil-related airway disorders, effect that was confirmed by *in vitro* and *in vivo* studies (Schiavi et al. 2018). All these results indicate that these molecules, when used in small doses, present strong immunomodulatory effects, characteristic that make them interesting to be applied as functional food ingredients. Besides, they adapt more easily to different food matrices than probiotics, as high cell counts should be maintained during storage in the last case, and parameters such as pH, water content, shelf life of the final product, temperature, exposure to oxygen, among others, are determinant.

15.4.6 Antibiofilm/Antibacterial Activity

Biofilms can be described as populations of microorganisms attached to a surface in response to environmental signals, such as exogenous stress, and allow them to escape from the host immune system (Rahbar Saadat et al. 2019). They can cause important contamination problems in diverse environments (hospitals, industries, homes), which can lead to health and economic negative consequences. Despite the findings of numerous strategies to control them, the exploration of novel and effective antibiofilm agents still continues (Merghni et al. 2017). In this sense, EPS are potential candidates as they may adhere to intestinal mucus and inhibit the adhesion of other bacteria by competition (Lynch et al. 2018b). There are some recent works in which this field has been developed, for instance, EPS from *L. paracasei* M7 presented antibiofilm activity against *Enterococcus faecalis, Bacillus subtilis, Bacillus cereus, Staphylococcus aureus, Klebsiella* sp., and *Pseudomonas aeruginosa* (Bhat et al. 2019). Furthermore, the EPS from *L. casei* NA-2 inhibited the biofilm formation of *B. cereus, S. aureus, S.* Typhimurium and *Escherichia coli* O157:H7 and was able to disperse *B. cereus, S. aureus, S.* Typhimurium and *E. coli* O157:H7 (Xu et al. 2020). Similarly, the EPS from *L. plantarum* WLPL04 inhibited the adhesion of *E. coli* O157:H7 to HT-29 cells in competition, replacement, and inhibition assays. Besides, the EPS exhibited strong inhibition against biofilm formation by *Pseudomonas aeruginosa* CMCC10104, *E. coli* O157:H7, *Salmonella* Typhimurium ATCC13311, and *Staphylococcus aureus* CMCC26003 (Liu et al. 2017b).

Regarding antibacterial activity in general, a novel EPS produced by *Lactobacillus* sp. Ca6 also demonstrated to inhibit *Bacillus*, *Staphylococcus*, and *Salmonella*. More interestingly, this EPS had wound healing activity *in vivo* (excision wound model in rats), accelerating the healing process (Trabelsi et al. 2017). In addition, Li et al. (2014b) proved by agar diffusion assay that EPS from *B. bifidum* WBIN03 and *L. plantarum* R315 have antibacterial activities against *Cronobacter sakazakii*, *E. coli*, *Listeria monocytogenes*, *S. aureus*, *Candida albicans*, *B. cereus*, *S.* Typhimurium and *Shigella sonnei*. It should be noticed that all these antibacterial or antibiofilm effects should be proved *in vivo*, considering the impact different food matrices may have on the results obtained.

15.5 Conclusion

An increasing number of EPS produced by *Lactobacillus* and *Bifidobacterium* strains have been isolated and characterized in the last few years, evidencing a high chemical diversity among them, which is causally related to the wide variety of properties that have already been described for them. In spite of their low yields, in some cases, and even at low concentrations, they are adequate to positively modify the rheology (enhanced texture and organoleptic characteristics) and functional properties of the final product, as observed for some low-fat cheeses. In this context, the use of HoPS and/or HePS from lactobacilli and bifidobacteria or the EPS-producing bacteria could have a promising future in their application as potential techno-functional ingredients for the design of novel foods. Additionally, these molecules are versatile and appropriate to be used in different food matrices because of their wide chemical and structural variability. For this reason, their integral characterization (structure, MW, presence of charged residues, type of branches, etc.) is essential, as recent studies strongly associated these characteristics with their rheological and health-promoting properties. In this way, by knowing the chemical characteristics of the EPS molecules to be added in a food matrix beforehand, the techno-functional properties of the final product could be inferred.

Recently, diverse health benefits were associated with EPS produced by bifidobacteria and lactobacilli, microorganisms widely associated with probiotic properties: immunomodulation, antioxidant role, microbiota modulation, among others. In most cases, these properties were extensively demonstrated *in vitro* and at high EPS concentrations, and, despite some interesting conclusions, their application on real systems such as food seems to be limited. For this reason, an interesting challenge for the future would be demonstrating, by clinical studies, the real effect of these metabolites on the intestinal environment when included at concentrations feasible to be used in a real food matrix.

References

Abid, Y., Casillo, A., Gharsallah, H. et al. (2018). Production and structural characterization of exopolysaccharides from newly isolated probiotic lactic acid bacteria. *International Journal of Biological Macromolecules* 108: 719–728. doi: 10.1016/j.ijbiomac.2017.10.155.

Adebayo-Tayo, B. and Fashogbon, R. (2020). *In vitro* antioxidant, antibacterial, *in vivo* immunomodulatory, antitumor and hematological potential of exopolysaccharide produced by wild type and mutant *Lactobacillus delbureckii* subsp. *bulgaricus*. *Heliyon* 6: e03268. doi: 10.1016/j.heliyon.2020.e03268.

Ahmed, H.I., Ransom-Jones, E., Sadiq, S. et al. (2020). Structural characterisation of two medium molecular mass exopolysaccharides produced by the bacterium *Lactobacillus fermentum* Lf2. *Carbohydrate Research* 488: 107909. doi: 10.1016/j.carres.2020.107909.

Ale, E.C., Batistela, V.A., Correa Olivar, G. et al. (2020). Statistical optimisation of the exopolysaccharide production by *Lactobacillus fermentum* Lf2 and analysis of its chemical composition. *International Journal of Dairy Technology* 73: 76–87. doi: 10.1111/1471-0307.12639.

Ale, E.C., Bourin, J.-B., Peralta, G.H. et al. (2019a). Functional properties of exopolysaccharide (EPS) extract from *Lactobacillus fermentum* Lf2 and its impact when combined with *Bifidobacterium animalis* INL1 in yoghurt. *International Dairy Journal* 96: 114–125. doi: 10.1016/j.idairyj.2019.04.014.

Ale, E.C., Perezlindo, M.J., Burns, P. et al. (2016a). Exopolysaccharide from *Lactobacillus fermentum* Lf2 and its functional characterization as a yogurt additive. *Journal of Dairy Research* 83: 487–492. doi: 10.1017/S0022029916000571.

Ale, E.C., Perezlindo, M.J., Pavón, Y. et al. (2016b). Technological, rheological and sensory characterizations of a yogurt containing an exopolysaccharide extract from *Lactobacillus fermentum* Lf2, a new food additive. *Food Research International* 90: 259–267. doi: 10.1016/j.foodres.2016.10.045.

Ale, E.C., Reinheimer, J., and Binetti, A.G. (2019b). Exopolysaccharides from *Lactobacillus*: biosynthesis, characterization, functional aspects and applications in food industry. In: The Many Benefits of Lactic Acid Bacteria (ed. J.G. LeBlanc and A. de Moreno de LeBlanc), 253–292. Nova Science Publishers

Ayyash, M., Abu-Jdayil, B., Hamed, F. et al. (2018). Rheological, textural, microstructural and sensory impact of exopolysaccharide-producing *Lactobacillus plantarum* isolated from camel milk on low-fat akawi cheese. *LWT - Food Science and Technology* 87: 423–431. doi: 10.1016/j.lwt.2017.09.023.

Ayyash, M., Abu-Jdayil, B., Itsaranuwat, P. et al. (2020). Characterization, bioactivities, and rheological properties of exopolysaccharide produced by novel probiotic *Lactobacillus plantarum* C70 isolated from camel milk. *International Journal of Biological Macromolecules* 144: 938–946. doi: 10.1016/j.ijbiomac.2019.09.171.

Bengoa, A.A., Dardis, C., Gagliarini, N. et al. (2020). Exopolysaccharides from *Lactobacillus paracasei* isolated from kefir as potential bioactive compounds for microbiota modulation. *Frontiers in Microbiology* 11: 58325411. doi: 10.3389/fmicb.2020.583254.

Bhat, B. and Bajaj, B.K. (2019). Hypocholesterolemic potential and bioactivity spectrum of an exopolysaccharide from a probiotic isolate *Lactobacillus paracasei* M7. *Bioactive Carbohydrates and Dietary Fibre* 19: 100191. doi: 10.1016/j.bcdf.2019.100191.

Birch, J., Van Calsteren, M.R., Pérez, S. et al. (2019). The exopolysaccharide properties and structures database: EPS-DB. Application to bacterial exopolysaccharides. *Carbohydrate Polymers* 205: 565–570. doi: 10.1016/j.carbpol.2018.10.063.

Castro-Bravo, N., Sánchez, B., Margolles, A. et al. (2018a). Biological activities and applications of bifidobacterial exopolysaccharides: from the bacteria and host perspective.

In: The Bifidobacteria and Related Organisms, 177–193. Elsevier. doi: 10.1016/B978-0-12-805060-6.00010-7.

Castro-Bravo, N., Wells, J.M., Margolles, A. et al. (2018b). Interactions of surface exopolysaccharides from *Bifidobacterium* and *Lactobacillus* within the intestinal environment. *Frontiers in Microbiology* 9: 2426. doi: 10.3389/fmicb.2018.02426.

Chen, X.Y., Levy, C., and Gänzle, M.G. (2016). Structure-function relationships of bacterial and enzymatically produced reuterans and dextran in sourdough bread baking application. *International Journal of Food Microbiology* 239: 95–102. doi: 10.1016/j.ijfoodmicro.2016.06.010.

Chen, Y.C., Wu, Y.J., and Hu, C.Y. (2019). Monosaccharide composition influence and immunomodulatory effects of probiotic exopolysaccharides. *International Journal of Biological Macromolecules* 133: 575–582. doi: 10.1016/j.ijbiomac.2019.04.109.

Dahech, I., Fakhfakh, J., Damak, M. et al. (2013). Structural determination and NMR characterization of a bacterial exopolysaccharide. *International Journal of Biological Macromolecules* 59: 417–422. doi: 10.1016/j.ijbiomac.2013.04.036.

Dertli, E., Colquhoun, I.J., Gunning, A.P. et al. (2013). Structure and biosynthesis of two exopolysaccharides produced by *Lactobacillus johnsonii* FI9785. *Journal of Biological Chemistry* 288: 31938–31951. 10.1074/jbc.M113.507418.

Di, W., Zhang, L., Wang, S. et al. (2017). Physicochemical characterization and antitumour activity of exopolysaccharides produced by *Lactobacillus casei* SB27 from yak milk. *Carbohydrate Polymers* 171: 307–315. doi: 10.1016/j.carbpol.2017.03.018.

Dilna, S.V., Surya, H., Aswathy, R.G. et al. (2015). Characterization of an exopolysaccharide with potential health-benefit properties from a probiotic *Lactobacillus plantarum* RJF4. *LWT - Food Science and Technology* 64: 1179–1186. doi: 10.1016/j.lwt.2015.07.040.

Dimopoulou, M., Bardeau, T., Ramonet, P. et al. (2016). Exopolysaccharides produced by *Oenococcus oeni*: from genomic and phenotypic analysis to technological valorization. *Food Microbiology* 53: 10–17. doi: 10.1016/j.fm.2015.07.011.

Domingos-Lopes, M.F.P., Nagy, A., Stanton, C. et al. (2017). Immunomodulatory activity of exopolysaccharide producing *Leuconostoc citreum* strain isolated from Pico cheese. *Journal of Functional Foods* 33: 235–243. doi: 10.1016/j.jff.2017.03.054.

El-Deeb, N.M., Yassin, A.M., Al-Madboly, L.A. et al. (2018). A novel purified *Lactobacillus acidophilus* 20079 exopolysaccharide, LA-EPS-20079, molecularly regulates both apoptotic and NF-κB inflammatory pathways in human colon cancer. *Microbial Cell Factories* 17: 29. doi: 10.1186/s12934-018-0877-z.

Fagunwa, O., Ahmed, H.I., Sadiq, S. et al. (2019). Isolation and characterization of a novel exopolysaccharide secreted by *Lactobacillus mucosae* VG1. *Carbohydrate Research* 484: 107781. doi: 10.1016/j.carres.2019.107781.

Gerwig, G.J. (2019). Structural analysis of exopolysaccharides from lactic acid bacteria. In: Lactic Acid Bacteria. Methods in Molecular Biology (ed. M. Kanauchi), 67–84. New York: Humana Press. doi: 10.1007/978-1-4939-8907-2_7.

Gibson, G.R., Hutkins, R., Sanders, M.E. et al. (2017). Expert consensus document: The International Scientific Association for Probiotics and Prebiotics (ISAPP) consensus statement on the definition and scope of prebiotics. *Nature Reviews Gastroenterology & Hepatology* 14: 491–502. doi: 10.1038/nrgastro.2017.75.

Gill, P.A., van Zelm, M.C., Muir, J.G. et al. (2018). Review article: short chain fatty acids as potential therapeutic agents in human gastrointestinal and inflammatory disorders. *Alimentary Pharmacology & Therapeutics* 48: 15–34. doi: 10.1111/apt.14689.

Hamet, M.F., Piermaria, J.A., and Abraham, A.G. (2015). Selection of EPS-producing *Lactobacillus* strains isolated from kefir grains and rheological characterization of the fermented milks. *LWT - Food Science and Technology* 63: 129–135. doi: 10.1016/j.lwt.2015.03.097.

Hidalgo-Cantabrana, C., Nikolic, M., López, P. et al. (2014a). Exopolysaccharide-producing *Bifidobacterium animalis* subsp. *lactis* strains and their polymers elicit different responses on immune cells from blood and gut associated lymphoid tissue. *Anaerobe* 26: 24–30. doi: 10.1016/j.anaerobe.2014.01.003.

Hidalgo-Cantabrana, C., Sánchez, B., Milani, C. et al. (2014b). Genomic overview and biological functions of exopolysaccharide biosynthesis in *Bifidobacterium* spp. *Applied and Environmental Microbiology* 80: 9–18. doi: 10.1128/AEM.02977-13.

Hill, C., Guarner, F., Reid, G. et al. (2014). Expert consensus document: the international scientific association for probiotics and prebiotics consensus statement on the scope and appropriate use of the term probiotic. *Nature Reviews Gastroenterology & Hepatology* 11: 506–514. doi: 10.1038/nrgastro.2014.66.

Huang, Z., Lin, F., Zhu, X. et al. (2020). An exopolysaccharide from *Lactobacillus plantarum* H31 in pickled cabbage inhibits pancreas α-amylase and regulating metabolic markers in HepG2 cells by AMPK/PI3K/Akt pathway. *International Journal of Biological Macromolecules* 143: 775–784. doi: 10.1016/j.ijbiomac.2019.09.137.

Ikeda, S., Murayama, D., Tsurumaki, A. et al. (2019). Rheological characteristics and supramolecular structure of the exopolysaccharide produced by *Lactobacillus fermentum* MTCC 25067. *Carbohydrate Polymers* 218: 226–233. doi: 10.1016/j.carbpol.2019.04.076.

Inturri, R., Molinaro, A., Di, F. et al. (2017). Chemical and biological properties of the novel exopolysaccharide produced by a probiotic strain of Bifidobacterium longum. *Carbohydrate Polymers* 174: 1172–1180. doi: 10.1016/j.carbpol.2017.07.039.

İspirli, H., Özmen, D., Yılmaz, M.T. et al. (2020). Impact of glucan type exopolysaccharide (EPS) production on technological characteristics of sourdough bread. *Food Control* 107: 106812. doi: 10.1016/j.foodcont.2019.106812.

Kant Lakra, A., Domdi, L., Mohd Tilwani, Y. et al. (2020). Physicochemical and functional characterization of mannan exopolysaccharide from *Weissella confusa* MD1 with bioactivities. *International Journal of Biological Macromolecules* 143: 797–805. doi: 10.1016/j.ijbiomac.2019.09.139.

Khan, T., Park, J.K., and Kwon, J.-H. (2007). Functional biopolymers produced by biochemical technology considering applications in food engineering. *Korean Journal of Chemical Engineering* 24: 816–826. doi: 10.1007/s11814-007-0047-1.

Knoshaug, E.P., Ahlgren, J.A., and Trempy, J.E. (2007). Exopolysaccharide expression in *Lactococcus lactis* subsp. *cremoris* Ropy352: evidence for novel gene organization. *Applied and Environmental Microbiology* 73: 897–905. doi: 10.1128/AEM.01945-06.

Korcz, E., Kerényi, Z., and Varga, L. (2018). Dietary fibers, prebiotics, and exopolysaccharides produced by lactic acid bacteria: potential health benefits with special regard to cholesterol-lowering effects. *Food & Function* 9: 3057–3068. doi: 10.1039/C8FO00118A.

Kšonžeková, P., Bystrický, P., Vlčková, S. et al. (2016). Exopolysaccharides of *Lactobacillus reuteri*: their influence on adherence of *E. coli* to epithelial cells and inflammatory response. *Carbohydrate Polymers* 141: 10–19. doi: 10.1016/j.carbpol.2015.12.037.

Lal, S., Kanhar, N.A., Kumar, P. et al. (2019). Production of exopolysaccharide by bifidobacteria and its viscometric analysis. *International Journal of Biosciences* 14 (5): 315–323. doi: 10.12692/ijb/14.5.315-323.

Laws, A., Gu, Y., Marshall, V. et al. (2001). Biosynthesis, characterisation, and design of bacterial exopolysaccharides from lactic acid bacteria. *Biotechnology Advances* 19: 597–625. doi: 10.1016/S0734-9750(01)00084-2.

Leo, F., Hashida, S., Kumagai, D. et al. (2007). Studies on a neutral exopolysaccharide of *Lactobacillus fermentum* TDS030603. *Journal of Applied Glycoscience* 54: 223–229. doi: 10.5458/jag.54.223.

Li, C., Li, W., Chen, X. et al. (2014a). Microbiological, physicochemical and rheological properties of fermented soymilk produced with exopolysaccharide (EPS) producing lactic acid bacteria strains. *LWT - Food Science and Technology* 57: 477–485. doi: 10.1016/j.lwt.2014.02.025.

Li, S., Huang, R., Shah, N.P. et al. (2014b). Antioxidant and antibacterial activities of exopolysaccharides from *Bifidobacterium bifidum* WBIN03 and *Lactobacillus plantarum* R315. *Journal of Dairy Science* 97: 7334–7343. doi: 10.3168/jds.2014-7912.

Li, W., Xia, X., Tang, W. et al. (2015). Structural characterization and anticancer activity of cell-bound exopolysaccharide (c-EPS) from *Lactobacillus helveticus* MB2-1. *Journal of Agricultural and Food Chemistry* 63 (13): 3454–3463. doi: 10.1021/acs.jafc.5b01086.

Liu, L., Li, H., Xu, R.-H. et al. (2017a). Expolysaccharides from *Bifidobacterium animalis* RH activates RAW 264.7 macrophages through toll-like receptor 4. *Food and Agricultural Immunology* 28 (1): 149–161. doi: 10.1080/09540105.2016.1230599.

Liu, Z., Zhang, Z., Qiu, L. et al. (2017b). Characterization and bioactivities of the exopolysaccharide from a probiotic strain of *Lactobacillus plantarum* WLPL04. *Journal of Dairy Science* 100: 6895–6905. doi: 10.3168/jds.2016-11944.

López, P., Monteserín, D.C., Gueimonde, M. et al. (2012). Exopolysaccharide-producing *Bifidobacterium* strains elicit different *in vitro* responses upon interaction with human cells. *Food Research International* 46: 99–107. doi: 10.1016/j.foodres.2011.11.020.

Lynch, K.M., Coffey, A., and Arendt, E.K. (2018a). Exopolysaccharide producing lactic acid bacteria: their techno-functional role and potential application in gluten-free bread products. *Food Research International* 110: 52–61. doi: 10.1016/j.foodres.2017.03.012.

Lynch, K.M., Zannini, E., Coffey, A. et al. (2018b). Lactic acid bacteria exopolysaccharides in foods and beverages: isolation, properties, characterization, and health benefits. *Annual Review of Food Science and Technology* 9: 155–176. doi: 10.1146/annurev-food-030117-012537.

Mende, S., Rohm, H., and Jaros, D. (2016). Influence of exopolysaccharides on the structure, texture, stability and sensory properties of yoghurt and related products. *International Dairy Journal* 52: 57–71. doi: 10.1016/j.idairyj.2015.08.002.

Merghni, A., Dallel, I., Noumi, E. et al. (2017). Antioxidant and antiproliferative potential of biosurfactants isolated from *Lactobacillus casei* and their anti-biofilm effect in oral *Staphylococcus aureus* strains. *Microbial Pathogenesis* 104: 84–89. doi: 10.1016/j.micpath.2017.01.017.

Min, W.-H., Fang, X.-B., Wu, T. et al. (2019). Characterization and antioxidant activity of an acidic exopolysaccharide from *Lactobacillus plantarum* JLAU103. *Journal of Bioscience and Bioengineering* 127: 758–766. doi: 10.1016/j.jbiosc.2018.12.004.

Nácher-Vázquez, M., Ballesteros, N., Canales, A. et al. (2015). Dextrans produced by lactic acid bacteria exhibit antiviral and immunomodulatory activity against salmonid viruses. *Carbohydrate Polymers* 124: 292–301. doi: 10.1016/j.carbpol.2015.02.020.

Oleksy, M. and Klewicka, E. (2018). Exopolysaccharides produced by *Lactobacillus* sp.: biosynthesis and applications. *Critical Reviews in Food Science and Nutrition* 58: 450–462. doi: 10.1080/10408398.2016.1187112.

Paik, W., Alonzo, F., and Knight, K.L. (2018). Probiotic exopolysaccharide protects against systemic *Staphylococcus aureus* infection, inducing dual-functioning macrophages that restrict bacterial growth and limit inflammation. *Infection and Immunity* 87: e00791–18. doi: 10.1128/IAI.00791-18.

Patrignani, F., Modesto, M., Michelini, S. et al. (2018). Technological potential of *Bifidobacterium aesculapii* strains for fermented soymilk production. *LWT - Food Science and Technology* 89: 689–696. doi: 10.1016/j.lwt.2017.11.048.

Polak-Berecka, M., Choma, A., Waśko, A. et al. (2015). Physicochemical characterization of exopolysaccharides produced by *Lactobacillus rhamnosus* on various carbon sources. *Carbohydrate Polymers* 117: 501–509. doi: 10.1016/j.carbpol.2014.10.006.

Prasanna, P.H.P., Bell, A., Grandison, A.S. et al. (2012a). Emulsifying, rheological and physicochemical properties of exopolysaccharide produced by *Bifidobacterium longum* subsp. *infantis* CCUG 52486 and *Bifidobacterium infantis* NCIMB 702205. *Carbohydrate Polymers* 90: 533–540. doi: 10.1016/j.carbpol.2012.05.075.

Prasanna, P.H.P., Grandison, A.S., and Charalampopoulos, D. (2012b). Screening human intestinal *Bifidobacterium* strains for growth, acidification, EPS production and viscosity potential in low-fat milk. *International Dairy Journal* 23: 36–44. doi: 10.1016/j.idairyj.2011.09.008.

Prasanna, P.H.P., Grandison, A.S., and Charalampopoulos, D. (2013). Microbiological, chemical and rheological properties of low fat set yoghurt produced with exopolysaccharide (EPS) producing *Bifidobacterium* strains. *Food Research International* 51: 15–22. doi: 10.1016/j.foodres.2012.11.016.

Püngel, D., Treveil, A., Dalby, M.J. et al. (2020). *Bifidobacterium breve* UCC2003 exopolysaccharide modulates the early life microbiota by acting as a potential dietary substrate. *Nutrients* 12: 948. doi: 10.3390/nu12040948.

Rahbar Saadat, Y., Yari Khosroushahi, A., and Pourghassem Gargari, B. (2019). A comprehensive review of anticancer, immunomodulatory and health beneficial effects of the lactic acid bacteria exopolysaccharides. *Carbohydrate Polymers* 217: 79–89. doi: 10.1016/j.carbpol.2019.04.025.

Riaz Rajoka, M.S., Jin, M., Haobin, Z. et al. (2018). Functional characterization and biotechnological potential of exopolysaccharide produced by *Lactobacillus rhamnosus* strains isolated from human breast milk. *LWT - Food Science and Technology* 89: 638–647. doi: 10.1016/j.lwt.2017.11.034.

Riaz Rajoka, M.S., Mehwish, H.M., Fang, H. et al. (2019). Characterization and anti-tumor activity of exopolysaccharide produced by *Lactobacillus kefiri* isolated from Chinese kefir grains. *Journal of Functional Foods* 63: 103588. doi: 10.1016/j.jff.2019.103588.

Rios-Covian, D., Cuesta, I., Alvarez-Buylla, J.R. et al. (2016). *Bacteroides fragilis* metabolises exopolysaccharides produced by bifidobacteria. *BMC Microbiology* 16: 150. doi: 10.1186/s12866-016-0773-9.

Salazar, N., Neyrinck, A.M., Bindels, L.B. et al. (2019). Functional effects of EPS-producing *Bifidobacterium* administration on energy metabolic alterations of diet-induced obese mice. *Frontiers in Microbiology* 10: 1809. doi: 10.3389/fmicb.2019.01809.

Şanli, T., Gursel, A., Şanli, E. et al. (2013). The effect of using an exopolysaccharide-producing culture on the physicochemical properties of low-fat and reduced-fat Kasar cheeses. *International Journal of Dairy Technology* 66 (4): 535–542. doi: 10.1111/1471-0307.12071.

Sasikumar, K., Kozhummal Vaikkath, D., Devendra, L. et al. (2017). An exopolysaccharide (EPS) from a *Lactobacillus plantarum* BR2 with potential benefits for making functional foods. *Bioresource Technology* 241: 1152–1156. doi: 10.1016/j.biortech.2017.05.075.

Schiavi, E., Plattner, S., Rodriguez-Perez, N. et al. (2018). Exopolysaccharide from *Bifidobacterium longum* subsp. *longum* 35624TM modulates murine allergic airway responses. *Beneficial Microbes* 9: 761–773. doi: 10.3920/BM2017.0180.

Shang, N., Xu, R., and Li, P. (2013). Structure characterization of an exopolysaccharide produced by *Bifidobacterium animalis* RH. *Carbohydrate Polymers* 91 (1): 128–134. doi: 10.1016/j.carbpol.2012.08.012.

Shao, L., Wu, Z., Zhang, H. et al. (2014). Partial characterization and immunostimulatory activity of exopolysaccharides from *Lactobacillus rhamnosus* KF5. *Carbohydrate Polymers* 107 (1): 51–56. doi: 10.1016/j.carbpol.2014.02.037.

Singh, P. and Saini, P. (2017). Food and health potentials of exopolysaccharides derived from lactobacilli. *Microbiology Research Journal International* 22: 1–14. doi: 10.9734/MRJI/2017/36935.

Sungur, T., Aslim, B., Karaaslan, C. et al. (2017). Impact of exopolysaccharides (EPSs) of *Lactobacillus gasseri* strains isolated from human vagina on cervical tumor cells (HeLa). *Anaerobe* 47: 137–144. doi: 10.1016/j.anaerobe.2017.05.013.

Swanson, K.S., Gibson, G.R., Hutkins, R. et al. (2020). The International Scientific Association for Probiotics and Prebiotics (ISAPP) consensus statement on the definition and scope of synbiotics. *Nature Reviews Gastroenterology & Hepatology* 17: 687–701. doi: 10.1038/s41575-020-0344-2.

Tang, X., Liu, R., Huang, W. et al. (2018). Impact of *in situ* formed exopolysaccharides on dough performance and quality of Chinese steamed bread. *LWT - Food Science and Technology* 96: 519–525. doi: 10.1016/j.lwt.2018.04.039.

Torino, M.I., de Valdez, G.F., and Mozzi, F. (2015). Biopolymers from lactic acid bacteria. Novel applications in foods and beverages. *Frontiers in Microbiology* 6: 1–16. doi: 10.3389/fmicb.2015.00834.

Trabelsi, I., Ktari, N., Ben Slima, S. et al. (2017). Evaluation of dermal wound healing activity and *in vitro* antibacterial and antioxidant activities of a new exopolysaccharide produced by *Lactobacillus* sp.Ca 6. *International Journal of Biological Macromolecules* 103: 194–201. doi: 10.1016/j.ijbiomac.2017.05.017.

van Hijum, S.A.F.T., Kralj, S., Ozimek, L.K. et al. (2006). Structure-function relationships of glucansucrase and fructansucrase enzymes from lactic acid bacteria. *Microbiology and Molecular Biology Reviews* 70: 157–176. doi: 10.1128/MMBR.70.1.157-176.2006.

Vitlic, A., Sadiq, S., Ahmed, H. et al. (2019). Isolation and characterisation of a high molecular mass beta-glucan from *Lactobacillus fermentum* Lf2 and evaluation of its immunomodulatory activity. *Carbohydrate Research* 476: 44–52. doi: 10.1016/j.carres.2019.03.003.

Wang, J., Fang, X., Wu, T. et al. (2020). *In vitro* immunomodulatory effects of acidic exopolysaccharide produced by *Lactobacillus planetarium* JLAU103 on RAW264.7 macrophages. *International Journal of Biological Macromolecules* 156: 1308–1315. doi: 10.1016/j.ijbiomac.2019.11.169.

Wang, J., Zhao, X., Tian, Z. et al. (2015a). Isolation and characterization of exopolysaccharide-producing *Lactobacillus plantarum* skt109 from Tibet kefir. *Polish Journal of Food and Nutrition Sciences* 65. doi: 10.1515/pjfns-2015-0023.

Wang, J., Zhao, X., Yang, Y. et al. (2015b). Characterization and bioactivities of an exopolysaccharide produced by *Lactobacillus plantarum* YW32. *International Journal of Biological Macromolecules* 74: 119–126. doi: 10.1016/j.ijbiomac.2014.12.006.

Wang, K., Li, W., Rui, X. et al. (2014). Characterization of a novel exopolysaccharide with antitumor activityfrom *Lactobacillus plantarum* 70810. *International Journal of Biological Macromolecules* 63: 133–139. doi: 10.1016/j.ijbiomac.2013.10.036.

Wang, K., Niu, M., Yao, D. et al. (2019a). Physicochemical characteristics and *in vitro* and *in vivo* antioxidant activity of a cell-bound exopolysaccharide produced by *Lactobacillus fermentum* S1. *International Journal of Biological Macromolecules* 139: 252–261. doi: 10.1016/j.ijbiomac.2019.07.200.

Wang, L., Wang, Y., Li, Q. et al. (2019b). Exopolysaccharide, isolated from a novel strain *Bifidobacterium breve* lw01 possess an anticancer effect on Head and Neck Cancer-genetic and biochemical evidences. *Frontiers in Microbiology* 10: 1044. doi: 10.3389/fmicb.2019.01044.

Wang, X., Shao, C., Liu, L. et al. (2017). Optimization, partial characterization and antioxidant activity of an exopolysaccharide from *Lactobacillus plantarum* KX041. *International Journal of Biological Macromolecules* 103: 1173–1184. doi: 10.1016/j.ijbiomac.2017.05.118.

Xiao, L., Ge, X., Yang, L. et al. (2020a). Anticancer potential of an exopolysaccharide from *Lactobacillus helveticus* MB2-1 on human colon cancer HT-29 cells via apoptosis induction. *Food & Function* 11: 10170–10181. doi: 10.1039/D0FO01345E.

Xiao, L., Han, S., Zhou, J. et al. (2020b). Preparation, characterization and antioxidant activities of derivatives of exopolysaccharide from *Lactobacillus helveticus* MB2-1. *International Journal of Biological Macromolecules* 145: 1008–1017. doi: 10.1016/j.ijbiomac.2019.09.192.

Xiu, L., Zhang, H., Hu, Z. et al. (2018). Immunostimulatory activity of exopolysaccharides from probiotic *Lactobacillus casei* WXD030 strain as a novel adjuvant *in vitro* and in vivo. *Food and Agricultural Immunology* 29: 1086–1105. doi: 10.1080/09540105.2018.1513994.

Xu, R., Shen, Q., Ding, X. et al. (2011). Chemical characterization and antioxidant activity of an exopolysaccharide fraction isolated from *Bifidobacterium animalis* RH. *European Food Research and Technology* 232: 231–240. doi: 10.1007/s00217-010-1382-8.

Xu, R., Shen, Q., Wu, R. et al. (2017). Structural analysis and mucosal immune regulation of exopolysaccharide fraction from *Bifidobacterium animalis* RH. *Food and Agricultural Immunology* 28: 1226–1241. doi: 10.1080/09540105.2017.1333578.

Xu, X., Peng, Q., Zhang, Y. et al. (2020). Antibacterial potential of a novel *Lactobacillus casei* strain isolated from Chinese northeast sauerkraut and the antibiofilm activity of its exopolysaccharides. *Food & Function* 11: 4697–4706. doi: 10.1039/D0FO00905A.

Xu, X., Qiao, Y., Peng, Q. et al. (2021). Antioxidant and immunomodulatory properties of partially purified exopolysaccharide from *Lactobacillus casei* isolated from Chinese northeast sauerkraut. *Immunological Investigations* doi: 10.1080/08820139.2020.1869777.

Xu, Y., Cui, Y., Yue, F. et al. (2019). Exopolysaccharides produced by lactic acid bacteria and bifidobacteria: structures, physiochemical functions and applications in the food industry. *Food Hydrocolloids* 94: 475–499. doi: 10.1016/j.foodhyd.2019.03.032.

Yan, S., Yang, B., Ross, R.P. et al. (2020). *Bifidobacterium longum* subsp. *longum* YS108R fermented milk alleviates DSS induced colitis via anti-inflammation, mucosal barrier maintenance and gut microbiota modulation. *Journal of Functional Foods* 73: 104153. doi: 10.1016/j.jff.2020.104153.

Yan, S., Yang, B., Zhao, J. et al. (2019). A ropy exopolysaccharide producing strain *Bifidobacterium longum* subsp. *longum* YS108R alleviates DSS-induced colitis by maintenance of the mucosal barrier and gut microbiota modulation. *Food & Function* 10: 1595–1608. doi: 10.1039/C9FO00014C.

Yildiz, H. and Karatas, N. (2018). Microbial exopolysaccharides: resources and bioactive properties. *Process Biochemistry* 72: 41–46. doi: 10.1016/j.procbio.2018.06.009.

Yilmaz, M.T., Dertli, E., Toker, O.S. et al. (2015). Effect of *in situ* exopolysaccharide production on physicochemical, rheological, sensory, and microstructural properties of the yogurt drink ayran: an optimization study based on fermentation kinetics. *Journal of Dairy Science* 98: 1604–1624. doi: 10.3168/jds.2014-8936.

You, J. and Yaqoob, P. (2012). Evidence of immunomodulatory effects of a novel probiotic, *Bifidobacterium longum* bv. *infantis* CCUG 52486. *FEMS Immunology and Medical Microbiology* 66: 353–362. doi: 10.1111/j.1574-695X.2012.01014.x.

You, X., Li, Z., Ma, K. et al. (2020). Structural characterization and immunomodulatory activity of an exopolysaccharide produced by *Lactobacillus helveticus* LZ-R-5. *Carbohydrate Polymers* 235: 115977. doi: 10.1016/j.carbpol.2020.115977.

Zarour, K., Goretti Llamas, M., Prieto, A. et al. (2017). Rheology and bioactivity of high molecular weight dextrans synthesised by lactic acid bacteria. *Carbohydrate Polymers* 174: 646–657. doi: 10.1016/j.carbpol.2017.06.113.

Zhang, L., Liu, C., Li, D. et al. (2013). Antioxidant activity of an exopolysaccharide isolated from *Lactobacillus plantarum* C88. *International Journal of Biological Macromolecules* 54: 270–275. doi: 10.1016/j.ijbiomac.2012.12.037.

Zhou, X., Hong, T., Yu, Q. et al. (2017). Exopolysaccharides from *Lactobacillus plantarum* NCU116 induce c-Jun dependent Fas/Fasl-mediated apoptosis via TLR2 in mouse intestinal epithelial cancer cells. *Scientific Reports* 7: 14247. doi: 10.1038/s41598-017-14178-2.

Zhou, Y., Cui, Y., and Qu, X. (2019). Exopolysaccharides of lactic acid bacteria: structure, bioactivity and associations: a review. *Carbohydrate Polymers* 207: 317–332. doi: 10.1016/j.carbpol.2018.11.093.

16

Characterization of Bacteriocins Produced by Lactic Acid Bacteria of Industrial Interest

Silvina Alicia Pujato, Andrea del Luján Quiberoni, and Daniela Marta Guglielmotti*

Instituto de Lactología Industrial (INLAIN, UNL-CONICET), Facultad de Ingeniería Química, Santiago del Estero 2829 (3000), Santa Fe, Argentina.
**Corresponding author: aquibe@fiq.unl.edu.ar*

16.1 Introduction

The ability of lactic acid bacteria (LAB) to ferment food and beverages has been used since ancient times because of their beneficial influence on nutritional, organoleptic, and shelf-life properties (Deegan et al. 2006; O'Connor et al. 2020). They cause rapid acidification of the raw material through the production of organic acids, mostly lactic and acetic acids; in addition, they can also produce other interesting metabolites such as ethanol, aroma compounds, bacteriocins, exopolysaccharides, hydrogen peroxide, fatty acids, and diverse enzymes. Many of the above mentioned compounds produced by LAB exert antimicrobial activities, which in fermented foods contribute significantly to food biopreservation (De Vuyst and Leroy 2007; Egan et al. 2016).

Bacteriocins encompass a large family of ribosomally synthesized antimicrobial peptides antagonistic against microorganisms related (competing bacteria) or not (spoilage or pathogenic bacteria) to the producer bacterium (Cotter et al. 2013). Even if both Gram positive and Gram negative bacteria are able to produce bacteriocins, those produced by LAB are of special interest since they have enormous potential in the biopreservation of foods (Dobson et al. 2012). In this sense, the use of bacteriocins from LAB in foods is considered safe for human health (food-grade) since they are produced by LAB generally used for or isolated from fermented foods (Perez et al. 2014). Among the LAB family the higher number of bacteriocin-producer strains belongs to *Lactobacillus* genera (recently reclassified, Zheng et al. 2020) (31% of the total bacteriocinogenic LAB), followed by *Enterococcus* (24%) and *Streptococcus* (17%). The amount of bacteriocins reported by strains belonging to other LAB genera is significantly lower, ranging between less than 1% to 8% for *Lactococcus* (8%), *Pediococcus* (7%), *Leuconostoc* (6%), *Carnobacterium* (4%), *Weissella* (2%) and *Bifidobacterium* (<1%).

16.2 Bacteriocin Classification: Brief Story of Its Evolution

Over the years, numerous approaches have been used to classify bacteriocins. In the early 1990s, bacteriocins were classified by Klaenhammer (1993) on the basis of common, mainly structural characteristics and mode of action into four groups: class I or

Biomolecules from Natural Sources: Advances and Applications, First Edition. Edited by Vijai Kumar Gupta, Satyajit D. Sarker, Minaxi Sharma, María Elida Pirovani, Zeba Usmani, and Chelliah Jayabaskaran.

lantibiotics (very small thermostable peptides of <5 kDa with lanthionine or derivatives in their structure); class II (small thermostable peptides of <10 kDa) divided into subclasses IIa (anti-*Listeria* peptides), IIb (two-peptide bacteriocins) and IIc (thiol-activated peptides); class III (large thermo-sensitive peptides of >30 kDa) and class IV (complex molecules of proteins added of lipids or carbohydrates). However, class IV was removed from the classification (Cleveland et al. 2001; Cotter et al. 2005) and Cotter et al. (2005) reduced it to only two major categories: class I or lantibiotics and class II or nonlanthionine-containing bacteriocins. The most remarkable change in this scheme was the reclassification of class III bacteriocins as bacteriolysins, considering them more as lytic enzymes rather than peptides (Perez et al. 2014). Lastly, Drider et al. (2006) divided bacteriocins into three groups, based on their genetics and biochemical properties: class I or lantibiotics, class II with the subclasses IIa (pediocin-like, anti-*Listeria* bacteriocins), IIb (two-peptide bacteriocins) and IIc (one peptide or non-pediocin-like bacteriocins) and class III (thermolabile large peptides or proteins). Regarding bacteriocins produced by LAB, all of them were classified as class IIa peptides, while those belonging to class IIc, IV, III and circular ones were less frequently found (Kassaa et al. 2019).

16.3 Application of Bacteriocins in the Food Industry

For some time now, concerns regarding the safety and quality of foods have increased and attention has been focused on new and more natural methods of food preservation (Drider et al. 2006). Many bacteriocin-producing LAB are generally recognized as safe (GRAS status) and they are naturally found and isolated from food material (Deegan et al. 2006). Thus, the use of bacteriocins as natural food preservatives satisfies consumer demands for high quality and safe food and beverages without the use of chemical preservatives (Silva et al. 2018). Bacteriocins from LAB are usually tolerant to high temperatures, have a wide range of pH, have no color, odor or taste, and are easily degraded by human proteases. Moreover, they barely influence human microbiota, and many of them are active against food pathogens such as *Bacillus cereus*, *Listeria monocytogenes*, *Staphylococcus aureus*, *Pseudomonas* sp. and *Escherichia coli*, among others (Daba and Elkhateeb 2020; O'Connor et al. 2020).

Bacteriocins can be applied in foods through diverse approaches: addition of the purified bacteriocin as food additive, addition of a bacteriocin-containing pre-fermented product, direct inoculation of the food with the bacteriocinogenic strain or as a protective biofilm or coating on the surface of foods (O'Connor et al. 2020; Silva et al. 2018) (Table 16.1).

16.4 Purified Bacteriocins as Food Additive

In theory, purified bacteriocins could be prepared as an ingredient and incorporated afterwards in the food. However, practically, this is difficult to accomplish since it requires regulatory approval. A list of important points must be checked before the use of bacteriocins in foods, beginning with the fact that the bacteriocin-producer strain must have food grade status (GRAS). In addition, the bacteriocin itself must fulfill a list of specific

Table 16.1 Bacteriocin usages in foods: advantages and disadvantages.

Method	Advantages	Disadvantages
Purified bacteriocin	Control of dose Use of small volumes No interferences from other metabolites Use in the pharmaceutical industry	Requires regulatory approval Expensive Activity will depend on chemical composition and physical conditions of the food
Bacteriocin-containing fermentate	Very low regulation for industrial use Very low cost Presence of other beneficial compounds	Use of large volumes
Bacteriocin-producer strain	Requires fewer regulations than purified bacteriocin (GRAS status strains) Less expensive Continuous release of bacteriocin Presence of other beneficial compounds	Presence of unwanted metabolites (such as CO_2) Negative interference with starter strains Presence of bacteria
Biofilm / coating	Use of lower quantity of bacteriocin Bacteriocin is protected from degradation by coating components	Only for food surface More suitable for solid foods Presence of bacteria

properties, such as broad spectrum of inhibition, good heat and pH stability, optimal solubility and stability for a specific food, being safe to human health and exhibiting beneficial effects as flavor improvement, and quality and/or safety enhancement of foods (Leroy and De Vuyst 2010). Nisin, a class Ia bacteriocin or lantibiotic produced by *Lactococcus lactis*, is the most studied and commercially important bacteriocin and it is currently approved commercially in more than 60 countries as a food additive. In fact, its commercial application is regulated by the European Food Safety Authority (EFSA), licensed as a food preservative (E234), and recognized as safe for application as a biopreservative in food products by the Joint Food and Agriculture Organization/World Health Organization (FAO/WHO) Expert Committee on Food Additives. Nisin is commercially available under the trade name Nisaplin^{TM} (Danisco), and it is a water soluble, thermally stable product that contains 2.5% nisin (Daba and Elkhateeb 2020; O'Connor et al. 2020; Todorov et al. 2020). Nisin has a broad spectrum of antimicrobial activity against many Gram-positive bacteria (including LAB), pathogens (*Staphylococcus* and *Listeria*) and spore-forming bacteria (*Bacillus* and *Clostridium*). Nisin was widely used in refrigerated milk to control the development of pathogenic *Listeria monocytogenes* and *Staphylococcus aureus*, as well as in diverse types of cheese to prevent late blowing caused by *Clostridium* (Silva et al. 2018).

16.5 Fermented Products Containing Bacteriocin

This antimicrobial product is obtained by growing the bacteriocin-producer strain in a food-grade matrix, such as milk or whey, where the bacteriocin and other antimicrobial metabolites (for example, lactic acid or diacetyl) will be produced. Bacteriocin-containing fermentates are commercially available and extensively used in the food industry. Some examples are MicroGARD® (DuPont™) and ALTA™ 2341 (Quest International), both containing the bacteriocin pediocin PA-1. Pediocins are class II bacteriocin produced by *Pediococcus* strains (*P. acidilactici* and *P. pentosaceus*) (Nieto-Lozano et al. 2010) and other LAB genera such as *Lactiplantibacillus plantarum* (Bernbom et al. 2009; Zheng et al. 2020). This bacteriocin was found to be effective against Gram-positive (*Listeria monocytogenes* and *Staphylococcus aureus*), Gram-negative bacteria (*E. coli* and *Pseudomonas*) and spores (Nieto-Lozano et al. 2010; Silva et al. 2018). MicroGARD pediocin-containing products are used in several types of foods including dairy products to delay/inhibit microbial development; for example, their use in cottage cheese allows increasing both shelf life and safety of the product (Makhal et al. 2015).

16.6 Inoculation of the Food with the Bacteriocin-Producer Strain

The bacteriocin-producing strain can be included in starter or adjunct cultures allowing the *in situ* production of the antimicrobial compound while the strain is developing, which is a regulatory free action and has the advantage of localized and continuous release of the antibacterial compound while the culture is multiplying. Another advantage would be economical because it can be less expensive than the use of purified bacteriocins (Balciunas et al. 2013; Pujato et al. 2014). Several bacteriocin-producer cultures are commercially available, such as BIOSAFE™ (Chr. Hansen), use for hard and semi-hard cheeses to avoid late blowing defects, containing the nisin A-producer strain, *Lactococcus lactis* subsp. *lactis* BS-10, or HOLDBAC™ protective cultures series, containing undefined bacteriocins from *Propionibacterium freudenreichii* subsp. *shermanii* DSM 706 and *Lacticaseibacillus rhamnosus* DSM 7061, (DuPont), used for a wide range of foods including diverse type of cheeses and other fermented dairy products, meats, fermented vegetables, and beverages (Chikindas et al. 2018; O'Connor et al. 2020).

16.7 Bioprotective Films or Coatings

A very interesting alternative is the use of purified bacteriocins as part of antimicrobial packaging films or coatings which can be applied on the surface of foods. Antimicrobial packaging film can inhibit microbial growth on food surfaces by direct contact of the package with the surfaces of foods by gradual release of bacteriocin from the film to the food surface.

Two procedures are frequently used to prepare packaging films with bacteriocins. One method incorporates bacteriocins directly into polymers; an example of this is the inclusion of nisin into biodegradable protein films (Padgett et al. 1998) and the incorporation of nisin into a polyethylene-based plastic film used to vacuum-pack beef carcasses (Siragusa et al. 1999). In the second method, bacteriocins are added to packaging films by coating or adsorbing bacteriocin to the polymer surface. This bioprotective packaging will prevent contamination and surface proliferation of pathogens or spoilage bacteria during handling, transportation and storage of food products. Moreover, this approach would allow the use of certainly much lower quantities of bacteriocin preparation in comparison to those needed for the entire food matrix (Balciunas et al. 2013; Silva et al. 2018).

16.8 Control of *Listeria monocytogenes* by a Bacteriocin-Producer Lactic Acid Bacteria

In the past few years, consumption of fermented food has had an abrupt growth due to the increasing interest of consumers for natural products and the boom in healthy diets.

Leuconostoc are heterofermentative lactic acid bacteria frequently used in the dairy industry. Many species of *Leuconostoc* are part of dairy starter cultures and they are also found in the dairy environment as non-starter lactic acid bacteria (NSLAB) (Plengvidhya et al. 2007). *Leuconostoc* is capable of producing low molecular weight organic compounds that provide aroma and flavor to the fermented products (Shin and Han 2015). Additionally, as with the majority of LAB, *Leuconostoc* may have a role in the prevention of growth and activity of spoilage and pathogenic microorganisms. Many *Leuconostoc* strains produced antilisterial, class II bacteriocins (Trias et al. 2008; Xiraphi et al. 2008), such as *Leuc. mesenteroides* (Trias et al. 2008; Xiraphi et al. 2008), *Leuc. gelidum* (Hastings et al. 1991), *Leuc. carnosum* (Felix et al. 1994) and *Leuc. pseudomesenteroides* (Sawa et al. 2010).

Listeria monocytogenes is an opportunistic pathogen of humans and animals able to survive and grow under various adverse environmental conditions, including refrigeration temperatures, high salt concentration and low pH, which makes it a potential hazard after the consumption of milk and dairy products, in particular soft cheeses, often implicated as the source of infection in severe outbreaks of listeriosis. *List. monocytogenes* is one of the most violent food pathogens, with a mortality rate of 20–30%, higher than almost all other food infections. Immunodepressed individuals including pregnant women, infants and aged persons, are more prone to suffer from infections (Cataldo et al. 2007). Over the years, diverse methods that include the use of antilisterial bacteriocins from LAB have been assayed to inhibit the growth of *List. monocytogenes* in foods, principally in dairy products. Many of these approaches prefer the addition of the bacteriocinogenic strain (as part of the starter culture) rather than the more complicated use of the purified bacteriocin (Dal Bello et al. 2012).

In previous studies carried out by our group (Cardamone et al. 2011; Pujato et al. 2014) interesting technological properties and specific antilisterial activity were evidenced in the supernatant of a *Leuc. citreum* strain named MB1. This strain presented two different genes (a-mb1 and u-mb1 of 130 and 300 bp, respectively) compatible with the active fraction of previous reported bacteriocins: leucocin A and UviB (Felix et al. 1994; Hastings et al. 1991; Kim et al. 2008; Sawa et al. 2010). At least one bacteriocin proved to be active, as demonstrated by the significant anti-*Listeria* activity of the cell-free supernatant obtained from *Leuc. citreum* MB1 (Pujato et al. 2014). In this sense, nucleotide sequence of a-mb1 gene (130 bp) shared only partial homology to the gene that codifies for leucocin A, while u-mb1 (300 bp) shared 100% homology with the sequence of a novel bacteriocin named UviB, firstly reported by Kim et al. (2008) as part of the complete genome of another *Leuc. citreum* strain. If the UviB protein is found to be active, it will be the first strain to report the activity of this bacteriocin since it was only reported in a complete genome.

In this work, optimal conditions (temperature and pH) for bacteriocin production, followed by the ability of *Leuc. citreum* MB1 to control the growth of *List. monocytogenes* by producing the antilisterial bacteriocin in a fermented-milk model were evaluated.

16.9 Effect of Temperature and pH on Bacteriocin Production

Bacteriocin production is influenced by diverse factors, principally temperature, pH, and composition of the medium. Generally, good bacterial growth is directly proportional to high bacteriocin production; although optimal cell growth is not always related to the maximum bacteriocin levels (Mataragas et al. 2003). Before investigating the antilisterial activity of *Leuc. citreum* MB1 in the fermented milk model, optimal temperature and pH conditions of the bacteriocinogenic strain to achieve maximum bacteriocin production were determined. *Leuc. citreum* MB1 aliquots were inoculated in MRS broth (pH 6.5) and incubated at 25 °C, 30 °C and 37 °C. In parallel, the bacteriocin-producer strain was grown in MRS broth adjusted to pH values of 5.0, 5.5, 6.0, and 7.0 and incubated at 30 °C. All samples were collected each hour during the first 10 h, and at 24 and 48 h of incubation. Maximum bacterial growth was verified at 30 °C, reflected in maximum viable cell counts (log CFU/ml) (Figure 16.1(a)). This result was predictable since 30 °C is the optimal temperature for *Leuconostoc* growth. However, the highest bacteriocin production—evaluated by measuring the inhibition halo resulted of inoculating cell-free supernatants obtained from *Leuc. citreum* MB1 on *List. monocytogenes* ATCC 15313 plates—was observed when *Leuc. citreum* MB1 was grown at 25 °C, which was a suboptimal temperature for this strain (Figure 16.1(b)). Mataragas et al. (2003) showed that the decrease of temperature (from 30 °C to 25 °C) below the optimum for bacterial growth improved bacteriocin production. This phenomenon could be due to better utilization of energy and essential metabolites at temperatures lower than optimal. At 9 h of incubation, bacteriocin production was similar both at 25 °C and 30 °C; though after that moment, higher bacteriocin activity was observed when the strain was incubated at 25 °C. On the contrary, both bacteriocin production and bacterial growth were visibly unfavoured at 37 °C.

Maximum bacteriocin production was demonstrated when the bacteriocin-producer strain was grown in MRS broth adjusted to pH 6 (between 5 h and 8 h of incubation at 30 °C), as shown by largest inhibition halos (Figure 16.1(d)). However, after 24 h incubation, the highest activity was revealed when the strain was grown in MRS broth adjusted to pH 5. The optimal pH for bacteriocin production was lower (pH 5 and 6) than that for maximum bacterial growth (pH 7) (Figure 16.1(c)). Similar results were obtained by other authors (Krier et al. 1998; Mataragas et al. 2003). High bacterial growth rates do not assure optimal bacteriocin production and on the contrary, it could be favoured by suboptimal bacterial growth conditions (Mataragas et al. 2003).

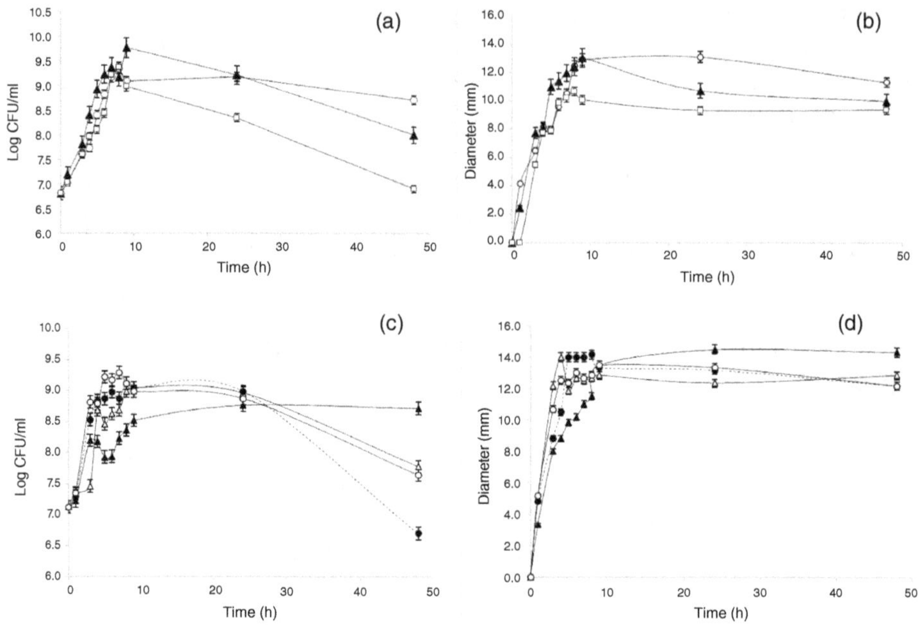

Figure 16.1 Growth kinetics and bacteriocin production of *Leuc. citreum* MB1. (a) and (b) show *Leuc. citreum* MB1 growth and bacteriocin activity against *List. monocytogenes* ATCC 15313, respectively, when the bacteriocin-producer strain was incubated at 25 °C (○), 30 °C (▲) and 37 °C. (c) and (d) show *Leuc. citreum* MB1 growth and bacteriocin activity, respectively, when the bacteriocin-producer strain was inoculated in MRS broth adjusted to pH 7.0 (○),6.0 (●), 5.5 (△), and 5.0 (▲). Values are the mean of three determinations.

16.10 Effect of Bacteriocin on Other Lactic Acid Bacteria

The incorporation of bacteriocins as biopreservative agents into model food systems has been shown to be an effective and natural method in the control of pathogenic and spoilage microorganisms, thus improving the safety of foods and contributing to the extension of their shelf-life. Nonetheless, the direct addition of bacteriocin-producing cultures into the food could be a more practical and economic option (Deegan et al. 2006; O'Sullivan et al. 2002). However, incompatibility issues between the bacteriocinogenic strain and other cultures of the starter used may occur. To avoid this inconvenience, some tests must be performed before the inclusion of the bacteriocin-producer strain. In this work, 16 strains belonging to diverse LAB of industrial importance species (*L. lactis, S. thermophilus, Lactobacillus helveticus, Lactobacillus delbrueckii, Lactiplantibacillus plantarum, Lacticaseibacillus paracasei, Leuc. pseudomesenteroides* and *Leuc. mesenteroides*) (Zheng et al. 2020) were tested for their sensitivity against bacteriocinogenic cell-free supernatants (CFS) obtained from *Leuc. citreum* MB1 (data not shown). In general, the CFS did not affect the normal growth of the tested LAB, excepting for *Leuconostoc* (*pseudomesenteroides* and *mesenteroides*) strains, which showed clear inhibition halos with both CFS and CFS adjusted to pH 7. The most sensitive strain was *Leuc. pseudomesenteroides* R707, followed by *Leuc. mesenteroides* L79-1 and D8, while *Leuc. mesenteroides* D4 revealed low sensitivity (as shown by very small inhibition halos obtained with CFS without pH adjustment) caused by organic acids present in the CFS (data not shown). In a similar manner, Budde et al. (2003) demonstrated that the antilisterial bacteriocin produced by a *Leuconostoc carnosum* strain inhibited diverse LAB, including a *Leuc. mesenteroides* strain.

16.11 Biocontrol of *Listeria monocytogenes* in a Fermented-Milk Model

Since the direct addition of bacteriocin-producing cultures as biopreservative agents in model food systems is an effective and natural method in the control of pathogenic microorganisms, the ability of *Leuc. citreum* MB1 to control *List. monocytogenes* ATCC 15313 growth was studied in a fermented-milk model. The selected fermentation conditions (30 °C and pH 6) were those that allowed the highest growth of *Leuc. citreum* and the highest production of bateriocin. Tubes containing skim milk were co-inoculated with *Leuc. citreum* MB1 and *List. monocytogenes* ATCC 15313 at 10^7 and 10^2 CFU ml^{-1}, respectively. Tubes inoculated with only *Leuconostoc* or *Listeria* were used as controls. One set of inoculated tubes was immediately incubated at 8 °C (refrigerated co-cultured milk), while other set of tubes was pre-incubated at 30 °C for 5 h and then kept at 8 °C for 28 days (fermented co-cultured milk). Viable cell counts in BHI (*Listeria*) or MRS (*Leuconostoc*) agar were carried out at the beginning and throughout the experiment. *List. monocytogenes* ATCC 15313 growth was retarded when it was co-inoculated with *Leuc. citreum* MB1, as viable cell counts of the pathogen were maintained below 10^2 CFU ml^{-1} during the first 3 days for both refrigerated and fermented co-cultured milk. *Listeria* grew slowly during the first 10 days of refrigerated incubation, after which the growth rose rapidly up to 10^6 CFU ml^{-1} at 28 days (Figure 16.2(a)). Conversely, fermented co-cultured milk allowed significant retard of the pathogen development even after 28 days, showing remarkably lower counts (6.7×10^4 CFU ml^{-1}) of the pathogen (Figure 16.2(b)). The ability of *Leuc. citreum* MB1 to control *List. monocytogenes* 15313 growth at 8 °C was greater in the fermented milk. The pre-incubation at 30 °C for 5 h certainly allowed rapid development of *Leuc. citreum* MB1 and the earlier production of high amounts of antimicrobial compounds which delayed the growth of the pathogenic strain.

In previous studies carried out by our group (Pujato et al. 2014), *List. monocytogenes* ATCC 15313 was grown (BHI broth, 37 °C, 48 h) in the presence of a concentrated bacteriocin-containing cell-free supernatant (CCFS) obtained from *Leuc. citreum* MB1. During 24 h of incubation, delay in *List. monocytogenes* growth was observed. The efficiency in preventing *List. monocytogenes* growth depended on both the initial number of pathogenic bacteria and the concentration of bacteriocin in the CCFS. Similar to that observed in the fermented milk, the inhibition of *List. monocytogenes* growth was greater when the CCFS was added after the pre-incubation of the pathogen. This effect could be attributable to an increased expression of the receptors linked to high metabolism of the

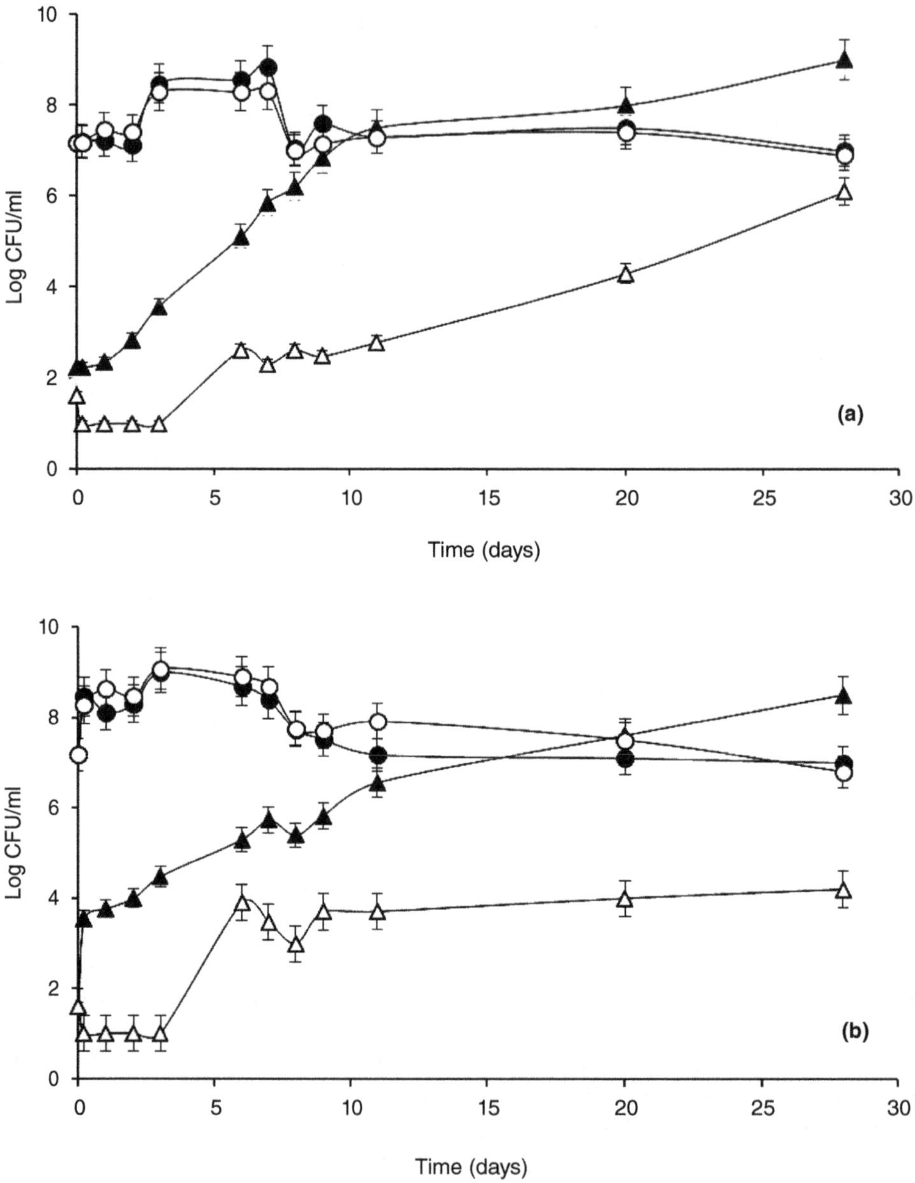

Figure 16.2 Co-culture growth of *Leuconostoc citreum* MB1 and *Listeria monocytogenes* ATCC 15313 in skimmed milk, at 10^6 CFU ml^{-1} (*Leuconostoc*) and 10^2 CFU ml^{-1} (*Listeria*) of initial cell loads. (a) Incubation at 8 °C for 28 days; (b) pre-incubated at 30 °C for 5 h, followed by incubation at 8 °C for 28 days. Symbols correspond to *Leuc. citreum* MB1 control (●), *Leuc. citreum* MB1 in co-culture (○), *List. monocytogenes* 15313 control (▲), and *List. monocytogenes* 15313 in co-culture (▲). Values are the mean of three determinations.

pathogenic strain (Kjos et al. 2011). Contrary to that observed in the fermented milk, the inhibitory effect disappeared after 24 h, probably because the antimicrobial compound lost its activity. As shown by the results, the use of CCFS could be a less efficient alternative in comparison to the addition of the bacteriocin-producing strain.

16.12 Conclusions

Bacteriocins have long been studied because of their huge potential as natural food preservatives that could be the perfect replacement of the extensively used but dangerous chemical preservatives. Application approaches of bacteriocin uses as biopreservatives in foods include the addition of the purified bacteriocin, which may be expensive and their commercialization as food additives requires regulatory approval. On the other hand, a bacteriocinogenic strain can be added to food so the bacteriocin will be produced *in situ*. Our results show that *Leuc. citreum* MB1 is effective in delaying *List. monocytogenes* growth in milk during refrigerated storage. Moreover, the highest production of bacteriocin was observed when *Leuc. citreum* MB1 was cultured at 25 °C and at pH values between 5 and 6, which are suboptimal conditions for bacterial growth. In turn, the incubation of diverse LAB with the bacteriocin-containing supernatant did not affect their normal growth, excepting for very closely related strains belonging to *Leuc. mesenteroides* and *Leuc. pseudomesenteroides* species. Direct incorporation of a bacteriocin-producer strain during the manufacture of fermented foods presents many advantages, since the metabolites produced during its development can contribute to the flavor and act as biopreservative agents as well. Furthermore, this is a more economical option in comparison to the use of purified bacteriocin and is subject to less regulatory controls.

References

Balciunas, E.M., Castillo Martinez, F.A., Todorov, S.D. et al. (2013). Novel biotechnological applications of bacteriocins: a review. *Food Control* 32: 134–142.

Bernbom, N., Jelle, B., Brogren, C.-H. et al. (2009). Pediocin PA-1 and a pediocin producing *Lactobacillus plantarum* strain do not change the HMA rat microbiota. *International Journal of Food Microbiology* 130: 251–257.

Budde, B.B., Hornbæk, T., Jacobsen, T. et al. (2003). *Leuconostoc carnosum* 4010 has the potential for use as a protective culture for vacuum-packed meats: culture isolation, bacteriocin identification, and meat application experiments. *International Journal of Food Microbiology* 83: 171–184.

Cardamone, L., Quiberoni, A., Mercanti, D.J. et al. (2011). Adventitious dairy *Leuconostoc* strains with interesting technological and biological properties useful for adjunct starters. *Dairy Science & Technology* 91: 457–470.

Cataldo, G., Conte, M.P., Chiarini, F. et al. (2007). Acid adaptation and survival of *Listeria monocytogenes* in Italian-style soft cheeses. *Journal of Applied Microbiology* 103: 185–193.

Chikindas, M.L., Week, R., Drider, D. et al. (2018). Functions and emerging applications of bacteriocins. *Current Opinion in Biotechnology* 49: 23–28.

Cleveland, J., Montville, T.J., Nes, I.F. et al. (2001). Bacteriocins: safe, natural antimicrobials for food preservation. *International Journal of Food Microbiology* 71: 1–20.

Cotter, P.D., Hill, C., and Ross, R.P. (2005). Bacteriocins: developing innate immunity for food. *Nature Reviews Microbiology* 3: 777–788.

Cotter, P.D., Ross, R.P., and Hill, C. (2013). Bacteriocins—A viable alternative to antibiotics? *Nature Reviews Microbiology* 11: 95–105.

Daba, G.M. and Elkhateeb, W.A. (2020). Bacteriocins of lactic acid bacteria as biotechnological tools in food and pharmaceuticals: current applications and future prospects. *Biocatalysis and Agricultural Biotechnology* 28: 101750. doi: 10.1016/j.bcab.2020.101750.

Dal Bello, B., Cocolin, L., Zeppa, G. et al. (2012). Technological characterization of bacteriocin producing *Lactococcus lactis* strains employed to control *Listeria monocytogenes* in cottage cheese. *International Journal of Food Microbiology* 153: 58–65.

De Vuyst, L. and Leroy, F. (2007). Bacteriocins from lactic acid bacteria: production, purification, and food applications. *Journal of Molecular Microbiology and Biotechnology* 13: 194–199.

Deegan, L.H., Cotter, P.D., Hill, C. et al. (2006). Bacteriocins: biological tools for biopreservation and shelf-life extension. *International Dairy Journal* 16: 1058–1071.

Dobson, A., Cotter, P.D., Ross, R.P. et al. (2012). Bacteriocin production: a probiotic trait? *Applied and Environmental Microbiology* 78: 1–6. doi: 10.1128/AEM.05576-11.

Drider, D., Fimland, G., Hechard, Y. et al. (2006). The continuing story of class IIa bacteriocins. *Microbiology and Molecular Biology Reviews* 70: 564–582.

Egan, K., Field, D., Rea, M.C. et al. (2016). Bacteriocins: novel solutions to age old spore-related problems? *Frontiers in Microbiology* 7: 461. doi: 10.3389/fmicb.2016.00461.

Felix, J.V., Papathanasopoulos, M.A., Smith, A.A. et al. (1994). Characterization of leucocin B-Ta11a: a bacteriocin from *Leuconostoc carnosum* Ta11a isolated from meat. *Current Microbiology* 29: 207–212.

Hastings, J.W., Sailer, M., Johnson, K. et al. (1991). Characterization of leucocin A-UAL 187 and cloning of the bacteriocin gene from *Leuconostoc gelidum*. *Journal of Bacteriology* 173: 7491–7500.

Kassaa, I.A.L., Rafei, R., Moukhtar, M. et al. (2019). LABiocin database: a new database designed specifically for Lactic Acid Bacteria bacteriocins. *International Journal of Antimicrobial Agents* 54: 771–779.

Kim, J.F., Jeong, H., Lee, J.-S. et al. (2008). Complete genome sequence of *Leuconostoc citreum* KM20. *Journal of Bacteriology* 190: 3093–3094.

Kjos, M., Nes, I.F., and Diep, D.B. (2011). Mechanisms of resistance to bacteriocins targeting the mannose phosphotransferase system. *Applied and Environmental Microbiology* 77: 3335–3342.

Klaenhammer, T.R. (1993). Genetics of bacteriocins produced by lactic acid bacteria. *FEMS Microbiology Reviews* 12: 39–86.

Krier, F., Revol-Junelles, A.M., and Germain, P. (1998). Influence of temperature and pH on production of two bacteriocins by *Leuconostoc mesenteroides* subsp. *mesenteroides* FR52 during batch fermentation. *Applied Microbiology and Biotechnology* 50: 359–363.

Leroy, F. and De Vuyst, L. (2010). Bacteriocins of lactic acid bacteria to combat undesirable bacteria in dairy products. *Australian Journal of Dairy Technology* 65: 143–149.

Makhal, S., Kanawjia, S.K., and Giri, A. (2015). Effect of microGARD on keeping quality of direct acidified Cottage cheese. *Journal of Food Science and Technology* 52: 936–943.

Mataragas, M., Metaxopoulos, J., Galiotou, M. et al. (2003). Influence of pH and temperature on growth and bacteriocin production by *Leuconostoc mesenteroides* L124 and *Lactobacillus curvatus* L442. *Meat Science* 64: 265–271.

Nieto-Lozano, J., Reguera-Useros, J., Peláez, C. et al. (2010). The effect of the pediocin PA-1 produced by *Pediococcus acidilactici* against *Listeria monocytogenes* and *Clostridium perfringens* in Spanish dry-fermented sausages and frankfurters. *Food Control* 21: 679–685.

O'Connor, P.M., Kuniyoshi, T.M., Oliveira, R.P.S. et al. (2020). Antimicrobials for food and feed; a bacteriocin perspective. *Current Opinion in Biotechnology* 61: 160–167.

O'Sullivan, L., Ross, R.P., and Hill, C. (2002). Potential of bacteriocin-producing lactic acid bacteria for improvements in food safety and quality. *Biochimie* 84: 593–604.

Padgett, T., Han, I.Y., and Dawson, P.L. (1998). Incorporation of food – grade antimicrobial compounds into biodegradable packaging films. *Journal of Food Protection* 61: 1330–1335.

Perez, R.H., Zendo, T., and Sonomoto, K. (2014). Novel bacteriocins from lactic acid bacteria (LAB): various structures and applications. *Microbial Cell Factories* 13 (Suppl 1): S3. doi: 10.1186/1475-2859-13-S1-S3.

Plengvidhya, V., Breidt, F., Lu, Z. et al. (2007). DNA fingerprinting of lactic acid bacteria in sauerkraut fermentations. *Applied and Environmental Microbiology* 73: 7697–7702.

Pujato, S.A., Quiberoni, A. del L., Candioti, M.C. et al. (2014). *Leuconostoc citreum* MB1 as biocontrol agent of *Listeria monocytogenes* in milk. *Journal of Dairy Research* 81: 137–145.

Sawa, N., Okamura, K., Zendo, T. et al. (2010). Identification and characterization of novel multiple bacteriocins produced by *Leuconostoc pseudomesenteroides* QU 15. *Journal of Applied Microbiology* 109: 282–291.

Shin, S.Y. and Han, N.S. (2015). *Leuconostoc* spp. as starters and their beneficial roles in fermented foods. In: Beneficial Microorganisms in Food and Nutraceuticals. Microbiology Monographs, vol 27 (ed. M.T. Liong). Cham: Springer. doi: 10.1007/978-3-319-23177-8_5.

Silva, C.C.G., Silva, S.P.M., and Ribeiro, S.C. (2018). Application of bacteriocins and protective cultures in dairy food preservation. *Frontiers in Microbiology* 9: 594. doi: 10.3389/fmicb.2018.00594.

Siragusa, G.R., Cutter, C.N., and Willett, J.L. (1999). Incorporation of bacteriocin in plastic retains activity and inhibits surface growth of bacteria on meat. *Food Microbiology* 16: 229–235.

Todorov, S.D., Kang, H.-J., Ivanova, I.V. et al. (2020). Bacteriocins from LAB and other alternative approaches for the control of *Clostridium* and Clostridiodes related gastrointestinal colitis. *Frontiers in Bioengineering and Biotechnology* 8: 581778. doi: 10.3389/fbioe.2020.581778.

Trias, R., Badosa, E., Montesinos, E. et al. (2008). Bioprotective *Leuconostoc* strains against *Listeria monocytogenes* in fresh fruits and vegetables. *International Journal of Food Microbiology* 127: 91–98.

Xiraphi, N., Georgalaki, M., Rantsiou, K. et al. (2008). Purification and characterization of a bacteriocin produced by *Leuconostoc mesenteroides* E131. *Meat Science* 80: 194–203.

Zheng, J., Wittouck, S., Salvetti, E. et al. (2020). A taxonomic note on the genus *Lactobacillus*: description of 23 novel genera, emended description of the genus *Lactobacillus* Beijerinck 1901, and union of *Lactobacillaceae* and *Leuconostocaceae*. *International Journal of Systematic and Evolutionary Microbiology* 70: 2782–2858.

Index

www.ingramcontent.com/pod-product-compliance
Lightning Source LLC
LaVergne TN
LVHW062203200926
848484LV00006B/1308
* 9 7 8 1 1 1 9 7 6 9 5 7 6 *